Handbook of Microbiology

Handbook of Microbiology

Editor: Drew Farmer

R CALLISTO REFERENCE

www.callistoreference.com

Callisto Reference,
118-35 Queens Blvd., Suite 400,
Forest Hills, NY 11375, USA

Visit us on the World Wide Web at:
www.callistoreference.com

ISBN: 978-1-64116-219-7 (Hardback)

Trademark Notice: Registered trademark of products or corporate names are used only for explanation and identification without intent to infringe.

Cataloging-in-Publication Data

Handbook of microbiology / edited by Drew Farmer.
   p. cm.
Includes bibliographical references and index.
ISBN 978-1-64116-219-7
1. Microbiology. 2. Microorganisms. 3. Biology. I. Farmer, Drew.
QR41.2 .H36 2019
579--dc23

TABLE OF CONTENTS

Preface

The study of microorganisms, including unicellular, multicellular and acellular microorganisms, is under the scope of microbiology. It is divided according to taxonomy into bacteriology, protozoology, mycology and phycology. Microbes are the causative agents for a number of human diseases. However, they are useful for a number of industrially beneficial processes such as production of vinegar, alcohol and dairy products. Various biopolymers such as polyamides, polysaccharides and polyesters are produced using microorganisms. These are also used for advanced medical and biomedical applications like drug delivery and tissue engineering, antibiotics production and as a carrier of DNA in complex organisms such as plants and animals. Microorganisms further play a critical role in the bioremediation of agricultural, domestic and industrial wastes in soils and marine environments. This book contains some path-breaking studies in the field of microbiology. It outlines the processes and applications of microbiology in detail. Coherent flow of topics, student-friendly language and extensive use of examples make this book an invaluable source of knowledge.

This book is a result of research of several months to collate the most relevant data in the field.

When I was approached with the idea of this book and the proposal to edit it, I was overwhelmed. It gave me an opportunity to reach out to all those who share a common interest with me in this field. I had 3 main parameters for editing this text:

1. Accuracy – The data and information provided in this book should be up-to-date and valuable to the readers.

2. Structure – The data must be presented in a structured format for easy understanding and better grasping of the readers.

3. Universal Approach – This book not only targets students but also experts and innovators in the field, thus my aim was to present topics which are of use to all.

Thus, it took me a couple of months to finish the editing of this book.

I would like to make a special mention of my publisher who considered me worthy of this opportunity and also supported me throughout the editing process. I would also like to thank the editing team at the back-end who extended their help whenever required.

Editor

Characterization of pneumococcal Ser/Thr protein phosphatase *phpP* mutant and identification of a novel PhpP substrate, putative RNA binding protein Jag

Aleš Ulrych, Nela Holečková, Jana Goldová, Linda Doubravová* ⓘ, Oldřich Benada, Olga Kofroňová, Petr Halada and Pavel Branny*

Abstract

Background: Reversible protein phosphorylation catalyzed by protein kinases and phosphatases is the primary mechanism for signal transduction in all living organisms. *Streptococcus pneumoniae* encodes a single Ser/Thr protein kinase, StkP, which plays a role in virulence, stress resistance and the regulation of cell wall synthesis and cell division. However, the role of its cognate phosphatase, PhpP, is not well defined.

Results: Here, we report the successful construction of a Δ*phpP* mutant in the unencapsulated *S. pneumoniae* Rx1 strain and the characterization of its phenotype. We demonstrate that PhpP negatively controls the level of protein phosphorylation in *S. pneumoniae* both by direct dephosphorylation of target proteins and by dephosphorylation of its cognate kinase, StkP. Catalytic inactivation or absence of PhpP resulted in the hyperphosphorylation of StkP substrates and specific phenotypic changes, including sensitivity to environmental stresses and competence deficiency. The morphology of the Δ*phpP* cells resembled the StkP overexpression phenotype and conversely, overexpression of PhpP resulted in cell elongation mimicking the *stkP* null phenotype. Proteomic analysis of the *phpP* knock-out strain permitted identification of a novel StkP/PhpP substrate, Spr1851, a putative RNA-binding protein homologous to Jag. Here, we show that pneumococcal Jag is phosphorylated on Thr89. Inactivation of *jag* confers a phenotype similar to the *phpP* mutant strain.

Conclusions: Our results suggest that PhpP and StkP cooperatively regulate cell division of *S. pneumoniae* and phosphorylate putative RNA binding protein Jag.

Keywords: Signal transduction, Protein phosphatase, Protein kinase, Cell division, *Streptococcus*, Phosphorylation, Jag

Background

Signal transduction via protein phosphorylation is one of the basic mechanisms that modulate numerous cellular processes in both prokaryotes and eukaryotes. Signal transduction in prokaryotes has been considered to occur primarily by two-component systems consisting of a histidine protein kinase and its cognate response regulator [1]. However, studies published in the last two decades have clearly demonstrated that this paradigm requires modification. Eukaryotic-type Ser/Thr protein kinases (ESTKs) as well as Ser/Thr phosphatases (ESTPs) operate in various bacterial species in parallel or overlapping signaling networks and regulate various cellular functions [2]. A distinct group of ESTKs which regulate cell cycle and cell division in many Gram-positive bacteria are conserved transmembrane proteins with a cytoplasmic kinase domain and repeated PASTA (penicillin-binding protein and Ser/Thr kinase-associated) domains in their extracellular region [2–6].

ESTKs are often co-expressed with their cognate phosphatases which are necessary for regulation of ESTK activity and quenching of signaling cascades; however, their physiological function in bacteria is still poorly

* Correspondence: linda@biomed.cas.cz; branny@biomed.cas.cz
Institute of Microbiology, v.v.i., Academy of Sciences of the Czech Republic, Vídeňská 1083, 142 20 Prague, Czech Republic

understood. The ESTPs associated with PASTA-possessing ESTKs are Mg^{2+}- or Mn^{2+}-dependent enzymes of the PPM family of Ser/Thr phosphatases, which share homology with the eukaryotic PP2C phosphatase [7]. Unlike ESTKs, only a few cognate ESTPs have been studied in detail, in part because several of them have been reported to be essential [8–10]. However, other detailed studies have demonstrated that knock-out mutants of phosphatase genes are viable and that ESTPs play a role in virulence, cell wall metabolism and cell segregation [11–16].

S. pneumoniae encodes a single PASTA-containing ESTK named StkP and a co-transcribed phosphatase, PhpP [8, 17]. Unlike PhpP, StkP has been extensively studied in past years, and its pleiotropic function in the regulation of different cellular processes has been described. StkP is a virulence determinant that is important for lung infection and bloodstream invasion in vivo and regulates pilus expression and bacterial adherence in vitro [8, 18]. StkP is essential for the resistance of S. pneumoniae to various stress conditions and competence development. Microarray analysis has revealed that StkP affects the transcription of a set of genes involved in cell wall metabolism, pyrimidine biosynthesis, DNA repair, iron uptake and oxidative stress response [8, 19]. StkP localizes to the division sites and plays important role in the regulation of cell division [20–22]. Cells with stkP mutations demonstrated disrupted cell wall synthesis and displayed elongated morphologies with multiple, often unconstricted, cell division septa, which suggest that StkP coordinates cell wall synthesis with cell division and thus helps pneumococcus to achieve its characteristic ovoid shape. Consistent with its role in cell division, StkP was found to phosphorylate several proteins involved in cell wall synthesis and cell division. The cell division proteins DivIVA [21, 23], LocZ (named also MapZ) [23–25] and the phosphoglucosamine mutase GlmM [17] are phosphorylated by StkP in vitro and in vivo. The cell division proteins FtsZ [22] and FtsA [20] and the cell wall biosynthesis enzyme MurC [26] are substrates of StkP in vitro; however, their phosphorylation by StkP in vivo has not been confirmed.

StkP is dephosphorylated by the cognate phosphatase PhpP, which is a PP2C-type Mn^{2+}-dependent enzyme. The PhpP catalytic domain contains 11 conserved signature motifs [27], and mutations of the highly conserved residues D192 and D231, which have been implicated in metal binding, completely abolish PhpP activity in vitro [17]. GFP-PhpP fusion protein is localized in the cytoplasm; however, the protein is often enriched in the mid-cell. The localization of PhpP to cell division sites depends on the presence of active StkP, indicating that both enzymes form a signaling couple in vivo [20]. Previously, phpP was reported to be essential for the viability of the unencapsulated Rx1 and R800 strains [10, 21]. According to global analysis performed by Thanassi et al. [28], both phpP and stkP genes were found to be essential; however, in the other global studies, phpP was not recognized as an essential gene [29, 30]. A recent report generated nonpolar markerless phpP knock-out mutants in two encapsulated pathogenic strains, S. pneumoniae D39 and 6A, indicating that PhpP is dispensable for pneumococcal survival [11]. Characterization of these mutants demonstrated the strain-specific role of PhpP in cell wall biosynthesis, adherence and biofilm formation. The StkP/PhpP signaling couple has been demonstrated to regulate the two-component system HK06/RR06, which modulates the expression of a major pneumococcal adhesin, CbpA [11].

In the present study, we show that the unencapsulated Rx1 phpP knock-out strain is viable. The morphology of both, the unencapsulated phpP null mutant and the phpP overexpression strain, clearly demonstrated that PhpP participates in the regulation of cell division and has an opposite regulatory effect to that of StkP. Our data suggest that PhpP modulates the level of protein phosphorylation in vivo both, through direct dephosphorylation of target proteins and dephosphorylation of its cognate kinase, StkP, resulting in coordination of cell wall synthesis and division in S. pneumoniae. Proteomic analysis of the $\Delta phpP$ strain revealed a novel StkP/PhpP substrate, Spr1851, a putative RNA-binding protein homologous to Jag protein of B. subtilis [31].

Results and discussion
PhpP is not essential for pneumococcal survival and catalyzes dephosphorylation of StkP and its substrates

Although phpP was reported to be essential in an $stkP^+$ genetic background [10], a nonpolar markerless phpP knock-out was generated in two encapsulated S. pneumoniae strains using the Janus cassette-based two-step negative selection strategy [11]. We attempted to use the same strategy to knock out phpP in the unencapsulated Rx1 strain as described in the Methods. We obtained viable $\Delta phpP$ transformants and characterized them further (see below). To exclude the possibility that the $\Delta phpP$ strain might carry an unlinked extragenic suppressor of the potentially lethal effect of loss of PhpP, the dose-response pattern for $\Delta phpP$ versus a wild type (WT) backcross was determined [32, 33]. Transformation by a single marker in pneumococcus displays a linear dependence on the dose of donor DNA (slope of regression curve equal to 1), whereas less efficient co-transformation by two markers follows a quadratic dependence on donor DNA dose (slope of regression curve equal to 2). Transfer of the $\Delta phpP$ mutation followed first order kinetics, and, therefore, viability of the $\Delta phpP$ strain does not depend on an extragenic suppressor mutation (Fig. 1a). In addition, the sequence of the neighboring genes *spr1579* and *stkP* was verified for the absence of mutation by DNA sequencing. The contradictory results reporting the essentiality [10, 21, 28] and non-essentiality [11, 29, 30] of the *phpP* gene may result from the different methods used for gene inactivation or from the genetic variability of the pneumococcal strains used. As reviewed in Massidda et al. [34], the genome of *S. pneumoniae* is very dynamic, and the number of genes found to be conditionally essential is dependent on the genetic background or the presence of capsule.

Using specific anti-PhpP (α-PhpP) and anti-StkP (α-StkP) antibodies, we confirmed that PhpP was deleted from the genome of the Δ*phpP* strain (Sp113), while the expression level of StkP was similar in both the Δ*phpP* and wild type strain (Fig. 1b). To evaluate the level of protein phosphorylation, we performed immunodetection with an anti-phospho-threonine (α-pThr) antibody. Thr phosphorylation in *S. pneumoniae* is largely dependent on StkP, and the majority of its substrates are membrane or membrane-associated proteins [23]. As previously reported, no Thr phosphorylated proteins were detected in the Δ*stkP* mutant (Sp10) (Fig. 1b). Immunodetection of phosphoproteins in the Δ*phpP* membrane fraction revealed a pattern similar to the StkP-dependent phosphoproteome [23]; however, we observed an increase in signal intensity corresponding to 192 ± 58.4 % of the wild type, indicating hyperphosphorylation of StkP substrates, including StkP itself (Fig. 1b). These data indicate that PhpP negatively regulates phosphorylation of StkP and its substrates.

To verify that the observed phosphorylation profile was the result of the *phpP* deletion, we constructed two complementation strains. First, we reverted the Δ*phpP* mutation back to the wild type genotype by transforming the wild type allele of the *phpP* gene into the Δ*phpP* strain (strain Sp222). The second complementation strain was prepared using ectopic expression of *phpP* under the control of a zinc-inducible promoter, P$_{czcD}$ (P$_{Zn}$) [35]. *phpP* was cloned into the pJWV25 plasmid under the control of the P$_{Zn}$ promoter and was inserted by double cross-over into the dispensable *bga* locus on the chromosome of the Δ*phpP* strain. The resulting strain, Sp120 (Δ*phpP bga*::P$_{Zn}$-*phpP*), was cultivated in the presence of different concentrations of ZnSO$_4$, and the expression of PhpP and the level of protein phosphorylation was monitored by immunoblotting. When expression of *phpP* was induced by addition of 0.2 mM ZnSO$_4$, we observed expression of PhpP and a phosphorylation signal intensity that correlated with the wild type (Fig. 1c, lane 6). Addition of 0.3 mM ZnSO$_4$ resulted in PhpP overexpression and a strong decrease in Thr phosphorylation (Fig. 1c, lane 7). The reverted strain WT$_R$ (Sp222) showed Thr phosphorylation levels that

were indistinguishable from the wild type strain (Fig. 1c, lane 2). These results demonstrate that complementation restores the wild type phosphorylation profile.

To demonstrate that PhpP directly dephosphorylates StkP substrates and not solely StkP, thus decreasing its activity, we performed in vitro dephosphorylation assays. We prepared pneumococcal strains Sp188 and Sp174 expressing the known StkP substrates LocZ and DivIVA, respectively, tagged with Flag-tag and isolated phosphorylated Flag-LocZ and DivIVA-Flag proteins from pneumococcal cell lysates. Purified proteins were incubated with recombinant His-PhpP as described in *Methods*, and protein dephosphorylation was monitored using immunodetection with the α-pThr antibody. As shown in Fig. 1d and e, the phosphorylation of DivIVA and LocZ significantly decreased with time, demonstrating that PhpP directly catalyzes dephosphorylation of both StkP substrates.

The *phpP* knock-out strain is sensitive to elevated temperature and oxidative stress

The *stkP* null mutant has an altered growth rate, and it is sensitive to environmental stresses [19], which highlights the importance of StkP in the resistance of pneumococcus to hostile environmental conditions in the host. The growth rate of the *phpP* knock-out strain was reduced in TSB medium (38 min doubling time) compared to the wild type strain (31 min doubling time). In addition, the mutant strain had a significantly prolonged lag phase and reached a lower final optical density (Fig. 2a), similar to the Δ*stkP* mutant strain. Further we examined the growth of the Δ*phpP* mutant in response to heat stress, osmotic stress and pH variation, as well as its viability after exposure to H$_2$O$_2$. The Δ*phpP*, Δ*stkP* and the wild type strains were inoculated in liquid medium and cultivated as described in detail in the *Methods*. Our experiments showed that unlike StkP, PhpP did not significantly affect the sensitivity to osmotic stress induced by high salt concentration or the tolerance to acidic or alkaline pH (data not shown). However, PhpP was important for normal growth at elevated temperatures: the doubling time and final density achieved were significantly affected when the Δ*phpP* strain was grown at 40 °C (Fig. 2b). In addition, we tested the resistance of the mutant strain to oxidative damage. When exposed to varying concentrations of H$_2$O$_2$, the Δ*phpP* strain, similar to the Δ*stkP* strain, displayed a lower survival rate than the wild type strain, indicating increased sensitivity to oxidative stress (Fig. 2c). In summary, the phenotype of the unencapsulated Rx1 derived Δ*phpP* mutant differs from the encapsulated strain 6A which displayed retarded growth under all stress conditions tested but also differs from the strain D39 which was affected only in the high-salt stress

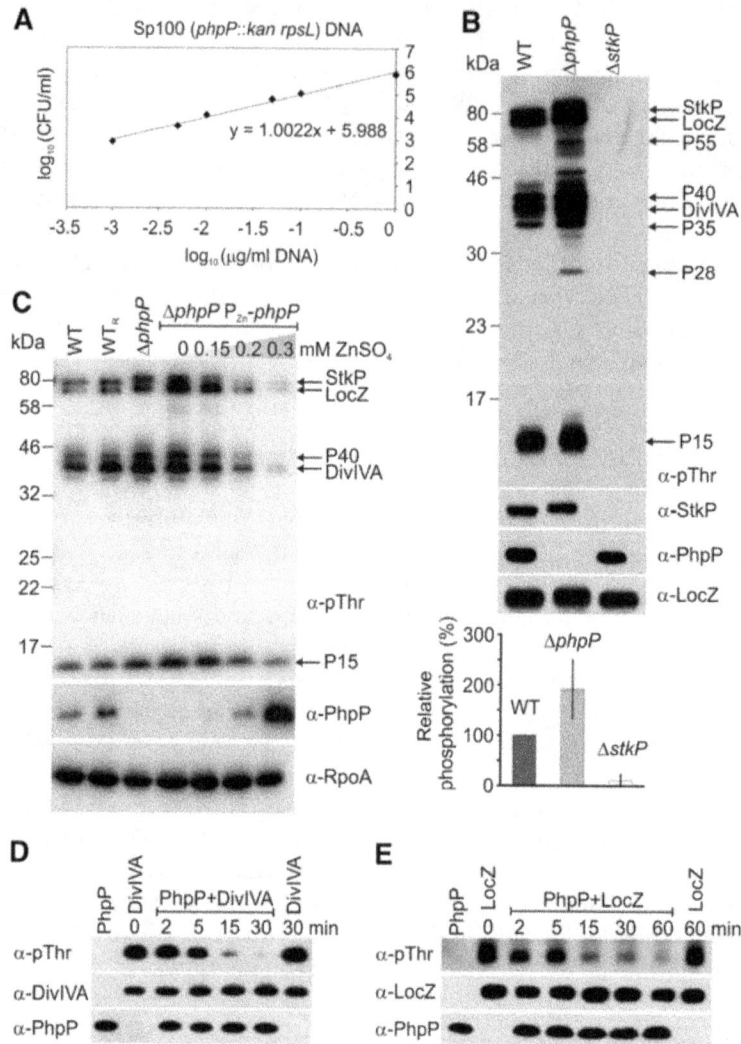

Fig. 1 PhpP regulates phosphorylation of StkP and its substrates. **a** Kinetics of transfer of *phpP* mutation. WT strain was transformed by Sp100 genomic DNA carrying Janus cassette inserted into *phpP* gene (*phpP::kan rpsL*). Number of kanamycin resistant transformants was plotted as a function of genomic DNA concentration in logarithmic scale. Transfer of tested marker follows linear kinetics with slope of line equal to 1 (y = ax + b, a = slope). **b** Phosphorylation profile of Δ*phpP* mutant. Membrane fraction from *S. pneumoniae* WT (Sp1), Δ*phpP* (Sp113) and Δ*stkP* (Sp10) was isolated and 30 μg of proteins was subjected to SDS-PAGE and immunoblotted with anti-phospho-threonine antibody (α-pThr) to detect phosphorylated proteins. The level of PhpP and StkP proteins was detected with anti-PhpP (α-PhpP) and anti-StkP antibody (α-StkP). Immunodetection of membrane protein LocZ was used as loading control. Arrows indicate position of StkP and its known (LocZ, DivIVA) and unknown (P15, P28, P35, P40, P55) substrates. Relative phosphorylation values represent mean ± SD. **c** Complementation of *phpP* mutation. Comparison of phosphoprotein pattern of wild type strain WT (Sp1), reverted wild type strain WT$_R$ (Sp222), Δ*phpP* strain (Sp113) and complementation strain Δ*phpP* P$_{Zn}$-*phpP* (Sp120) cultivated in C + Y medium in the presence or absence of inducer (ZnSO$_4$). Total protein lysates (30 μg) were subjected to SDS-PAGE and immunoblotted with α-pThr antibody to detect phosphorylated proteins. The total amount of PhpP was detected using α-PhpP antibody. Immunodetection of α-subunit of RNA polymerase (α-RpoA) was used as loading control of total cell lysate. Arrows indicate position of StkP substrates. **d** PhpP dephosphorylates DivIVA. **e** PhpP dephosphorylates LocZ. Purified DivIVA-Flag and Flag-LocZ were incubated with His-PhpP in vitro and reaction was stopped at indicated times. Samples were subjected to SDS-PAGE and immunoblotted with α-pThr antibody to visualize dephosphorylation in time. PhpP, LocZ and DivIVA were detected with specific antibodies as described above. To exclude the spontaneous decay, phosphorylated form of both proteins, DivIVA-Flag and Flag-LocZ, was incubated for 30 and 60 min, respectively, in phosphatase reaction buffer without addition of His-PhpP

conditions [11]. These results suggest that genetic background significantly affects the demonstration of *phpP* mutation although we cannot exclude the role of polysaccharide capsule itself. Considering that the Rx1 derived *stkP* mutant strain was sensitive to elevated temperature, acidic pH, osmotic and oxidative stress [19], we did not confirm the opposite effect of PhpP. Our data suggest that unbalanced activity of both, PhpP and StkP, is critical for bacterial physiology, and the adaptive response to environmental stress is not cooperatively regulated by the PhpP/StkP signaling couple.

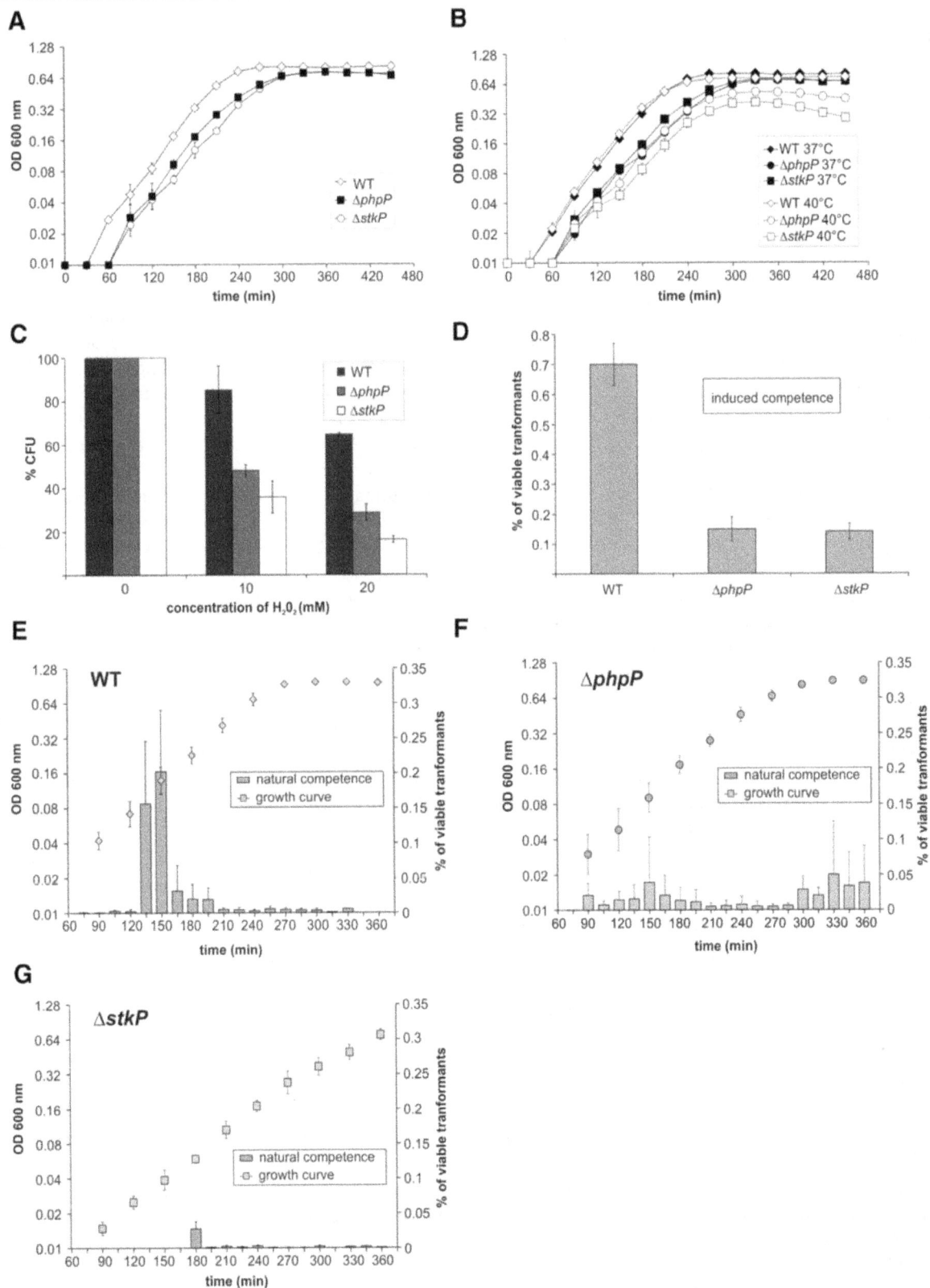

Fig. 2 (See legend on next page.)

The Δ*phpP* strain displays decreased competence for genetic transformation

Competence for genetic transformation is powerful mechanism for generating genetic diversity and acquiring antibiotic resistance. Natural competence in *S. pneumoniae* is a transient event regulated by a quorum-sensing system and occurs via a peptide pheromone signal (e.g., competence-stimulating peptide (CSP)). Previous studies demonstrated the importance of StkP for competence development [10, 19]. Here, we tested the capability of the Δ*phpP* mutant strain to develop natural and induced competence in conditions optimal for competence development [36]. Induced competence was defined as the transformation efficiency in response to the addition of synthetic CSP. Similar to the Δ*stkP* strain, the *phpP* null mutant strain weakly developed induced competence, and the transformation efficiency was low compared to wild type (Fig. 2d). Natural competence was defined as the transformation efficiency in the absence of added CSP and was monitored during the growth. As shown in Fig. 2e, that wild type strain developed natural competence during the early exponential phase of growth (OD_{600} 0.08–0.16), which is observed as a peak in viable transformants obtained by transformation with control DNA. On the other hand, two low peaks of competence, one during the exponential phase and the second upon the entry into the stationary phase, were detected during growth of the *phpP* knock-out strain (Fig. 2f). However, the transformation efficiency was about fivefold lower than that observed for the wild type, and therefore, the strain is competence deficient, similar to the Δ*stkP* strain (Fig. 2 g). In the Δ*stkP strain* the reduced transformation efficiency may be the result of a weak induction of DNA uptake and processing genes [19]. However, the molecular mechanism responsible for the transformation deficiency in the Δ*stkP* mutant remains unclear. To date, none of the proteins that play a direct role in competence development have been identified as a substrate of StkP. Our results suggest that StkP and PhpP do not function as antagonists in the control of competence regulation. Therefore, we cannot exclude that competence deficiency of both mutants is an indirect consequence of the pleiotropic effects of *phpP* and *stkP* mutations on pneumococcal physiology.

PhpP regulates cell division in pneumococcus

The newly established role of StkP in cell division prompted us to investigate the morphology of the *phpP* knock-out strain and the potential role of PhpP in the regulation of cell division. Although no morphologic changes were observed in the encapsulated Δ*phpP* mutants [11], phase contrast microscopy of the Rx1 derived Δ*phpP* strain revealed that the phenotype differed from that of the wild type strain. Cell size measurement using the automated MicrobeTracker software confirmed that the Δ*phpP* cells were significantly smaller (median cell length 1.48 ± 0.22 μm; median cell width 0.64 ± 0.08 μm) than the wild type cells (median cell length 1.58 ± 0.25 μm; median cell width 0.66 ± 0.04 μm) (Fig. 3a, c) and showed phenotype similar to the StkP-overexpressing strain. To demonstrate that the observed phenotype was the result of the *phpP* deletion, we analyzed the reverted strain WT_R (Sp222) and the complementation strain Δ*phpP* P_{Zn}-*phpP* (Sp120). Cell morphology (Fig. 3a) and cell size (Fig. 3c) of the reverted strain were not different from the wild type strain. The analysis of the complementation strain Sp120 cultivated in the presence of a growing concentration of the inducer ($ZnSO_4$) demonstrated that increasing expression of PhpP correlated with the decrease in protein phosphorylation (Fig. 1c) and an increase in cell length (Fig. 3b, c). In the presence of 0.3 mM $ZnSO_4$, the median cell length reached 1.9 ± 0.33 μm (Fig. 3b, c), clearly indicating that overexpression of PhpP results in a phenotype similar to that observed with the StkP-depleted strain (strain Sp10, median cell length 2.11 ± 0.32 μm) (Fig. 3a, c).

To obtain further insight into the morphological changes induced by inactivation of *phpP*, we analyzed the mutant strain Sp113 by electron microscopy. Figure 3d shows a scanning electron microscopy image of the wild type (Sp1) compared with the Δ*phpP* (Sp113) and Δ*stkP* (Sp10) strains. As reported previously, the Δ*stkP* strain produces long cells with unconstricted septa. On the other hand, we observed cell size heterogeneity in Δ*phpP* mutant cells, with numerous smaller cells; however, their shape appeared to be normal in general. Additionally, transmission electron microscopy did not reveal significant abnormalities (thicker cell walls)

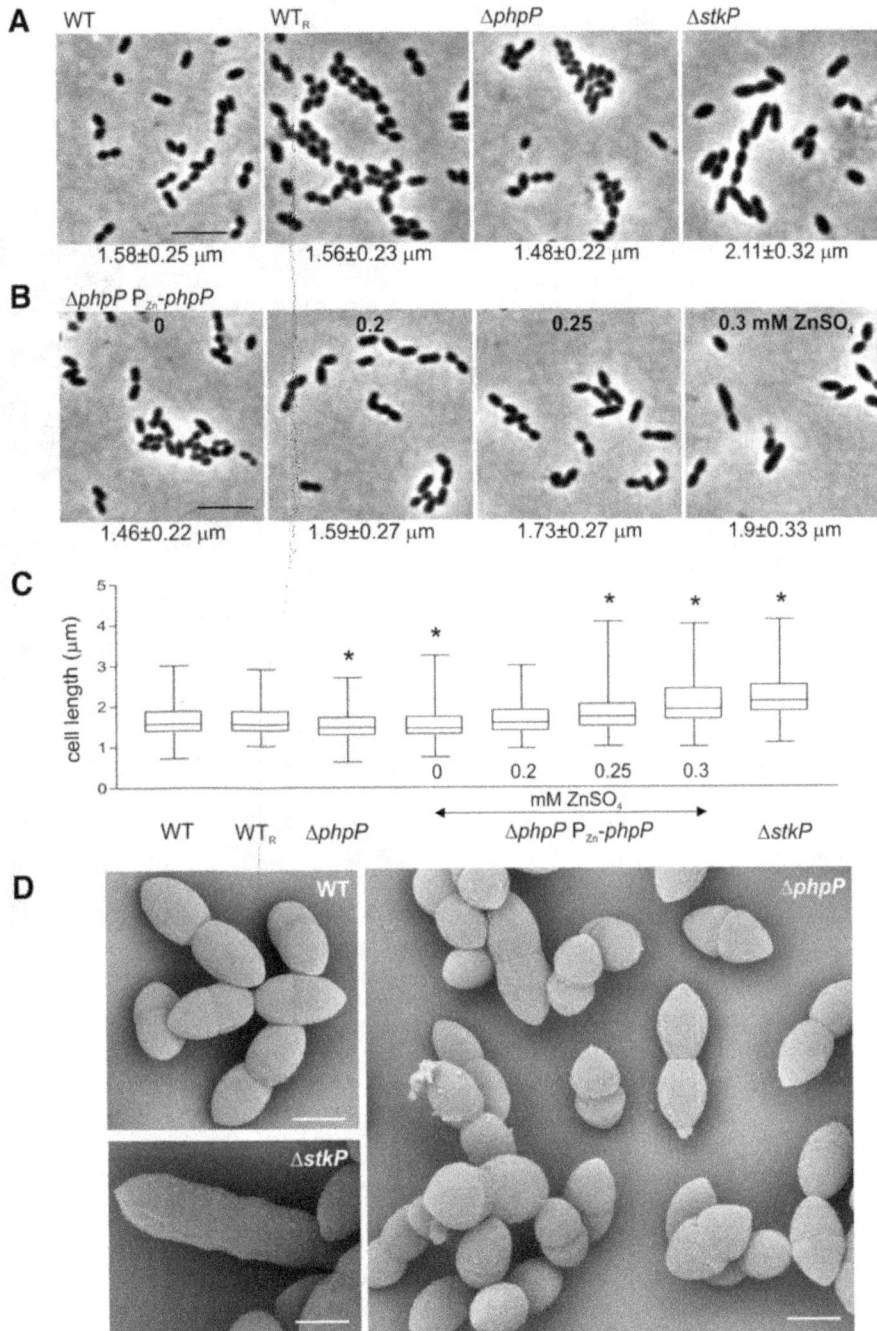

Fig. 3 PhpP is involved in regulation of cell division in *S. pneumoniae*. **a** Morphology and cell length analysis of WT strain (Sp1), reverted strain WT_R (Sp222) and Δ*phpP* (Sp113) or Δ*stkP* (Sp10) strains grown in C + Y medium to mid-exponential phase. **b** Cell length depends on expression of PhpP in complementation strain Δ*phpP* P_Zn-*phpP* (Sp120). Micrographs of complementation strain cultivated in the presence of 0, 0.2, 0.25 and 0.3 mM $ZnSO_4$ in C + Y medium. Cell length in panel A and B is expressed as a median value ± MAD (*n* = 300). Bar 5 μm. **c** Cell length analysis. Cell length parameters measured with MicrobeTracker software were analyzed and plotted in box-and-whiskers graph. Mann-Whitney *U* test: * cell length of mutant strain is significantly different from that of the WT strain *P* < 0.0001. 300 cells were scored per sample. **d** Scanning electron microscopy of WT strain (Sp1), Δ*stkP* strain (Sp10) and Δ*phpP* strain (Sp113) cultivated in TSB medium. Magnifications are the same for all panels. Bar 0.5 μm. a-d: Representative data for three independent experiments are shown

associated with the loss of *phpP* (data not shown) which were observed in the encapsulated Δ*phpP* strain [11].

Analysis of 600 cells showed that 24.2 % of the Δ*phpP* cells formed chains longer than 4 cells in contrast with 2.5 % of chaining cells in the wild type. However, we did not observe aggregation and abnormally long chains which were detected in the encapsulated mutant strains [11]. Regulation of chain length in streptococci depends on wall-associated autolytic activity. Therefore we tested the expression of genes encoding the peptidoglycan hydrolases *pcsB*, *lytA* and *lytB*, which may affect cell separation, using qRT-PCR, but we did not detect any differences in transcript levels in the Δ*phpP* and wild type strain (data not shown); thus, the reason for the increased chain formation remains unknown.

To characterize the role of PhpP in cell division in more detail, we investigated the localization of nascent PG synthesis sites in live Δ*phpP* cells (Sp113) stained with fluorescently labeled vancomycin (Van-FL), a marker of nascent peptidoglycan synthesis (PG) (Fig. 4). Labeling was observed predominantly at current and future cell division sites in the mutant cells, a pattern similar to that observed in the wild type cells. However, 4.5 % of mutant cells (58/1300) showed disturbed Van-FL labeling (Fig. 4) indicating that minority of cells display perturbed cell wall synthesis. When we induced overexpression of PhpP in complementation strain Sp120 (Δ*phpP* P$_{Zn}$-*phpP*) by the addition of 0.3 mM ZnSO$_4$, we observed significant elongation of cells, and Van-FL staining revealed that the cells often contained multiple unconstricted division septa, which is a distinct feature of *stkP*-depleted cells (Fig. 4). Further we investigated localization of cell division proteins LocZ/MapZ, FtsA and DivIVA but we did not find significant differences between mutant and wild type cells (data not shown).

Our data clearly show that PhpP plays an opposing role to StkP in regulation of cell division which was not recognized in the previous study by Agarwal et al. [11]. We hypothesize that the morphological differences between Δ*phpP* mutants derived from different strains may be largely caused by the presence or absence of capsule. However, the cell division defect caused by the depletion of *phpP* is less severe than the abnormalities observed either in the absence of StkP or in the presence of the excess of PhpP suggesting that hyperphosphorylation of StkP substrates is better tolerated than the absence of phosphorylation.

Conserved residues D192 and D231 are essential for PhpP activity *in vivo*

PhpP contains two conserved aspartate residues, D192 and D231, which are directly involved in metal ions binding and are essential for the activity of eukaryotic

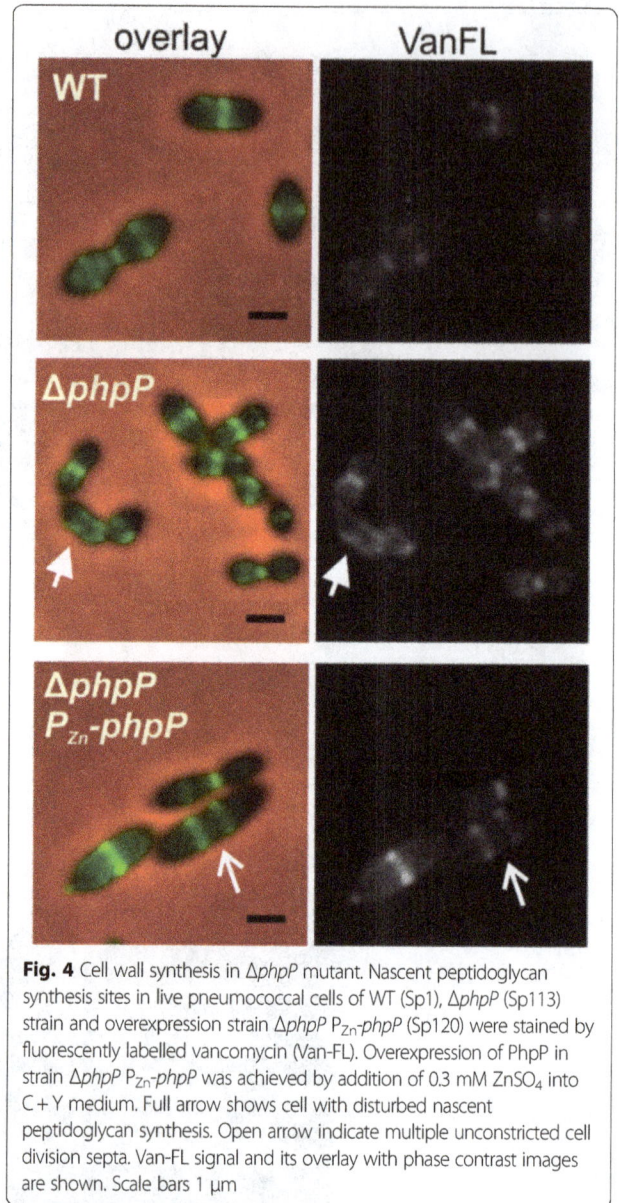

Fig. 4 Cell wall synthesis in Δ*phpP* mutant. Nascent peptidoglycan synthesis sites in live pneumococcal cells of WT (Sp1), Δ*phpP* (Sp113) strain and overexpression strain Δ*phpP* P$_{Zn}$-*phpP* (Sp120) were stained by fluorescently labelled vancomycin (Van-FL). Overexpression of PhpP in strain Δ*phpP* P$_{Zn}$-*phpP* was achieved by addition of 0.3 mM ZnSO$_4$ into C + Y medium. Full arrow shows cell with disturbed nascent peptidoglycan synthesis. Open arrow indicate multiple unconstricted cell division septa. Van-FL signal and its overlay with phase contrast images are shown. Scale bars 1 μm

PP2C phosphatases [37]. Previously, we reported that substitution of D192 and D231 for alanine abolished PhpP activity in vitro [17]. Here, we investigated the importance of D192 and D231 for enzymatic activity and localization of PhpP in vivo. We constructed strains expressing PhpP mutant alleles D192A and D231A fused to GFP under an inducible P$_{Zn}$ promoter in the Δ*phpP* genetic background. Strains expressing GFP-PhpP-WT (Sp140), the D192A allele (Sp292) and the D231A allele (Sp293) were cultivated in C + Y medium with or without zinc, and PhpP expression and the phosphorylation pattern were detected using specific antibodies. Expression of PhpP-WT and PhpP-D192A was similar; however, the expression of PhpP-D231A was lower as indicated by immunodetection with the α-GFP antibody

(Fig. 5a). Immunodetection with the α-pThr antibody showed that increasing the expression of the wild type allele resulted in a decrease in overall phosphorylation intensity. On the other hand, increased expression of PhpP-D192A or D231A upon addition of zinc did not affect the phosphorylation intensity in strains Sp292 and Sp293, respectively, indicating that both alleles are catalytically inactive. Phase contrast microscopy revealed that the morphology of strain Sp140 expressing GFP-PhpP-WT changed depending on zinc concentration, and cell length increased (Fig. 5b). On the other hand, the morphology of strains expressing the mutant alleles of PhpP did not change (Fig. 5b) indicating inability to complement the mutant phenotype. Cell size analysis confirmed these observations (Fig. 5c). PhpP-WT expression in the Sp140 strain led to increases in cell length up to 2.11 ± 0.36 μm when 0.3 mM $ZnSO_4$ was added to the medium, while cell length remained unchanged in strains Sp292 (PhpP-D192A) (1.49 ± 0.2 μm) and Sp293 (PhpP-D231A) (1.43 ± 0.23 μm).

Previously, we demonstrated that the protein phosphatase PhpP is localized in the cytoplasm, but it is significantly enriched at the midcell during the early exponential phase of growth, and this localization depends on the presence of active StkP [20]. To determine localization of catalytically inactive GFP-PhpP, we cultivated strains Sp140, Sp292 and Sp293 in medium supplemented with 0.2 mM $ZnSO_4$ until the early exponential phase (OD_{600} 0.2) and examined live cells using fluorescence microscopy. PhpP-WT was clearly associated with the cell division septum in 23 % (176/766) of the cells and showed cytoplasmic distribution in 77 % of cells (Fig. 5d). PhpP-D192A was enriched at the midcell in 19 % (152/800) of the cells; however, the GFP signal in these cells was more diffuse (Fig. 5d). To quantify the difference, we measured the fluorescence intensity profiles in the cells at the first stage of cell division (predivisional cells). We confirmed different distributions of the GFP-PhpP-WT and GFP-PhpP-D192A signals along the cell axis, which indicates that PhpP-D192A is more abundant in the cytoplasm (Fig. 5d). Interestingly, PhpP-D231A was localized exclusively in the cytoplasm (Fig. 5d). These data suggest that mutant alleles of PhpP not only loose the catalytic activity but also lose the ability to co-localize with cell division apparatus.

Jag protein (Spr1851) is a previously unknown substrate of StkP and PhpP

The elevated level of Thr phosphorylation in the *phpP* null mutant helped us to detect phosphorylated membrane proteins previously unrecognized in wild type lysates. Because the phosphorylated proteins designated P35 and P40 migrate close to phosphorylated DivIVA, we generated strain Δ*phpP*Δ*divIVA* (designated Sp169), which would enable

appropriate separation of the new substrates. To identify these StkP substrates we extracted proteins from the membrane fraction using trifluorethanol (TFE) [38] and separated them using two-dimensional (2D) SDS-PAGE as described in the *Methods*. The protein spot corresponding to P40 was successfully resolved, and its phosphorylation was confirmed by immunoblotting (Fig. 6a). The protein spot was excised, digested by trypsin and identified using MALDI-TOF mass spectrometry as Spr1851, a homolog of Jag/SpoIIIJ-associated protein from *B. subtilis*. We named the product of the *spr1851* gene Jag_Spn (Fig. 6b). Jag_Spn contains an N-terminal Jag_N domain and KH domain followed by an R3H domain at the C-terminus (Conserved Domain Database (CDD) [39]) (Fig. 6b). The Jag_N domain located at the N-terminus of bacterial Jag proteins is a conserved stretch of 50 amino acids without a defined function (CDD). The KH domain is a single-stranded nucleic acid-binding domain that mediates RNA target recognition in proteins that regulate gene expression in eukaryotes and prokaryotes (reviewed in [40]). The R3H motif is present in proteins from a diverse range of organisms that includes Eubacteria, green plants, fungi and various groups of metazoans, and it is predicted to bind ssDNA or ssRNA in a sequence-specific manner. Jag homologues are conserved in bacteria, especially in *Firmicutes*, and their domain architecture suggests that they bind RNA; however, their function is unknown.

To verify the phosphorylation of Jag_Spn we constructed a *jag* null mutant named Sp295 using the Janus cassette strategy described in the *Methods*. We detected phosphorylated proteins in whole cell lysates of the wild type, Δ*jag* and Δ*stkP* strains and compared them with strain Δ*phpP*Δ*divIVA* to better distinguish different phosphoprotein bands. Immunoblotting showed that a phosphoprotein corresponding to 40 kDa (P40) is present in the wild type and Δ*phpP*Δ*divIVA* strain but absent in the Δ*jag* strain, which indicates that Spr1851/Jag corresponds to StkP substrate P40 (Fig. 6c).

To verify further phosphorylation of Jag we prepared the complementation strain Sp304 expressing Jag with a Flag-tag at the C-terminus (Jag-Flag) under the inducible P_{Zn} promoter in the Δ*jag* genetic background. Figure 6d shows that addition of zinc induced expression of Jag-Flag in strain Sp304 (Δ*jag* P_{Zn}-*jag-flag*), and immunodetection with the α-pThr antibody confirmed that the protein is phosphorylated. A second, faster migrating band of Jag was detected with the α-Flag antibody when $ZnSO_4$ was added at concentrations of 0.25 and 0.3 mM. This protein band interacted weakly with the α-pThr antibody, which indicated that this form of Jag is also phosphorylated.

Jag is phosphorylated on Thr89

Spr1851/Jag was previously found to be phosphorylated on Thr89 in a global study published by Sun et al. [41]. Thr89

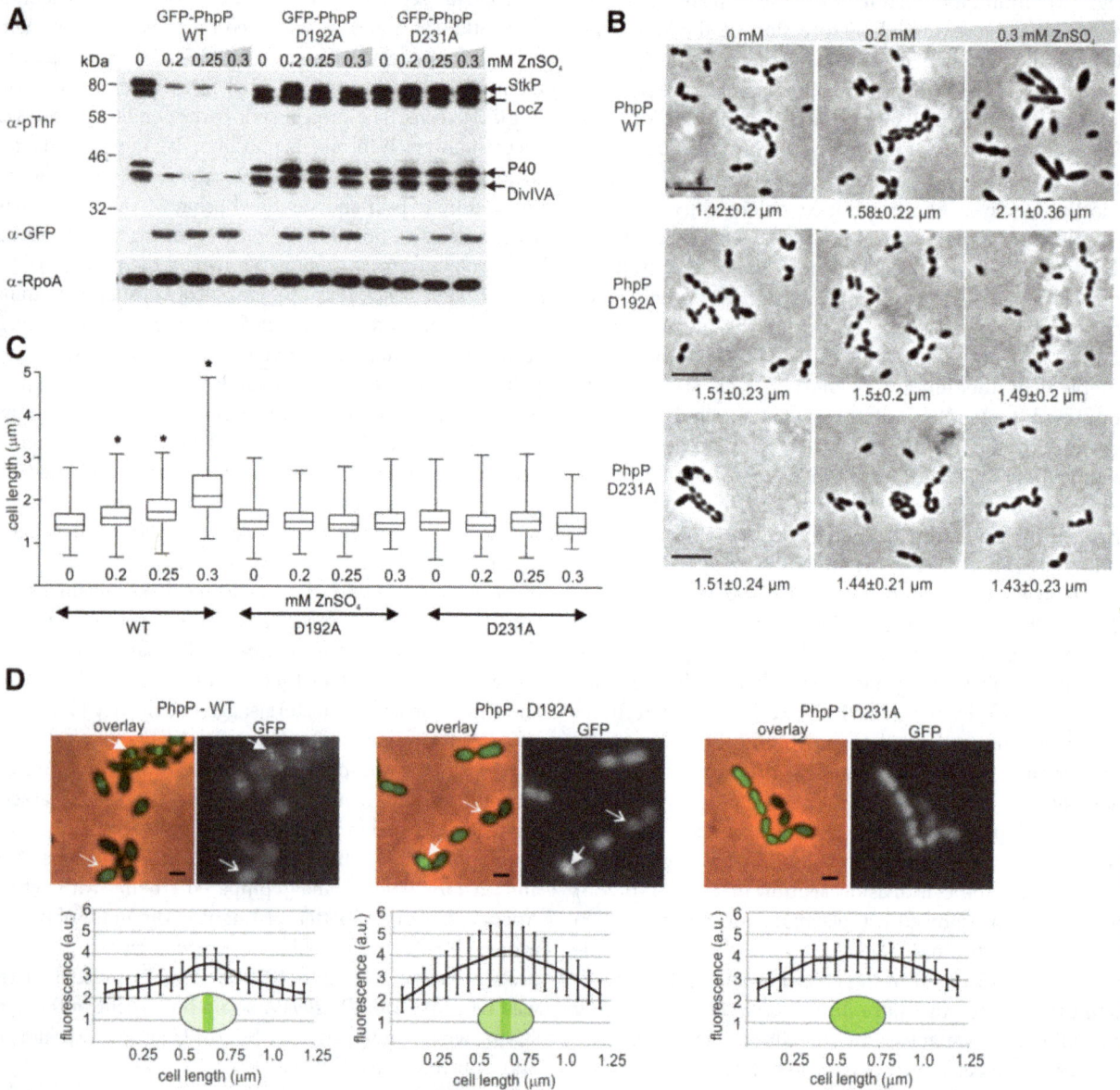

Fig. 5 D192 and D231 are essential for PhpP catalytic activity *in vivo*. **a** Phosphorylation pattern after induction of expression of GFP-PhpP-WT (Sp140), GFP-PhpP-D192A (Sp292) and GFP-PhpP-D231A (Sp293) in Δ*phpP* genetic background. Total cell lysates from cultures grown in C + Y medium in the presence or absence of $ZnSO_4$ were separated by SDS-PAGE and immunoblotted with α-pThr antibody to document protein phosphorylation. α-GFP antibody was used to show expression level of GFP-PhpP and immunodetection of RpoA was used as a loading control. Position of StkP and its substrates is indicated by arrows. **b** Morphology of strains expressing GFP-PhpP-WT (Sp140), GFP-PhpP-D192A (Sp292) and GFP-PhpP-D231A (Sp293) in Δ*phpP* genetic background. Pneumococcal strains were cultivated in C + Y medium supplemented with $ZnSO_4$. Phase contrast images show cell morphology in the presence of 0.2 and 0.3 mM $ZnSO_4$ and median cell lengths ± MAD (*n* = 300) corresponding to each image are shown below. Bar, 5 μm. **c** Cell length analysis. Cell length parameters were analyzed and plotted in box-and-whiskers graph. Mann-Whitney *U* test: * cell length in the presence of inducer is significantly different from uninduced conditions (0 mM $ZnSO_4$) *P* < 0.0001. 300 cells were scored per sample. **d** Localization of PhpP. Strains expressing GFP-PhpP-WT (Sp140), GFP-PhpP-D192A (Sp292) and GFP-PhpP-D231A (Sp293) were cultivated in C + Y medium supplemented with 0.2 mM $ZnSO_4$. GFP signal and overlay of phase contrast and GFP signal are shown. Enrichment of PhpP at midcell of cells at first stage of cell division (predivisional cells) is indicated by full arrow; cells showing cytoplasmic localization of PhpP are indicated by open arrow. Bar, 1 μm. Predivisional cells (*n* = 20) showing either midcell enrichment of PhpP (WT and D192A) or cytoplasmic localization (D231A) were selected to quantify distribution of GFP-PhpP along the cell axis. Fluorescence intensity (arbitrary units) versus cell length is plotted in corresponding graphs (error bars show SD)

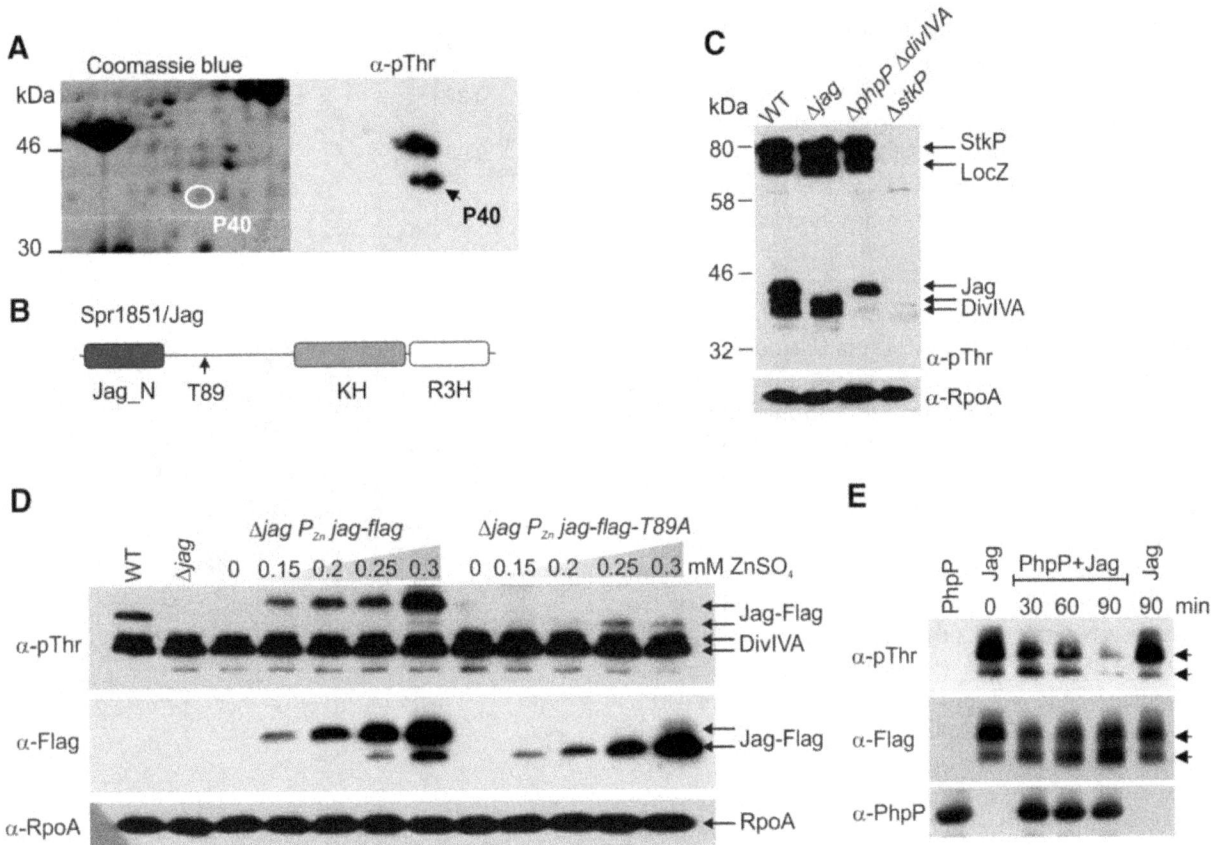

Fig. 6 Spr1851/Jag is a new substrate of StkP and PhpP. **a** Identification of Jag. Membrane fraction isolated and extracted by TFE/chloroform from strain Sp169 (ΔphpP ΔdivIVA) was resolved on 2D SDS-PAGE, immunoblotted with α-pThr antibody and matched with parallel gel stained with Coomassie Blue. Spot corresponding to P40 was excised, analyzed by MALDI-MS and identified as Spr1851. **b** Schematic structure of Spr1851/Jag. Jag_N: conserved domain found at the N-terminus of Jag proteins; T89: phosphorylated threonine 89; KH: single-stranded RNA binding domain; R3H: putative single-stranded nucleic acid binding domain. **c** Deletion of jag. Whole cell lysates of WT (Sp1), Δjag (Sp295), ΔphpP Δdi-vIVA (Sp169) and ΔstkP (Sp10) were separated by SDS-PAGE and immunoblotted with α-pThr antibody to detect protein phosphorylation. Immunodetection of α-subunit of RNA polymerase (α-RpoA) was used as a loading control. Arrows indicate position of StkP and its substrates. **d** Phosphorylation of T89. Total protein lysates (30 μg) of WT (Sp1), Δjag (Sp295), complementation strain Δjag PZn-jag-flag (Sp304) and Δjag PZn-jag-flag-T89A (Sp302) were separated by SDS-PAGE and immunoblotted with α-pThr antibody to detect protein phosphorylation. Expression of Jag-Flag was monitored by immunodetection with anti-Flag antibody (α-Flag) and immunodetection of RpoA (α-RpoA) was used as a loading control. Arrows indicate position of proteins. **e** Dephosphorylation of Jag. Purified Jag-Flag was incubated with His-PhpP in vitro and reaction was stopped at given time. Samples were subjected to SDS-PAGE and immunoblotted with α-pThr antibody to visualize dephosphorylation in time. PhpP-His was detected with α-PhpP antibody and Jag-Flag was detected with α-Flag antibody. To exclude the spontaneous decay, phosphory-lated form of Jag-Flag was incubated for 90 min in phosphatase reaction buffer without addition of His-PhpP. Arrows indicate position of two differentially phosphorylated forms of Jag

is located in a region that does not show significant homology to any conserved domain; however, Thr89 is conserved in many streptococcal species (KEGG, Kyoto Encyclopedia of Genes and Genomes). To verify phosphorylation of Thr89, we mutagenized this residue to unphosphorylatable alanine and constructed strain Sp302 expressing the phosphoablative allele jag-flag-T89A under the P_{Zn} promoter in the Δjag genetic background. Upon induction of expression, we detected production of Jag-Flag-T89A migrating faster than the major form of Jag-Flag-WT. This suggests that alteration of mobility is related to the unphosphorylated state (Fig. 6d). Immunodetection with the α-pThr antibody showed a significant decrease in

phosphorylation; however, the signal was not completely lost, and Jag-T89A reacted weakly with the α-pThr antibody when expressed at higher levels (0.25–0.3 mM ZnSO₄) (Fig. 6d, lanes 11, 12). These results confirm that Thr89 is indeed a phosphoacceptor residue in vivo; however, another still unidentified secondary phosphorylation site is present in Jag.

Jag is dephosphorylated by PhpP

To demonstrate that PhpP directly dephosphorylates Jag, we isolated phosphorylated Jag-Flag from cell lysates of strain Sp304 via affinity chromatography and performed in vitro dephosphorylation reactions as described in the

Methods. Phosphorylation of Jag was monitored by immunodetection with the α-pThr antibody, and the results showed that the loss of the phosphorylation signal was time dependent (Fig. 6e). This experiment confirmed that PhpP directly dephosphorylates Jag. Two different forms of Jag were detected upon incubation with PhpP, correlating with our finding that Jag most likely contains more than one phosphorylated residue (Fig. 6d).

Characterization of the Δ*jag* phenotype

To obtain insight into Jag function in pneumococcus, we characterized the phenotype of a Δ*jag* mutant. The Δ*jag* mutant showed retarded growth in TSB medium, with a longer doubling time (32 min) than the wild type (29 min). The mutant had a significantly longer lag phase and reached stationary phase at a lower optical density (Fig. 7a). Phase contrast microscopy indicated that mutant cells are smaller, a phenotype reminiscent of the Δ*phpP* mutant. Cell size analysis confirmed that the median cell length (1.33 ± 0.19 μm) and cell width (0.59 ± 0.01 μm) of the mutant were significantly smaller than the median cell length and width of the wild type (1.57 ± 0.2 μm and 0.67 ± 0.09 μm) ($P < 0.0001$; Mann-Whitney rank sum test). Scanning electron microscopy further supported these data; however, no significant abnormalities in cell shape and morphology were observed (Fig. 7b). We also determined cell size of the complementation strain Sp304 (Δ*jag* P_{Zn}-*jag*-*flag*) upon addition of increasing zinc concentrations. Induced expression of Jag-WT resulted in complementation characterized by increasing cell length (Fig. 7c). This rescue of the phenotype confirmed the relationship between inactivation of *jag* and decreased cell dimensions. The cells reached wild type cell length at a concentration of approximately 0.25 mM ZnSO₄ (1.67 ± 0.25 μm). The cell length increased further upon addition of 0.3 mM ZnSO₄ until it reached 1.76 ± 0.26 μm, suggesting that overexpression of Jag led to significant cell elongation. These data suggest that Jag plays a role in pneumococcal cell division and helps to maintain proper cell shape.

The *jag* homologue in *B. subtilis* forms a bicistronic operon with the *spoIIIJ* gene [31], which corresponds to pneumococcal *spr1852* encoding the YidC1/Oxa1 membrane protein insertase. This gene cluster is widely conserved (KEGG). YidC homologues are required for the insertion and/or proper folding of integral membrane proteins (reviewed in [42]). Most Gram-positive bacteria encode two YidC paralogues, YidC1 and YidC2, which correspond to Spr1852 and Spr1790, respectively, in *S. pneumoniae*. The role of YidC homologues in *S. pneumoniae* has not been described; however, in *S. mutans*, disruption of YidC2 results in a loss of genetic competence, decreased membrane-associated ATPase activity

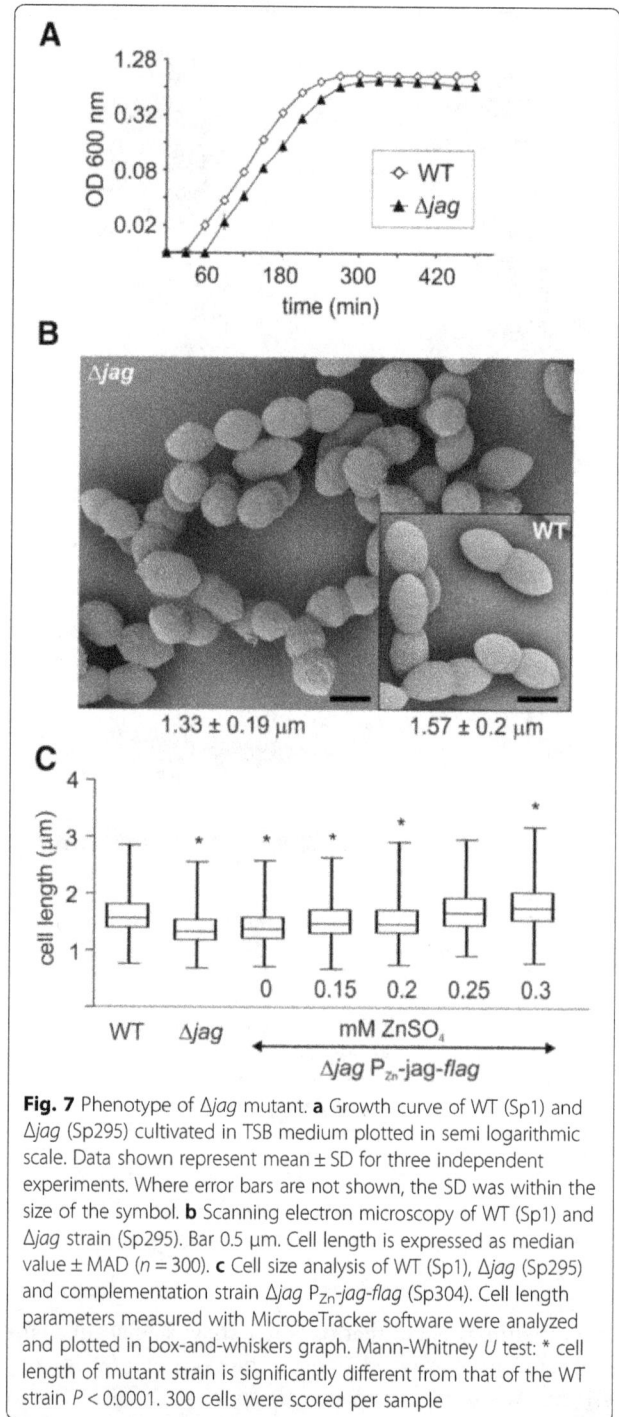

Fig. 7 Phenotype of Δ*jag* mutant. **a** Growth curve of WT (Sp1) and Δ*jag* (Sp295) cultivated in TSB medium plotted in semi logarithmic scale. Data shown represent mean ± SD for three independent experiments. Where error bars are not shown, the SD was within the size of the symbol. **b** Scanning electron microscopy of WT (Sp1) and Δ*jag* strain (Sp295). Bar 0.5 μm. Cell length is expressed as median value ± MAD ($n = 300$). **c** Cell size analysis of WT (Sp1), Δ*jag* (Sp295) and complementation strain Δ*jag* P_{Zn}-*jag*-*flag* (Sp304). Cell length parameters measured with MicrobeTracker software were analyzed and plotted in box-and-whiskers graph. Mann-Whitney *U* test: * cell length of mutant strain is significantly different from that of the WT strain $P < 0.0001$. 300 cells were scored per sample

and stress sensitivity. Loss of YidC1 has less severe defects, with little observable effect on growth or stress sensitivity [43]. Although the two insertases have different physiological functions, both of them contribute to biofilm formation and cariogenicity in rats [43].

The Jag association with the membrane and likely cotranscription with *yidC1* suggest that both proteins might be functionally linked. It is tempting to speculate that Jag$_{Spn}$ plays an indirect role in targeting of the integral membrane proteins. Given the Δjag phenotype and its phosphorylation by StkP, which regulates cell division in pneumococcus, Jag$_{Spn}$ might specifically affect targeting of cell division proteins. YidC homologues are involved in cell division processes in different bacteria. The well-studied YidC1/SpoIIIJ in *B. subtilis* is required for sporulation [31]. Interestingly, the cell division proteins FtsQ and FtsEX have been found to be substrates of YidC in *E. coli* and *Shigella*, respectively [44, 45]. A recent report also showed that YidC assists in the biogenesis of penicillin-binding proteins (PBP) in *E. coli*, and in the absence of YidC, two critical PBPs, PBP2 and PBP3, are not correctly folded, and their substrate-binding capacity is reduced, although the total amount of protein in the membrane is not affected [46].

Conclusions

Streptococcus pneumoniae has a characteristic ovoid shape, which is most likely achieved by the concerted action of two peptidoglycan biosynthetic machineries: peripheral and septal [47, 48]. We previously proposed a model in which Ser/Thr protein kinase StkP coordinates cell wall synthesis and cell division in *S. pneumoniae* [20]. Here, we demonstrate that the cognate Ser/Thr protein phosphatase PhpP is not essential as published previously [10, 21, 28] and plays an opposing role in cell division to that of StkP. Overexpression of PhpP, which leads to dephosphorylation of StkP substrates, mimics the *stkP* null phenotype and the dividing cells are elongated and contain multiple unconstricted cell division septa. In the absence of *phpP* we observe enhanced autophosphorylation of StkP and hyperphosphorylation of StkP substrates. We show that PhpP regulates not only activity of StkP but dephosphorylates directly StkP substrates. The morphology of $\Delta phpP$ cells resembles StkP overexpression, and the cells do not achieve the size of the wild type, most likely due to insufficient elongation of cells or premature constriction of the Z-ring. We hypothesize that PhpP and StkP co-ordinately regulate the shift from peripheral to septal cell wall synthesis through phosphorylation of several substrates, including cell division proteins. In contrast, we did not confirm a straightforward regulatory impact of PhpP on the other functions of StkP. Characterization of the *phpP* null mutant revealed that like the *stkP* null mutant, it is more sensitive to elevated temperature, oxidative stress and that both mutant strains have reduced competence for genetic transformation. These results suggest that unbalanced activity of each of these enzymes is critical for bacterial physiology. We cannot exclude the possibility that PhpP may also have broader substrate specificity and may dephosphorylate phosphoproteins other than StkP substrates. Detection of proteins specifically

phosphorylated in the $\Delta phpP$ strain allowed us to identify new substrate modified by StkP/PhpP couple. The product of gene *spr1851* called Jag$_{Spn}$ is a putative RNA binding protein phosphorylated on Thr89. Jag proteins are widely conserved in bacteria and their role is unknown. Phenotype of the Δjag mutant suggests that Jag$_{Spn}$ is involved in cell division and maintaining proper cell shape of *S. pneumoniae*.

Methods

Bacterial strains and growth conditions

The bacterial strains used in this study are listed in Table 1. *E. coli* DH5α used as a general purpose cloning host and *E. coli* BL21 used for protein expression were cultured in Luria–Bertani (LB) broth at 37 °C. The wild type *S. pneumoniae* strain Rx1 and its corresponding mutants were grown statically at 37 °C in Brain-heart Infusion (BHI) medium, Tryptone Soya Broth (TSB) medium, semi-synthetic C medium supplemented with 0.1 % yeast extract (C + Y) [49] or in Casein Tryptone (CAT) medium supplemented with 0.2 % glucose and 1/30 volume 0.5 M K_2HPO_4, pH 7.5 [49]. DNA from the strain CP1016 (rif-23) was used as the donor DNA for the competence assays [50]. The following antibiotics were added when necessary at the indicated concentrations (in µg ml^{-1}): rifampin (Rif), 1; kanamycin (Kan), 200; streptomycin (Sm), 500; tetracycline (Tet), 2.5; erythromycin (Erm), 1 (for *S. pneumoniae*); ampicillin (Amp), 100; kanamycin (Kan), 50; erythromycin (Erm), 100 (for *E. coli*).

Plasmid construction

Plasmids used in this study are listed in Table 1 and oligonucleotides are listed in Additional file 1: Table S1. To construct plasmid pZn-PhpP, *phpP* was amplified with primers JG19 and JG20 using WT chromosomal DNA as a template. The P_{czcD} (P_{Zn}) promoter was amplified with primers LN123 and JG21 from a template plasmid pJWV25 [35]. Both PCR products were used as a template in a fusion PCR with primers LN123 and JG20. The final PCR product P_{Zn}-*phpP* was cloned into the KpnI and NotI restriction sites of the plasmid pJVW25. Plasmid pZn-jag-flag was constructed as follows: *jag* gene was amplified with primer pairs AU77 and AU79 (containing Flag sequence and NotI restriction site) using wild type chromosomal DNA as a template. The P_{Zn} promoter was amplified with primers LN123 and AU76 from a template plasmid pJWV25. Both PCR products were used as a template in a fusion PCR with primers LN123 and AU79. The final PCR product was cloned into the EcoRI and NotI restriction sites of the plasmid pJVW25. To generate pZn-flag-locZ, the P_{Zn} promoter was amplified with primers LN123 and NS1 from a template pJWV25 and *locZ* was amplified with

Table 1 Bacterial strains and plasmids used in this study

Strain/plasmid	Genotype or description	Source
Strains		
S. pneumoniae		
Sp1 (Rx1)	unencapsulated wild-type, *str1; hexA*	[54]
Sp10	*Cm, stkP::cm*	[17]
Sp26	*Cm, divIVA::cm*	[55]
Sp57	*locZ::lox72*	[25]
Sp100	*Kan, phpP::kan rpsL*	This work
Sp113	*ΔphpP*	This work
Sp120	*Tet, ΔphpP bga::P$_{Zn}$-phpP*	This work
Sp140	*Tet, ΔphpP bga::P$_{Zn}$-gfp-phpP*	This work
Sp161	*Kan, ΔphpP divIVA::kan rpsL*	This work
Sp169	*ΔphpP ΔdivIVA*	This work
Sp174	*Erm, ΔdivIVA pMU-P96-divIVA-flag*	This work
Sp188	*Erm, ΔlocZ pMU-P96-flag-locZ*	This work
Sp220	*Kan, phpP::kan rpsL* (reverted from Sp113)	This work
Sp222	wild-type (reverted from Sp113)	This work
Sp292	*Tet, ΔphpP bga::P$_{Zn}$-gfp-phpP-D192A*	This work
Sp293	*Tet, ΔphpP bga::P$_{Zn}$-gfp-phpP-D231A*	This work
Sp295	*Δjag*	This work
Sp302	*Tet, Δjag bga::P$_{Zn}$-jag-flag-T89A*	This work
Sp304	*Tet, Δjag bga::P$_{Zn}$-jag-flag*	This work
E. coli		
DH5α	*F- Φ80lacZΔM15 Δ(lacZYA-argF) U169 recA1 endA1 hsdR17 (rk-, mk+) phoA supE44 λ- thi-1 gyrA96 relA1*	Invitrogen
BL21	*F- ompT gal [dcm][lon] hsdSB (rB- mB-) (DE3)*	Novagen
Plasmids		
pJWV25	*Amp, tet, bgaA, P$_{Zn}$-gfp+*	[35]
pZn-PhpP	*Amp, tet, bgaA, P$_{Zn}$-phpP*	This work
pZn-flag-locZ	*Amp, tet, bgaA, P$_{Zn}$-flag-locZ*	This work
pJWV25-phpP	*Amp, tet, bgaA, P$_{Zn}$-gfp-phpP*	[20]
pZn-gfp-phpP-D192A	*Amp, tet, bgaA, P$_{Zn}$-gfp-phpP-D192A*	This work
pZn-gfp-phpP-D231A	*Amp, tet, bgaA, P$_{Zn}$-gfp-phpP-D231A*	This work
pZn-jag-flag-T89A	*Amp, tet, bgaA, P$_{Zn}$-jag-flag-T89A*	This work
pZn-jag-flag	*Amp, tet, bgaA, P$_{Zn}$-jag-flag*	This work
pEXphpP-D231A	*Kan, phpP-D231A*	[17]
pMU1328	*Erm,* empty vector	[51]
pMU-P96-divIVA-flag	*Erm, P96-divIVA-flag*	This work
pMU-P96-flag-locZ	*Erm, P96-flag-locZ*	This work
Janus cassette	*Kan, kan-rpsL$^+$*	[52]

Amp ampicillin resistance marker, *cm* chloramphenicol resistance marker, *kan* kanamycin resistance marker, *tet* tetracycline resistance marker, *erm* erythromycin resistance marker

primers NS2 (containing Flag sequence) and LN155 using wild type chromosomal DNA as a template. Both PCR products were used as a template in a fusion PCR with primers LN123 and LN155 and the final PCR product was digested and cloned into the EcoRI and NotI restriction sites of the plasmid pJVW25.

To construct plasmid pMU-P96-flag-locZ, the P96 promoter was amplified with primer LN231 and LN215 from a template plasmid pMU1328 [51]. The *flag-locZ* fragment was amplified with primers NS3 and NS4 using pZn-flag-locZ as a template. Both PCR products were used as a template in a fusion PCR with primers LN231 and NS4. The resulting PCR fragment was cloned into EcoRI and SalI sites of pMU1328 vector. To generate plasmid pMU-P96-divIVA-flag, the P96 promoter was amplified with primers LN214 and LN215 from a template pMU1328. The *divIVA* gene was amplified with primers LN218 and LN229 (containing Flag sequence) using wild type chromosomal DNA as a template. Both PCR fragments were fused in a fusion PCR with primer pair LN214/LN229 and the acquired PCR fragment was inserted into BamHI and SalI sites of pMU1328. All constructs were verified by DNA sequencing.

Site directed mutagenesis

To introduce specific mutations in the *phpP* and *jag* genes we used the QuickChange mutagenesis kit (Stratagene) according to manufacturer's instructions. T89A mutation was introduced into plasmid pZn-jag-flag using primer pair AU80/AU81 to generate plasmid pZn-jag-flag-T89A. D192A mutation was introduced into pJWV25-phpP plasmid using primer pair AU69/AU70 to generate plasmid pZn-gfp-phpP-D192A. Plasmid pZn-gfp-phpP-D231A was constructed as follows: gene *phpP-D231A* was amplified by PCR using plasmid pEXphpP-D231A as a template and primers AU67 and AU68 containing SpeI and NotI restriction site, respectively. PCR product was cloned into pJWV25 generating pZn-gfp-phpP-D231A. All constructs were verified by DNA sequencing.

Construction of pneumococcal strains

pJWV25 derived strains expressing proteins under control of P_{Zn} promoter were prepared by transformation of *S. pneumoniae* competent cells with the corresponding pJWV25 derived plasmids previously linearized by digestion with PvuI. Tetracycline resistant transformants were obtained by a double-crossover recombination event between the chromosomal *bgaA* gene of the parental strain and *bgaA* regions located on the plasmids as described previously [35]. Following plasmids were used for construction of corresponding strains: pZn-PhpP: Sp120;

pJWV25-phpP: Sp140; pZn-gfp-phpP-D192A: Sp292; pZn-gfp-phpP-D231A: Sp293; pZn-jag-flag: Sp304; pZn-jag-flag-T89A: Sp302. Strain Sp174 was prepared by transformation of strain Sp26 (Δ*divIVA*) with plasmid pMU-P96-divIVA-flag. Strain Sp188 was prepared by transformation of strain Sp57 (Δ*locZ*) with pMU-P96-flag-locZ.

Strain Sp113 (Δ*phpP*) was constructed as described by Agarwal et al. [11], using a Janus cassette (kanamycin resistance gene followed by the recessive *rpsL* gene)-based two-step negative selection strategy [52]. In the first step 1100 bp and 1037 bp fragments corresponding to the upstream and downstream flanking regions of the *phpP* gene were amplified from the wild type chromosomal DNA with JG24/JG25 and JG26/JG27 primer pairs, respectively. The Janus cassette (1333 bp) amplified by JG28/JG29 primers from the Janus casette DNA fragment was attached to the *phpP* flanking regions by fusion PCR using primers JG24 and JG27. The resulting PCR fragment was used for the transformation of the *S. pneumoniae* strain Rx1, and Kan^R/Sm^S transformants (Sp100, *phpP::kan rpsL*) were selected. The PCR fragments, consisting of the upstream and downstream flanking region of the *phpP* gene, were amplified by JG24/JG31 and JG27/JG30 primer pairs, respectively, and fused by overlap extension using primers JG24/JG27. The resulting fragment was transformed into the strain Sp100 to gain Sp113 (Sm^R/Kan^S). Reverted strain Sp222 (WT$_R$) was constructed as follows: in the first step, Δ*phpP* strain (Sp113) was transformed by PCR fragment consisting of upstream and downstream flanking regions of the *phpP* gene fused with Janus cassette, as described in the previous section and Sm^S/Kan^R transformants (strain Sp220, *phpP::kan rpsL*) were selected. The PCR fragments, consisting of the upstream and downstream flanking region of the *phpP* gene, were amplified by JG24/JG68 and JG27/JG67 primer pairs, respectively, and fused by overlap extension with *phpP* gene (amplified by primers JG65 and JG66) using the primers JG24/JG27. The resulting fragment was transformed into the strain Sp220 to obtain reverted strain Sp222 (Sm^R/Kan^S).

Strain Sp295 (Δ*jag*) was constructed using a Janus cassette strategy [52]. In the first step, upstream and downstream flanking regions of the *jag* gene were amplified from the wild type chromosomal DNA with AU57/AU58 and AU59/AU60 primer pairs, respectively. The Janus cassette (1333 bp) amplified by JG28/JG29 primers from the Janus casette DNA fragment was attached to the *jag* gene flanking regions by fusion PCR using primers AU57 and AU60. The resulting PCR fragment was used for the transformation of the *S. pneumoniae* strain Rx1, and Sm^S/Kan^R transformants (*jag::kan rpsL*) were selected. The PCR fragments, consisting of the upstream and downstream flanking region of the *jag* gene, were amplified by AU57/AU74 and AU60/AU75 primer pairs, respectively, and fused by overlap extension using the primers AU57/AU60. The resulting fragment was transformed into the *jag::kan rpsL* strain and Sm^R/Kan^S transformants were selected (strain Sp295).

Strain Sp169 ($\Delta phpP$ $\Delta divIVA$) was constructed as follows: the upstream and downstream flanking regions of the *divIVA* gene were amplified from the wild type chromosomal DNA with JG57/JG58 and JG59/JG60 primer pairs, respectively. Both flanking regions were attached to the amplified Janus cassette (see above) by fusion PCR using primers JG57 and JG60. The resulting PCR fragment was used for the transformation of the $\Delta phpP$ strain (Sp113), and Sm^S/Kan^R transformants were selected to obtain Sp161 ($\Delta phpP$ *divIVA::kan rpsL*). To construct deletion of *divIVA* gene without selectable marker, the PCR fragments, consisting of the upstream and downstream flanking region of the *divIVA* gene, were amplified by JG57/JG62 and JG60/JG61 primer pairs, respectively, and fused by overlap extension using the primers JG57/JG60. Resulting fragment was transformed into Sp161 strain to yield Sm^R/Kan^S strain named Sp169.

Western blot analysis and immunodetection

Cells were grown in C + Y medium with or without the addition of an appropriate concentration of $ZnSO_4$ to an OD_{600} 0.4, harvested and resuspended in 1 ml of precooled lysis buffer containing 25 mM Tris (pH 7.5), 100 mM NaCl, Benzonase (Merck), and protease inhibitors (Roche). Cells were disintegrated using glass beads in a FastPrep homogenizer (ThermoScientific). Cell debris was pelleted by a centrifugation at $5\,000 \times g$. The total cell lysate was further fractionated by centrifugation at $100\,000 \times g$ for 1 h at 4 °C and the cytoplasmic and membrane fractions were obtained. The protein concentration was determined using a bicinchoninic acid (BCA) protein estimation kit (Pierce). An aliquot of 30 μg of each protein fractions or total cell lysate was diluted in $1 \times$ SDS loading buffer and boiled for 10 min. After SDS-PAGE separation, proteins were transferred to a PVDF membrane by Western blotting. Phosphorylated proteins were detected using anti-phosphothreonine polyclonal rabbit antibody (Cell Signalling, 9381S, LOT 22). StkP, PhpP, LocZ and RpoA were detected using specific custom made polyclonal rabbit sera derived from rabbits immunized with corresponding purified full length His-tagged proteins (Apronex, Czech Republic) and were used as described previously [20, 23, 25]. Specificity of anti-StkP and anti-PhpP antibody is documented by loss of reactivity in corresponding mutant strains (Fig. 1b; [20]). Specificity of anti-LocZ antibody is documented by loss of reactivity in

$\Delta locZ$ strain [25]. Specificity of antibody against housekeeping gene product RpoA is documented by reactivity with pure protein [23]. Flag-tagged and GFP-tagged proteins were detected with anti-Flag rabbit (Sigma-Aldrich, F7425, LOT 064M4757V) and anti-GFP mouse (Santa Cruz Biotech, sc-9996, LOT A1111) antibody. Protein abundance was measured using ECL detection substrate (Pierce) and signal was developed using G:Box Chemi XRQ instrument (SynGene) or by exposition on medical X-ray film (Agfa). To quantify protein phosphorylation in cell lysates the immunoblot was scanned by G:Box Chemi XRQ instrument to obtain linear range of exposure and signal was analyzed by Quantity One software, version 4.6.3 (Bio-Rad). Data represent mean ± standard deviation (SD) from three independent experiments and were normalized to the total protein level.

Trifluoroethanol/chloroform extraction, two-dimensional (2D) SDS-PAGE and mass spectrometry

Protein membrane fractions were extracted by trifluoroethanol/chloroform mixture as described previously [38]. Resulting aqueous and insoluble fractions were solubilized in lysis buffer (7 M urea, 2 M thiourea, 2 % Triton X-100, 0,5 % amido sulfobetaine-14 (ASB14), 1 % Ampholytes 3–10, 50 mM DTT) and purified with 2-D Clean-Up Kit (GE Healthcare). The protein concentration was determined using 2-D Quant kit (GE Healthcare). 2D SDS-PAGE was performed on IPG strips pH 4–7 NL (Amersham Biosciences) as described previously [23]. Gels were either stained with colloidal Coomassie Brilliant Blue G-250 (CBB G-250) or electroblotted. The protein spots selected for mass spectrometric analysis were destained using 50 mM 4-ethylmorpholine acetate (pH 8.1) in 50 % acetonitrile (MeCN) and in-gel digested overnight with trypsin (100 ng; Promega) in a cleavage buffer containing 25 mM 4-ethylmorpholine acetate. The resulting peptides were extracted to 40 % MeCN/0.1 % TFA and measured on an Ultraflex III MALDI-TOF mass spectrometer (BrukerDaltonics) in a mass range of 700–4000 Da. For protein identification the peptide mass spectra were searched against SwissProt or NCBI (National Center for Biotechnology Information) bacterial database using an in-house Mascot search engine. The identity of protein candidates was confirmed using MS/MS analysis.

Competence assays

S. pneumoniae was induced to competence using competence stimulating peptide (CSP) as described previously [36], with minor modifications. Briefly, an exponential culture of cells cultivated in BHI medium was diluted 1:20 in BHI supplemented with 0.2 % BSA (Bovine Serum Albumin) and 1 mM $CaCl_2$ and pH adjusted to 7.8. The recipient strain was activated by the addition of CSP (250 ng ml^{-1}) and incubated for 10 min at room temperature. The *rif-23* donor DNA (1 μg ml^{-1}) was then added and DNA uptake was obtained by 20 min incubation at room temperature. The mixture was then diluted 1:10 in BHI medium and incubated at 37 °C for 2 h. Serial dilutions of transformed cultures were plated, and transformation efficiencies were calculated as the ratio of the viable counts on plates with and without rifampin. To generate natural competence profiles of the wild type and mutant strains, method according to Echenique et al. [36] was used with several modifications. Briefly, stocks of bacteria grown in BHI medium to an OD$_{600}$ of 0.5 were diluted 100-fold in the same medium supplemented with 0.2 % BSA and 1 mM CaCl$_2$ and pH adjusted to 7.8 and grown at 37 °C. Samples were withdrawn at 15-min intervals, diluted 10-fold into BHI medium containing *rif-23* donor DNA, and incubated for 30 min at 30 °C. Further incubation was carried out at 37 °C for 90 min before plating serial dilutions with and without rifampin. Transformation efficiencies were calculated as the ratio of the viable counts on plates with and without rifampin.

Growth and environmental stress tolerance

To generate the growths curves pneumococcal strains were inoculated (6.8 × 10^5 CFU ml^{-1}) in TSB medium, cultivated statically and the growth was monitored every 30 min by measuring OD$_{600}$ for period of 7 to 8 h. The tolerance of pneumococcal strains to environmental stress was examined in a manner similar to that described previously [19]. To investigate heat stress resistance cultures were inoculated into TSB medium prewarmed to 37 and 40 °C. The acid tolerance of all strains was monitored by measuring the growth curve in TSB medium adjusted to pHs 6.5 and 7.5. The alkaline tolerance was monitored at pH 8.0. To test the tolerance to osmotic stress, bacteria were first grown to early exponential phase (OD$_{600}$ 0.2) and then inoculated into prewarmed TSB medium with or without 400 mM NaCl. The sensitivity of cells to H$_2$O$_2$ was tested by exposing exponential cultures (OD$_{600}$ 0.4) grown in CAT medium at 37 °C to 10 mM and 20 mM H$_2$O$_2$ for 15 min. Viable cell counts were determined by plating serial dilutions of cultures onto agar plates before and after exposure to H$_2$O$_2$. The results were expressed as percentages of survival.

Protein purification and dephosphorylation assay

Recombinant His-PhpP was purified as described previously [17]. To purify Flag-tagged proteins form *S. pneumoniae* the strains Sp174 (Δ*divIVA pMU-P96-divIVA-flag*), Sp188 (Δ*locZ pMU-P96-flag-locZ*) and Sp304 (Δ*jag bga::* P$_{Zn}$-*jag-flag*) were grown statically at 37 °C in C + Y

medium supplemented with 0.25 mM ZnSO$_4$. Total cell lysates were prepared as described above and Flag-tagged proteins were purified by affinity chromatography using ANTI-Flag M2 Affinity Gel (Sigma-Aldrich) according to the manufacturer's instructions. Dephosphorylation assay was performed basically as described previously [17]. Briefly, Flag fusion-proteins of interest (Flag-LocZ, DivIVA-Flag or Jag-Flag) bounded on the M2 affinity gel were mixed with 55 μl of reaction buffer and 4 μg of purified His-PhpP and incubated at 37 °C. Phosphatase reaction was terminated by the addition of 5× SDS-PAGE sample buffer at different time intervals (0–90 min). Samples were boiled, subjected to SDS-PAGE and immunoblotted as described above.

Electron microscopy

Samples for electron microscopy were prepared as described elsewhere [25] except the dried samples were sputter coated with 3 nm of platinum in a Q150T ES sputter coater (Quorum Technologies Ltd.). The final samples were examined in a FEI Nova NanoSem 450 scanning electron microscope (FEI Czech Republic s.r.o.) at 5 kV using Circular Backscatter Detector and backscattered electrons.

Fluorescence microscopy

Fluorescence microscopy was performed basically as described before [20]. Cells were grown statically at 37 °C in C + Y medium, and the expression of the GFP fusion proteins was induced by adding desired concentration of ZnSO$_4$. To stain the unfixed cells with fluorescently labelled vancomycin (VanFL) (Molecular Probes) the pneumococcal cultures were grown to OD$_{600}$ 0.2 in C + Y medium, and the samples were labelled with 0.1 μg ml^{-1} of Van-FL/vancomycin (50:50) mixture for 5 min at 37 °C before examination. A quantity of 2 μl of the culture was spotted onto a microscope slide and covered with a 1 % PBS agarose slab. The samples were observed using an Olympus Cell^R IX 81 microscope equipped with an Olympus FV2T Digital B/W Fireware Camera and 100× oil immersion objective (N.A. 1.3) (phase contrast). The images were modified for publication using Cell^R Version 2.0 software, ImageJ (http://rsb.info.nih.gov/ij/) and CorelDRAW X7 (Corel Corporation). Fluorescence intensity line scans were acquired using ImageJ and plotted as a function of cell length measured with MicrobeTracker Suite [53].

Cell size analysis

The phase-contrast images were analyzed using automated MicrobeTracker software [53] and cell size parameters were evaluated by the Mann-Whitney rank sum test and plotted using GraphPad Prism 3.0. $P <$ 0.0001 was considered as statistically significant. Cell size throughout the text is indicated as the median cell size ± median absolute deviation (MAD).

Abbreviations
2D SDS-PAGE: Two-dimensional sodium dodecyl sulfate polyacrylamide gel electrophoresis; CDD: Conserved domain database; CSP: Competence-stimulating peptide; ESTK: Eukaryotic-type serine/threonine protein kinases; ESTP: Eukaryotic-type serine/threonine phosphatases; GFP: Green fluorescent protein; Jag_N: Jag N-terminal domain; Jag_{Spn}: S. pneumoniae Jag; KEGG: Kyoto encyclopedia of genes and genomes; KH: Ribonucleoprotein K homology domain; MAD: Median absolute deviation; MALDI-TOF: Matrix-assisted laser desorption/ionization- time of flight mass spectrometry; P35: Protein 35 kDa; P40: Protein 40 kDa; PASTA: Penicillin-binding protein and Ser/Thr kinase-associated; PBP: Penicillin-binding protein; PG: Peptidoglycan; PP2C: Protein phosphatase 2C; PPM: Protein phosphatases $Mg2+/Mn2+$ dependent; pThr: phospho-threonine; P_{Zn}: Zinc-inducible promoter; qRT-PCR: quantitative real-time polymerase chain reaction; R3H: Arginine-x-x-x-Histidine domain; SD: Standard deviation; TFE: Trifluorethanol; VanFL: Fluorescently labeled vancomycin; WT: Wild type

Acknowledgements
We sincerely thank Orietta Massidda for critical reading of the manuscript and valuable suggestions. We thank Don Morrison for the kan-rpsL (Janus) cassette and Monica Moschioni for plasmid pMU1328. PH, OB and OK gratefully acknowledge the support of Operational Program Prague – Competitiveness project (CZ.2.16/3.1.00/24023) supported by EU. This work used instruments provided by C4Sys infrastructure.

Funding
This work was supported by Grant Czech Science Foundation Grants P302/12/0256 and P207/12/1568 to P.B., by Grant LH 12055 of the Ministry of Education, Youth and Sports of the Czech Republic, and by Institutional Research Concept RVO 61388971.

Authors' contributions
LD, JG and AU designed the study. LD and PB supervised the project. JG and AU prepared and analyzed mutant strains and performed immunodetection. JG and NH tested gene essentiality. NH tested growth, stress sensitivity and performed statistical analysis. NH, JG and LD performed phase contrast and fluorescence microscopy. PB and LD performed competence assays. AU and NH purified proteins and performed in vitro assays. OB and OK performed electron microscopy. PH carried out mass spectrometric analysis. LD, AU, NH, JG and PB wrote the manuscript. All authors discussed the results and implications and commented on the manuscript at all stages. All authors read and approved the final manuscript.

Competing interests
The authors declare that they have no competing interest.

References
1. Jung K, Fried L, Behr S, Heermann R. Histidine kinases and response regulators in networks. Curr Opin Microbiol. 2012;15(2):118–24.
2. Dworkin J. Ser/Thr phosphorylation as a regulatory mechanism in bacteria. Curr Opin Microbiol. 2015;24:47–52.
3. Jones G, Dyson P. Evolution of transmembrane protein kinases implicated in coordinating remodeling of gram-positive peptidoglycan: inside versus outside. J Bacteriol. 2006;188(21):7470–6.
4. Manuse S, Fleurie A, Zucchini L, Lesterlin C, Grangeasse C. Role of eukaryotic-like serine/threonine kinases in bacterial cell division and morphogenesis. FEMS Microbiol Rev. 2016;40(1):41–56.
5. Pereira SF, Goss L, Dworkin J. Eukaryote-like serine/threonine kinases and phosphatases in bacteria. Microbiol Mol Biol Rev. 2011;75(1):192–212.
6. Yeats C, Finn RD, Bateman A. The PASTA domain: a beta-lactam-binding domain. Trends Biochem Sci. 2002;27(9):438.
7. Shi L, Potts M, Kennelly PJ. The serine, threonine, and/or tyrosine-specific protein kinases and protein phosphatases of prokaryotic organisms: a family portrait. FEMS Microbiol Rev. 1998;22(4):229–53.
8. Echenique J, Kadioglu A, Romao S, Andrew PW, Trombe MC. Protein serine/threonine kinase StkP positively controls virulence and competence in Streptococcus pneumoniae. Infect Immun. 2004;72(4):2434–7.
9. Jin H, Pancholi V. Identification and biochemical characterization of a eukaryotic-type serine/threonine kinase and its cognate phosphatase in Streptococcus pyogenes: their biological functions and substrate identification. J Mol Biol. 2006;357(5):1351–72.
10. Osaki M, Arcondeguy T, Bastide A, Touriol C, Prats H, Trombe MC. The StkP/PhpP signaling couple in Streptococcus pneumoniae: cellular organization and physiological characterization. J Bacteriol. 2009;191(15):4943–50.
11. Agarwal S, Agarwal S, Pancholi P, Pancholi V. Strain-specific regulatory role of eukaryote-like serine/threonine phosphatase in pneumococcal adherence. Infect Immun. 2012;80(4):1361–72.
12. Banu LD, Conrads G, Rehrauer H, Hussain H, Allan E, van der Ploeg JR. The Streptococcus mutans serine/threonine kinase, PknB, regulates competence development, bacteriocin production, and cell wall metabolism. Infect Immun. 2010;78(5):2209–20.
13. Beltramini AM, Mukhopadhyay CD, Pancholi V. Modulation of cell wall structure and antimicrobial susceptibility by a Staphylococcus aureus eukaryote-like serine/threonine kinase and phosphatase. Infect Immun. 2009;77(4):1406–16.
14. Burnside K, Lembo A, de Los Reyes M, Iliuk A, BinhTran NT, Connelly JE, et al. Regulation of hemolysin expression and virulence of Staphylococcus aureus by a serine/threonine kinase and phosphatase. PLoS One. 2010;5(6), e11071.
15. Rajagopal L, Clancy A, Rubens CE. A eukaryotic type serine/threonine kinase and phosphatase in Streptococcus agalactiae reversibly phosphorylate an inorganic pyrophosphatase and affect growth, cell segregation, and virulence. J Biol Chem. 2003;278(16):14429–41.
16. Sajid A, Arora G, Singhal A, Kalia VC, Singh Y. Protein phosphatases of pathogenic bacteria: role in physiology and virulence. Annu Rev Microbiol. 2015;69:527–47.
17. Novakova L, Saskova L, Pallova P, Janecek J, Novotna J, Ulrych A, et al. Characterization of a eukaryotic type serine/threonine protein kinase and protein phosphatase of Streptococcus pneumoniae and identification of kinase substrates. FEBS J. 2005;272(5):1243–54.
18. Herbert JA, Mitchell AM, Mitchell TJ. A Serine-threonine kinase (StkP) regulates expression of the pneumococcal pilus and modulates bacterial adherence to human epithelial and endothelial cells In vitro. PLoS One. 2015;10(6), e0127212.
19. Saskova L, Novakova L, Basler M, Branny P. Eukaryotic-type serine/threonine protein kinase StkP is a global regulator of gene expression in Streptococcus pneumoniae. J Bacteriol. 2007;189(11):4168–79.
20. Beilharz K, Novakova L, Fadda D, Branny P, Massidda O, Veening JW. Control of cell division in Streptococcus pneumoniae by the conserved Ser/Thr protein kinase StkP. Proc Natl Acad Sci U S A. 2012;109(15):E905–13.
21. Fleurie A, Cluzel C, Guiral S, Freton C, Galisson F, Zanella-Cleon I, et al. Mutational dissection of the S/T-kinase StkP reveals crucial roles in cell division of Streptococcus pneumoniae. Mol Microbiol. 2012;83(4):746–58.
22. Giefing C, Jelencsics KE, Gelbmann D, Senn BM, Nagy E. The pneumococcal eukaryotic-type serine/threonine protein kinase StkP co-localizes with the cell division apparatus and interacts with FtsZ in vitro. Microbiology. 2010;156(Pt 6):1697–707.
23. Novakova L, Bezouskova S, Pompach P, Spidlova P, Saskova L, Weiser J, et al. Identification of multiple substrates of the StkP Ser/Thr protein kinase in Streptococcus pneumoniae. J Bacteriol. 2010;192(14):3629–38.
24. Fleurie A, Lesterlin C, Manuse S, Zhao C, Cluzel C, Lavergne JP, et al. MapZ marks the division sites and positions FtsZ rings in Streptococcus pneumoniae. Nature. 2014;516(7530):259–62.
25. Holeckova N, Doubravova L, Massidda O, Molle V, Buriankova K, Benada O, et al. LocZ is a New cell division protein involved in proper septum placement in streptococcus pneumoniae. MBio. 2014;6(1):e01700–14.

26. Falk SP, Weisblum B. Phosphorylation of the *Streptococcus pneumoniae* cell wall biosynthesis enzyme MurC by a eukaryotic-like Ser/Thr kinase. FEMS Microbiol Lett. 2013;340(1):19–23.

27. Bork P, Brown NP, Hegyi H, Schultz J. The protein phosphatase 2C (PP2C) superfamily: detection of bacterial homologues. Protein Sci. 1996;5(7):1421–5.

28. Thanassi JA, Hartman-Neumann SL, Dougherty TJ, Dougherty BA, Pucci MJ. Identification of 113 conserved essential genes using a high-throughput gene disruption system in *Streptococcus pneumoniae*. Nucleic Acids Res. 2002;30(14):3152–62.

29. Bijlsma JJ, Burghout P, Kloosterman TG, Bootsma HJ, de Jong A, Hermans PW, et al. Development of genomic array footprinting for identification of conditionally essential genes in *Streptococcus pneumoniae*. Appl Environ Microbiol. 2007;73(5):1514–24.

30. Song JH, Ko KS, Lee JY, Baek JY, Oh WS, Yoon HS, et al. Identification of essential genes in *Streptococcus pneumoniae* by allelic replacement mutagenesis. Mol Cells. 2005;19(3):365–74.

31. Errington J, Appleby L, Daniel RA, Goodfellow H, Partridge SR, Yudkin MD. Structure and function of the *spoIIIJ* gene of *Bacillus subtilis*: a vegetatively expressed gene that is essential for sigma G activity at an intermediate stage of sporulation. J Gen Microbiol. 1992;138(12):2609–18.

32. Kent JL, Hotchkiss RD. Kinetic analysis of multiple, linked recombinations in pneumococcal transformation. J Mol Biol. 1964;9:308–22.

33. Piotrowski A, Burghout P, Morrison DA. *spr1630* is responsible for the lethality of *clpX* mutations in *Streptococcus pneumoniae*. J Bacteriol. 2009; 191(15):4888–95.

34. Massidda O, Novakova L, Vollmer W. From models to pathogens: how much have we learned about *Streptococcus pneumoniae* cell division? Environ Microbiol. 2013;15(12):3133–57.

35. Eberhardt A, Wu LJ, Errington J, Vollmer W, Veening JW. Cellular localization of choline-utilization proteins in *Streptococcus pneumoniae* using novel fluorescent reporter systems. Mol Microbiol. 2009;74(2):395–408.

36. Echenique JR, Chapuy-Regaud S, Trombe MC. Competence regulation by oxygen in *Streptococcus pneumoniae*: involvement of *ciaRH* and *comCDE*. Mol Microbiol. 2000;36(3):688–96.

37. Das AK, Helps NR, Cohen PT, Barford D. Crystal structure of the protein serine/threonine phosphatase 2C at 2.0 A resolution. EMBO J. 1996;15(24): 6798–809.

38. Zuobi-Hasona K, Crowley PJ, Hasona A, Bleiweis AS, Brady LJ. Solubilization of cellular membrane proteins from *Streptococcus mutans* for two-dimensional gel electrophoresis. Electrophoresis. 2005;26(6):1200–5.

39. Marchler-Bauer A, Derbyshire MK, Gonzales NR, Lu S, Chitsaz F, Geer LY, et al. CDD: NCBI's conserved domain database. Nucleic Acids Res. 2015; 43(Database issue):D222–6.

40. Nicastro G, Taylor IA, Ramos A. KH-RNA interactions: back in the groove. Curr Opin Struct Biol. 2015;30:63–70

41. Sun X, Ge F, Xiao CL, Yin XF, Ge R, Zhang LH, et al. Phosphoproteomic analysis reveals the multiple roles of phosphorylation in pathogenic bacterium *Streptococcus pneumoniae*. J Proteome Res. 2010;9(1):275–82.

42. Hennon SW, Soman R, Zhu L, Dalbey RE. YidC/Alb3/Oxa1 family of insertases. J Biol Chem. 2015;290(24):14866–74.

43. Palmer SR, Crowley PJ, Oli MW, Ruelf MA, Michalek SM, Brady LJ. YidC1 and YidC2 are functionally distinct proteins involved in protein secretion, biofilm formation and cariogenicity of *Streptococcus mutans*. Microbiology. 2012; 158(Pt 7):1702–12.

44. Gray AN, Li Z, Henderson-Frost J, Goldberg MB. Biogenesis of YidC cytoplasmic membrane substrates is required for positioning of autotransporter IcsA at future poles. J Bacteriol. 2014;196(3):624–32.

45. Scotti PA, Urbanus ML, Brunner J, de Gier JW, von Heijne G, van der Does C, et al. YidC, the *Escherichia coli* homologue of mitochondrial Oxa1p, is a component of the Sec translocase. EMBO J. 2000;19(4):542–9.

46. de Sousa Borges A, de Keyzer J, Driessen AJ, Scheffers DJ. The *Escherichia coli* membrane protein insertase YidC assists in the biogenesis of penicillin binding proteins. J Bacteriol. 2015;197(8):1444–50.

47. Higgins ML, Shockman GD. Model for cell wall growth of *Streptococcus faecalis*. J Bacteriol. 1970;101(2):643–8.

48. Zapun A, Vernet T, Pinho MG. The different shapes of cocci. FEMS Microbiol Rev. 2008;32(2):345–60.

49. Lacks S, Hotchkiss RD. A study of the genetic material determining an enzyme in Pneumococcus. Biochim Biophys Acta. 1960;39:508–18.

50. Auzat I, Chapuy-Regaud S, Le BG, Dos SD, Ogunniyi AD, Le TI, et al. The NADH oxidase of *Streptococcus pneumoniae*: its involvement in competence and virulence. Mol Microbiol. 1999;34(5):1018–28.

51. Lo SM, Hilleringmann M, Barocchi MA, Moschioni M. A novel strategy to over-express and purify homologous proteins from *Streptococcus pneumoniae*. J Biotechnol. 2012;157(2):279–86.

52. Sung CK, Li H, Claverys JP, Morrison DA. An *rpsL* cassette, janus, for gene replacement through negative selection in *Streptococcus pneumoniae*. Appl Environ Microbiol. 2001;67(11):5190–6.

53. Sliusarenko O, Heinritz J, Emonet T, Jacobs-Wagner C. High-throughput, subpixel precision analysis of bacterial morphogenesis and intracellular spatio-temporal dynamics. Mol Microbiol. 2011;80(3):612–27.

54. Morrison DA, Lacks SA, Guild WR, Hageman JM. Isolation and characterization of three new classes of transformation-deficient mutants of *Streptococcus pneumoniae* that are defective in DNA transport and genetic recombination. J Bacteriol. 1983;156(1):281–90.

55. Fadda D, Pischedda C, Caldara F, Whalen MB, Anderluzzi D, Domenici E, et al. Characterization of *divIVA* and other genes located in the chromosomal region downstream of the *dcw* cluster in *Streptococcus pneumoniae*. J Bacteriol. 2003;185(20):6209–14.

2

Whole genome sequencing analyses of *Listeria monocytogenes* that persisted in a milkshake machine for a year and caused illnesses in Washington State

Zhen Li[1], Ailyn Pérez-Osorio[1], Yu Wang[2], Kaye Eckmann[1], William A. Glover[1], Marc W. Allard[2], Eric W. Brown[2] and Yi Chen[2*] (iD)

Abstract

Background: In 2015, in addition to a United States multistate outbreak linked to contaminated ice cream, another outbreak linked to ice cream was reported in the Pacific Northwest of the United States. It was a hospital-acquired outbreak linked to milkshakes, made from contaminated ice cream mixes and milkshake maker, served to patients. Here we performed multiple analyses on isolates associated with this outbreak: pulsed-field gel electrophoresis (PFGE), whole genome single nucleotide polymorphism (SNP) analysis, species-specific core genome multilocus sequence typing (cgMLST), lineage-specific cgMLST and whole genome-specific MLST (wgsMLST)/outbreak-specific cgMLST. We also analyzed the prophages and virulence genes.

Results: The outbreak isolates belonged to sequence type 1038, clonal complex 101, genetic lineage II. There were no pre-mature stop codons in *inlA*. Isolates contained *Listeria* Pathogenicity Island 1 and multiple internalins. PFGE and multiple whole genome sequencing (WGS) analyses all clustered together food, environmental and clinical isolates when compared to outgroup from the same clonal complex, which supported the finding that *L. monocytogenes* likely persisted in the soft serve ice cream/milkshake maker from November 2014 to November 2015 and caused 3 illnesses, and that the outbreak strain was transmitted between two ice cream production facilities. The whole genome SNP analysis, one of the two species-specific cgMLST, the lineage II-specific cgMLST and the wgsMLST/outbreak-specific cgMLST showed that *L. monocytogenes* cells persistent in the milkshake maker for a year formed a unique clade inside the outbreak cluster. This clustering was consistent with the cleaning practice after the outbreak was initially recognized in late 2014 and early 2015. Putative prophages were conserved among prophage-containing isolates. The loss of a putative prophage in two isolates resulted in the loss of the *Asc*I restriction site in the prophage, which contributed to their *Asc*I-PFGE banding pattern differences from other isolates.

Conclusions: The high resolution of WGS analyses allowed the differentiation of epidemiologically unrelated isolates, as well as the elucidation of the microevolution and persistence of isolates within the scope of one outbreak. We applied a wgsMLST scheme which is essentially the outbreak-specific cgMLST. This scheme can be combined with lineage-specific cgMLST and species-specific cgMLST to maximize the resolution of WGS.

Keywords: Listeriosis, Ice cream, Outbreak, Whole genome sequencing, Core genome multilocus sequence typing

* Correspondence: yi.chen@fda.hhs.gov
[2]Center for Food Safety and Applied Nutrition, Food and Drug
Administration, College Park, MD, USA
Full list of author information is available at the end of the article

Background

Listeria monocytogenes is a Gram-positive, facultative intracellular bacterium that causes high mortality food-borne illnesses through contaminated food products [1]. *L. monocytogenes* exists in different environments due to its hardiness in harsh conditions, such as a wide pH range, high salt concentrations and ability to grow and persist at refrigeration temperatures [2]. These unique characteristics have made *L. monocytogenes* one of the major threats to the food industry and public health. Several listeriosis outbreaks occurred in United States recently, linked to dairy products and fresh produce [3–5]. Ice cream-associated outbreaks are rarely reported. However, two epidemiologically unrelated outbreaks were linked to contaminated ice cream in recent years. A 2010–2015 multistate listeriosis outbreak was linked to contaminated ice cream manufactured in the southern United States [6]. In late 2014, a different listeriosis outbreak in Washington State, unrelated to the 2010–2015 multistate outbreak, occurred in a hospital (Hospital X) in the Pacific Northwest of the United States, involving patients hospitalized for other medical conditions prior to exposure to milkshakes made from contaminated ice cream mixes manufactured in a company (Company A) [7]. Following the investigation of the Washington State outbreak, intensive cleaning and sanitizing were conducted in the facility and hospital kitchen, although cleaning of the soft serve shake freezer took extra efforts, because milkshake was made inside the machine and disassembly was required for thorough cleaning [7]. In November 2015, another patient from Hospital X, hospitalized for other conditions prior to exposure to *L. monocytogenes*, was linked to contaminated milkshakes by pulsed-field gel electrophoresis (PFGE) [8]. Hospital X was using a different brand of ice cream mix from the 2014 outbreak, which was tested negative for *L. monocytogenes*; but isolates recovered from the milkshake samples and swab samples from the milkshake machine matched the outbreak-associated isolates collected in 2014 [8], confirming that this third patient was also associated with this outbreak.

Single nucleotide polymorphism (SNP)-based and multi-locus sequence typing (MLST) allele-based whole genome sequence (WGS) analyses have been utilized to support the findings of the listeriosis outbreak investigations and offer various advantages over PFGE [5, 6, 9]. SNP-based analyses could target SNPs in the whole genome (i.e. entire genome including coding and noncoding regions) or the core genome (i.e. coding regions that are present in a set of strains). A whole genome MLST (wgMLST) scheme, targeting a specific pan-genome of 4797 loci defined based on over 150 publicly available reference genomes of *L. monocytogenes* [10], was implemented in PulseNet [9]. An alternative way to perform whole genome-based MLST is to target the entire coding loci that are specific to a set of closely-related isolates (e.g, those of the same outbreak strain). This scheme may target loci unique to these isolates, which are not included in any pre-defined pan-genome locus set.

Four *L. monocytogenes* species-specific core genome MLST (cgMLST) schemes have been developed [4, 11–13]. Further, lineage-specific cgMLST schemes for 3 genetic lineages of *L. monocytogenes* were developed to improve the discriminatory power [4]. The objective of this study is to determine whether the results of whole genome SNP analysis, whole genome-specific MLST (wgsMLST)/outbreak-specific cgMLST, lineage-specific and species-specific cgMLST analyses were consistent with PFGE, and could support epidemiological evidence to delineate the Hospital X - acquired outbreak.

Results

Isolates selected for WGS analysis are listed in Table 1. The outbreak-associated isolates had sequence type (ST) 1038, belonging to clonal complex (CC) 101, a genetic lineage II clonal group [4]. The non-outbreak isolate CFSAN028854 (as discussed below) had ST5, which belonged to CC5, a serotype 1/2b or 3b clonal group [4], and thus it is not illustrated in the phylogenetic trees. The outbreak isolates contained *Listeria* pathogenicity island (LIPI)-1, internalin A, B, C, E, F, H, J and P. They did not contain LIPI-3 or LIPI-4. There were no premature stop codons (PMSC) in *inlA*.

The two clinical isolates collected in November and December 2014 exhibited two PFGE profiles, *Asc*I-P1/*Apa*I-P1 and *Asc*I-P2/*Apa*I-P1 (Fig. 1). Isolates from ice cream products manufactured by Company A and environmental samples from Company A facility areas, and isolates of unopened ice cream products and machine-dispensed products from Hospital X, collected after the outbreak recognition in 2014, exhibited *Asc*I-P1/*Apa*I-P1, *Asc*I-P2/*Apa*I-P1, *Asc*I-P3/*Apa*I-P1 and *Asc*I-P4/*Apa*I-P2. One environmental isolate was collected in March 2015 from Company A and one environmental isolate was collected in April 2015 from Company B who purchased dairy ingredients from Company A; and they both exhibited *Asc*I-P1/*Apa*I-P1. After the identification of the case-patient in November 2015, isolates were collected from the ice cream that remained in and were dispensed from the milkshake maker in Hospital X and environmental samples from different areas (e.g., side walls of internal parts and nozzle assembly) of the milkshake maker; and they exhibited *Asc*I-P1/*Apa*I-P1, the same as the 2015 clinical isolate. *Asc*I-P1/*Apa*I-P1, *Asc*I-P2/*Apa*I-P1, *Asc*I-P3/*Apa*I-P1 were rare PFGE profiles in PulseNet; prior to this outbreak only one isolate in 2010, with no epidemiological link to this outbreak, exhibited *Asc*I-P2/*Apa*I-P1. Overall, 27 of 29 food and environmental isolates had indistinguishable

Table 1 List of *L. monocytogenes* isolates analyzed in this study

Isolate ID	Collection time	Source Type	AscI-PFGE pattern	ApaI-PFGE pattern	SRA ID or GenBank Accession
PNUSAL001207	November, 2014	Clinical	AscI-P1	ApaI-P1	SRR1745448
PNUSAL001241	December, 2014	Clinical	AscI-P2	ApaI-P1	SRR1745474
CFSAN028842	December, 2014	Ice cream/Hospital X	AscI-P1	ApaI-P1	SRR3130313
CFSAN028843	December, 2014	Ice cream/Hospital X	AscI-P1	ApaI-P1	SRR3091402
CFSAN028844	December, 2014	Ice cream/Hospital X	AscI-P1	ApaI-P1	SRR3130327
CFSAN028845	December, 2014	Ice cream/Hospital X	AscI-P1	ApaI-P1	SRR3066080
CFSAN028846	December, 2014	Ice cream/Hospital X	AscI-P1	ApaI-P1	SRR3130329
CFSAN028847	December, 2014	Ice cream/Hospital X	AscI-P1	ApaI-P1	SRR3091403
CFSAN028848	December, 2014	Ice cream/Company A	AscI-P1	ApaI-P1	SRR3091404
CFSAN028849	December, 2014	Ice cream/Company A	AscI-P1	ApaI-P1	SRR3091405
CFSAN028850	December, 2014	Ice cream/Company A	AscI-P1	ApaI-P1	SRR3130331
CFSAN028851	December, 2014	Ice cream/Company A	AscI-P1	ApaI-P1	SRR3091406
CFSAN028852	December, 2014	Environmental/Company A	AscI-P1	ApaI-P1	SRR3130333
CFSAN028853	December, 2014	Environmental/Company A	AscI-P1	ApaI-P1	SRR3130335, MAKW00000000.1
CFSAN028855	December, 2014	Environmental/Company A	AscI-P1	ApaI-P1	SRR3130404
CFSAN028856	December, 2014	Environmental/Company A	AscI-P1	ApaI-P1	SRR3130406
CFSAN028857	December, 2014	Environmental/Company A	AscI-P3	ApaI-P1	SRR3130409
CFSAN028858	December, 2014	Environmental/Company A	AscI-P1	ApaI-P1	SRR3130413
CFSAN028859	December, 2014	Environmental/Company A	AscI-P1	ApaI-P1	SRR3130415
CFSAN028860	December, 2014	Environmental/Company A	AscI-P1	ApaI-P1	SRR3130350
CFSAN028861	December, 2014	Environmental/Company A	AscI-P2	ApaI-P1	SRR3130375
CFSAN029502	December, 2014	Environmental/Company A	AscI-P1	ApaI-P1	SRR3130341
CFSAN030692	March, 2015	Environmental/Company A	AscI-P1	ApaI-P1	SRR1974103
CFSAN032836	April, 2015	Environmental/Company B	AscI-P1	ApaI-P1	SRR2035442
CFSAN043359	November, 2015	Ice cream/Hospital X	AscI-P1	ApaI-P1	SRR3052035
CFSAN043360	November, 2015	Ice cream/Hospital X	AscI-P1	ApaI-P1	SRR3053137
CFSAN043361	November, 2015	Ice cream/Hospital X	AscI-P1	ApaI-P1	SRR3086932
CFSAN043362	November, 2015	Environmental/Hospital X	AscI-P1	ApaI-P1	SRR3086935
CFSAN043363	November, 2015	Environmental/Hospital X	AscI-P1	ApaI-P1	SRR3086936
CFSAN043364	November, 2015	Environmental/Hospital X	AscI-P1	ApaI-P1	SRR3052036
PNUSAL001911	November, 2015	Clinical	AscI-P1	ApaI-P1	SRR2994642
CFSAN028854	December, 2014	Environmental/Company A	AscI-P4	ApaI-P2	SRR3130337
CFSAN004336	NA	Food	NA	NA	SRR1818032

PFGE profiles from the 3 clinical isolates. Two other environmental isolates had PFGE profiles not observed in any clinical isolates.

The whole genome SNP analysis clustered all food, environmental and clinical isolates in 2014 and 2015, except CFSAN028854 (*Asc*I-P4/*Apa*I-P2); and separated them from the outgroup, CFSAN004336, a CC101 strain not associated with the outbreak (Fig. 2). Isolates exhibiting the two clinical PFGE profiles (*Asc*I-P1/*Apa*I-P1 had *Asc*I-P2/*Apa*I-P1) and an isolate exhibiting *Asc*I-P3/*Apa*I-P1 were clustered together (Fig. 2). This is consistent with the epidemiological finding that Company A was the likely

source of the outbreak in Hospital X in November 2014. Considering samples from unopened containers of ice cream mixes used to make ice cream/milkshakes in Hospital X in November 2015 were not from Company A and were tested negative for *L. monocytogenes*, and that swab samples throughout the hospital kitchen surfaces were tested negative for *L. monocytogenes* [8], *L. monocytogenes* isolates linked to the illnesses in 2014 likely persisted in the ice cream/milkshake machine of Hospital X through November 2015 and contaminated products that were consumed by the case-patient identified in November 2015. Company B, who purchased ingredients from

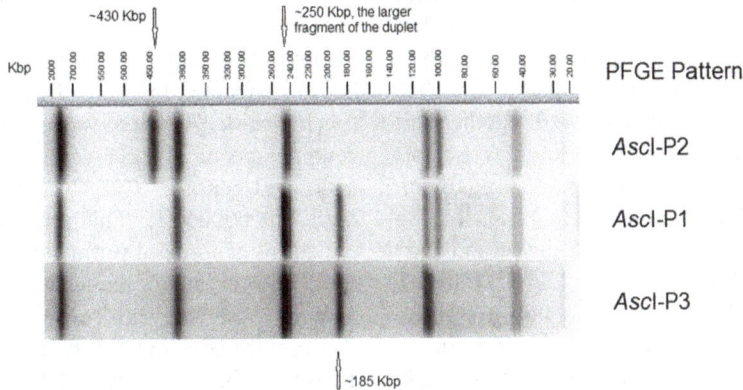

Fig. 1 Three *Asc*I-PFGE banding patterns observed among outbreak-associated isolates

Company A, yielded an environmental isolate that was clustered together with Company A isolates. Thus, the whole-genome SNP analysis was able to trace the spread of the outbreak strain over more than one facility. A second SNP analysis containing only outbreak-associated isolates identified 59 polymorphic loci and revealed that the pairwise SNP distances among all isolates were 0 to 18 (median of 7). The food, environmental and clinical isolates collected in Hospital X in November 2015 formed a distinct clade (Fig. 2). Two SNPs specifically distinguished the November 2015 isolates from other isolates, one synonymous SNP (nucleotide A in the November 2015 isolates and nucleotide T in other isolates) in an ABC transporter ATP-binding protein (AFY11_ 00690 of the reference genome) and one non-synonymous SNP (nucleotide T in the November 2015 isolates and nucleotide C in other isolates) in 50S ribosomal protein L4 (AFY11_ 15190 of the reference genome).

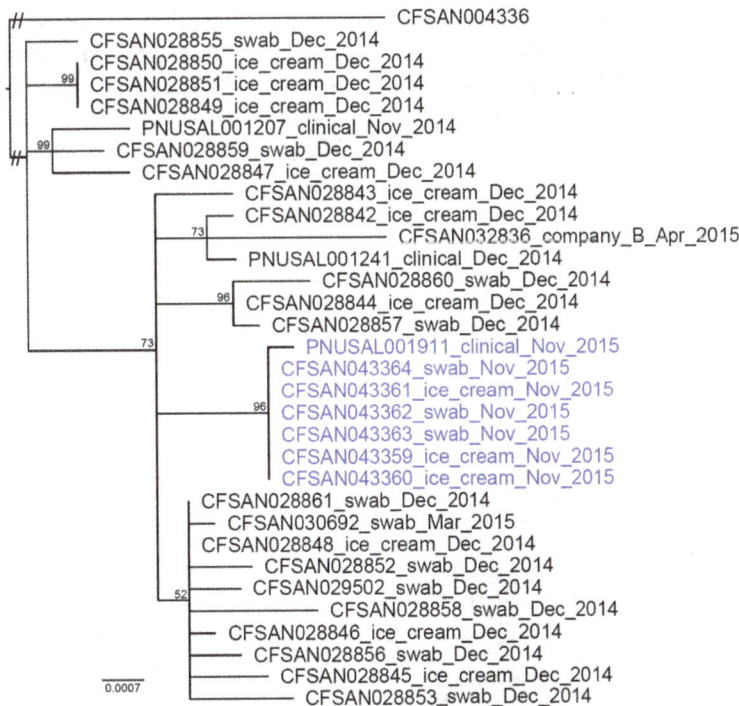

Fig. 2 Maximum likelihood phylogeny of outbreak-associated isolates based on single nucleotide polymorphisms (SNPs) identified by the Center for Food Safety and Applied Nutrition (CFSAN) SNP Pipeline using CFSAN004336 for comparison. The tree is rooted at midpoint. Isolate ID is followed by sample type and collection year and abbreviation of month. The isolates that persisted in the milkshake maker until November 2015 and the clinical isolate collected in November 2015 are in *blue color*. The isolates collected in March and April 2015 were from ice cream processing facilities, not the hospital milkshake maker

A species-specific cgMLST scheme targeting 1827 genes (hereinafter designated as 1827-cgMLST) generated a phylogeny congruent with the SNP-based WGS phylogeny. Outbreak-associated food, environmental and clinical isolates in 2014 and 2015 were clustered together and separated from the outgroup (Fig. 3a). Outbreak isolates differed by 0 to 12 (median, 6) alleles. Among them, isolates collected in 2014 differed by up to 12 alleles; the two 2014 clinical isolates differed by 10 alleles and they differed from the 2015 clinical isolate by 7 and 11 alleles. The isolates that persisted in the hospital milkshake machine until November 2015 and the clinical isolate collected in November 2015 also formed a distinct clade inside the outbreak cluster. Two alleles specifically distinguished the November 2015 isolates from other isolates, lmo2631 (encoding 50S ribosomal protein L4) and lmo2751 (encoding ABC transporter ATP-binding protein). Another species-specific cgMLST scheme targeting 1748 genes (hereinafter designated as 1748-cgMLST) also clustered together outbreak-associated isolates collected in 2014 and 2015 (Fig. 3b). With this scheme, outbreak isolates differed by 0 to 12 (median, 5) alleles. Among them, isolates collected in 2014 differed by up to 10 alleles; the two 2014 clinical isolates differed by 8 alleles and they differed from the 2015 clinical isolate by 4 and 8 alleles. However, the outbreak-associated isolates collected in November 2015 did not form a distinct clade inside the outbreak cluster, because lmo2631 and lmo2751 were not in the gene set targeted by 1748-cgMLST. The minimum spanning tree (Fig. 4) based on 1827-cgMLST also revealed that the November 2015 isolates formed its own clade. This tree does not show a clear central allele profile

because the majority of the isolates have their unique allele profiles.

Due to the difference in clustering outbreak-associated isolates collected in November 2015 by the two species-specific cgMLST schemes, we further developed a whole genome-specific MLST scheme (wgsMLST) using the annotated genome (CFSAN028853, GenBank Accession MAKW00000000.1). This scheme targeted 3017 loci in the entire genome of outbreak-associated isolates, which could be alternatively named as outbreak-specific cgMLST because those loci were core to the outbreak isolates. We also performed a previously developed lineage-II specific cgMLST scheme targeting 2342 loci [4]. The results of these two schemes corroborated those of the 1827-cgMLST and whole genome SNP analyses: food, environmental and clinical isolates collected in November 2015 formed a distinct clade inside the outbreak cluster (trees not shown). Both schemes indeed contained the genes encoding the 50S ribosomal protein L4 and the ABC transporter ATP-binding protein.

Two putative complete prophages in the reference genome CFSAN028853 were predicted by PHAST/PHASTER [14, 15]: prophage 1 (31.6Kbp, position 163,457 to 195,099 of contig 11) and prophage 2 (34 Kbp, position 1 to 34,017 of contig 15). BLAST analysis showed that all outbreak isolates contained a conserved prophage 1 (>93% query coverage, >99% sequence identity). BLAST analysis further showed that all outbreak isolates exhibiting AscI-P1 and AscI-P3 contained a conserved prophage 2 (>98% coverage, >99% sequence identity). The outbreak isolates exhibiting AscI-P2 (PNUSAL001241 and CFSAN028861) did not contain prophage 2 (<3% BLAST coverage). The

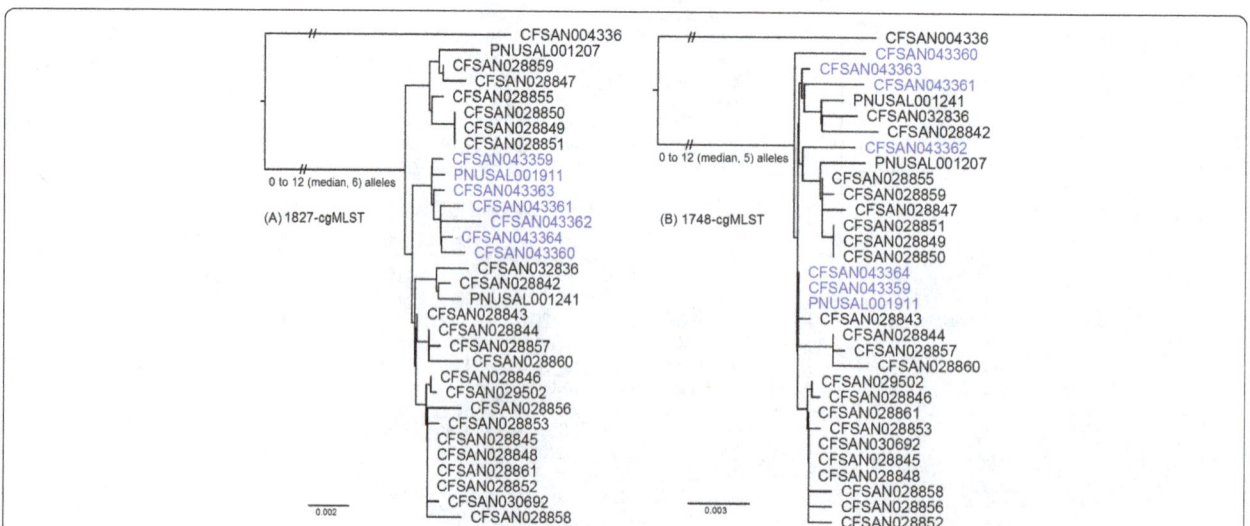

Fig. 3 Neighbor-joining phylogeny of outbreak-associated isolates based on two cgMLST schemes, **a** 1827-cgMLST and **b** 1748-cgMLST, using CFSAN004336 for comparison. The trees are rooted at midpoint. The isolates that persisted in the milkshake maker until November 2015 and the clinical isolate collected in November 2015 are in *blue color*. The isolates collected in March and April 2015 were from ice cream processing facilities, not the hospital milkshake maker. The minimum, maximum and median of pairwise allele differences of outbreak isolates are indicated near the root

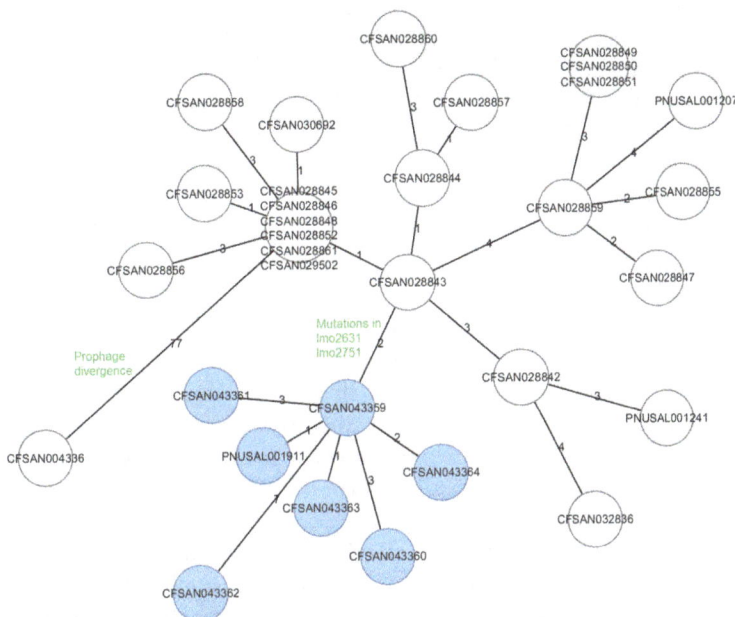

Fig. 4 Minimum spanning tree of outbreak-associated isolates based on 1827-cgMLST using CFSAN004336 for comparison. The allele differences between nodes are listed by the connection lines. The isolates collected in November 2015 are marked in *light blue color*. The prophage divergence and SNPs in lmo2631 and lmo2751 are indicated in the figure

non-outbreak isolate (CFSAN028854) from Company A environment aligned with prophage 1 and prophage 2 for 18% and 71% BLAST coverage, respectively, indicating significant prophage divergence from outbreak isolates. The outgroup CFSAN004336 aligned with prophage 1 and prophage 2 for 87% and 14% BLAST coverage, respectively. The reference genome is a draft genome, and thus multiple fragments that belonged to other putative prophages were also predicted by the software but either they were incomplete prophages or we could not assemble the complete prophages from this draft genome (data not shown). Prophage 2 contained an *Asc*I restriction site (position 9598 to 9605 of contig 15 of reference genome), thus the loss of prophage 2 contributed to the *Asc*I-PFGE pattern change from *Asc*I-P1 or *Asc*I-P3 (all other outbreak isolates) to *Asc*I-P2 (PNUSAL001241 and CFSAN028861). Loss of an *Asc*I restriction site would lead to the combination of two restriction fragments (~185 Kbp and ~250 Kbp) and loss of the prophage 2 would result in a 34 Kbp deletion. Thus the combined fragment should be around 400 Kbp, which was slightly different from the ~430 Kbp fragment (Fig. 1). Thus other DNA variations (e.g., replacement of prophage 2 with a similar-sized prophage through recombination) could co-cause this PFGE banding pattern change. We were not able to identify this variation using the draft genome.

Discussion

A 7-gene MLST scheme has been used to define major clonal groups of *L. monocytogenes*, designated as CC or singleton [16] and it was recently demonstrated that the this definition was generally compatible with WGS clustering [4, 11]. Isolates associated with this outbreak had ST1038, belonging to CC101 which was commonly isolated in the mid-1950s and its prevalence had since decreased [17]. However, it likely has re-emerged recently, evidenced by its isolation in 31 clinical cases in Lombardy, Italy between 2006 and 2010 [17], as well as its association with a 2012 U.S. outbreak linked to ricotta salata cheese products manufactured in Italy [18]. Thus, CC101 has been involved in at least two outbreaks to date and represents an epidemic clone. *inlA* encodes internalin A, which is involved in the invasion of human intestinal epithelia cells and could play an important role in *Listeria* virulence [19]. Premature stop codons (PMSCs) in *inlA* lead to truncated protein in some *Listeria* strains and were linked to attenuated virulence of those strains in mammalian hosts and thus it has been [19]. Moreover, PMSCs are mostly found in lineage II strains and isolated more frequently from food and food production environment than from human listeriosis cases [19]. Our finding is consistent with previous observation that outbreak-associated strains generally do not contain PMSCs [20]. To date, 4 *Listeria* pathogenicity islands (LIPI) have been characterized as virulence factors: LIPI-1 is conserved across the entire species of *L. monocytogenes* [21]; LIPI-2 is specific to *L. ivanovii* [22]; LIPI-3 is mostly present in lineage I isolates [23]; and the very recently identified LIPI-4 is present in several lineage I CCs and a few lineage II, III and IV strains [11]. This explains why we observed only LIPI-1 in isolates associated with this outbreak. In comparison, the 2010–2015 multistate ice cream outbreak strains (CC5) contained LIPI-1, but not LIPI-3 or LIPI-4 [6] and the recent isolates (singleton ST382) from the stone fruit, caramel apple and leafy green salad outbreaks contained all LIPI-1, LIPI-3 and LIPI-4 [24].

In this study, WGS could not only trace the transmission of the outbreak strain between two facilities, but also reveal its persistence in the soft serve ice cream/milkshake machine for one year. Furthermore, WGS was able to cluster all outbreak-associated isolates, despite their difference in *Asc*I-PFGE profiles. The milkshake maker had a regular sanitation schedule, and extra sanitizing effort after the outbreak, including disassembly of the machine at least twice, was taken after the outbreak recognition [8]. Thus, it is probable that while the majority of the *L. monocytogenes* cells contaminating the milkshake maker in 2014 were eliminated during sampling and sanitizing; a few cells escaped and/or survived, contaminated other areas of the milkshake maker later, and eventually made their presence in the final product during operation. This also concurred with our results that the food, environmental and clinical isolates collected in November 2015 formed its own WGS clade and thus had a recent common ancestor.

The high resolution of WGS has been demonstrated in various studies by differentiating outbreak isolates from non-outbreak isolates, especially those matched by PFGE [5, 6]. In addition, WGS can achieve more than just discrimination between unrelated isolates. For example, it can be used to study the microevolution of different isolates in the same outbreak setting, identify genotypes that may be specific to different product varieties or production lines in the same facility [6], and ultimately shed some light on root cause analysis. For another example, combination of core genome and accessory genome was needed to elucidate the epidemiology of isolates persistent in a food processing facility for 12 years [4]. Comparison between outbreak and non-outbreak isolates could provide potential candidates for future functional genomics analyses on virulence. Comparison among strains persistent in an environment could identify potential candidates for studying the evolution and mechanisms of *L. monocytogenes* persistence. WGS analyses targeting the entire genome could certainly reveal more potential genetic markers. For example, in this study, whole genome-based analyses, lineage-specific cgMLST and 1827-cgMLST identified unique variants in the genes encoding ABC transporter ATP-binding protein (lmo2751 in EGD-e) and 50S ribosomal protein L4 (lmo2631 in EGD-e) of the persistent cells, which contributed to the formation of the distinct clade containing these persistent cells in the phylogenetic trees; while 1748-cgMLST did not yield that

clustering because of the absence of these two genes in its gene set. The 1827-cgMLST scheme and the 1748-cgMLST scheme share 1324 common loci (i.e., there are 503 unique loci in the 1827-cgMLST scheme and 424 in the 1748-cgMLST scheme), thus, despite their sufficient discriminatory power to distinguish epidemiologically unrelated isolates, differences in results are expected when they are applied to study the microevolution within the scope of an outbreak. We also evaluated another cgMLST scheme targeting 1701 core genes of *L. monocytogenes* [12] which contained lmo2751 and lmo2631, and it generated a phylogeny congruent with that by 1827-cgMLST and SNP analyses: the November 2015 isolates formed a distinct clade within the outbreak cluster (tree not shown). It is possible that different isolates collected in November 2015 simultaneously accumulated SNPs in both genes during evolution from their 2014 ancestor; however, it is also possible that their 2014 ancestor already had these two unique SNPs, which remained unchanged during persistence in the milkshake maker until November 2015. The expression of 50S ribosomal protein L4 in *L. monocytogenes* is affected by alkali-tolerance response, which may be critical for this pathogen to survive in human gastrointestinal tract and during food processing [25]. Mutation in this gene has been linked to antimicrobial resistance in other bacteria [26, 27]. The ABC transporter ATP-binding protein encoded by lmo2751 is upregulated when *L. monocytogenes* was exposed to bacteriocin pediocin [28] or during growth within murine macrophages [29]. The exact roles of these two proteins in the survival and evolution of *L. monocytogenes* in food processing environment remain to be investigated.

Generally, a cgMLST scheme targets the entire population of *L. monocytogenes* [4, 11–13] and the core loci comprise ~60% of a typical coding genome. The species-specific schemes are suitable for evolutionary analysis and nomenclature. A standardized nomenclature system based on cgMLST could be beneficial in surveillance studies because isolates from different environment, food commodities and geographical locations analyzed in different studies can be easily compared to suggest possible links. However, vigorous collaborative validation needs to be performed on multiple elements in this nomenclature system: the centralized database, the set of cgMLST targets, the platform(s) to run the analysis, the parameter(s) and algorithm to call allele differences, the mechanism to deal with missing regions due to draft sequencing, and the threshold to define cgMLST or cluster types. Minimum spanning tree based on a 7-gene MLST has been used to define clonal complexes [16], in which the majority of the strains share the same sequence type, and that sequence type serves as the central allele profile to define single locus variants. This approach might not be suitable for cgMLST since isolates could easily differ by one allele in cgMLST and thus no clear central allele profile can be easily identified, as illustrated in this outbreak (Fig. 4). An approach to define cgMLST cluster type is to set an allele threshold among isolates, but different studies have proposed different thresholds [4, 11, 12]. In addition, such approach is not perfect because the entry order of submission of a set of isolates could potentially affect the assignment of cgMLST cluster types of those isolates [30]. Nonetheless, species-specific cgMLST schemes are generally satisfactory in differentiating epidemiologically unrelated isolates [4, 11, 12]. However, when WGS is used to differentiate among a set of closely related isolates and/or to study the microevolution of isolates within the scope of one outbreak, a flexible definition of cgMLST could be explored to fully utilize the high resolution of WGS [31]. cgMLST schemes specific to individual lineages of *L. monocytogenes* contain loci comprising ~80% of a typical coding genome [4]. In this study, we developed a wgsMLST, which was essentially an outbreak-specific cgMLST, to analyze the microevolution among outbreak isolates. In contrast to wgMLST which contains a pre-defined set of pan-genome loci [10], this wgsMLST targets the entire set of coding loci of an outbreak isolate, and thus could target any novel loci that are not in the pan-genome pre-defined based on a set of previously published genomes. This is similar to the whole genome SNP-based approach that maximizes the resolution by selecting one outbreak isolate as the reference to analyze other outbreak isolates [32].

Several previous studies using SNP-based WGS analysis employed a complete genome closed by PacBio® technology as the reference [5, 24]. However, using PacBio® may not be practical in every outbreak investigations. Here, we explored the use of a CLC Genomics Workbench-assembled draft genome (CFSAN028853) as the reference for the CFSAN SNP Pipeline and produced a WGS phylogeny that supported PFGE and epidemiological evidence. We also tried CFSAN028853 assembled by SPAdes assembler 3.9.0 [33], and the WGS analysis generated the same phylogeny with minor changes of the lengths of several tree branches (data not shown). CLC and SPAdes both map raw reads to initial assembly for error correction; however, error in the final assembly could still occur. We mapped raw reads of CFSAN028853 to both CLC assembly and SPAdes assembly using the CFSAN SNP Pipeline and found that raw reads were consistent with CLC assembly but differed from the SPAdes assembly by 6 SNPs. As a result, between the final SNP matrices generated using CLC and Spades assemblies the pairwise SNP differences of several isolate pairs differed by 1–2 SNPs. Thus, we believe the CLC assembly was more accurate than the SPAdes assembly in this case, although we found that SPAdes can be more accurate in

other cases (data not shown). Nonetheless, using either assembly as the reference yielded the same WGS clustering. The wgsMLST was defined using the annotated, CLC-assembled draft genome of CFSAN028853. A completely closed CFSAN028853 genome could probably have revealed additional coding regions as wgsMLST loci and thus further improved the resolution of wgsMLST, although this improved resolution may not be critical for the purpose of outbreak investigation.

AscI-PFGE banding pattern changes due to prophage variations have been observed among isolates associated with a few outbreaks. In this outbreak, a prophage loss led to the loss of an *AscI* restriction site in the prophage, resulting in combination of two *AscI* fragments, although other unidentified DNA variations could co-affect the PFGE banding pattern change from *AscI*-P1/*AscI*-P3 to *AscI*-P2. DNA variations underlying the difference between *AscI*-P1 and *AscI*-P3 were not identified. Among isolates of several other outbreaks, gain/loss of prophage in a PFGE fragment caused the fragment to shift to a different position in the gel [6, 34, 35]. In one outbreak, gain/loss of 3 prophages occurred among different isolates, and *AscI* restriction analysis of the completely closed genome allowed the precise determination of the genome positions of all *AscI* fragments, which unambiguously identified the gain/loss of the specific prophage in the specific *AscI* fragment [6]. Previous analyses have shown that reference-based reads mapping and SNP calling in repetitive regions and insertion regions of prophages could yield inconclusive SNPs, which were usually present in high density ($\geq$3 SNPs in 1000 bp) [6, 24]. Recombination events could also generate high - density SNPs [6]. Thus, when analyzing a group of closely related isolates (e.g., $\leq$50 SNPs in 3×10^6 bp) associated with the same outbreak, CFSAN SNP Pipeline offers the option to apply a filter to remove these high-density SNPs from the final SNP matrices. In this study, the 4 removed high - density variant regions contained SNPs only between PNU-SAL001241/CFSAN028861 and other isolates, and these high - density variant regions were all in insertion sites of putative prophages: the prophage 2 described above, and 3 fragments from other incomplete prophage(s) (data not shown).

The patients in Hospital X involved in the Washington State outbreak under discussion in this study and patients in a hospital (Hospital Y) involved in the 2010–2015 multistate outbreak [6] consumed milkshakes prepared from contaminated ice cream products, and this highlights the potential of ice cream as a vehicle for listeriosis infection, given that *L. monocytogenes* could grow in milkshakes, especially when the milkshakes go through temperature abuse during serving. However, *L. monocytogenes* in milkshakes prepared from the multistate outbreak-associated ice cream had a relatively long log phase (9 h) and a slow growth rate of 0.186 CFU/log/h at room temperature, which could be attributed to low level of initial amount of naturally occurring *L. monocytogenes* [36] and/or the variety and levels of competing microflora present in the ice cream samples [37]. The ice cream products associated with the Washington State outbreak were not available for such analysis. The milkshake maker used in Hospital Y involved in the multistate outbreak was a drink mixer employing simple propellers and were relatively easy to clean [36], and the environmental testing in the hospital kitchen, including the drink mixer, did not yield *L. monocytogenes* [38]. The milkshake maker used in Hospital X was a soft serve shake freezer which held ice cream mix and made milkshake inside the machine to serve through the dispensing nozzle, and had reusable parts of reservoirs, pipes and mixers that made contact with ice cream [8]; thus it was more difficult to clean, which could explain why *L. monocytogenes* was able to survive the sampling and cleaning. The occurrence of these two outbreaks, involving patients with weakened immune systems, could contribute to our understanding of the risk associated with *L. monocytogenes* contamination in ice cream.

Conclusions

WGS analyses clustered epidemiologically related isolates, and clarified the microevolution and persistence of isolates within the scope of one outbreak. A flexible definition of core genome MLST, targeting a species, a genetic lineage or an outbreak, could be explored to offer different levels of resolution based on the set of strains investigated and the purpose of the analysis.

Methods

Isolates

L. monocytogenes isolates were collected from ice cream products and environmental samples from Company A as well as patients and ice cream products from Hospital X during the outbreak investigation in 2014. An environmental isolate from Company A and an environmental isolate from Company B were collected in early 2015 as Company B purchased dairy ingredients from Company A [7, 8]. Isolates were also collected in November 2015 from the patient, the ice cream that remained in and were dispensed from the milkshake maker in Hospital X, and the environmental samples from different areas (e.g., side walls of internal parts and nozzle assembly) of the milkshake maker. Sequences are publicly available at NCBI Sequence Read Archive (SRA) of the Genome-Trakr database [39] with SRA ID listed in Table 1.

Whole genome SNP analysis

WGS was performed using Illumina MiSeq platform (Illumina, San Diego, USA) as previously described

[40]. Reference-based whole genome SNP analysis was performed using SNP pipeline (version 0.7.0) developed by the Center for Food Safety and Applied Nutrition (CFSAN), as previously described [32, 40, 41]. CLC Genomics Workbench 9.0.1 (Qiagen, Hilden, Germany) was used to assemble CFSAN028853 (200× coverage, GenBank Accession MAKW00000000.1) to serve as the reference. A *L. monocytogenes* ST101 strain, CFSAN004336 (Table 1), which belonged to the same CC101 as the outbreak-associated isolates, was used as the outgroup to demonstrate that WGS can separate this strain from the outbreak-associated isolates. A second WGS analysis was performed on only the outbreak-associated isolates for precise SNP calling because accurate SNP calling by reference-based methods may be affected by ascertainment bias when these methods were applied to slightly more diverse isolates [42, 43]. Briefly, raw reads from each isolate were mapped to the reference genome using default settings of Bowtie2 version 2.2.9 [44]. The BAM file was sorted using Samtools version 1.3.1 [45], and a pileup file for each isolate was produced. These files were then processed using VarScan2 version 2.3.9 [46] to identify high quality variant sites using the mpileup2snp option. A Python script was used to parse the.vcf files and construct an initial SNP matrix. For these closely related isolates, the SNP Pipeline applied a filter to exclude variant sites in high density variant regions (≥3 variant sites in ≤1000 bp of any one genome) since they may be the result of recombination or low quality sequencing/mapping and/or be associated with repetitive elements. Four regions, 42 bp (containing 16 variant sites), 387 bp (14 variant sites), 95 bp (4 variant sites) and 102 bp (9 variant sites), all in prophage insertion areas, were filtered out. The 9 variant sites in the 102 bp region were also within 500 bp of the start of a reference genome contig, which typically had lower quality of mapping and assembly due to less reads mapped to contig ends. GARLI was used to create topologies based on the SNP matrix [47].

Species-specific and lineage II-specific cgMLST analyses

Two species-specific cgMLST schemes, developed using strain EGD-e (GenBank Accession NC_003210.1) as the reference, were used to analyze the isolates: the 1827-cgMLST targeting 1827 core genes [4] and the 1748-cgMLST targeting 1748 core genes [11]. Core genome loci of both schemes were incorporated into Ridom Seq-Sphere + (Ridom GmbH, Münster, Germany) and cgMLST was performed using default settings as previously described [4]. Neighbor-joining and minimum spanning trees were constructed from the allele profiles of all the isolates. A previously developed lineage II-specific cgMLST, targeting 2342 core genes of lineage II, was also performed [4].

wgsMLST/outbreak-specific cgMLST

The reference genome used for SNP analysis, CFSAN028853 (GenBank Accession MAKW00000000.1), was used as the only genome to define a wgsMLST scheme using the cgMLST target definer (version 3.5.0) function of SeqSphere+ (Ridom GmbH, Germany) with default parameters as previously described [12]. The software collected all coding loci of CFSAN028853 and filtered out those that could be generated by assembly/annotation errors, resulting in 3107 loci. These loci were core to the outbreak isolates and thus could be alternatively called as an outbreak-specific cgMLST.

In silico MLST, prophage, virulence gene presence and *inlA* premature stop codon analyses

In silico MLST analysis was performed using the tools in the BIGSdb-Lm database (http://bigsdb.pasteur.fr/listeria/listeria.html). Putative prophages were identified from individual contigs of the reference genome (CFSAN028853) by PHAST/PHASTER [14, 15], and the predicted complete phages were compared to other isolates using BLAST. A threshold of ≥60% query coverage with ≥80% sequence identity [48] of BLAST alignment indicated the presence of a CFSAN028853 prophage in a genome. The presence of major virulence genes [11] was examined using BLAST. The *inlA* sequences of all isolates were examined for the presence of premature stop codons.

Pulsed-field gel electrophoresis (PFGE)

PFGE was performed using the PulseNet standard protocol [49].

Abbreviations

CC: Clonal complex; cgMLST: Core genome multilocus sequence typing; PFGE: Pulsed-field gel electrophoresis; PMSC: Premature stop codon; SNP: Single nucleotide polymorphism; ST: Sequence type; WGS: Whole genome sequencing; wgsMLST: Whole genome-specific multilocus sequence typing

Acknowledgements

This project was supported in part by an appointment to the Research Participation Program at the Center for Food Safety and Applied Nutrition administration by the Oak Ridge Institute for Science and Education through an interagency agreement between the U.S. Department of Energy and the U.S. Food and Drug Administration.

Funding

This project was supported in part by an appointment to the Research Participation Program at the Center for Food Safety and Applied Nutrition administration by the Oak Ridge Institute for Science and Education through an interagency agreement between the U.S. Department of Energy and the U.S. Food and Drug Administration.

Authors' contributions

ZL, AP, WG, YC: designed and coordinated the study; ZL, YC: wrote the manuscript; ZL, YW, YC: performed bioinformatics analyses and edited the manuscript; ZL, KE: carried out laboratory work and edited the manuscript; MA, EB: provided scientific advisement and edited the manuscript. All authors read and approved the manuscript.

Competing interests

The authors declare that they have no competing interests.

Author details

[1]Washington State Department of Health, Public Health Laboratories, Shoreline, Washington, USA. [2]Center for Food Safety and Applied Nutrition, Food and Drug Administration, College Park, MD, USA.

References

1. Kathariou S. *Listeria monocytogenes* Virulence and pathogenicity, a food safety perspective. J Food Prot. 2002;65(11):1811–29.
2. Ferreira V, Wiedmann M, Teixeira P, Stasiewicz MJ. *Listeria monocytogenes* Persistence in food-associated environments: epidemiology, strain characteristics, and implications for public health. J Food Prot. 2014;77(1):150–70.
3. Silk BJ, Date KA, Jackson KA, Pouillot R, Holt KG, Graves LM, et al. Invasive listeriosis in the foodborne diseases active surveillance network (FoodNet), 2004-2009: further targeted prevention needed for higher-risk groups. Clin Infect Dis. 2012;54(Suppl 5):S396–404.
4. Chen Y, Gonzalez-Escalona N, Hammack TS, Allard MW, Strain EA, Brown EW. Core genome multilocus sequence typing for identification of globally distributed clonal groups and differentiation of outbreak strains of *Listeria monocytogenes*. Appl Environ Microbiol. 2016;82(20):6258–72.
5. Chen Y, Burall LS, Luo Y, Timme R, Melka D, Muruvanda T, et al. *Listeria monocytogenes* In stone fruits linked to a multistate outbreak: enumeration of cells and whole-genome sequencing. Appl Environ Microbiol. 2016; 82(24):7030–40.
6. Chen Y, Luo Y, Curry P, Timme R, Melka D, Doyle M, et al. Assessing the genome level diversity of *Listeria monocytogenes* from contaminated ice cream and environmental samples linked to a listeriosis outbreak in the United States. PLoS One. 2017;12(2):e0171389.
7. Rietberg K, Lloyd J, Melius B, Wyman P, Treadwell R, Olson G, et al. Outbreak of *Listeria monocytogenes* infections linked to a pasteurized ice cream product served to hospitalized patients. Epidemiol Infect. 2016;144(13):2728–31.
8. Mazengia E, Kawakami V, Rietberg K, Kay M, Wyman P, Skilton C, et al. Hospital-acquired listeriosis linked to a persistently contaminated milkshake machine. Epidemiol Infect. 2017:1–7.
9. Jackson BR, Tarr C, Strain E, Jackson KA, Conrad A, Carleton H, et al. Implementation of nationwide real-time whole-genome sequencing to enhance listeriosis outbreak detection and investigation. Clin Infect Dis. 2016;
10. *Listeria monocytogenes* whole genome sequence typing [http://www.applied-maths.com/news/listeria-monocytogenes-whole-genome-sequence-typing] Assessed May 5th, 2017.
11. Moura A, Criscuolo A, Pouseele H, Maury MM, Leclercq A, Tarr C, et al. Whole genome-based population biology and epidemiological surveillance of *Listeria monocytogenes*. Nat Microbiol. 2016;2:16185.
12. Ruppitsch W, Pietzka A, Prior K, Bletz S, Fernandez HL, Allerberger F, et al. Defining and evaluating a core genome multilocus sequence typing scheme for whole-genome sequence-based typing of *Listeria monocytogenes*. J Clin Microbiol. 2015;53(9):2869–76.
13. Pightling AW, Petronella N, Pagotto F. The *Listeria monocytogenes* Core-genome sequence Typer (LmCGST): a bioinformatic pipeline for molecular characterization with next-generation sequence data. BMC Microbiol. 2015;15:224.
14. Arndt D, Grant JR, Marcu A, Sajed T, Pon A, Liang Y, et al. PHASTER: a better, faster version of the PHAST phage search tool. Nucleic Acids Res. 2016; 44(W1):W16–21.
15. Zhou Y, Liang Y, Lynch KH, Dennis JJ, Wishart DS: PHAST: a fast phage search tool. *Nucleic acids research* 2011, 39(Web Server issue):W347–352.
16. Ragon M, Wirth T, Hollandt F, Lavenir R, Lecuit M, Le Monnier A, et al. A new perspective on *Listeria monocytogenes* evolution. PLoS Pathog. 2008; 4(9):e1000146.

17. Haase JK, Didelot X, Lecuit M, Korkeala H. Group LmMS, Achtman M: the ubiquitous nature of *Listeria monocytogenes* clones: a large-scale multilocus sequence typing study. Environ Microbiol. 2014;16(2):405–16.

18. Bergholz TM, den Bakker HC, Katz LS, Silk BJ, Jackson KA, Kucerova Z, et al. Determination of evolutionary relationships of outbreak-associated *Listeria monocytogenes* strains of serotypes 1/2a and 1/2b by whole-genome sequencing. Appl Environ Microbiol. 2015;82(3):928–38.

19. Nightingale KK, Ivy RA, Ho AJ, Fortes ED, Njaa BL, Peters RM. Wiedmann M: inlA premature stop codons are common among *Listeria monocytogenes* isolates from foods and yield virulence-attenuated strains that confer protection against fully virulent strains. Appl Environ Microbiol. 2008;74(21): 6570–83.

20. Van Stelten A, Simpson JM, Chen Y, Scott VN, Whiting RC, Ross WH, et al. Significant shift in median guinea pig infectious dose shown by an outbreak-associated *Listeria monocytogenes* epidemic clone strain and a strain carrying a premature stop codon mutation in inlA. Appl Environ Microbiol. 2011;77(7):2479–87.

21. Vazquez-Boland JA, Kuhn M, Berche P, Chakraborty T, Dominguez-Bernal G, Goebel W, et al. *Listeria* pathogenesis and molecular virulence determinants. Clin Microbiol Rev. 2001;14(3):584–640.

22. Dominguez-Bernal G, Muller-Altrock S, Gonzalez-Zorn B, Scortti M, Herrmann P, Monzo HJ, et al. A spontaneous genomic deletion in *Listeria ivanovii* identifies LIPI-2, a species-specific pathogenicity island encoding sphingomyelinase and numerous internalins. Mol Microbiol. 2006;59(2):415–32.

23. Cotter PD, Draper LA, Lawton EM, Daly KM, Groeger DS, Casey PG, et al. Listeriolysin S, a novel peptide haemolysin associated with a subset of lineage I *Listeria monocytogenes*. PLoS Pathog. 2008;4(9):e1000144.

24. Chen Y, Luo Y, Pettengill J, Timme R, Melka D, Doyle M, et al. Singleton sequence type 382, an emerging clonal group of *Listeria monocytogenes* associated with three multistate outbreaks linked to contaminated stone fruit, caramel apples, and leafy green salad. J Clin Microbiol. 2017;55(3):931–41.

25. Giotis ES, Muthaiyan A, Blair IS, Wilkinson BJ, McDowell DA. Genomic and proteomic analysis of the alkali-tolerance response (AITR) in *Listeria monocytogenes* 10403S. BMC Microbiol. 2008;8:102.

26. Holzel CS, Harms KS, Schwaiger K, Bauer J. Resistance to linezolid in a porcine *Clostridium perfringens* strain carrying a mutation in the rplD gene encoding the ribosomal protein L4. Antimicrob Agents Chemother. 2010; 54(3):1351–3.

27. Zaman S, Fitzpatrick M, Lindahl L, Zengel J. Novel mutations in ribosomal proteins L4 and L22 that confer erythromycin resistance in *Escherichia coli*. Mol Microbiol. 2007;66(4):1039–50.

28. Laursen MF, Bahl MI, Licht TR, Gram L, Knudsen GM. A single exposure to a sublethal pediocin concentration initiates a resistance-associated temporal cell envelope and general stress response in *Listeria monocytogenes*. Environ Microbiol. 2015;17(4):1134–51.

29. Schultze T, Hilker R, Mannala GK, Gentil K, Weigel M, Farmani N, et al. A detailed view of the intracellular transcriptome of *Listeria monocytogenes* in murine macrophages using RNA-seq. Front Microbiol. 2015;6:1199.

30. Core genome MLST cluster type [http://www.seqsphere.de/ug/Core_Genome_MLST_Cluster_Type.html] Assessed May 5th, 2017.

31. Mellmann A, Bletz S, Boking T, Kipp F, Becker K, Schultes A, et al. Real-time genome sequencing of resistant bacteria provides precision infection control in an institutional setting. J Clin Microbiol. 2016;54(12):2874–81.

32. Pettengill JB, Luo Y, Davis S, Chen Y, Gonzalez-Escalona N, Ottesen A, et al. An evaluation of alternative methods for constructing phylogenies from whole genome sequence data: a case study with *Salmonella*. PeerJ. 2014;2:e620.

33. Bankevich A, Nurk S, Antipov D, Gurevich AA, Dvorkin M, Kulikov AS, et al. SPAdes: a new genome assembly algorithm and its applications to single-cell sequencing. J Comput Biol. 2012;19(5):455–77.

34. Gilmour MW, Graham M, Van Domselaar G, Tyler S, Kent H, Trout-Yakel KM, et al. High-throughput genome sequencing of two *Listeria monocytogenes* clinical isolates during a large foodborne outbreak. BMC Genomics. 2010;11:120.

35. Lomonaco S, Verghese B, Gerner-Smidt P, Tarr C, Gladney L, Joseph L, et al. Novel epidemic clones of *Listeria monocytogenes*, United States, 2011. Emerg Infect Dis. 2013;19(1):147–50.

36. Chen Y, Allard E, Wooten A, Hur M, Sheth I, Laasri A, et al. Recovery and growth potential of *Listeria monocytogenes* in temperature abused milkshakes prepared from naturally contaminated ice cream linked to a listeriosis outbreak. Front Microbiol. 2016;7:764.

37. Ottesen A, Ramachandran P, Reed E, White JR, Hasan N, Subramanian P, et al. Enrichment dynamics of *Listeria monocytogenes* and the associated microbiome from naturally contaminated ice cream linked to a listeriosis outbreak. BMC Microbiol. 2016;16(1):275.

38. Chen Y, Burall L, Macarisin D, Pouillot R, Strain E, De Jesus A, et al. Prevalence and level of *Listeria monocytogenes* in ice cream linked to a listeriosis outbreak in the United States. J Food Prot. 2016;79(11):1828–32.

39. Allard MW, Strain E, Melka D, Bunning K, Musser SM, Brown EW. Timme R: the PRACTICAL value of food pathogen traceability through BUILDING a whole-genome sequencing network and database. J Clin Microbiol. 2016;

40. Chen Y, Burall LS, Luo Y, Timme R, Melka D, Muruvanda T, et al. Isolation, enumeration and whole genome sequencing of *Listeria monocytogenes* in stone fruits linked to a multistate outbreak. Appl Environ Microbiol. 2016;

41. Davis S, Pettengill JB, Luo Y, Payne J, Shpuntoff A, Rand H, Strain A. CFSAN SNP Pipeline: an automated method for constructing SNP matrices from next-generation sequence data. PeerJ Computer Science. 2015;1:e20. https://doi.org/10.7717/peerj-cs.20.

42. Bertels F, Silander OK, Pachkov M, Rainey PB, van Nimwegen E. Automated reconstruction of whole-genome phylogenies from short-sequence reads. Mol Biol Evol. 2014;31(5):1077–88.

43. Pightling AW, Petronella N, Pagotto F. Choice of reference-guided sequence assembler and SNP caller for analysis of *Listeria monocytogenes* short-read sequence data greatly influences rates of error. BMC Res Notes. 2015;8:748.

44. Langmead B, Trapnell C, Pop M, Salzberg SL. Ultrafast and memory-efficient alignment of short DNA sequences to the human genome. Genome Biol. 2009;10(3):R25.

45. Li H, Handsaker B, Wysoker A, Fennell T, Ruan J, Homer N, et al. Genome project data processing S: the sequence alignment/map format and SAMtools. Bioinformatics. 2009;25(16):2078–9.

46. Koboldt DC, Chen K, Wylie T, Larson DE, McLellan MD, Mardis ER, et al. VarScan: variant detection in massively parallel sequencing of individual and pooled samples. Bioinformatics. 2009;25(17):2283–5.

47. Zwickl DJ. Genetic algorithm approaches for the phylogenetic analysis of large biological sequence datasets under the maximum likelihood criterion: The University of Texas at Austin; 2006.

48. Schmitz-Esser S, Muller A, Stessl B, Wagner M. Genomes of sequence type 121 *Listeria monocytogenes* strains harbor highly conserved plasmids and prophages. Front Microbiol. 2015;6:380.

49. Graves LM, Swaminathan B. PulseNet standardized protocol for subtyping *Listeria monocytogenes* by macrorestriction and pulsed-field gel electrophoresis. Int J Food Microbiol. 2001;65(1–2):55–62.

3

Bacillus species (BT42) isolated from Coffea arabica L. rhizosphere antagonizes Colletotrichum gloeosporioides and Fusarium oxysporum and also exhibits multiple plant growth promoting activity

Tekalign Kejela[1,2]*, Vasudev R. Thakkar[3] and Parth Thakor[3]

Abstract

Background: Colletotrichum and Fusarium species are among pathogenic fungi widely affecting Coffea arabica L., resulting in major yield loss. In the present study, we aimed to isolate bacteria from root rhizosphere of the same plant that is capable of antagonizing Colletotrichum gloeosporioides and Fusarium oxysporum as well as promotes plant growth.

Results: A total of 42 Bacillus species were isolated, one of the isolates named BT42 showed maximum radial mycelial growth inhibition against Colletotrichum gloeosporioides (78%) and Fusarium oxysporum (86%). BT42 increased germination of Coffee arabica L. seeds by 38.89%, decreased disease incidence due to infection of Colletotrichum gloeosporioides to 2.77% and due to infection of Fusarium oxysporum to 0 ($p < 0.001$). The isolate BT42 showed multiple growth-promoting traits. The isolate showed maximum similarity with Bacillus amyloliquefaciens.

Conclusion: Bacillus species (BT42), isolated in the present work was found to be capable of antagonizing the pathogenic effects of Colletotrichum gloeosporioides and Fusarium oxysporum. The mechanism of action of inhibition of the pathogenic fungi found to be synergistic effects of secondary metabolites, lytic enzymes, and siderophores. The major inhibitory secondary metabolite identified as harmine (β-carboline alkaloids).

Keywords: Biocontrol, Colletotrichum gloeosporioides, Fusarium oxysporum, Plant growth promoting rhizobacteria, Coffea arabica L

Background

The word coffee comes from the name of the place in Ethiopia called "Kaffa". "Kaffa" means the plants of God [1]. Coffee classified under the family of *Rubiaceae* in the genus *Coffea*. There are many species of coffee, but the two most widely cultivated are *C. arabica* L. and *C. canephora (robusta)*. Southwestern and southeastern Ethiopia considered as the origin of *C. arabica* L.

(Arabica coffee) [2]. Of the total world production of coffee, *C. arabica* L. takes the lion's share, which is 66% and *C. canephora* only of 34%. Although coffee produced in few countries, it is the most traded agricultural products around the globe after oil. According to a 2014 report by the International Coffee Organization, the top six coffee producing countries in our globe are Brazil, Vietnam, Colombia, Indonesia, Ethiopia, and India. In Ethiopia, it is mostly exported cash crop that accounts for 69% of all agriculturally export commodities and it was estimated that at least 15 million of Ethiopian population depend directly or indirectly on coffee production

* Correspondence: ftekakej@gmail.com
[1]Department of Biology, Faculty of Natural and Computational Sciences, Mettu University, Mettu, Ethiopia
[2]Present Address: BRD school of Biosciences, Sardar Patel University, Vallabh Vidyanagar 388120, India
Full list of author information is available at the end of the article

[1]. Similarly, there are around 250,000 coffee growers in India; 98% of them are small-scale growers.

One of the challenges in coffee production industry is the impact of the coffee pathogen, especially pathogenic fungi, which results in reduced production and low quality of coffee seeds [3]. The yield loss due to the fungal pathogens, especially *Colletotrichum species,* and *Fusarium species* have been repeatedly reported from coffee growing areas [4–7]. Coffee berry disease caused by *Colletotrichum khawae* is causing a major yield loss in coffee growing areas of Ethiopia [8, 9]. *C. khawae* and *C. gloeosporioides* are the most abundantly found pathogens in diseased coffee seeds [9]. *C. gloeosporioides* also listed as one of important coffee pathogens by the coffee board of India. Apart from the coffee plant, *C. gloeosporioides Penz* possesses a broad host range (470 genera of plants) and ranked as the most devastating plant pathogen in the genus *Colletotrichum* [10, 11]. *Fusarium species* also cause serious impact on the coffee production industry. Coffee wilt disease caused by *Gibberella xylarioides* (anamorph: *Fusarium xylarioides*) causes approximately 3360 t of coffee yield losses each year in Ethiopia [12]. This production loss causes a great economic loss around the world, for example, Ethiopia loses an estimated 3.7 million American dollars every year.

Chemical pesticides used currently to control coffee pathogens needed to spray 7–8 times annually which is laborious and expensive. Furthermore, the extensive use of chemical pesticides also contributes to emerging pesticide resistant pathogens. The uses of chemical pesticides and fertilizers have also a negative impact on the indigenous microbial community by disturbing the natural distribution of microbial niche. The coffee cultivated with no or less application of chemical pesticides have more consumer acceptance. In addition, the use of environmentally friendly and sustainable way of disease controlling system gained major attention in recent years. In this view, rhizosphere is the ideal place to search potential rhizobacteria that are capable of promoting plant growth and suppressing the phytopathogens.

Extensive studies of the use of plant growth promoting rhizobacteria (PGPR) for disease control and plant growth promotion in the coffee plant have not been reported. It is necessary and useful to evaluate and document indigenous beneficial microbe isolated from coffee and test them against coffee pathogens. In the current study, several bacteria from the rhizosphere of *C. arabica* L. were isolated and the potent bacterium antagonistic to *C. gloeosporioides and F. oxysporum was* chosen, which also showed multiple plant growth promoting activity. The potent isolate showed maximum similarity with *Bacillus amyloliquefaciens* by 16 s rRNA gene sequencing and by blasting this sequence against reference sequences found in the international nucleotide database using the program called BLASTn.

Results

Isolation of Bacillus species

From the rhizosphere of *Coffea arabica* L. 42 pure *Bacillus species* were isolated. These isolates were gram-positive, catalase-positive, spore-forming, rod-shaped and able to survive at 80 °C (Fig. 1). It forms central/sub terminal/ellipsoidal endospores. The bacterium grew at the temperature range of 15–50 °C, the optimum temperature of 30–42 °C, optimum pH of 7 in aerobic condition.

The antagonistic effect of Bacillus species isolates against C. gloeosporioides and F. oxysporum

Bacteria isolates were studied against two fungal pathogens *C. gloeosporioides* and *F. oxysporum* for radial mycelial growth inhibition. Sixteen *Bacillus* species isolates showed greater than 40% mycelial growth inhibition. Among these isolates, BT42 showed maximum radial mycelial growth inhibition against *C. gloeosporioides* (78%) and *F. oxysporum* (86%) (Fig. 2). Therefore, BT42 selected for in vitro and vivo studies.

Identification of isolate

The isolate BT42 selected for identification because it showed highest mycelial growth inhibition of *C. gloeosporioides* and *F. oxysporum* when compared to other isolates. The 16S rRNA gene amplified from the genomic DNA of BT42 (1.5 kb) sequenced and analyzed by nucleotide Blast analysis (BLASTn). The BT42 (NCBI

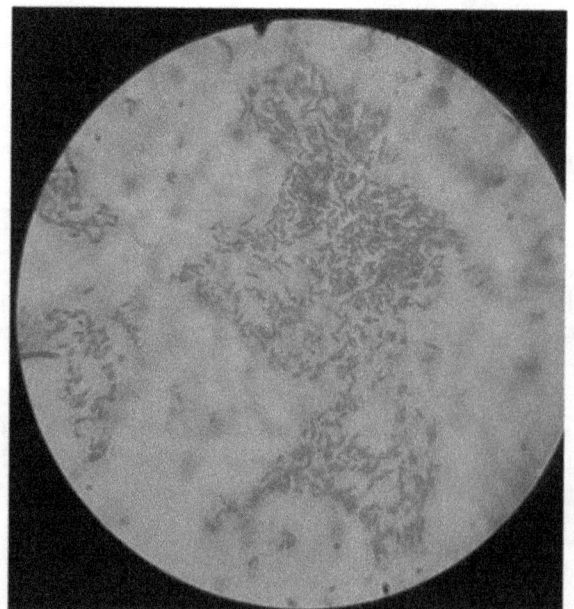

Fig. 1 Gram staining of cells of *Bacillus sp* BT42 isolated from *Coffea arabica* L. rhizosphere

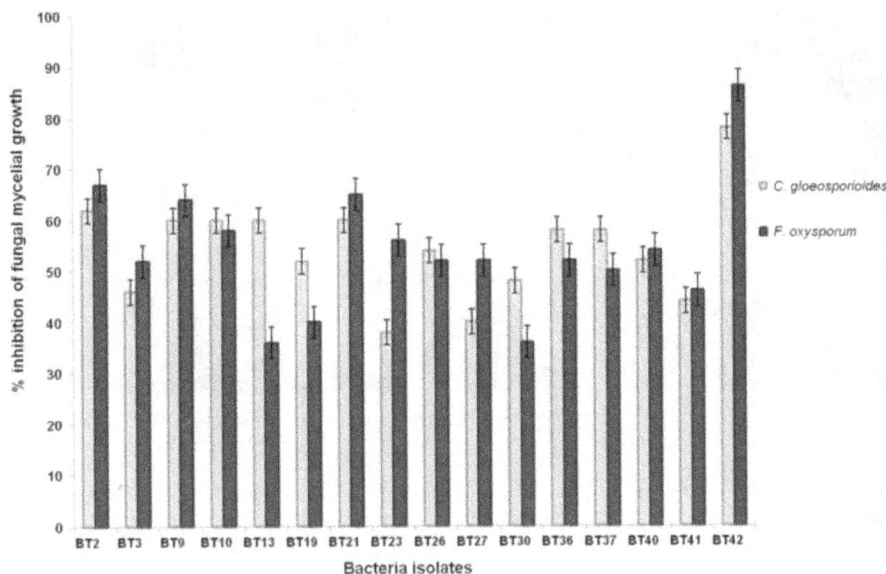

Fig. 2 In vitro mycelial growth inhibition of *C. gloeosporioides* and *F. oxysporum* by *Bacillus species* isolated from *Coffea arabica* L. rhizosphere. Error bars represent ± Standard Deviation (SD). Values are means of three replicates

accession number KT220617) showed maximum similarity with *Bacillus amyloliquefaciens* (Fig. 3). When compared to Ez taxon database it showed 86.55% similarity with *Bacillus amyloliquefaciens* subsp. plantarum FZB42.

Plant growth promoting characteristics of Bacillus species BT42

BT42 produced 14.56 ± 0.862 µg/ml IAA in the medium supplemented with L-tryptophan, produced ammonia, solubilized 6.36 ± 0.48 µg/ml tri-calcium phosphate, formed 37.5 ± 0.56 mm of holo zone by solubilizing insoluble zinc oxides (Fig. 4), produced 75.90 ± 1.24% siderophore units

in the iron free succinate medium, grew on the media supplemented with 1-aminocyclopropane-1-carboxylic acid (ACC) as a sole nitrogen source, and formed robust pellicles when grown in LB broth at the liquid air interference.

Bioassay
Effect of BT42 on seed germination and disease incidence
To study the effect of BT42 on germination of *C. arabica* L. seeds, suspension of overnight grown rhizobacteria (0.5 McFarland Standard) was coated on the seeds of *C. arabica* L and the effect on germination were observed and recorded. Percentage germination of *C. arabica* L.

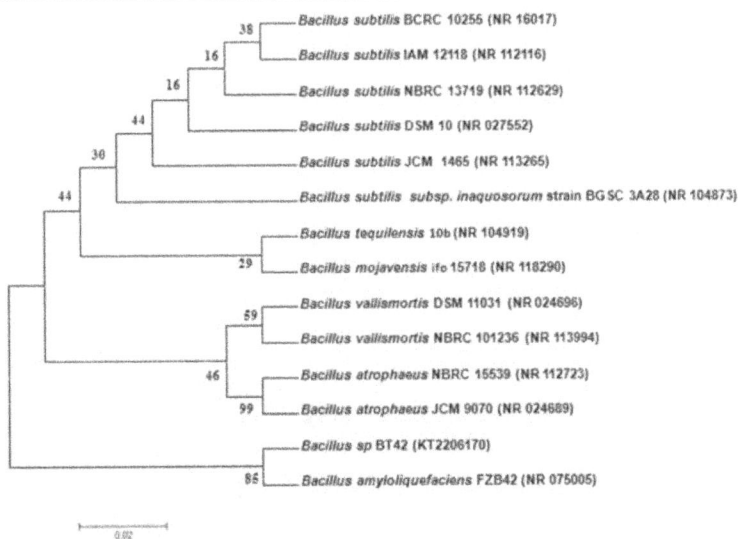

Fig. 3 Neighbor joining tree of isolated *Bacillus* sp. BT42 and closely related species. Bootstrap values based on 1,000 replications

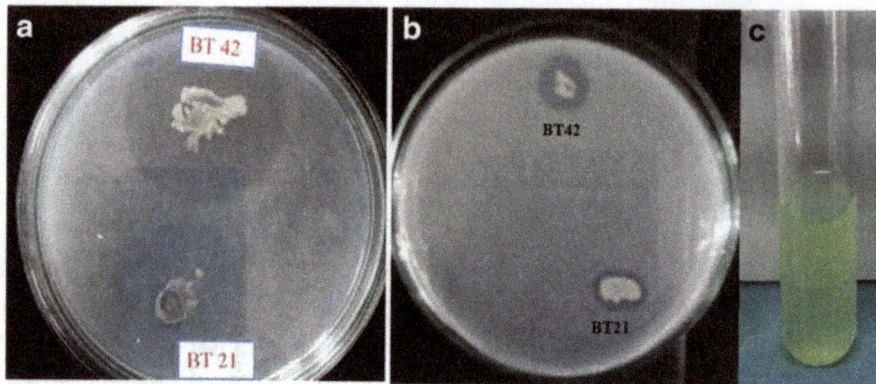

Fig. 4 Zinc solubilization (**a**), Tricalcium phosphate solubilization (**b**) and Siderophore production (**c**) by Bacillus species(BT42)

improved from 50% of untreated seeds to higher in the presence of rhizobacteria (Table 1). None of the *C. arabica* L. seeds infected with *C. gloeosporioides* could germinate (Fig. 5a), but in the presence of BT42, there was an improvement in germination percentage and decrease in the disease incidence. The disease incidence was as high as 91.67% when the seeds infected by *C. gloeosporioides* spore suspension alone. Disease incidence reduced by 88.9% when the BT42 was simultaneously surface coated with *C. gloeosporioides* (Table 1).

Mechanisms of inhibition

Lytic enzyme production

BT42 formed a clear zone around its colony on the medium supplemented with colloidal chitin, laminarin, tween 80 and skim milk, indicating extracellular production of chitinase, β-1,3 glucanase, protease, and lipase respectively. The correlation of diameter of hole zone of chitinase with mycelial growth inhibition of *C. gloeosporioides* ($r = 0.905$, $P < 0.05$) and *F. oxysporum* ($r = 0.780$, $P < 0.05$) were positive. Similarly, the correlation of diameter of hole zone

of β-1,3 glucanase with mycelial growth inhibition of *C. gloeosporioides* ($r = 0.604$, $P < 0.01$) and *F. oxysporum* ($r = 0.802$, $P < 0.01$) were also positive (Fig. 6).

Production of Antifungal compounds

To investigate the production of antifungal metabolite/s, 5 ml of growing culture of BT42 (in three replicates) was taken at different time intervals and its supernatant was tested for antifungal activity against *C. gloeosporioides* and *F. oxysporum* by agar well diffusion method. The highest antifungal activity was found in 48 h old culture grown at 30 °C under shaking condition of 150 rpm (Fig. 7).

Stability of antifungal compounds

The ethyl acetate extract of 48-hour-old cell-free culture was checked for its antifungal activity at different temperatures and it was found stable for more than 3 months at 4 °C, for 1 h at 100 °C and for 3 months at room temperature.

Separation and purification of antifungal compound

The ethyl acetate extract was separated on the silica thin layer chromatography (TLC) plate using isopropanol: ammonia: water (10:1.5:1) as the solvent system (Fig. 8a). Five bands with *Rf* values 0.33, 0.50, 0.61, 0.72 and 0.87 were detected. Compounds corresponding to a metabolite in each band named from bottom to top as C1, C2, C3, C4, and C5 were assayed for antifungal activity against *C. gloeosporioides* and *F. oxysporum*. Among the five bands, the metabolite named as C1 showed significant mycelial growth inhibition of *C. gloeosporioides* (Fig. 8b) and *F. oxysporum* (Fig. 8c) and was selected further for characterization and identification. The purity of the band showing maximum antifungal activity (that is C1, *Rf* 0.33) was checked again using different solvent systems on TLC, which showed a single band.

Table 1 Effects of selected rhizobacteria isolates on germination of *C. arabica* L. seeds and on disease incidence caused by *C. gloeosporioides* and *F. oxysporum*

Bioassay	Parameters	
	Germination %	DI (%)
Control (untreated seeds)	50 ± 5.33^a	0^a
C. gloeosporioides infected	0^b	91.67 ± 8.33^b
BT42 treated	88.89 ± 4.81^c	0^a
BT42 + *C. gloeosporioides*	72.22 ± 2.83^d	2.77 ± 4.81^a
Control (untreated seeds)	50 ± 5.33^a	0^a
F. oxysporum infected seeds	11.11 ± 4.81^b	88.89 ± 9.62^b
BT42 treated	88.89 ± 4.81^c	0^a
BT42 + *F. oxysporum*	80.56 ± 12.72^c	0^a

Values followed by dissimilar letters in each column indicate significance difference (one-way ANOVAs, Duncan's test). Values are means of three replicates

Fig. 5 Effect of inoculation of BT42 on coffee seed germination. *C. gloeosporioides* infected (**a**), BT42 treated (**b**), BT42+ *C. gloeosporioides* (**c**) and untreated (**d**)

Characterization and identification of potent compound

Purified compound C1 was subjected to spectral scan analysis from 100–1100 nm to identify its absorbance maxima, which was found to be 205 nm. The compound C1 was found to be soluble in methanol and sparingly soluble in water. Liquid chromatography-mass spectrometry (LC-MS) data of C1 showed the retention time 4.9 min by the photodiode anode (PDA) detector. The same fragment was subjected to mass spectrometry (MS) analysis. MS analysis clearly indicated that purified C1 compound has the molecular weight 212.10 (Fig. 9a). For the investigation of the numbers of carbon, hydrogen, oxygen and nitrogen atoms, absolute intensity of M + 1 peak and natural

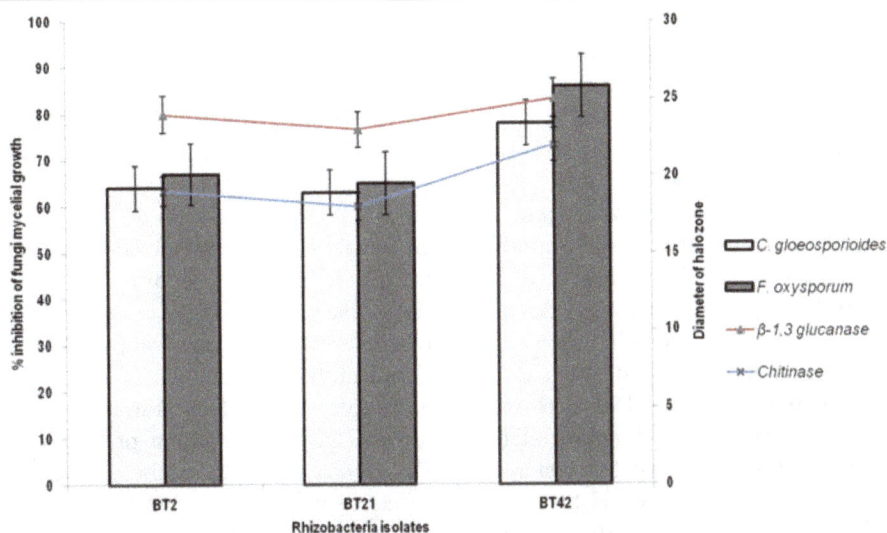

Fig. 6 Percent inhibition of mycelial growth of *C. gloeosporioides* and *F.oxysporum* (*Bar graph*) and production of lytic enzymes (*line graph*) by selected *Bacillus species* isolates

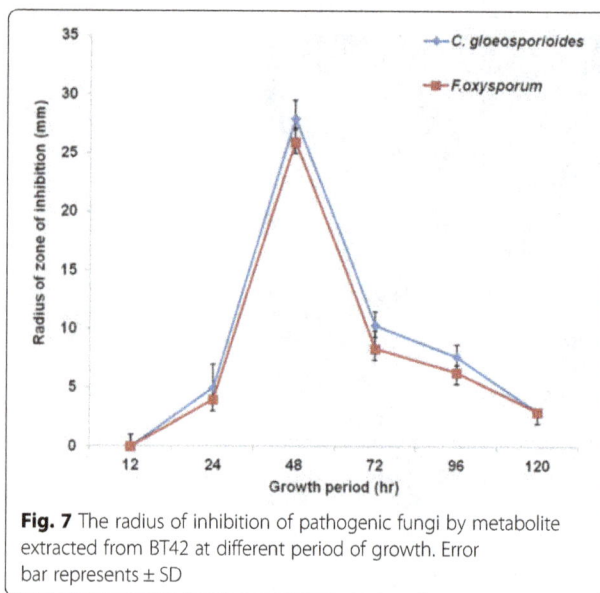

Fig. 7 The radius of inhibition of pathogenic fungi by metabolite extracted from BT42 at different period of growth. Error bar represents ± SD

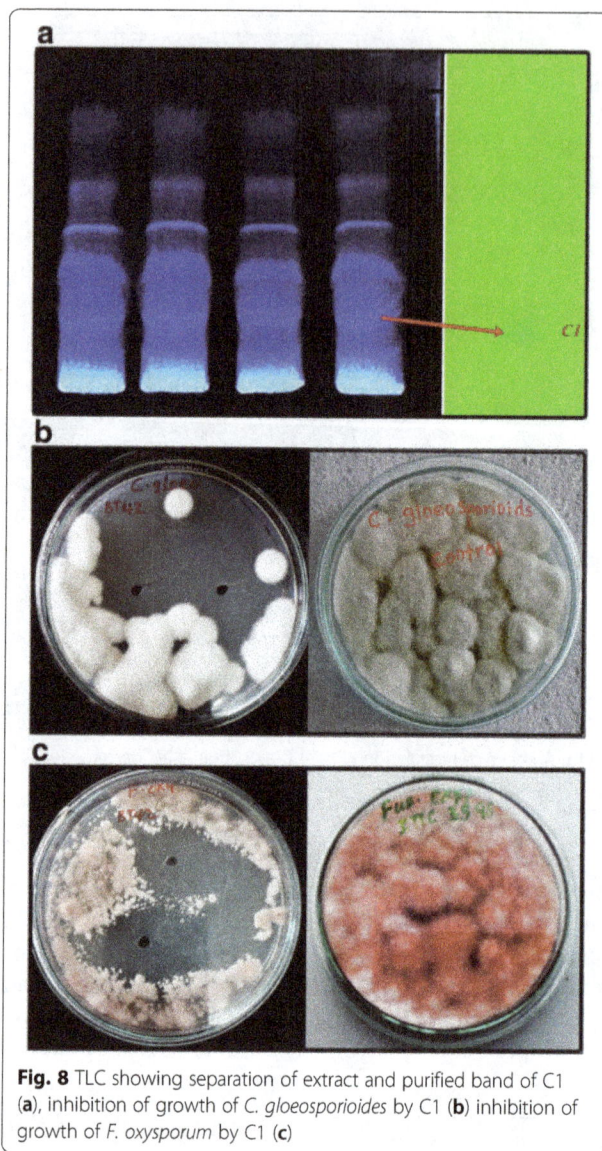

Fig. 8 TLC showing separation of extract and purified band of C1 (**a**), inhibition of growth of *C. gloeosporioides* by C1 (**b**) inhibition of growth of *F. oxysporum* by C1 (**c**)

abundance of isotopes were considered. Based on the calculation we found that there were 13 carbons, 12 hydrogens, 2 nitrogens and 1 oxygens atom. From m/z cloud the value of mass (i.e. 212.09) (Fig. 9b) is exactly matching with the compound harmine (CAS registry Number. 442-51-3) and its MS spectrum was matched to NIST database. Fourier transform infrared spectroscopy (FTIR) data of the compound C1 showed the presence of functional groups. The frequency at 3405 cm^{-1} indicates the presence of -NH Indole stretching, while aromatic C-H stretching at 3044 cm^{-1}, -CH of alkane stretching at 2958 cm^{-1}, C = N stretching at 1665 cm^{-1}, -CH of -CH$_3$ bending at 1357 cm^{-1}, asymmetrical C-O-C stretching at 1077 and 1231 cm^{-1}, C-N of Indole at 1100 cm^{-1}, aromatic C = C stretching at 1590 and 1448 cm^{-1}, -CH bending of aromatic ring at 864 cm^{-1} were observed (Fig. 10). Based on LC-MS data and FTIR assignment, the suggested formula for the compound is $C_{13}H_{12}N_2O$ and its structural formula showed in Fig. 11.

Further confirmation of the compound was carried out by qualitative tests using the general reagent for alkaloids (Dragendroff's reagent) and it was found to be positive and specifically for the presence of indole derivates using glyoxylic-sulphuric acid test. Accordingly, the solution containing the C1 showed a positive test for the presence of alkaloid and showed a positive test for the presence of indole derivative, forming a purple to violet ring at the junction of two distinct phases when glyoxylic acid and the test solution mixed in the presence of conc. H_2SO_4. The secondary metabolite harmine could be the principal reason for inhibition of growth of *C. gloeosporioides* and *F. oxysporum*.

Discussion

In the present study, we have isolated potent *Bacillus species* (BT42) that is able to inhibit *C. gloeosporioides* (78%) and *F. oxysporum* (86%). In a review of similar studies, *Bacillus sp.* strain RMB7, which has broad range antifungal activity showed 71% and 78% mycelial growth inhibition of *F. oxysporum* and *C. gloeosporioides* respectively [13]. BT42 produces known extracellular lytic enzymes (chitinase and β-1, 3-glucanase), which are significantly correlated with fungi mycelial growth inhibition. In a similar study conducted elsewhere, significant mycelial growth inhibition of *G. xylarioides* (anamorph: *Fusarium xylarioides*) by the chitinase producing *Bacillus* species isolate named as JU5444 was also reported [12]. Apart from chitinase and β-1,3-glucanase, BT42 produced protease and lipase that might be involved in the inhibition of the mycelial growth of *C. gloeosporioides and*

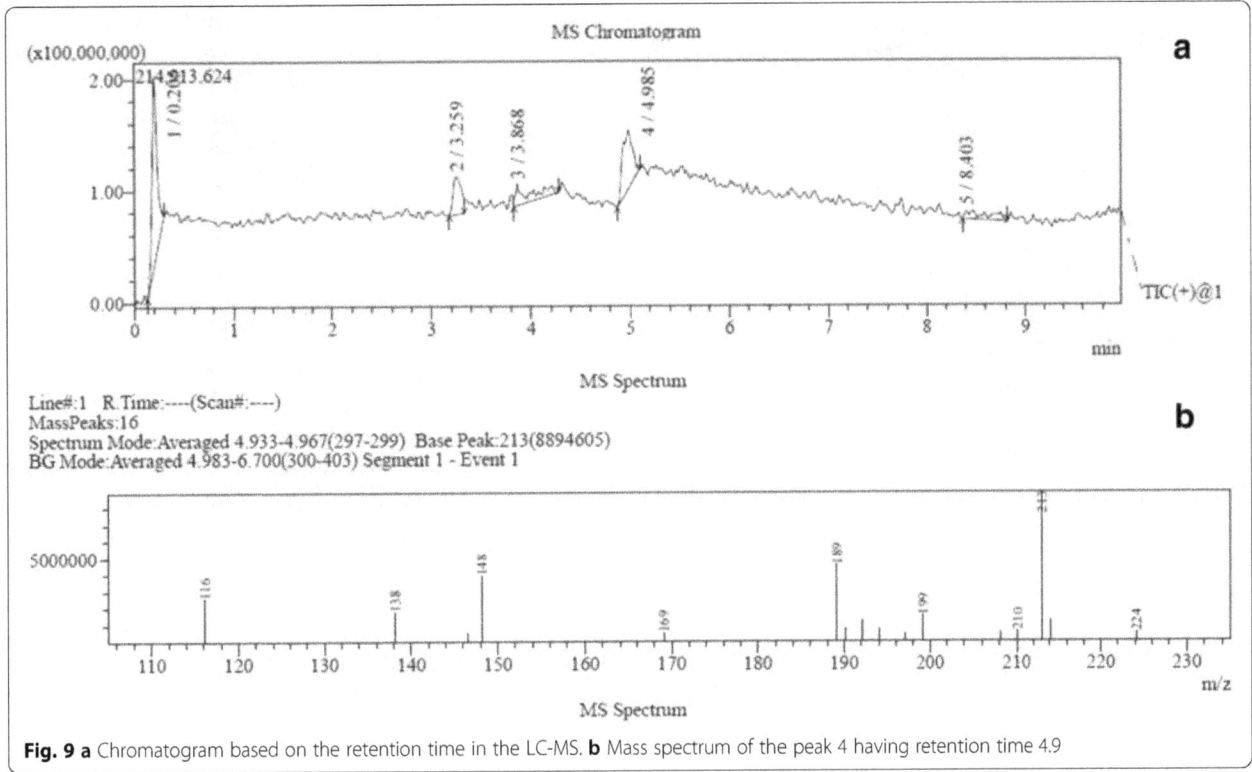

Fig. 9 a Chromatogram based on the retention time in the LC-MS. **b** Mass spectrum of the peak 4 having retention time 4.9

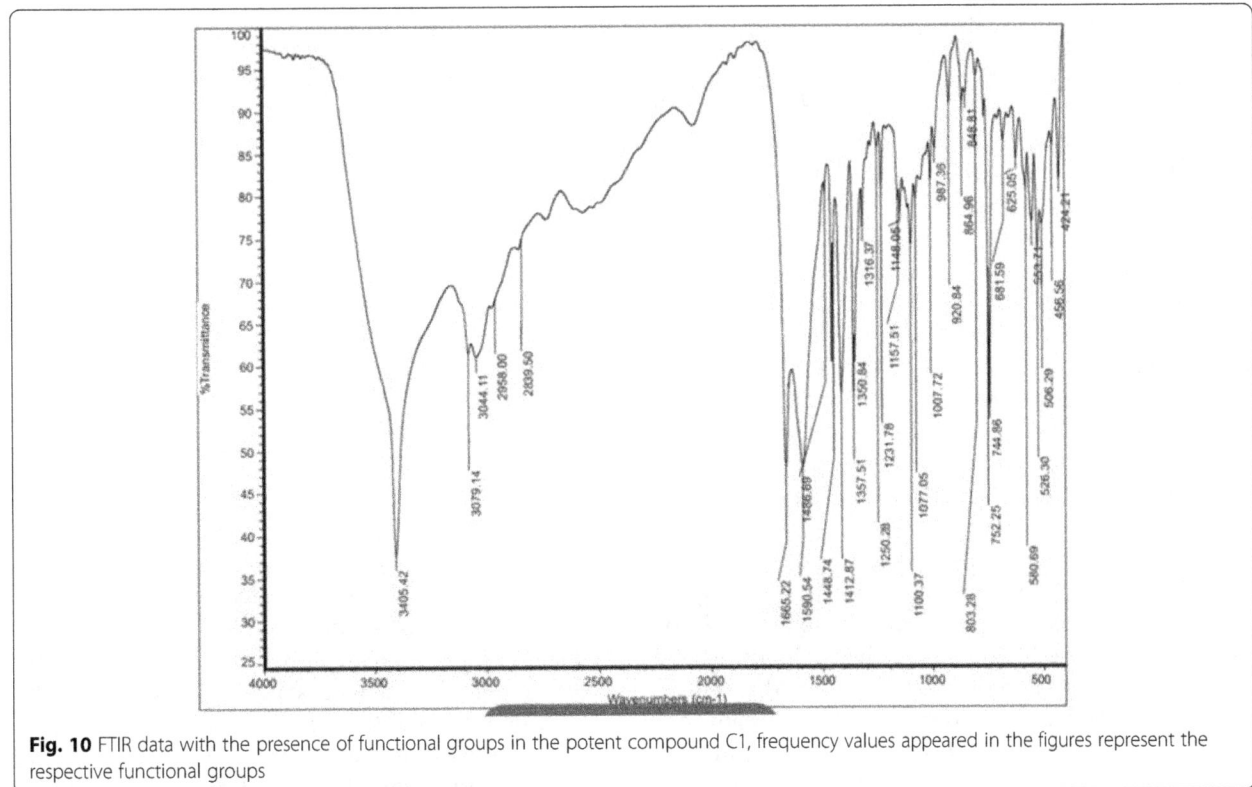

Fig. 10 FTIR data with the presence of functional groups in the potent compound C1, frequency values appeared in the figures represent the respective functional groups

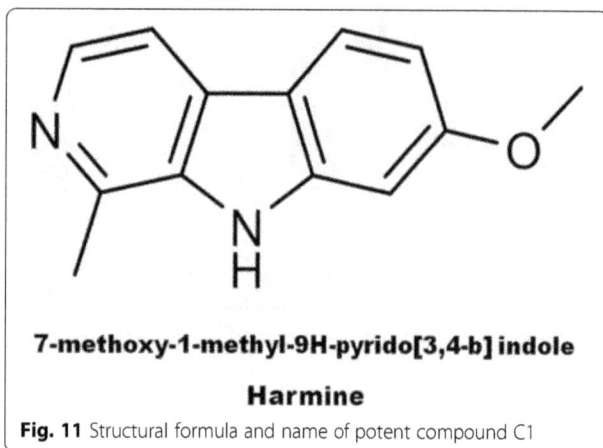

7-methoxy-1-methyl-9H-pyrido[3,4-b] indole

Harmine

Fig. 11 Structural formula and name of potent compound C1

F. oxysporum. Earlier reports indicated that protease [14, 15] and lipase production [12] by bacteria are associated with inhibition of mycelial growth of different fungi.

Since the statistical correlation of growth inhibition of pathogenic fungi and lytic enzyme produced were positive but mostly less than 0.9 r, more than one mechanism of inhibition of pathogenesis was suspected. The production of siderophore by BT42 might also be involved in the inhibition of fungi. Bacteria produce Siderophores during iron starvation. The previous report indicates that siderophores produced by *Bacillus subtilis* have got potential to inhibit the growth of *F. oxysporum* [16]. *Bacillus amyloliquefaciens* FZB42, which has maximum similarity with BT42, harbors genes responsible for the synthesis of siderophores [17].

Apart from production of lytic enzymes and siderophores, from the culture filtrate of BT42, we also isolated and identified antifungal compound, which is confirmed as Harmine [18]. Harmine is β-carboline alkaloid that was first isolated from *Peganum harmala* L [19, 20]. Harmine is rarely reported from bacteria and it is mostly extracted from higher plants [19, 21]. Harmine production in bacteria was limited to a few species including *Enterococcus faecium, Myxobacter* and *Pseudomonas* species [22–25]. However, the recent report by Saad and Zakaria demonstrated that *Bacillus flexus* isolated from fresh water produced harmine as a mechanism of inhibition of toxic cyanobacteria and this was considered as the first report from *Bacillus* species [26]. In our present study, we identified harmine from the cultural extract of *Bacillus species* (BT42) isolated from root rhizosphere of *C. arabica* L., which is the first report of its type. This is the first report of inhibition of pathogens of *C. arabica* L., *C. gloeosporioides* and *F. oxysporum* by plant growth promoting rhizobacterium by an alkaloid harmine. The inhibitory activity of β-carboline alkaloids against different fungi that include *C. gloeosporioides* and *F. oxysporum* was recently described although the β-carboline alkaloids were extracted from the plant [21].

To sum up, the highest inhibition of *C. gloeosporioides* and *F. oxysporum* mycelial growth by the *Bacillus species* (BT42) is due to the synergistic effect of multiple mechanisms, which could be explained as the production of lytic enzymes and harmine, which is produced as an extracellular secondary metabolite.

Apart from biocontrol activity, *Bacillus species* (BT42) also exhibited plant growth promoting characteristics (IAA production, ammonia production, phosphate solubilization and zinc solubilization) that directly involved in plant growth promotion and indirectly in the pathogen suppression. The production of IAA by rhizobacteria is so important in supporting plant growth and development by interfering with the endogenous IAA produced by the plant and serve in defense responses [27–29]. Studies showed that *Bacillus amyloliquefaciens* is known to produce a substantial amount of IAA [17, 30]. Another important trait displayed by *Bacillus species* (BT42) is the ability to produce an enzyme ACC deaminase. The production of ACC deaminase by the BT42 is also vital in plant-microbe interaction; that enables the host plant to withstand stress due to drought and flooding by decreasing the endogenous ethylene levels. Studies showed that *Bacillus amyloliquefaciens* has the versatile potential for both plant growth and biocontrol activity of different phytopathogenic fungi, which is similarly strengthened by our present study [17, 31–33].

The high disease incidence caused by *C. gloeosporioides* and *F. oxysporum* were significantly reduced in the presence of BT42 due to the potential biocontrol activity of this bacterium. In BT42 treated *C. arabica* L seeds, germination occurred much better than that of untreated, which indicates that the isolate facilitated germination of the *C. arabica* L. seeds by the extracellular production of PGPR traits that supported the germination.

Conclusions

In conclusion, we have isolated a *Bacillus species* (BT42), which shows the versatility of direct and indirect plant growth promoting traits. BT42 showed maximum similarity with *Bacillus amyloliquefaciens* in NCBI BLASTn result and shows maximum similarity with the same bacterium by Ez taxon database. This bacterium could effectively inhibit pathogens of *C. arabica* L. The mechanisms of inhibition of *C. gloeosporioides* and *F. oxysporum* were found to be the production of lytic enzymes, siderophores as well as antifungal compounds. The major antifungal compound was identified as harmine (a member of β-carboline alkaloids), which is not previously reported as a mechanism of action of PGPR. To this end, the BT42 isolate can be a potential candidate to be used as a biocontrol of *C. gloeosporioides* and *F. oxysporum* and also as biofertilizer.

Methods

Sampling site and sample collection

Rhizosphere soil samples of *Coffea arabica* L. were collected from two different fields at Khusalnagar, Karnataka, South India. These sites are located between $12^0 28'05.7''N$ and $75^0 57'51.7''E$. Soil samples were collected after consent was obtained from coffee farm owners.

Isolation and Identification of bacterial isolate

Isolation of rhizobacteria from root rhizosphere of *C. arabica* L. was performed as previously stated [12]. The pure isolates were further confirmed by standard microbiological techniques [34]. Pure isolates were stored at -20 °C with 50% glycerol for further study. Primers fD1 (forward, 5′-AGAGTTTGATCCTGGCTC AG-3′) and rP2 (reverse, 5′-ACGGCTACCTTGTTACG ACTT-3′), were obtained from Eurofins (India) and used for the amplification of *16S rRNA* gene [35]. A PCR mixture (20 µl) consisting of: 2.5 µl of 10X Mg^{2+} buffer containing 15 mM Mg^{2+}, 1 µl of Taq polymerase enzyme (1U/µl), 3 µl of dNTP mixture (3 mM), 1 µl of each primer (10 pM), 2 µl of template DNA (50–100 ng/µl) and 9.5 µl of Milli-Q water was prepared. DNA thermocycler was used for the amplification of the DNA at 94 °C for 3 min, followed by 32 cycles of 30 s at 94 °C, 15 s at 54 °C and 1 min at 72 °C with an extension of 72 °C for 5 min. One µl of PCR product along with standard DNA ladder (1.5 kb) were loaded on a 0.8% agarose gel containing 3 µl of ethidium bromide in 1X TAE buffer and electrophoresed at 100 V for 35 min. The PCR product checked and visualized using a UV transilluminator. The PCR product purified and sequenced at Eurofins Genomics India Pvt Ltd, Bangalore, India.

Using the NCBI website, Basic Local Alignment Search Tool (BLASTn) used for checking the 16S rRNA gene sequences with comparative sequences of reference strains. Mega 6 software used for the construction of phylogenetic tree after the sequences aligned. To carry out phylogenetic analysis, sequences of 16S rRNA of thirteen reference strains that are important for comparison downloaded from NCBI database http://www.ncbi.nlm.nih.gov.

In vitro antagonistic study of bacteria against pathogens

In vitro antagonistic study of the bacteria was carried out against two fungal pathogens *Colletotrichum gloeosporioids sp.coffee* (ITCC 7131) and *Fusarium oxysporum* (ITCC 3595). Both the cultures were obtained from Indian Type Culture Collection (ITCC), Division of Plant pathology, Indian agricultural research institute, New Delhi 110012 (India).

Potato dextrose agar used for dual culture during the antagonistic studies of bacteria against fungal pathogens. Percent fungal radial growth inhibition was calculated as stated below [12].

$$\text{Fungal mycelial growth inhibition} = [(C\text{-}T)/C] \times 100$$

Where,

T = fungal radial mycelial growth during dual culture (Bacteria + fungus)

C = fungal radial mycelial growth (without antagonistic bacteria).

Experimental design and bioassay

The experiment involved only one factor (the rhizobacteria antagonist). In the experiment *Bacillus sp* BT42, used as it showed the highest reduction of radial mycelial growth of both *C. gloeosporioides* and *F. oxysporum*. Healthy *Coffea arabica* L. seeds collected from selected healthy coffee plants. First, the exocarp of the coffee fruits was removed and kept in water at 30 °C for 24 h with the other fruit part (mesocarp and endocarp) then mesocarp was removed by washing them and left on a tray to dry [36]. Surface sterilization of coffee seeds was carried out as reported elsewhere [37]. Two hundred sixteen surface sterilized coffee seeds randomly distributed in four groups: untreated (36 seeds), rhizobacteria treated (36 seeds), rhizobacteria + fungi (72 seeds) and fungi (72 seeds) for the experiment. The bacterium grown in nutrient broth under shaking condition of 150 rpm for 18–20 h (1 X 10^8 cells/ml) at 30 °C were used for coating seeds. The surface sterilized seeds inoculated in the liquid culture of bacteria and dried in aseptic condition under laminar flow hood for 30 min. In the same manner, spore suspensions of *C. gloeosporioides* (1 X 10^4 spores/ml) inoculated and dried under laminar flow hood for 30 min. Similarly, bacteria and fungi surface coated on the seed of *C. arabica* L., in the same manner, one after the other. Surface sterilized seeds without any treatment taken as controls.

Seeds in each group (bacterial treated, fungi infected, fungi + bacteria treated and untreated) placed on pre-sterilized Whatman filter paper in separate Petri plate (three replicates). Plates with the coffee seeds incubated at $30 + 1$ °C, dampness of the seeds maintained by spraying 1–1.5 ml distilled water on filter paper as necessary and any physiological changes in each group of seeds inspected and recorded daily. The experiment carried out in completely randomized design under controlled conditions. The experiment performed in three replicates.

Studies of plant growth promoting traits

Qualitative and quantitative test for Siderophore production

Rhizobacterium was grown in iron-free medium (K_2HPO_4, 6.0 g L^{-1}; KH_2PO_4, 3.0 g L^{-1}; $MgSO_47H_2O$,0.2 g L^{-1};$(NH_4)_2SO_4$,1.0 g L^{-1}; and Succinic acid 4.0 g L^{-1}, pH 7.0) and incubated for 48 h at 30 °C with constant shaking of

150 rpm [38]. After 48 h of incubation, the fermented broth centrifuged at 10,000 rpm for 15 min and supernatant were taken and checked for the presence of siderophore. For estimation of siderophore, 0.5 ml of supernatant was mixed with CAS reagent and absorbance was calculated at 630 nm [38]. Percent siderophore units calculated using the formula:

$$\% \text{ siderophore units } = (Ar\text{-}As)/Ar \times 100$$

Where, Ar = absorbance of reference at 630 nm (CAS reagent) and As = absorbance of the sample at 630 nm.

Test for phytohormones production

For the production of IAA, purified bacterium isolate was grown in Luria-Bertani (LB) broth under shaking condition (150 rpm) around the clock at 30 °C. Overnight grown culture centrifuged at 10,000 rpm for 15 min and the supernatant collected. To the supernatant (app. 2 ml) two drops of O-phosphoric acid added; the manifestation of pink color indicates IAA production by the rhizobacteria isolates. Quantitative estimation of IAA was done colorimetrically [39].

For the production of Gibberellic acid, bacterium isolate was grown in 100 ml nutrient broth at 30 °C for 48 h. The growth of bacterium was monitored by measuring turbidity at 600 nm. For extraction of Gibberellic acid, 50 ml of the culture was taken and centrifuged at 7500 rpm for 10 min. The supernatant was collected and pH was adjusted to 2.5 using 37% HCl. Supernatant extracted using ethyl acetate in 1:1 volume ratio [40]. Gibberellic acid quantitatively estimated by using a UV spectrophotometer at 254 nm [41].

Test for ammonia production

Rhizobacterium isolate grown in peptone water for 4 days at 30 °C in 50 ml test tubes. In each test tube containing bacterial isolates, 1 ml of Nessler's reagent added. The appearance of a faint yellow color is evidence of weak reaction and deep yellow to brownish color was confirmation of strong reaction [42].

Test for zinc and phosphate solubilization

Phosphate solubilizing ability of bacterium isolate was determined by the inoculation of overnight grown bacterial isolates on pre-solidified specific medium for phosphate solubilization test [43]. The rhizobacterium incubated for 96 h at 30 °C and any clear zone around the rhizobacterium colonies indicated phosphate solubilization. Quantitative determination of phosphate solubilizing activity was performed calorimetrically [44]. Zinc solubilization was performed by plate assay using modified Pikovskaya agar [45]. The rhizobacteria isolates were inoculated into a medium consisting of: ammonium sulfate (1 g L^{-1}),

dipotassium hydrogen phosphate (0.2 g L^{-1}), glucose (10.0 g L^{-1}), magnesium sulfate (0.1 g L^{-1}), potassium chloride (0.2 g L^{-1}), Yeast (0.2 g L^{-1}), distilled water (1000 ml), pH 7.0 and 0.1% insoluble zinc compounds (ZnO, ZnCO$_3$ and ZnS). The rhizobacteria isolate grown in this medium for 48 h at 28 °C. The clear zone around the colony indicated solubilization of insoluble zinc compounds.

Qualitative test for 1-Aminocyclopropane-1-carboxylate (ACC) deaminase

ACC deaminase production was checked using Dworkin and Foster (DF) minimal salts medium [46]. To the pre-solidified DF minimal salts medium 3 mM ACC solution sprayed and allowed to dry in aseptic condition for 10 min then bacterial isolates inoculated. After 48 h of incubation at 30 °C, any growth of bacteria on the media was considered as ACC deaminase production [47].

Studies of mechanisms of inhibition of pathogenic fungi
Lytic enzymes production

Test for Production of β-1, 3 glucanases, and chitinase Test for the production of β-1, 3 glucanases was performed using laminarin as the only carbon source for growth of bacteria. Accordingly, the isolates were inoculated on media containing Na$_2$HPO$_4$ (6 g L^{-1}), KH$_2$PO$_4$ (3 g L^{-1}), NH$_4$Cl (0.5 g L^{-1}), yeast extract (0.05 g L^{-1}), Agar (15 g L^{-1}) and 0.05% laminarin, (Sigma) and incubated at 30 °C for 48 h. After 48 h of incubation, the clear zones obtained. To visualize clearly, plates were flooded with a mixture of 0.666% KI and 0.333% Iodine and isolates showing yellow clear zone around the bacterial colony confirmed as the production of β-1, 3 glucanases.

For the checking of chitinase production, colloidal chitin was prepared from chitin powder (Hi-Media) for the preparation of the solid media. The compositions of solid media were colloidal chitin 1% (w/v), Na$_2$HPO$_4$ (6 g L^{-1}), NaCl (0.5 g L^{-1}), KH$_2$PO$_4$ (3 g L^{-1}); NH$_4$Cl (1 g L^{-1}), yeast extract (0.05 g L^{-1}) and agar (15 g L^{-1}). The bacteria isolates were checked for their production of chitinase by observation of clear zone around the colonies after five days incubation at 30 ± 1 °C [48].

Test for Production of Protease and Lipase Test for protease production was done using the protease specific medium as earlier described [49]. Similarly, the rhizobacteria isolates were checked for lipase enzyme using lipase media [2]. This media contains calcium chloride 0.1 g, Peptone 10 g, sodium chloride 5 g, Agar 15 g, distilled water 1 Liter, 10 ml sterile Tween 20. The bacterial isolates streaked on this medium and incubated at 27 °C for 48 h, the clear zone around the bacterial colonies show the activity of lipase enzyme.

Production of antifungal compound from isolate BT42

The test for antifungal activity of the culture filtrate was done as described elsewhere with modifications [50, 51]. Briefly, to check the antifungal activity of culture filtrate from strain BT42, a single colony of the isolate inoculated into Luria-Bertani (LB) broth and incubated for 120 h under shaking condition of 150 rpm at 30 °C. During the incubation period, 5 ml samples taken from flasks at different time intervals and centrifuged at 12,500 rpm for 10 min at 4 °C. The cell pellet removed and supernatant filtered through a membrane filter (25 mm) to remove any suspended cell. The collected sample tested for the antifungal activity by agar well diffusion method using 20 µl of the culture filtrate and sterile broth as a control on Potato dextrose agar spread with a spore suspension of pathogenic fungi (1 X 10⁴ spores/ml). The plates incubated at 28 °C for 3 days.

Extracellular metabolites extracted from the 48 h grown culture filtrate using ethyl acetate. Accordingly, an equal amount of ethyl acetate added to the culture filtrate and both phases collected and concentrated to dryness. Active compounds obtained from both the phases subjected to the antifungal activity bioassay after dissolving them in methanol

Purification, identification and characterization of antifungal compound

The crude extract dissolved in methanol was separated by TLC on silica gel plates (20 × 20 cm, 0.5 mm thick, G), developed in Isopropanol: ammonia: water (10:1.5:1, v/v) as the mobile phase. The TLC plate was visualized under UV transilluminator. Each specific band corresponding to specific metabolite eluted by carefully scraping from the TLC plate, suspended in methanol and checked for antifungal activity. The antifungal compound was once again subjected to TLC using the same as well as different solvent systems (chloroform: methanol 90:10; benzene: acetic acid 95:05; ammonia: methanol: chloroform 0.5:1:8.5) as stated above to check its purity. The purified compound characterized and identified using FTIR and LC-MS analysis. The alkaloid nature of the purified antifungal compound was also confirmed by a qualitative test using Dragendorff's reagent and glyoxylic-sulphuric acid test specific for indole derivatives as described elsewhere [52, 53].

Statistical analysis

IBM SPSS Statistics software version 19 was used to analyze the data related to correlations of fungal radial mycelial growth inhibition and lytic enzyme production. Similarly, one-way ANOVAs was used to compare the mean difference between treatments, and the level of significance was set at $P < 0.05$.

Abbreviations
ACC: 1-Aminocyclopropane-1-carboxylate; BLAST: Basic Local Alignment Search Tool; C. arabica: Coffea arabica; C. gloeosporioides: Colletotrichum gloeosporioides; CAS: Chrome Azul S; DF: Dworkin and Foster; F. oxysporum: Fusarium oxysporum; FTIR: Fourier transform infrared spectroscopy; IAA: Indole-3-acetic acid; ITCC: Indian Type Culture Collection; LC-MS: Liquid chromatography-mass spectrometry; MS: Mass spectrometry; PDA: Photodiode anode; PGPR: Plant growth promoting rhizobacteria; TLC: Thin layer chromatography

Acknowledgement
We would like to thank BRD School of Biosciences, Sardar Patel University for lab facilities and the Indian Council for Cultural Relations (ICCR), as a sponsor of the scholarship for the first author. Mr. Parth Thakor is thankful to DST Inspire program for the fellowship. Authors are also thankful to Mr. Sampark Thakkar and Prof. Arabinda Ray from PDPIAS, Charusat for the critical evaluation of the LC-MS and FTIR portion of the manuscript. We thank DST, New Delhi for the assistance in general and for the PURSE central facility LC-MS sponsored under PURSE program grant vide sanction letter Do.No.SR/59/Z-23/2010/43 dated 16th March 2011.

Funding
No specific fund was obtained for this study; however, the study was supported by BRD School of Biosciences, Sardar Patel University, Vallabh Vidyanagar, Gujarat, India.

Authors' contributions
TK involved in designing of the study, experimentations, data analysis and interpretation, and write-up of the manuscript. VRT designed the study, supervised and guided the experimental process, and prepared the manuscript for publication. PT involved in LC-MS and FTIR data analysis and interpretation and manuscript writing. All the authors read and approved the final manuscript.

Competing interests
The authors declare that they have no competing interests.

Author details
¹Department of Biology, Faculty of Natural and Computational Sciences, Mettu University, Mettu, Ethiopia. ²Present Address: BRD school of Biosciences, Sardar Patel University, Vallabh Vidyanagar 388120, India. ³BRD School of Biosciences, Sardar Patel University, Vadtal Road, Satellite Campus, Post Box No.39, Vallabh Vidyanagar 388120, Gujarat, India.

References
1. Amamo AA. Coffee production and marketing in Ethiopia. Eur J Bus Manag. 2014;6(37):109–22.
2. Muleta D. Microbial Inputs in Coffee (Coffea arabica L.) Production Systems, Southwestern Ethiopia. Implications for Promotion of Biofertilizer and Biocontrol Agents. Doctoral thesis,Swedish University of Agricultural Sciences/Uppsala.; 2007.
3. Admasu W, Sahile S, Kibret M. Assessment of potential antagonists for anthracnose (Colletotrichum gloeosporioides) disease of mango (Mangifera indica L.) in North Western Ethiopia (Pawe). Arch Phytopathol Plant Prot. 2014;47(18):2176–86. doi:10.1080/03235408.2013.870110.
4. Hindorf H, Omondi CO. A review of three major fungal diseases of Coffea arabica L. in the rainforests of Ethiopia and progress in breeding for resistance in Kenya. J Adv Res. 2011;2(2):109–20. doi:10.1016/j.jare.2010.08.006.

5. Kilambo DL, Mabagala RB, Varzea VMP, Haddad F, Loureiro A, Teri JM. Characterization of *Colletotrichum kahawae* strains in Tanzania. Int J Microbiol Res. 2013;5(2):382–9.

6. Muleta D, Assefa F, Börjesson E, Granhall U. Phosphate-solubilising rhizobacteria associated with Coffea arabica L. in natural coffee forests of southwestern Ethiopia. J Saudi Soc Agric Sci. 2013. doi:10.1016/j.jssas.2012.07.002.

7. Nguyen THP, Säll T, Bryngelsson T, Liljeroth E. Variation among *Colletotrichum gloeosporioides* isolates from infected coffee berries at different locations in Vietnam. Plant Pathol. 2009;58(5):898–909. doi:10.1111/j.1365-3059.2009.02085.x.

8. Derso E, Waller JM. Variation among *Colletotrichum* isolates from diseased coffee berries in Ethiopia. Crop Prot. 2003;22:561–5. doi:10.1016/S0261-2194(02)00191-6.

9. Rutherford MA, Phiri N. Pests and Diseases of Coffee in Eastern Africa: A Technical and Advisory Manual. Wallingford, UK: CAB International; 2006.

10. Sharma M, Kulshrestha S. Colletotrichum gloeosporioides: an anthracnose causing pathogen of fruits and vegetables. Biosci, Biotech Res Asia. 2015;12:1233–46.

11. Martínez EP, Hío JC, Osorio LA, Torres MF, Erika P. Identification of *Colletotrichum species* causing anthracnose on Tahiti lime, tree tomato, and mango. Agron Colomb. 2009;27(2):211–8.

12. Tiru M, Muleta D, Berecha G, Adugna G. Antagonistic Effects of Rhizobacteria Against Coffee Wilt Disease Caused by *Gibberella xylarioides*. Asian J Plant Pathol. 2013;7(3):109–22. doi:10.3923/ajppaj.2013.109.122.

13. Ali S, Hameed S, Imran A, Iqbal M, Lazarovits G. Genetic, physiological and biochemical characterization of Bacillus sp.strain RMB7 exhibiting plant growth promoting and broad spectrum antifungal activities. Microb Cell Fact. 2014;13:144.

14. Chaiharn M, Chunhaleuchanon S, Kozo A, Lumyong S. Screening of rhizobacteria for their plant growth promoting activities. KMITL Sci Technol J. 2008;8(1):18–23.

15. Liao CY, Chen MY, Chen YK, et al. Characterization of three *Colletotrichum acutatum* isolates from *Capsicum sp.* Eur J Plant Pathol. 2012;133(3):599–608. doi:10.1007/s10658-011-9935-7.

16. Patil S, Bheemaraddi CM, Shivannavar TC, Gaddad MS. Biocontrol activity of siderophore producing *Bacillus subtilis* CTS-G24 against wilt and dry root rot causing fungi in chickpea. IOSR J Agric Vet Sci. 2014;7(9):63–8.

17. Chen XH, Koumoutsi A, Scholz R, et al. Comparative analysis of the complete genome sequence of the plant growth–promoting bacterium *Bacillus amyloliquefaciens* FZB42. Nat Biotechnol. 2007;25(9):1007–14. doi:10.1038/nbt1325.

18. Silverstein RM, Webster FX, Kiemle DJ. Spectrometric Identification of Organic Compounds. New York: John Wiley & Sons; 2005. p. 1–550.

19. Berrougui H, Isabelle M, Cloutier M, Hmamouchi M, Khalil A. Protective effects of Peganum harmala L. extract, harmine and harmaline against human low-density lipoprotein oxidation. J Pharm Pharmacol. 2006;58(7):967–74. doi:10.1211/jpp.58.7.0012.

20. de Meester C. Genotoxic potential of beta-carbolines: a review. Mutat Res. 1995;339(3):139–53.

21. Li Z, Chen S, Zhu S, Luo J, Zhang Y, Weng Q. Synthesis and fungicidal activity of β-Carboline alkaloids and their derivatives. Molecules. 2015;20(8):13941–57. doi:10.3390/molecules200813941.

22. Aassila H, Bourguet-Kondracki ML, Rifai S, Fassouane A, Guyot M. Identification of harman as the antibiotic compound produced by a tunicate-associated bacterium. Mar Biotechnol (NY). 2003;5(2):163–6. doi:10.1007/s10126-002-0060-7.

23. Bohlendorf B, Forche E, Bedorf N, et al. Antibiotics from Gliding Bacteria, LXXIII Indole and Quinoline Derivatives as Metabolites of Tryptophan in Myxobacteria. European J Org Chem. 1996. doi:10.1002/jlac.199619960108.

24. Kodani S, Imoto A, Mitsutani A, Murakami M. Isolation and identification of the antialgal compound, harmane (1-methyl-β-carboline), produced by the algicidal bacterium, Pseudomonas sp.K44-1. J Appl Phycol. 2002;14(2):109–14.

25. Zheng L, Chen H, Han X, Lin W, Yan X. Antimicrobial screening and active compound isolation from marine bacterium NJ6-3-1 associated with the sponge Hymeniacidon perleve. World J Microbiol Biotechnol. 2005;21(2):201–6. doi:10.1007/s11274-004-3318-6.

26. Alamri SA, Mohamed ZA. Selective inhibition of toxic cyanobacteria by B-carboline-containing bacterium *Bacillus flexus* isolated from Saudi freshwaters. Saudi J Biol Sci. 2013;20(4):357–63. doi:10.1016/j.sjbs.2013.04.002.

27. Mohite B. Isolation and characterization of indole acetic acid (IAA) producing bacteria from rhizospheric soil and its effect on plant growth. J Sci Plant Nutr. 2013. doi:10.4067/S0718-95162013005000051.

28. Weselowski B, Nathoo N, Eastman AW, Macdonald J, Yuan Z-C. Isolation, identification and characterization of Paenibacillus polymyxa CR1 with potentials for biopesticide, biofertilization, biomass degradation and biofuel production. BMC Microbiol. 2016. doi:10.1186/s12866-016-0860-y.

29. Shao J, Li S, Zhang N, et al. Analysis and cloning of the synthetic pathway of the phytohormone indole-3-acetic acid in the plant-beneficial Bacillus amyloliquefaciens SQR9. Microb Cell Fact. 2015;14(1):130. doi:10.1186/s12934-015-0323-4.

30. Ait A, Noreddine K, Chaouche K, Ongena M, Thonart P. Biocontrol and plant growth promotion characterization of *Bacillus Species* isolated from Calendula officinalis Rhizosphere. Indian J Microbiol. 2013;53(4):447–52. doi:10.1007/s12088-013-0395-y.

31. Yu GY, Sinclair JB, Hartman GL, Bertagnolli BL. Production of iturin A by *Bacillus amyloliquefaciens* suppressing Rhizoctonia solani. Soil Biol Biochem. 2002;34(7):955–63. doi:10.1016/S0038-0717(02)00027-5.

32. Jiang C-H, Wu F, Yu Z-Y, et al. Study on screening and antagonistic mechanisms of *Bacillus amyloliquefaciens* 54 against bacterial fruit blotch (BFB) caused by Acidovorax avenae subsp. citrulli. Microbiol Res. 2015;170:95–104. doi:10.1016/j.micres.2014.08.009.

33. Shahzad R, Waqas M, Khan AL, et al. Seed-borne endophytic *Bacillus amyloliquefaciens* RWL-1 produces gibberellins and regulates endogenous phytohormones of Oryza sativa. Plant Physiol Biochem. 2016; 106(September):236–43. doi:10.1016/j.plaphy.2016.05.006.

34. Ahmad F, Ahmad I, Khan MS. Screening of free-living rhizospheric bacteria for their multiple plant growth promoting activities. Microbiol Res. 2008; 163(2):173–81. http://dx.doi.org/10.1016/j.micres.2006.04.001.

35. Weisburg WG, Barns SM, Pellettier DA, Lane DJ. 16S ribosomal DNA amplification for phylogenetic study. J Bacteriol. 1991;173(2):697–703.

36. Rosa SDVF da, Mcdonald MB, Veiga AD, Vilela FdeL, Ferreira IA. Staging coffee seedling growth : a rationale for shortenning the coffee seed germination test. Seed Sci Technol. 2010;38:421–431.

37. Shanmugam V, Kanoujia N. Biological management of vascular wilt of tomato caused by *Fusarium oxysporum f.sp.* lycospersici by plant growth-promoting rhizobacteria mixture. Biol Control. 2011;57(2):85–93. doi:10.1016/j.biocontrol.2011.02.001.

38. Sayyed RZ, Badgujar MD, Sonawane HM, Mhaske MM, Chincholkar SB. Production of microbial iron chelators (siderophores) by *fluorescent Pseudomonads*. Indian J Biotechnol. 2005;4(4):484–90.

39. Tsavkelova EA, Cherdyntseva TA, Klimova SY, Shestakov AI, Botina SG, Netrusov AI. Orchid-associated bacteria produce indole-3-acetic acid, promote seed germination, and increase their microbial yield in response to exogenous auxin. Arch Microbiol. 2007;188(6):655–64. doi:10.1007/s00203-007-0286-x.

40. Cho KY, Sakurai A, Kamiya Y, Takahashi N, Tamura S. Effects of the new plant growth retardants of quaternary ammonium iodides on gibberellin biosynthesis in Gibberella fujikuroi. Plant Cell Physiol. 1979;20(1):75–80.

41. Kumar PKR, Lonsane BK. Immobilized growing cells of Gibberella fujikuroi P-3 for production of gibberellic acid and pigment in batch and semi-continuous cultures. Appl Microbiol Biotechnol. 1986;28:537–42.

42. Cappuccino JC, Sherman N. Microbiology; a laboratory manual. New York: Benjamin/Cumming Pub. Co; 1992. p. 125–79.

43. Pikovskaya RI. Mobilization of phosphorus in soil in connection with the vital activity of some microbial species. Mikrobiologiya. 1948;17:362–70.

44. King EJ. The colorimetric determination of phosphorus. Biochem J. 1932;26:292.

45. Saravanan VS, Subramoniam SR, Raj SA. Assessing in vitro solubilization potential of different zinc solubilizing bacterial (ZSB) isolates. Brazilian J Microbiol. 2004;35(1-2):121–5. doi:10.1590/S1517-83822004000100020.

46. Ramamoorthy V, Viswanathan R, Raguchander T, Prakasam V, Samiyappan R. Induction of systemic resistance by plant growth promoting rhizobacteria in crop plants against pests and diseases. Crop Prot. 2001;20:1–11.

47. Penrose DM, Glick B. Methods for isolating and characterizing ACC deaminase containing plant growth promoting rhizobacteria. Physiol Plant. 2002;118:10–5.

48. Renwick A, Campbell R, Coe S. Assessment of in vivo screening systems for potential biocontrol agents of Gaeumannomyces graminis. Plant Pathol. 1991;40:524–32.

49. Smibert RM, Krieg NR. Phenotypic characterization. In: Gerhardt P, Murray RGE, Wood WA, Krieg NR, editors. Methods for General and Molecular Bacteriology. Washington DC: American Society of Microbiology; 1994. p. 607–54.

50. Kumar A, Saini S, Wray V, Nimtz M, Prakash A, Johri BN. Characterization of an antifungal compound produced by Bacillus sp. strain A(5) F that inhibits *Sclerotinia sclerotiorum*. J Basic Microbiol. 2012;52(6):670–8. doi:10.1002/jobm.201100463.

51. Vilarinho BR, Silva JP, Pomella AWV, Marcellino LH. Antimicrobial and plant growth-promoting properties of the cacao endophyte *Bacillus subtilis* ALB629. J Appl Microbiol. 2014;116:1584–92. doi:10.1111/jam.12485.

52. Paech K, Tracey MV. Modern Methods of Plant Analysis/Moderne Methoden der Pflanzenanalyse. 1st edition, Biemann K, Boardman NK, Breyer B, et al., editors. Berlin, Heidelberg: Springer; 1962.p.1–509. doi:10.1007/978-3-642-45993-1.

53. Crotti AEM, Paul J, Gates JLCL, NPL. Based characterization of β -carbolines-mutagenic constituents of thermally processed meat. Mol Nutr Food Res. 2010;54:433–9. doi:10.1002/mnfr.200900064.

4

In vitro assessment of *Pediococcus acidilactici* Kp10 for its potential use in the food industry

Sahar Abbasiliasi[1,2], Joo Shun Tan[3], Fatemeh Bashokouh[4], Tengku Azmi Tengku Ibrahim[4,5], Shuhaimi Mustafa[1,2], Faezeh Vakhshiteh[4], Subhashini Sivasamboo[1] and Arbakariya B. Ariff[1*]

Abstract

Background: Selection of a microbial strain for the incorporation into food products requires in vitro and in vivo evaluations. A bacteriocin-producing lactic acid bacterium (LAB), *Pediococcus acidilactici* Kp10, isolated from a traditional dried curd was assessed in vitro for its beneficial properties as a potential probiotic and starter culture. The inhibitory spectra of the bacterial strain against different gram-positive and gram-negative bacteria, its cell surface hydrophobicity and resistance to phenol, its haemolytic, amylolytic and proteolytic activities, ability to produce acid and coagulate milk together with its enzymatic characteristics and adhesion property were all evaluated in vitro.

Results: *P. acidilactici* Kp10 was moderately tolerant to phenol and adhere to mammalian epithelial cells (Vero cells and ileal mucosal epithelium). The bacterium also exhibited antimicrobial activity against several gram-positive and gram-negative food-spoilage and food-borne pathogens such as *Listeria monocytgenes* ATCC 15313, *Salmonella enterica* ATCC 13311, *Shigella sonnei* ATCC 9290, *Klebsiella oxytoca* ATCC 13182, *Enterobacter cloaca* ATCC 35030 and *Streptococcus pyogenes* ATCC 12378. The absence of haemolytic activity and proteinase (trypsin) and the presence of a strong peptidase (leucine-arylamidase) and esterase-lipase (C4 and C8) were observed in this LAB strain. *P. acidilactici* Kp10 also produced acid, coagulated milk and has demonstrated proteolytic and amylolactic activities.

Conclusion: The properties exhibited by *P. acidilactici* Kp10 suggested its potential application as probiotic and starter culture in the food industry.

Keywords: *Pediococcus acidilactici* Kp10, Probiotic, Starter culture, Adhesion property, Proteolytic, Food-borne pathogens, Food industry

Background

The importance of proper selection of the bacterial strains for incorporation in food products is related to the considerable variations of the beneficial properties among different strains. Lactic acid bacteria (LAB) which are used worldwide have been focused in recent years for a variety of fermented foods production [1].

LAB play an important role in improving the nutritional and keeping qualities of foods by virtue of the organic acids produced during fermentation of the raw materials [2]. At the industrial scale, short fermentation duration is preferred in order to increase the plant output as well as to reduce microbial contamination. The use of LAB as a starter culture in food fermentation will increase the fermentation rates and also will improve product quality [3] due to LAB versatile metabolic characteristics such as acidification and proteolytic activities and ability to synthesize metabolites such as bacteriocin [4, 5]. Thus, the isolation and characterization of new strains of LAB for broader industrial applications is currently of industrial importance.

LAB species presence in traditional foods of Southeast Asian countries have not been extensively investigated and there is every likelihood that some species could be

---
* Correspondence: arbarif@upm.edu.my
[1]Department of Microbiology, Faculty of Biotechnology and Biomolecular Sciences, Universiti Putra Malaysia, 43400 UPM Serdang, Selangor, Malaysia
Full list of author information is available at the end of the article

of commercial potential [1]. With the realization that there is a need to identify new strains with useful characteristics, in our previous study we had identified and characterized the LAB strain with ability to produce bacteriocin-like inhibitory substances (BLIS) for potential applications in the food industry. The isolate, *P. acidilactici* Kp10, could be a potential probiotic as it exerted beneficial and positive effects on the intestinal flora which included tolerance to bile salts (0.3%) and acidic conditions (pH 3), produced β-galactosidase, stable in a wide range of pH (2–9) and not resistant to vancomycin. Most interesting, the LAB strain showed the highest level of BLIS activity against *Listeria monocytogenes*, a virulent food pathogenic bacterium. To further substantiate its probiotic potential and application as a starter culture the present study further evaluated in vitro other physicochemical properties of *P. acidilactici* Kp10 which include inhibitory spectra of activities against different gram-positive and gram negative bacteria, cell surface hydrophobicity, resistance to phenol, haemolytic, amylolytic and proteolytic activities, ability to produce acid and coagulate milk and enzymatic characterization along with its adhesive properties.

Methods

Microorganism and maintenance

Isolation and characterization of the bacterium, *P. acidilactici*Kp10, used in this study were as described previously [1]. The culture was maintained on agar slopes at 4 °C and prior to its use in the present study the culture was sub-cultured twice in M17 broth (Merck, Darmstadt, Germany).

Determination of probiotic properties

Inhibitory activity

The inhibitory activities of *P. acidilactici* Kp10 against different gram-positive and gram-negative bacteria (*Listeria monocytogenes* ATCC 15313, *Salmonella enterica* ATCC 13311, *Shigella sonnei* ATCC 9290, *Klebsiella oxytoca* ATCC 13182, *Enterobacter cloaca* ATCC 35030, *Streptococcus pyogenes* ATCC 12378) were determined according to the method as described in our previous study. Briefly, antimicrobial activity of *P. acidilactici* Kp10 was assessed by the agar well diffusion method using cell-free culture supernatants (CFCS). *P. acidilactici* Kp10 was grown in M17 broth at 30 °C for 24 h and the cultures were centrifuged at 12,000 g for 20 min at 4 °C (rotor model 1189, Universal 22R centrifuge, Hettich AG, Switzerland).

One hundred μL of the CFCS was placed into 6-mm wells of agar plates previously seeded with 1% (v/v) actively growing test strains. The plates were incubated at 37 °C for 24 h for the growth of test strains. After 24 h, the growth inhibition zones were measured, and the antimicrobial activity (AU mL^{-1}) was calculated as described previously [6].

Adhesion of *P. acidilactici* Kp10 on mammalian epithelial cells

Adhesion of *P. acidilactici* Kp10 to vero cells

Assessment of the adhesion of *P. acidilactici* Kp10 to Vero cells (African green monkey kidney cell line, ATCC CCL81) was performed by the method as described previously [7] with some modifications. Vero cells were cultured in Roswell Park Memorial Institute Medium (RPMI; Gibco, Grand Island, NY, USA) supplemented with 10% (v/v) fetal calf serum, 100 U/mL penicillin and 100 mg/mL streptomycin (Sigma, Switzerland). The cell lines were maintained in a humidified incubator (Binder, Tuttlingen, Germany) at 37 °C in atmosphere of 5% CO_2 and 95% air. Cells with 80–85% confluence were washed three times with sterile phosphate-buffered saline (PBS: NaCl, 0.8, K_2HPO_4, 0.121, KH_2PO_4, 0.034, pH 7.2) and transferred (10^5 cells/mL) onto cover slips placed in six-well plates containing fresh culture medium. The plates were incubated at 37 °C in an atmosphere of 5% CO_2 and 95% air. Cell monolayers (10^5 cells/mL) on glass cover slips were washed three times with PBS. Prior to the adhesion test, overnight culture of *P. acidilactici* Kp10 was harvested and washed three times with PBS and centrifuged for 10 min at 3000×g. The bacterial cells (1×10^9 CFU/mL in PBS) were resuspended in 1 mL of Dulbecco's modified Eagle medium (DMEM) and transferred to the washed monolayer cells on cover slips, placed in six-well plates and incubated at 37 °C in an atmosphere of 5% CO_2 and 95% air for 1 h.

For scanning electron microscopy (SEM) examination, the cells were fixed in 2.5% glutaraldehyde in 0.1 M sodium cacodylate buffer for 4–6 h and washed thrice in sodium cacodylate buffer. Samples were then postfixed in 1% aqueous osmium tetroxide, dehydrated in ascending grades of acetone concentrations (30, 50, 75, 80, 95 and 100%) critically point-dried and sputter coated with gold palladium.

Adhesion of *P. acidilactici* Kp10 to ileal mucosal epithelium

The method of Mäyrä-Mäukinen & Gyllenberg, [8] with slight modifications was employed to evaluate the adhesion of *P. acidilactici* Kp10 to ileal mucosal epithelium. Samples of goat ileum, obtained immediately after slaughter from a local abattoir were washed in PBS to remove the ingesta from the mucosal surface. The samples were transported back to the laboratory in cooled PBS and incubated in cell suspension of *P. acidilactici* Kp10 (10^9 CFU/mL PBS) at 37 °C for 30 min. The samples were then prepared for scanning electron microscopy as described above.

Auto-aggregation and co-aggregation assays

The procedure as described by Polak-Berecka et al., [9] with some modifications was used to determine the specific cell–cell interactions using auto-aggregation and co-aggregation assays. Cells harvested at the stationary phase were collected by centrifugation (5000×g for 10 min at room temperature), washed twice and resuspended in PBS (pH 7.2). For both assays, the culture suspension was standardized to OD $_{600\ nm}$ = 1.0 (2×10^8 CFU/mL). For auto-aggregation assay, 5 mL of bacterial suspension was vortexed for 10 s and incubated at 37 °C for 2 h. Absorbance of the supernatant was measured at 600 nm using a spectrophotometer (Perkin Elmer, Lambda 25, USA). The auto-aggregation coefficient (AC) was calculated according to Eq. 1 [10]:

$$AC_t(\%) = [1-(OD_{2h}/OD_i)] \times 100 \tag{1}$$

where, OD_i is the initial optical density of the microbial suspension at 600 nm.

For the co-aggregation assay an equal volume (2 mL, 2×10^8 CFU/mL) of P. acidilactici Kp10 and pathogenic bacterium (L. monocytogenes ATCC 15313) cultures were mixed, vortexed for 10 s and incubated at 37 °C for 2 h. Each control tubes contained 4 mL of each bacterial suspension. The supernatants were measured at $OD_{600\ nm}$ and co-aggregation was calculated according to Eq. 2 [11]:

$$Co\text{-}aggregation(\%) = [1-OD_{mix}/(OD_{strain} + OD_{pathogen})/2] \times 100 \tag{2}$$

where, OD_{mix} is the optical density of the mixture of P. acidilactici Kp10 and L. monocytogenes at 600 nm, OD_{strain} is the optical density of P. acidilactici Kp10 at 600 nm and $OD_{pathogen}$ is the optical density of L. monocytogenes at 600 nm. Experiments were conducted in triplicates on two separate occasions.

Adhesion of P. acidilactici Kp10 cell to solvents
Adhesion of P. acidilactici Kp10 cell to solvents was assayed according to the method as described previously [12] with some modifications. Three tubes each containing 3 mL of P. acidilactici Kp10 cell (grown in M17 broth at 37 °C for 18 h) suspension in PBS (pH 7.2) at 10^8 CFU/mL, were each mixed with 1 mL of xylene, chloroform, ethylene acetate and n-hexadecane. The mixture was then vortexed for 1–2 min and allowed to stand for 5–10 min to allow separation of the mixture into two phases. The aqueous phase was measured at 600 nm using a spectrophotometer (Perkin Elmer, Lambda 25, USA). Bacterial affinities to solvents (BATS) with different physicochemical properties (hydrophobicity and electron donor–electron acceptor interactions) were expressed using Eq. 3:

$$BATS(\%) = (1-A_{10\ min}/A_{0\ min}) \times 100 \tag{3}$$

Where, A_{10min} is the absorbance at t = 10 min and A_{0min} is the absorbance at t = 0 min.

In a separate experiment, Congo red dye method was used to further investigate the cell surface hydrophobicity of P. acidilactici Kp10. Agar plates were initially prepared by mixing 2% (w/v) NaCl in de Man, Rogosa and Sharpe (MRS) medium (Merck, Darmstadt, Germany), followed by the addition of sterile 0.03% (w/v) Congo red to the mixture. The bacterial strain was then cross-streaked and incubated at 37 °C for 24 h. The colonies stained red were hydrophobic whereas the colorless colonies were considered as non-hydrophobic [13].

Survivability studies on tolerance to phenol
Study on the tolerance of P. acidilactici Kp10 to phenol was performed by inoculating the cultures in M17 broth with and without phenol. The samples (100 μL) were then spread-plated onto MRS agar and incubated at 37 °C for 24 h. Bacterial survivability was enumerated using the formula as described previously [14].

Transmission electron microscopy (TEM) for detection of the S-layer
Cell suspensions of P. acidilactici Kp10 and Lactobacillus crispatus DSM 20584 (used as a control) were centrifuged at 5000×g for 10 min. The supernatants were pipetted and the pellets fixed in 2.5% glutaraldehyde in 0.1 M sodium cacodylate buffer for 4 to 6 h. The samples were then centrifuged and the supernatants pipetted to remove the fixative. A few drops of horse serum were added to each of the pellets. The coagulated pellets were then diced into 1 mm pieces. Following three washings with sodium cacodylate buffer the samples were post-fixed in 1% aqueous osmium tetroxide and dehydrated in ascending grades of acetone concentrations (30, 50, 75, 80, 95 and 100%). Samples were then infiltrated overnight with an equal mixture (1:1) of resin and acetone. The samples were infiltrated with 100% resin in the following morning and dropped into resin-filled, pre-labeled BEEM capsules and polymerized at 60 °C for 16 h. Ultrathin sections on copper grids were stained with uranyl acetate and lead citrate and examined under the TEM. Cross sections of bacterial cells were examined to detect the S-layer in the cell wall of both strains.

Haemolytic activity
The haemolytic activity of P. acidilactici Kp10 was determined by growing the bacterial strain in M17 agar at 37 °C for 18 h, and then streaked onto Columbia Agar plates containing 5% v/v of sheep blood (BioMérieux, Hazelwood, MO, USA). The plates were incubated at 37 °C overnight. Haemolytic reactions were recorded by the presence of a clear zone (β-haemolysis), green zone (α-haemolysis) or the absence of zone (γ-haemolysis) around the colonies [15].

Determination of starter culture properties

Enzymatic characterization

API ZYM strips (API Identification Systems, bioMérieux, France), according to the manufacturer's instructions, were used to determine the enzymatic characteristics of *P. acidilactici* Kp10. The strips were incubated at 37 °C for 4 h, and the reagents were then added. The color intensity was assessed according to the manufacturer's color chart. The test was performed in triplicates.

Acidification and coagulation activities

Effect of acidification and coagulation activities of *P. acidilactici* Kp10 was assayed by its inoculation into 10% skim milk at 1% level which incubated at 30 °C. The activities were evaluated by observation for commencement of clotting followed by pH measurement after 72 h [16].

Qualitative proteolytic activity and starch hydrolysis

P. acidilactici Kp10 culture was streaked on M17 agar for 24–48 h. Heavy inoculum of the culture was then streaked on skim milk agar and M17-starch agar and incubated at 37 °C for 24–48 h. Clear zone surrounding colonies on skim milk agar indicated proteolytic activity. To detect the hydrolysis of starch, M17-starch agar was topped with iodine solution [17]. *L. monocytogenes* ATCC 15313 and *E. coli* ATCC 25922 were used as negative controls.

Results and discussion

The inhibitory activity of the probiotic strain plays an important role in competing with other microorganisms in the gastrointestinal tract (GIT) protecting the latter from being colonized by food-borne pathogens. The inhibitory spectra of *P. acidilactici* Kp10 against different gram-positive and gram-negative bacteria in the present study showed an antagonistic effect of the growth of gram-positive and gram-negative pathogenic microorganisms. The potential probiotic bacterial strain in this study demonstrated an inhibitory activity against *L. monocytogenes* ATCC 15313, *S. enterica* ATCC 13311, *Sh. sonnei* ATCC 9290, *K. oxytoca* ATCC 13182, *E. cloaca* ATCC 35030, *St. pyogenes* ATCC 12378 (Table 1). There was significant difference ($P < 0.05$) between the inhibitory spectrum of *P. acidilactici* Kp10 against *L. monocytogenes* 15313 and five other strains while no significant differences ($P > 0.05$) was observed in inhibitory spectrum of Kp10 against these five strains. To date there are limited reports concerning the inhibitory effects of LAB on gram-negative bacteria due to the structure of their bacterial cell envelopes which is much more complex compared to that of gram-positive bacteria [18]. Their resistance to many antimicrobial agents is attributed to an effective permeable barrier of lipopolysaccharide layer of the outer membrane.

Table 1 Inhibitory spectrum of *P. acidilactici* Kp10 against gram-positive and gram-negative bacteria

Microorganism	Zone diameter (mm)
L. monocytgenes ATCC 15313	21 ± 0.1^a
S. enterica ATCC 13311	11 ± 0.05^b
Sh. sonnei ATCC 9290	11 ± 0.8^b
K. oxytoca ATCC 13182	11 ± 0.03^b
E. cloaca ATCC 35030	11 ± 0.5^b
S. pyogenes ATCC 12384	11 ± 0.7^b
P. acidilactici Kp10	0

Data are mean values ± SD ($n = 3$)
Values with different superscript letters (a and b) are significantly different ($P < 0.05$)

*P. acidilactici*Kp10 inhibited the growth of *L. monocytogenes* which is an important food-borne pathogen (Fig. 1). This observation could infer that *P. acidilactici* Kp10 has the potential to be used as a probiotic microorganism to overcome some major challenges facing the food industry and regulatory agencies. In addition, Kp10 was resistant to its own BLIS as indicated by the absence of activity around the well (Fig. 1). All bacteriocin producing isolates could protect themselves from the adverse effect of their own bacteriocins by the production of an immune protein commonly linked to the C-terminal domain of the bacteriocin [19]. Our finding is in agreement with the earlier reports which stated that bacteriocin producer could protect itself from the adverse effect of its own antimicrobial compounds by a defense system which is expressed concomitantly with the antimicrobial peptide(s) [20, 21]. Some bacteriocinogenic strains have no receptors which would then absorb their own bacteriocins thus rendering the bacteriocin ineffective against their own producer strain. Bacteriocin action and bacteriocin resistance were demonstrated to be contributed by the cell wall as well as its membrane lipid composition. As shown in Fig. 1, two zones of inhibition were observed. During the initial phase of incubation there was high antimicrobial activity which was demonstrated by an inner clear zone. During incubation there was an accompanying increase in pH of the substrate whence the antimicrobial range of activity was approaching its optimum. The antimicrobials further inhibit the growth of the microorganism in the area of the peripheral zone where the concentration of antimicrobials are lower than that presence in the central area [22]. However, it could be result of the presence of more than one bacteriocin.

Adhesion of *P. acidilactici* Kp10 to Vero cells and goat ileum mucosal epithelium as observed under the scanning electron microscope (SEM) are shown in Fig. 2a and b. To our knowledge previous reports on the adhesion of LAB were tested in rats intestine [23], columnar epithelial cells of pigs and calves [8] and ileum of

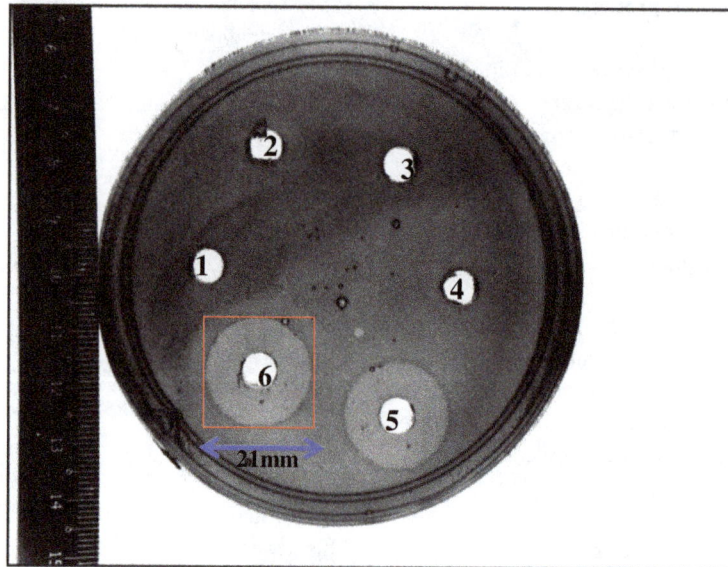

Fig. 1 Antimicrobial activity of *P. acidilactici* Kp10 against *L. monocytogenes* ATCC 15313 determined by agar well diffusion method (1 and 2: water; 3 and 4: media; 5 and 6: CFCS of *P. acidilactici* Kp10)

Landrace pigs [10]. The objective of this part of our study was to test qualitatively the colonization of LAB onto epithelial cells. As probiotic could be used in both human and animals we therefore examined LAB colonization in an animal species which have not been previously reported and in this case the goat. Human epithelial cells were not used as these cells were not easily available from our perspective. The goat being a ruminant is thus a species which is most remotely related to the human; however surprisingly our results demonstrated that LAB are capable of colonizing the goat epithelium which further augment our claim that *P. acidilactici* Kp10 is applicable to both human and animals. Colonization with extended transit time is most critical for optimal expression of general and specific physiological functions of probiotic microorganisms. Probiotic strains invariably should demonstrate the ability to adhere to the surface mucosal epithelial cells, an important requirement with reference

to effective colonization [24]. Cell adhesion which involve contact between the cell membrane of the bacteria and that of the mucosal epithelium is no doubt a complex process. There were a number of constrains in the evaluation of bacterial adhesion capability in vivo especially in humans. These constrains had prompted a number in vitro studies to be undertaken instead which were directed towards screening bacterial strains with adhering potentials.

For the beneficial effect of probiotics to manifest, there is a need to achieve an adequate mass through aggregation. In a number of ecological niches auto-aggregation, which are cell aggregation between microorganisms of similar strain or co-aggregation, aggregation of genetically different strain, are of considerable importance [25]. LAB with aggregation ability and hydrophobicity cell surface could be more capable to adhere to intestinal epithelial cells. It has been reported that some LAB can prevent

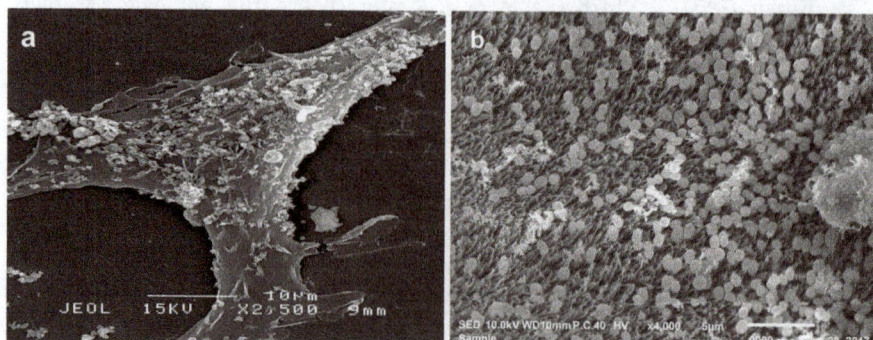

Fig. 2 SEM showing adhesion of *P. acidilactici* Kp10 to the surface of: **a** Vero cells, and **b** mucosal epithelium of goat ileum

adherence of pathogens to intestinal mucosa either by forming a barrier via auto-aggregation or by co-aggregation with the pathogens [26–28]. Invariably cell adherence properties are aggregation ability related.

Auto-aggregation of probiotics appeared to be necessary for the adhesion to intestinal epithelial cells. In addition, the ability to co-aggregate with pathogens may form a barrier which prevents colonization by pathogens. Adherence of bacterial cells is usually related to cell surface characteristics [29, 30]. Hydrophobicity, one of cell surface physico-chemical characteristics could affect auto-aggregation and adhesion of bacteria to different surfaces [25]. It was reported that auto-aggregation of LAB is associated with their adhesion ability [28].

The co-aggregation ability could allow LAB strains to inhibit the growth of pathogens in the gastrointestinal and urogenital tracts [31]. Furthermore, LAB strains have a major influence on the micro-environment around the pathogens and in the process of co-aggregation increase the concentration of antimicrobial substances secreted [26, 32]. Additionally, co-aggregation of inhibitor-producing LAB with the pathogens could possibly constitute an important host defense mechanism in the urogenital and GIT. The ability of LAB to co-aggregate with gut pathogens could potentially be a probiotic property of the microorganism [25].

Thus, the potential of *P. acidilactici* Kp10 as a probiotic strain was evaluated for its auto-aggregation and co-aggregation ability with a foodborne pathogenic bacterium, *L. monocytogenes*. *P. acidilactici* Kp10 had higher auto-aggregation values (35.2%) compared to that of *L. monocytogenes* ATCC 15313 (24.7%). *P. acidilactici* Kp10 had a co-aggregation ability with *L. monocytogenes* ATCC 15313 of about 46% (Table 2). Our results concurred with that reported previously [11] for *P. acidilactici* KACC 12307 which had auto-aggregation and co-aggregation values of 35.2 and 46%, respectively. It was also reported that probiotics had higher auto-aggregation abilities than the pathogens [26, 33].

Cell surface hydrophobicity is another physico-chemical property that facilitates first contact between microorganisms and host cells. This non-specific initial interaction is weak and reversible and precedes the subsequent adhesion process mediated by more specific mechanisms involving cell-surface proteins and lipoteichoic acids [34–36]. Thus the contribution of hydrophobicity to adhesion capacity could probably be due to the lack of correlation between hydrophobicity and bacterial adhesion [37–39].

Affinity for chloroform, an acidic and monopolar solvent, reflected the reducing (alkalic) nature of the bacterium. However, its affinity to ethylacetate, an alkalic and monopolar solvent, reflected the oxidizing (acidic) nature of the bacterium. Furthermore, affinity towards apolar solvents (hexadecane and xylene) demonstrated the hydrophobic nature of the bacterium. High hydrophobicity is linked to glycoproteins on the bacterial surface while low hydrophobicity is linked to the presence of polysaccharides on the bacterial surface [40].

The adhesion ability of *P. acidilactici* Kp10 to four different solvents (chloroform, xylene, ethylacetate and n-hexadecane) are summarized in Table 3. *P. acidilactici* Kp10 has a strong affinity (46.97%) for xylene, indicating the cells were hydrophobic. The Lewis acid-base characteristics of the cell surface of *P. acidilactici* Kp10 was assessed by its adhesion to chloroform and ethyl acetate. The results showed that *P. acidilactici* Kp10 had a stronger/higher affinity to chloroform (12.42%), an acidic solvent and electron acceptor compared to that of ethyl acetate (5.67%) a basic solvent and electron donor. *P. acidilactici* Kp10 showed a low hydrophobicity (14.55%) for n-hexadecane and positive to Congo red by the presence of red colonies on the agar plate, indicating that it has the hydrophobic structures in its cell wall (Fig. 3).

Some aromatic amino acids derived from dietary or endogenously produced proteins that can be deaminated by gut bacteria leading to the formation of phenolic compounds [41]. These compounds exert a bacteriostatic effect against some bacterial strains. The survivability test of probiotics in the intestine refers to their resistance to 0.4% phenol, a catabolic product of aromatic amino acids with bacteriostatic activity [14]. The tolerance of *P. acidilactici* Kp10 to phenol for 24 h is shown in Table 4. Growth of the bacterium was not markedly inhibited as the bacterial strain could still grow in the presence of 0.1% phenol during the incubation.

Table 2 Aggregation abilities of *P. acidilactici* Kp10 and *L. monocytogenes* ATCC 15313

	P. acidilactici Kp10	*L. monocytogenes* ATCC 15313
Auto-aggregation (%)	35.2 ± 0.07^a	24.7 ± 0.1^b
	P. acidilactici Kp10 with *L. monocytogenes* ATCC 15313	
Co-aggregation (%)	46 ± 0.6	

Mean (± standard deviation) of results from three separate experiments
Values with different superscript letters (a and b) are significantly different (*P* < 0.05)

Table 3 Adhesion of *P. acidilactici* Kp10 to xylene, chloroform, ethyl acetate and n-hexadecane

Solvent	Xylene	Chloroform	Ethyl acetate	n-hexadecane
Adhesion (%)	46.97 ± 0.01^a	12.42 ± 0.01^c	5.67 ± 0.04^d	$14.55\% \pm 0.1^b$

Mean (± standard deviation) of results from three separate experiments
Values with different superscript letters (a, b, c, d) are significantly different (*P* < 0.05)

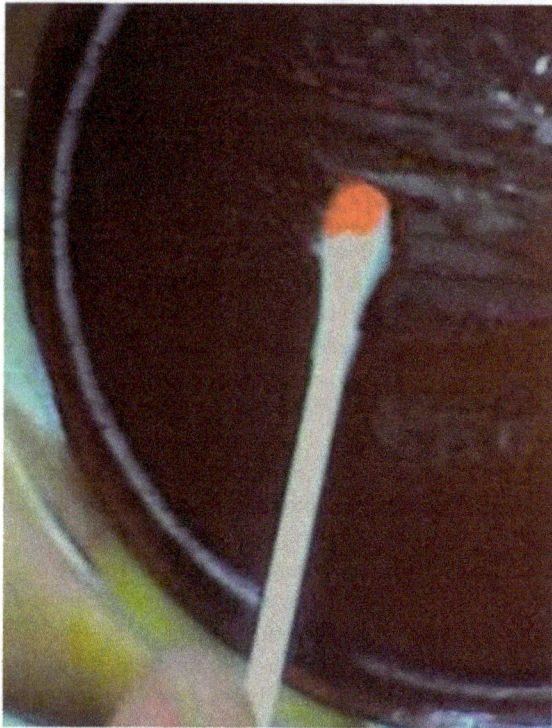

Fig. 3 Cell surface hydrophobicity of *P. acidilactici* Kp10 with Congo red dye

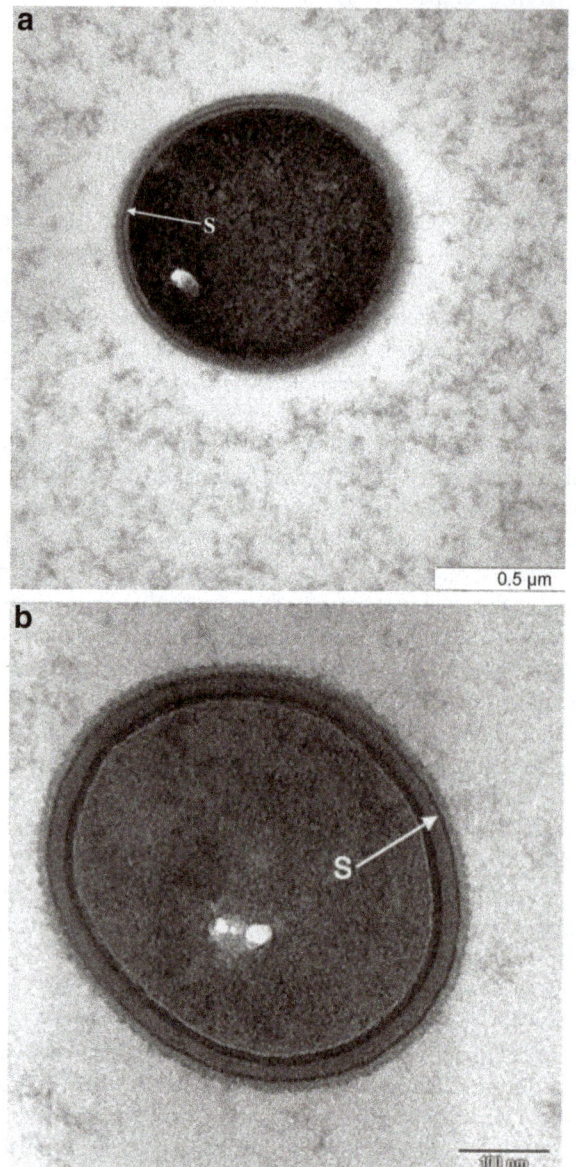

Fig. 4 TEM of a cross-section of (**a**) *P. acidilactici* Kp10 and (**b**) *Lb. crispatus* DSM 20584 cells showing the S-layer (*arrow*) in the cell wall of the bacterium

Results showed that *P. acidilactici* Kp10 was moderately tolerant to phenol. A similar result was also reported for *Lb. plantarum* Lp-115 [28]. Bacteria that are tolerant to phenols may have better chances of survival in the GIT. Some LAB strains such as *Lb. acidophilus* DC601, *Lb. gasseri* BO3, *Lb. paracasei* BO52 are tolerant to high phenol concentrations (0.4 to 0.5%) [14, 42], although the physiology of these bacteria are closely related to *P. acidilactici* Kp10.

Transmission electron micrographs of the *P. acidilactici* Kp10 and *Lb. crispatus* DSM 20584 (DSM: Deutsche Sammlung von Mikroorganismen un Zellkulturen GmbH/Braunschweig, Germany) are shown in Fig. 4a and b, respectively. From the micrographs, it can be seen that S-layers were presence in the cell wall of both strains. In

P. acidilactici Kp10, the S-layer was located in the middle of the thick cell wall. However, the S-layer of *Lb. crispatus* DSM 20584 was located more superficially in the bacterial cell wall. S-layer or crystalline surface layer is a common feature of eubacteria and archaebacteria [43]. The structure is composed of identical subunits consisting of a single protein species linked to each other as well as to the supporting cell wall, also known as specific hydrophobic cell surface proteins [44]. The biological functions of the S-layer in eubacteria include protection, cell adhesion and surface recognition [45]. The S-layer protein from *Lb. crispatus* JCM 5810 was also involved in adhesion [46] and

Table 4 Tolerance of *P. acidilactici* Kp10 cells to phenol

M17+ % of phenol	Viable counts[a] (Log_{10} CFU/ mL)		
	T_0	T_{24}	Inhibition[b]
Blank (without phenol)	5.09 ± 0.01	7.56 ± 0.0	−2.47
0.1	5.04 ± 0.0	6.47 ± 0.0	−1.43
0.2	5.04 ± 0.15	4 .75 ± 0.13	0.29
0.3	5.06 ± 0.06	4.11 ± 0.08	0.95
0.4	5.07 ± 0.25	3.48 ± 0.0	1.59

[a]Log mean counts of three trials (mean ± S.E)
[b]Inhibition = log_{10}(initial population) − log_{10}(final population)

the inhibition of adhesion of *E. coli* to the basement membrane of mucosal epithelium [47]. With reference to the function of the S-layer it could be a contributing factor in the adhesion of *P. acidilactici* Kp10 to Vero cells and the intestinal mucosa of goat ileum as observed in the present study. A more conclusive identification of this structure could be obtained by generating an antibody against the specific hydrophobic cell surface protein and gold-labeling the antibody [48].

The absence of pathogenicity traits such as the absence of haemolytic activity in cultures, as observed in this study, suggested the suitability of application of *P. acidilactici* Kp10 in foods [49]. The absence of haemolytic activity is considered a safety prerequisite for the selection of a probiotic strain [50]. *P. acidilactici* Kp10 exhibited γ-haemolytic activity (no haemolysis) when grown in Columbia blood agar. Similar observations were reported in *Lb. paracasei subsp. paracasei, Lactobacillus spp.* and *Lb. casei* isolated from dairy products which showed γ-haemolysis except of few that showed α-haemolysis [51]. Most of the LAB strains (69 from 71 strains) have been reported as γ-haemolytic (i.e. no haemolysis) [52].

Application of the commercial API-ZYM is for the selection of strains as potential starter cultures based on superior enzyme profiles especially peptidases and esterases. The test system is also applicable in the determining accelerated maturation and flavor development of fermented products [53]. Esterase in particular from LAB may be involved in the development of fruity flavors and quality improvement in dairy and meat products such as cheese, cured bacon and fermented sausages [54]. Enzymatic activities of *P. acidilactici* Kp10 as evaluated by the semi-quantitative API-ZYM system is shown in Table 5. *P. acidilactici* Kp10 exhibited a very low level of alkaline phosphatase, a lipolytic enzyme. Kp10 demonstrated strong peptidase (leucine-arylamidase) and esterase-lipase (C4 and C8) activities. Proteinases (trypsin) activity is however absent in Kp10. The above are two possible desirable traits for the production of typical flavor. Similar results have been reported on the use of LAB as a starter culture and potential technological implications by increasing desirable flavor in seafood products [55–57].

Acidification is an important technological and functional property in the selection of LAB as a starter culture [58]. It was found that *P. acidilactici* Kp10 acidified the skim milk used by lowering the pH to 5.3 apart from showing strong coagulating activities. The potential of LAB strains for application as a starter or adjunct cultures in the production of fermented products is demonstrated by their ability to coagulate milk. Results showed that *P. acidilactici* Kp10 exhibited proteolytic activity which is in agreement with the reports published by [59] and [60] for other LAB. LAB are weakly proteolytic

Table 5 Enzyme activities of *P. acidilactici* Kp10

	Enzyme	Production
1	Control	−
2	Alkalinephosphatase	+
3	Esterase (C4)	++
4	Esteraselipase (C8)	++
5	Lipase (C14)	−
6	Leucinearylamidas	++
7	Valinearylamidase	−
8	Cystinearylamidas	−
9	Trypsin	−
10	α-chymotrypsin	++++
11	Acidphosphatase	+
12	Naphthol-AS-Bl-phosphohydrolase	++
13	α-galactosidase	++++
14	β-galactosidase	++++
15	β-glucuronidase	−
16	α-glucosidase	++++
17	β-glucosidase	++++
18	N-acetyl-b-glucosaminidase	−
19	α-mannosidase	−
20	α-fucosidase	−

'+' refers to positive reaction; '-' refers to negative reaction

compared with other groups of bacteria such as *Bacillus, Proteus, Pseudomonas* and *Coliforms* [61] but the bacterial strains do cause a significant degree of proteolysis in many fermented dairy products [62]. LAB is capable of hydrolyzing oligopeptides into small peptides and amino acids as it possess a very comprehensive proteinase/peptidase system [63]. Many dairy starter cultures are proteolytic thus bioactive peptides can be generated and used in the manufacturing of fermented dairy products. To prepare an experimental starter the technological properties of LAB should include growth, acidifying, proteolytic and amylolytic activities [64].

P. acidilactici Kp10 showed positive results for amylolactic activity. Amylases produced by amylolytic LAB (ALAB) facilitate hydrolysis and fermentation of starch to lactic acid in a single step process [65]. ALAB can thus be utilized in commercial production of lactic acid from starchy materials and in reducing the viscosity of starchy complementary foods [66, 67]. Apart from altering the microstructure of starch, ALAB could also modify the amylography and viscosity of starch. α-amylases of ALAB has the ability of partially hydrolyzing raw starch and as such this microorganism could ferment different types of amylaceous raw materials viz. wheat, potato and different starchy substrates [68]. Taking into consideration the global importance and availability of starchy biomass, production of amylases and lactic acid from starch present two potential industrial applications of ALAB. Bulk production of amylases through microbial fermentation could beneficially be utilized in starch degradation which could supply 25–33% of the global enzyme market [69]. Direct conversion of starchy materials to lactic acid by LAB with ability in secreting amylolytic enzymes in a single-step production process is preferred at industrial scale. This approach will eliminate the two-step process, which include enzymatic saccharification for stach hydrolysis followed with LAB fermentation to convert sugar to lactic acid, Production cost could be substantially reduced with a sing-step process to ensure it is economically viable.

Conclusion

Results from the present study provided ample evidences to claim that *P. acidilactici* Kp10 is a potential probiotic and starter culture. However the data generated were based purely on in vitro studies. In order to claim that this microorganism is categorically a probiotic strain, the survivability and ability to express its probiotic potential in the gastrointestinal environment is also the important criterion to be considered. The environment in the gastrointestinal tract is not only different from that of in vitro, there are also a number of the interacting factors that have major influences on the survivability and its probiotic characteristics. The robust environment at industrial scale may not be favourable to the performance and the capability of the selected probiotic strain. To support the recommendation of using *P. acidilactici* Kp10 in food industry, a comprehensive study to identify their comparative advantages is required. However, this is not the objective of this paper. The results obtained from the present in vitro studies gave ample evidences to indicate that *P. acidilactici* Kp10 is a promising probiotic and starter culture potential. However a comprehensive in vivo investigations are required to categorically substantiate its true potential.

Abbreviations

AC: Auto-aggregation coefficient; ALAB: Amylolytic LAB; BATS: Bacterial cell adhesion to solvent; BLIS: Bacteriocin-like inhibitory substances; DMEM: Dulbecco's modified eagle medium; GIT: Gastrointestinal tract; LAB: Lactic acid bacteria; MRS: de Man, Rogosa and Sharpe; OD: Optical density; PBS: Phosphate-buffered saline; RPMI: Roswell park memorial institute medium; SEM: Scanning electron microscope; TEM: Transmission electron microscope

Acknowledgement

The authors wish to specifically thank to Prof. Dr. Abdul Rahman Omar, Laboratory of Vaccine and Therapeutics, Institute of Bioscience, University Putra Malaysia for his contribution on the use of cell lines in this study.

Funding

This study was financially supported by research fund from the Ministry of Higher Education Malaysia under Prototype Research Grant Scheme (PRGS) and the reference number is PRGS/2/2015/SG05/UPM/01/2.

Authors' contributions

SA participated in the project conception, carried out all the experimental work, analyzed and interpreted the data and wrote the manuscript. ABA was corresponding author, designed and supervised the entire project. All authors contributed to the design and interpretation of experimental results, as well as editing and revising the manuscript. All authors have read and approved the final manuscript.

Competing interests

The authors declare that they have no competing interests.

Author details

[1]Department of Microbiology, Faculty of Biotechnology and Biomolecular Sciences, Universiti Putra Malaysia, 43400 UPM Serdang, Selangor, Malaysia. [2]Bioprocessing and Biomanufacturing Research Centre, Faculty of Biotechnology and Biomolecular Sciences, Universiti Putra Malaysia, 43400 UPM Serdang, Selangor, Malaysia. [3]School of Industrial Technology, Universiti Sains Malaysia, 11800 George Town, Penang, Malaysia. [4]Institute of Bioscience, Universiti Putra Malaysia, 43300 Serdang, Selangor, Malaysia. [5]Faculty of Veterinary Medicine, Universiti Putra Malaysia, 43400 UPM Serdang, Selangor, Malaysia.

References

1. Abbasiliasi S, Joo Shun T, Ibrahim TAT, Ramanan RN, Vakhshiteh F, Mustafa S. Isolation of *Pediococcus acidilactici* Kp10 with ability to secrete bacteriocin-like inhibitory substance from milk products for applications in food industry. BMC Microbiol. 2012;12(1):260.
2. Rhee SJ, Lee JE, Lee CH. Importance of lactic acid bacteria in Asian fermented foods. Microb Cell Factories. 2011;10:55–68.
3. Visessanguan W, Benjakul S, Smitinont T, Kittikun C, Thepkasikul P, Panya A. Changes in micro- biological, biochemical and physico-chemical properties of nham inoculated with different inoculum levels of Lactobacillus curvatus. LWT-Food Sci Technol. 2006;39:814–26.
4. Quarantelli A, Righi F, Agazzi A, Invernizzi G, Ferroni M, Chevaux E. Effects of the administration of *Pediococcus acidilactici* to laying hens on productive performance. Vet Res Commun. 2008;32:359–61.
5. Guerra NP, Bernárdez PF, Méndez J, Cachaldora P, Pastrana Castro L. Production of four potentially probiotic lactic acid bacteria and their evaluation as feed additives for weaned piglets. Animal Feed Sci Technol. 2007;134(1):89–107.
6. Abbasiliasi S, Tan J, Kadkhodaei S, Nelofer R, Tengku Ibrahim TA, Mustafa S, Ariff AB. Enhancement of BLIS production by *Pediococcus acidilactici* kp10 in optimized fermentation conditions using an artificial neural network. RSC Adv. 2016;6(8):6342–9.
7. Gopal PK, Prasad J, Smart J, Gill HS. In vitro adherence properties of *Lactobacillus rhamnosus* DR20 and *Bifidobacterium lactis* DR10 strains and their antagonistic activity against an enterotoxigenic *Escherichia coli*. Int J Food Microbiol. 2001;67(3):207–16.
8. Mäyrä-Mäkinen A, Manninen M, Gyllenberg H. The adherence of lactic acid bacteria to the columnar epithelial cells of pigs and calves. J Appl Microbiol. 1983;55:241–5.
9. Polak-Berecka M, Waśko A, Paduch R, Skrzypek T, Sroka-Bartnicka A. The effect of cell surface components on adhesion ability of *Lactobacillus rhamnosus*. Antonie Van Leeuwenhoek 2014, 106:751–762.
10. Kos B, Suskovic J, Vukovic S, Simpraga M, Frece J, Matosic S. Adhesion and aggregation ability of probiotic strain *Lactobacillus acidophilus* M92. J Appl Microbiol. 2003;94(6):981–7.
11. Xu H, Jeong HS, Lee HY, Ahn J. Assessment of cell surface properties and adhesion potential of selected probiotic strains. Lett Appl Microbiol. 2009; 49(4):434–42.
12. Beena AK, Anupa A. A study on the probiotic aspects of *Lactobacillus* isolated from raw milk of Vechur. Int J Sci Res. 2015;4(12):192–3.
13. Sharma KK, Soni SS, Meharchandani S. Congo red dye agar test as an indicator test for detection of invasive bovine *Escherichia coli*. Veterinarski Arhiv. 2006; 76(4):363–6.
14. Xanthopoulos V, Litopoulou-Tzanetaki E, Tzanetakis N. Characterization of *Lactobacillus* isolates from infant faeces as dietary adjuncts. Food Microbiol. 2000;17(2):205–15.

15. De Vuyst L, Foulquie Moreno MR, Revets H. Screening for enterocins and detection of hemolysin and vancomycin resistance in enterococci of different origins. Int J Food Microbiol. 2003;84:299–318.

16. Chettri R, Tamang JP. Functional properties of tungrymbai and bekang, naturally fermented soybean foods of North East India. Int J Fermented Foods. 2014;3:87–103.

17. Pailin T, Kang DH, Schmidt K, Fung DYC. Detection of extracellular bound proteinase in EPS-producing lactic acid bacteria cultures on skim milk agar. J Appl Microbiol. 2001;133:45–9.

18. Chung H-J. Control of foodborne pathogens by bacteriocin-like substances from Lactobacillus spp. in combination with high pressure processing. USA: The Ohio State University; 2003.

19. Bharti V, Mehta A, Singh S, Jain N, Ahirwal L, Mehta S. Bacteriocin: A novel approach for preservation of food. Int J Pharm Pharm Sci. 2015;17(9):20–9.

20. oglu Gulahmadov SG, Batdorj B, Dalgalarrondo M, Chobert JM, oglu Kuliev AA, Haertlé T. Characterization of bacteriocin-like inhibitory substances (BLIS) from lactic acid bacteria isolated from traditional Azerbaijani cheeses. Euro Food Res Technol. 2006;224(2):229–35.

21. Koponen O. Studies of producer self-protection and nisin biosynthesis of Lactococcus lactis. Helsinki: University of Helsinki; 2004.

22. Korkeala H, Pekkanen TJ. The testing of the antibiotic sensitivity of bacteria on an agar medium: the problem of a double zone of inhibition. Acta Path Micro Im B. 1977;85:174–6.

23. Anggraeni D. Attachment study of lactic acid bacteria originated from human breast milk. Bogor: Bogor Agricultural University; 2010.

24. Duary RK, Rajput YS, Batish VK, Grover S. Assessing the adhesion of putative indigenous probiotic lactobacilli to human colonic epithelial cells. Ind J Medic Res. 2011;134(5):664.

25. Balakrishna A. In vitro Evaluation of Adhesion and Aggregation Abilities of Four Potential Probiotic Strains Isolated from Guppy (Poecilia reticulata). Braz Arch Biol Technol. 2013;56:793–800.

26. Li Q, Liu X, Dong M, Zhou J, Wang Y. Aggregation and adhesion abilities of 18 lactic acid bacteria strains isolated from traditional fermented f ood. Int J Agri Policy Res. 2015;3:84–92.

27. Vlková E, Rada V, Smehilova M, Killer J. Auto- aggregation and co-aggregation ability in bifidobacteria and clostridia. Folia Microbiol. 2008;53:263–9.

28. Collado MC, Meriluoto J, Salminen S. Adhesion and aggregation properties of probiotic and pathogen strains. Eur Food Res Technol. 2008;226:1065–73.

29. Bibiloni R, Perez PF, Garrote GL, Disalvo EA, De Antoni GL. Surface characterization and adhesive properties of bifidobacteria. Method Enzymol. 2001;336:411–27.

30. Canzi E, Guglielmetti S, Mora D, Tamagnini T, Parini C. Conditions affecting cell surface properties of human intestinal bifidobacteria. Antonie Van Leeuwenhoek. 2005;88:207–19.

31. Botes M, Loos B, van Reenen CA, Dicks LM. Adhesion of the probiotic strains Enterococcus mundtii ST4SA and Lactobacillus plantarum 423 to Caco-2 cells under conditions simulating the intestinal tract, and in the presence of antibiotics and anti-inflammatory medicaments. Arch Microbiol. 2008;190:573–84.

32. Kaewnopparat S, Dangmanee N, Kaewnopparat N, Srichana T, Chulasiri M, Settharaksa S. In vitro probiotic properties of Lactobacillus fermentum SK5 isolated from vagina of a healthy woman. Anaerobe. 2013;22:6–13.

33. Pan X, Chen F, Wu T, Tang H, Zhao Z. The acid, bile tolerance and antimicrobial property of Lactobacillus acidophilus NIT. Food Control. 2009;20:598–602.

34. Granato D, Perotti F, Masserey I, Rouvet M, Golliard M, Servin A. Cell surface-associated lipoteichoic acid acts as an adhesion factor for attachment of Lactobacillus johnsonii La1 to human enterocyte-like Caco-2 cells. Appl Environ Microbiol. 1999;65(3):1071–7.

35. Rojas M, Ascencio F, Conway PL. Purification and characterization of a surface protein from Lactobacillus fermentum 104R that binds to porcine small intestinal mucus and gastric mucin. Appl Environ Microbiol. 2002; 68(5):2330–6.

36. Roos S, Jonsson H. A high-molecular-mass cell-surface protein from Lactobacillus reuteri 1063 adheres to mucus components. Microbiol. 2002;148(2):433–42.

37. Kim HJ, Camilleri M, McKinzie S, Lempke MB, Burton DD, Thomforde GM. A randomized controlled trial of a probiotic, VSL# 3, on gut transit and symptoms in diarrhoea-predominant irritable bowel syndrome. Alimentary Pharmacol Ther. 2003;17(7):895–904.

38. Vinderola CG, Reinheimer JA. Lactic acid starter and probiotic bacteria: a comparative "in vitro" study of probiotic characteristics and biological barrier resistance. Food Res Int. 2003;36(9):895–904.

39. Van Loosdrecht MC, Lyklema J, Norde W, Schraa G, Zehnder AJ. The role of bacterial cell wall hydrophobicity in adhesion. Appl Environ Microbiol. 1987; 53(8):1893–7.

40. Bellon-Fontaine MN, Rault J, Van Oss CJ. Microbial adhesion to solvents: a novel method to determine the electron-donor/electron-acceptor or Lewis acid-base properties of microbial cells. Colloids Surface B. 1996;7(1):47–53.

41. Šušković J, Brkić B, Matošić S, Marić V. Lactobacillus acidophilus M92 as potential probiotic strain. Milchwissenschaft. 1997;52(8):430–5.

42. Shehata MG, El Sohaimy SA, El-Sahn MA, Youssef MM. Screening of isolated potential probiotic lactic acid bacteria for cholesterol lowering property and bile salt hydrolase activity. Ann Agri Sci. 2016;61:65–75.

43. Messner P, Sleytr UB. Crystalline bacterial cell-surface layers. Adv Microbial Physiol. 1992;33:213–75.

44. MacKenzie DA, Jeffers F, Parker ML, Vibert-Vallet A, Bongaerts RJ, Roos S. Strain-specific diversity of mucus-binding proteins in the adhesion and aggregation properties of Lactobacillus reuteri. Microbiol. 2010;156(11):3368–78.

45. Gruber K, Sleytr UB. Influence of an S-layer on surface properties of Bacillus stearothermophilus. Arch Microbiol. 1991;156(3):181–5.

46. Sillanpää J, Martínez B, Antikainen J, Toba T, Kalkkinen N, Tankka S. Characterization of the collagen-binding S-layer protein CbsA of Lactobacillus crispatus. J Bacteriol. 2000;182(22):6440–50.

47. Horie M, Ishiyama A, Fujihira-Ueki Y, Sillanpää J, Korhonen TK, Toba T. Inhibition of the adherence of Escherichia coli strains to basement membrane by Lactobacillus crispatus expressing an S-layer. J Appl Microbiol. 2002;92(3):396–403.

48. Johnson-Henry KC, Hagen KE, Gordonpour M, Tompkins TA, Sherman PM. Surface-layer protein extracts from Lactobacillus helveticus inhibit enterohaemorrhagic Escherichia coli O157: H7 adhesion to epithelial cells. Cell Microbiol. 2007;9(2):356–67.

49. Embarek PK, Jeppesen VF, Huss HH. Antibacterial potential of Enterococcus faecium strains to inhibit Clostridium botulinum in sous-vide cooked fish fillets. Food Microbiol. 1994;11:525–36.

50. FAO/WHO, 2002. Joint FAO/WHO Working group report on drafting guidelines for the evaluation of probiotics in food London, Ontario, Canada, April 30 and May 1, 2002.

51. Maragkoudakis PA, Zoumpopoulou G, Miaris C, Kalantzopoulos G, Pot B, Tsakalidou E. Probiotic potential of Lactobacillus strains isolated from dairy products. Int Dairy J. 2006;16(3):189–99.

52. Argyri AA, Zoumpopoulou G, Karatzas K-LG, Tsakalidou E, Nychas G-JE, Panagou EZ, Tassou CC. Selection of potential probiotic lactic acid bacteria from fermented olives by in vitro tests. Food Microbiol. 2013;33:282–91.

53. Tamang JP, Tamang B, Schillinger U, Guigas C, Hlzapfel WH. Functional properties of lactic acid bacteria isolated from ethnic fermented vegetables of the Himalayas. Int J Food Microbiol. 2009;135:28–33.

54. Gobbetti M, Smacchi E, Corsetti A. Purification and characterisation of a cell surface-associated esterase from Lactobacillus fermentum DT41. Int Dairy J. 1997;7:13–21.

55. Thapa N, Pal J, Tamang JP. Phenotypic identification and technological properties of lactic acid bacteria isolated from traditionally processed fish products of the eastern Himalayas. Int J Food Microbiol. 2006;107:33–8.

56. Thapa N, Pal J, Tamang JP. Microbial diversity in ngari, hentak and tungtap, fermented fish products of north-east India. World J Microbiol Biotechnol. 2004;20:599–607.

57. Nanasombat S, Phunpruch S, Jaichalad T. Screening and identification of lactic acid bacteria from raw seafoods and Thai fermented seafood products for their potential use as starter cultures. Songklanakarin J Sci Technol. 2012;34(3): 255–62.

58. De Vuyst L. Technology aspects related to the application of functional starter cultures. Food Technol Biotechnol. 2000;38(2):105–12.

59. Tulini FL, Hymery N, Haertlé T, Blay GL, De Martinis ECP. Screening for antimicrobial and proteolytic activities of lactic acid bacteria isolated from cow, buffalo and goat milk and cheeses marketed in the southeast region of Brazil. J Dairy Res. 2015:1–10.

60. Biswas SR, Ray P, Johnson MC, Ray B. Influence of growth conditions on the production of a bacteriocin, pediocin AcH, by Pediococcus acidilactici H. Appl Environmental Microbiol. 1991;57:1265–7.

61. Khairul Islam M, Abdul Alim Al-Bari M, Shakhawat Hasan M, Alam Khan M, Kudrat-E-Zahan M, Anwar Ul Islam M. Synergistic inhibitory activities and enhancing antibiotic sensitivities of lactobacilli from rajshahi traditional curd. J Adv Bio Biotechnol. 2015;3(1):1–11.

62. Kivanc M, Yilmaz M, Çakir E. Isolation and identification of lactic acid bacteria from boza, and their microbial activity against several reporter strains. Turk J Bio. 2011;35:313–24.

63. Nespolo CR, Brandelli A. Production of bacteriocin-like substances by lactic acid bacteria isolated from regional ovine cheese. Braz J Microbiol. 2010;41: 1009–18.

64. Madrau MA, Mangia NP, Murgia MA, Sanna MG, Garau G, Leccis L, Caredda M, Deiana P. Employment of autochthonous microflora in Pecorino Sardo cheese manufacturing and evolution of physicochemical parameters during ripening. Int Dairy J. 2006;16:876–85.

65. Reddy G, Altaf M, Naveena BJ, Venkateshwar M, Kumar EV. Amylolytic bacterial lactic acid fermentation-a review. Biotechnol Adv. 2008;26:22–34.

66. Songré-Ouattara LT, Mouquet-Rivier C, Humblot C, Rochette I, Diawara B, Guyot JP. Ability of selected lactic acid bacteria to ferment a pearl millet-soybean slurry to produce gruels for complemen- tary foods for young children. J Food Sci. 2010;75:261–9.

67. Mukisa IM, Byaruhanga YB, Aijuka M, Schüller RB, Sahlstrøm S, Langsrud T, Narvhus JA. Influence of cofermentation by amylolytic Lactobacillus plantarum and Lactococcus lactis strains on the fermentation process and rheology of sorghum porridge. Appl Environ Microbiol. 2012;78(15):5220–8.

68. Putri WDR, Haryadi DW, Marseno-Cahyanto MN. Effect of biodegradation by lactic acid bacteria on physical properties of cassava starch. Int Food Res J. 2011;18:1149–54.

69. Fossi BT, Tavea F, Jiwoua C, Ndjouenkeu R. Simultaneous production of raw starch degrading highly thermostable α-amylase and lactic acid by Lactobacillus fermentum 04BBA19. Afr J Biotechnol. 2011;10(34):6564–74.

5

Comparison of biomarker based Matrix Assisted Laser Desorption Ionization-Time of Flight Mass Spectrometry (MALDI-TOF MS) and conventional methods in the identification of clinically relevant bacteria and yeast

Ali Kassim[1]*, Valentin Pflüger[3], Zul Premji[1], Claudia Daubenberger[2] and Gunturu Revathi[1]

Abstract

Background: MALDI-TOF MS is an analytical method that has recently become integral in the identification of microorganisms in clinical laboratories. It relies on databases that majorly employ pattern recognition or fingerprinting. Biomarker based databases have also been developed and there is optimism that these may be superior to pattern recognition based databases. This study compared the performance of ribosomal biomarker based MALDI-TOF MS and conventional methods in the identification of selected bacteria and yeast.

Methods: The study was a cross sectional study identifying clinically relevant bacteria and yeast isolated from varied clinical specimens submitted to a clinical laboratory. The identification of bacteria using conventional Vitek 2™ automated system, serotyping and MALDI-TOF MS was performed as per standard operating procedures. Comparison of sensitivities were then carried out using Pearson Chi-Square test and p-value of <0.05 was considered statistically significant. Secondary outcomes analyzed included the major and minor error rates.

Results: Of the 383 isolates MALDI-TOF MS and conventional methods identified 97.6 and 95.7% ($p = 0.231$) to the genus level and 97.4 and 88.0% ($p = 0.000$) to the species level respectively. Biomarker based MALDI-TOF MS was significantly superior to Vitek 2™ in the identification of Gram negative bacteria and Gram positive bacteria to the species level. For the Gram positive bacteria, significant difference was observed in the identification of Coagulase negative *Staphylococci* ($p = 0.000$) and *Enterococcus* ($p = 0.008$). Significant difference was also observed between serotyping and MALDI-TOF MS ($p = 0.005$) and this was attributed to the lack of identification of *Shigella* species by MALDI-TOF MS. There was no significant difference observed in the identification of yeast however some species of *Candida* were unidentified by MALDI-TOF MS.

Conclusion: Biomarker based MALDI-TOF MS had good performance in a clinical laboratory setting with high sensitivities in the identification of clinically relevant microorganisms.

Keywords: MALDI-TOF MS, PAPMID™, VITEK 2™, SARAMIS™

* Correspondence: alliassimo@gmail.com
[1]Aga Khan University Hospital, Nairobi, Kenya
Full list of author information is available at the end of the article

Background

Matrix Assisted Laser Desorption Ionization-Time of Flight Mass Spectrometry (MALDI-TOF MS) is an analytical method developed in mid-1980s that has evolved rapidly to fingerprint spectra for various microorganisms including bacteria and fungi by analyzing protein profiles [1–5]. Databases were subsequently developed and adopted in clinical microbiology laboratories for the identification of clinically relevant microorganisms. These databases employ the concept of pattern recognition or fingerprinting where mass spectra obtained from a bacteria or yeast is compared to the existing spectra in the databases to find the closest match [6, 7].

The biomarker approach to identification of bacteria uses the specific proteins found within the bacterial cells. Ribosomal proteins have turned out to be one of the ideal biomarkers because they are abundant, highly conserved and encoded by chromosomal genes. They also have molecular masses that fall within the 4 to 30 kDa range of MALDI-TOF MS [8]. Despite being highly conserved there are inter-species and inter-strain differences that can be employed in typing and sub-typing of microorganisms. Using ribosomal biomarkers Suarez et al. were able to group various strains of Neisseria meningitidis into six subgroups that corresponded to sequence types and/or clonal complexes [9].

The Putative Assigned Protein Masses for Identification Database (PAPMID™) (Mabritec AG, Switzerland) is a biomarker based database that comprises molecular masses of ribosomal proteins calculated from partial or whole bacterial genome sequences. This database has been shown to supplement pattern recognition reference databases like the SARAMIS™ database [8]. Ziegler et al. found that it performed as well as 16S rRNA sequencing in the correct identification of root nodule bacteria [8]. This approach has also been shown to differentiate strains of Acinetobacter Genomic species 13BJ/14TU that are intrinsically resistant to polymixins from Acinetobacter haemolyticus [10]. This database therefore has the ability to be an easily accessible and affordable alternative to gene sequencing especially for resource poor settings in the developing world. We set out to compare the sensitivity of biomarker based MALDI-TOF MS to conventional methods like Vitek 2 and serotyping in the identification of bacteria and yeast from a clinical microbiology laboratory.

Methods

The study is a cross sectional study carried out at the Aga Khan University Hospital, Nairobi, Kenya (AKUH, N) and Mabritec laboratory, Riehen, Switzerland. Ethical approval was granted by the AKUH, N's Research and Ethics Committee (Ref 2016/REC-06). Clinically relevant bacteria and yeasts identified from clinical specimens submitted to the AKUH, N laboratories were included in the study. The specimens were given special codes and delinked from patient identifiers throughout the study. Only specimens classified as 'UN3373 Biological Substances Cat B' were shipped under 'Dry Ice UN 1845' for the MALDI TOF analysis at Mabritec AG. Blinding was maintained throughout the various stages of the study.

Processing of samples using Vitek 2™ and serotyping

The processing of the clinical specimen and identification of bacteria using conventional Vitek 2™ automated system was done at AKUH, N. Standard operating procedures (SOPs) in processing and culture of these specimens were strictly adhered to and the organisms were then put through the Vitek 2™ automated system for the final biochemical identification. Serotyping was employed for the identification of some isolates including Streptococcus, Salmonella and Shigella species.

Processing of samples for MALDI TOF MS analyses

Freshly cultured isolates were spotted in duplicates directly onto MALDI TOF target plates. The spots were then overlaid with 1 ul of 25% formic acid and allowed to air dry. They were then overlaid with 1 ul of matrix solution consisting of 40 g of Alpha–cyano-4-hydroxycinnamic acid (CHCA; Sigma-Aldrich, Buchs, Switzerland) in 33% ethanol, 33% deionized water, 33% acetonitrile (ACN) (Sigma-Aldrich) and 3% trifluoroacetic acid (TFA). For the preparation of yeast, a formic acid suspension protocol was used instead of direct smear. A colony of yeast was picked using a 1 ul plastic inoculation loop and suspended in 20 ul of 25% formic acid. One microliter of this suspension was then spotted onto the MALDI plate, allowed to dry and then overlaid with the matrix. The matrix was then allowed to dry in room air.

The MALDI plates were loaded onto the Axima™ Confidence (Shimadzu-Biotech Corp., Kyoto, Japan) mass spectrometer and mass spectra obtained in positive linear mode at a frequency of 50 Hz and within mass range of 3000 Da to 20,000 Da. Each MALDI plate was externally calibrated using a spectra of reference strain of Escherichia coli DH5α (Invitrogen, Carlsbad, USA) that was also spotted onto the plates.

Data acquisition and analysis using SARAMIS™ and PAPMID™

Empiric spectra for each spot was acquired and an average of 50 to 100 protein mass fingerprints were processed using the Launchpad™ 2.8 software (Shimadzu-Biotech). The spectra were then analyzed using the Saramis™ database and matched with the SuperSpectra™ to look for the closest match. The spectra was then compared to the PAPMID™ database to look for matches in species or strain specific ribosomal biomarkers. The closest match was taken to be the identification of the microorganism.

Data analysis

Data collected were entered into Excel worksheets and analyzed using SPSS version 23.0 (IBM; Armonk, New York, USA). Comparison of sensitivities were then carried out using Pearson Chi-Square test and P-value of <0.05 was considered statistically significant. Secondary outcomes analyzed include the major and minor error rates reported in percentages.

Results

The 383 isolates recruited included 222 Gram negative bacteria, 131 Gram positive bacteria and 30 yeast. Of all the isolates, biomarker based MALDI-TOF MS identified 97.6% correctly to the genus level while the conventional methods identified 95.7% to the genus level with a p-value of 0.231. At the species level, 358 isolates were

analyzed. Of these, MALDI-TOF MS identified 97.4% correctly while conventional methods identified only 88.0% correctly with a significant p-value of 0.000.

In Table 1 below the sensitivities of 195 Gram negative bacteria identified using Vitek 2 and biomarker based MALDI-TOF MS are shown. Of these, 100 and 92.3% (p = 0.000) were correctly identified to the genus level while 100 and 88.2% (p = 0.000) were correctly identified to the species level by MALDI-TOF MS and Vitek 2™ respectively. Vitek 2 correctly identified 48 out of 55 *E. coli* isolates to both the genus and species levels while MALDI TOF MS identified all correctly with a significant p-value of 0.006. Vitek 2™ misidentified four isolates of *E. coli* as *Serratia liquefaciens* and *Serratia fonticola*. The other three isolates of *E. coli* were misidentified as *Klebsiella pneumoniae*, *Pseudomonas aeruginosa* and *Moraxella* species. One isolate of *E. coli* was identified correctly by both PAPMID™ database and Vitek 2™ while SARAMIS™ database misidentified it as *Shigella sonnei*.

Table 1 Comparison of the sensitivities of Vitek 2 and MALDI-TOF MS in the identification of Gram negative bacteria to the species and genus levels

Organism (*n*)	Number (%) of isolates with Correct identification to the species level:			Number (%) of isolates with Correct identification to the genus level:		
	Vitek 2	MALDI-TOF MS	P value	Vitek 2	MALDI-TOF MS	P value
E coli (55)	48 (87.3)	55 (100.0)	0.006	48 (87.3)	55 (100.0)	0.006
Klebsiella pneumoniae (33)	32 (97.0)	33 (100.0)	0.314	32 (97.0)	33 (100.0)	0.314
Klebsiella oxytoca (4)	3 (75.0)	4 (100.0)	0.285	4 (100.0)	4 (100.0)	
Pseudomonas aeruginosa (14)	14 (100.0)	14 (100.0)		14 (100.0)	14 (100.0)	
Pseudomonas mendocina (1)	0 (0.0)	1 (100.0)		1 (100.0)	1 (100.0)	
Pseudomonas pseudoalcaligenes (1)	0 (0.0)	1 (100.0)		1 (100.0)	1 (100.0)	
Acinetobacter baumannii (18)	18 (100.0)	18 (100.0)		18 (100.0)	18 (100.0)	
Acinetobacter genomospecies 13BJ/14TU (1)	0 (0.0)	1 (100.0)		1 (100.0)	1 (100.0)	
Acinetobacter ursingii (1)	1 (100.0)	1 (100.0)		1 (100.0)	1 (100.0)	
Citrobacter freundii (6)	5 (83.3)	6 (100.0)	0.296	6 (100.0)	6 (100.0)	
Citrobacter koseri (3)	3 (100.0)	3 (100.0)		3 (100.0)	3 (100.0)	
Enterobacter cloacae (12)	8 (66.7)	12 (100.0)	0.028	9 (75.0)	12 (100.0)	0.064
Enterobacter aerogenes (5)	5 (100.0)	5 (100.0)		5 (100.0)	5 (100.0)	
Enterobacter gergoviae (1)	1 (100.0)	1 (100.0)		1 (100.0)	1 (100.0)	
Haemophilus influenzae (4)	3 (75.0)	4 (100.0)	0.285	3 (75.0)	4 (100.0)	
Morganella morganii (8)	7 (87.5)	8 (100.0)	0.302	8 (100.0)	8 (100.0)	
Proteus mirabilis (7)	7 (100.0)	7 (100.0)		7 (100.0)	7 (100.0)	
Proteus penneri/vulgaris (7)	6 (85.7)	7 (100.0)	0.299	7 (100.0)	7 (100.0)	
Providencia rettgeri (1)	1 (100.0)	1 (100.0)		1 (100.0)	1 (100.0)	
Stenotrophomonas maltophilia (10)	7 (70.0)	10 (100.0)	0.06	7 (70.0)	10 (100.0)	0.06
Serratia marcescens (1)	1 (100.0)	1 (100.0)		1 (100.0)	1 (100.0)	
Vibrio alginolyticus (1)	1 (100.0)	1 (100.0)		1 (100.0)	1 (100.0)	
Elizabethkingia meningoseptica (1)	1 (100.0)	1 (100.0)		1 (100.0)	1 (100.0)	
Total (195)	172 (88.2)	195 (100.0)	0.000	180 (92.3)	195 (100.0)	0.000

Of the Gram positive bacteria, 111 were identified routinely using Vitek 2™ while the rest were routinely identified using serotyping. Table 2 below shows the sensitivities of Gram positive bacteria identified using Vitek 2™ and MALDI-TOF MS. Of these, 100% were correctly identified to both the genus and species levels by MALDI-TOF MS while 99.1 and 83.8% were correctly identified to the genus and species levels respectively by Vitek 2™ with a significant *P* value of 0.000. All *Staphylococcus* species were identified correctly to the genus level by both Vitek 2™ and MALDI-TOF MS. However, Vitek 2™ correctly identified only 77.1% of Coagulase negative *Staphylococcus* to the species level while MALDI-TOF MS identified all correctly. All *Enterococcus* species were correctly identified to the species level by MALDI-TOF MS while Vitek 2™ identified only 72.7% correctly to the species level.

All the 18 *Salmonella* isolates were correctly identified by both MALDI-TOF MS and serotyping to the genus level. However, no comparison was performed at the species level. All 9 *Shigella* isolates were misidentified as *E. coli* by MALDI-TOF MS. All 20 Gram positive bacteria, mainly *Streptococcus* species, were identified correctly to the genus level by both MALDI-TOF MS

and Serotyping while only one was misidentified to the species level by serotyping. Table 3 below shows the sensitivities of serotyping and MALDI-TOF MS in the identification of some Gram negative and Gram positive bacteria routinely identified using serotyping.

Table 4 below shows the comparison of sensitivities of biomarker based MALDI-TOF MS and Vitek 2™ in the identification of yeasts. Of the 30 yeast isolates, 23 were correctly identified to the genus level by both MALDI-TOF MS and Vitek 2™ while 23 and 22 were identified correctly to the species level by MALDI-TOF MS and Vitek 2™ respectively. Six isolates of *Candida haemulonii* and 1 isolate of *Candida guillermondii* that were unidentified by MALDI-TOF MS were excluded from analysis.

Overall, MALDI-TOF MS had 2.6% minor errors while conventional methods had 12.0% (*p* = 0.000). The major errors were noted to be at 2.4 and 4.3% (*p* = 0.231) for MALDI-TOF MS and conventional methods respectively. No errors were made by MALDI-TOF in the identification of Gram positive bacteria in comparison to 16.2% (*p* = 0.000) minor errors and 0.9% (*p* = 0.316) major errors noted for Vitek 2™. Vitek 2™ had 11.8%

Table 2 Comparison of the sensitivities of Vitek 2 and MALDI-TOF MS in the identification of Gram positive bacteria to the species and genus levels

Organism (n)	Number (%) of isolates with Correct identification to the species level:			Number (%) of isolates with Correct identification to the genus level:		
	Vitek 2	MALDI-TOF MS	P value	Vitek 2	MALDI-TOF MS	P value
Staph aureus (33)	33 (100.0)	33 (100.0)		33 (100.0)	33 (100.0)	
CoNS (48)	37 (77.1)	48 (100.0)	0.000	48 (100.0)	48 (100.0)	
Staph capitis (3)	3 (100.0)	3 (100.0)		3 (100.0)	3 (100.0)	
Staph cohnii (1)	1 (100.0)	1 (100.0)		1 (100.0)	1 (100.0)	
Staph epidermidis (10)	9 (90.0)	10 (100.0)	0.305	10 (100.0)	10 (100.0)	
Staph hemolyticus (6)	3 (50.0)	6 (100.0)	0.046	6 (100.0)	6 (100.0)	
Staph hominis (2)	2 (100.0)	2 (100.0)		2 (100.0)	2 (100.0)	
Staph saprophyticus (23)	18 (78.3)	23 (100.0)	0.018	23 (100.0)	23 (100.0)	
Staph sciuri (1)	1 (100.0)	1 (100.0)		1 (100.0)	1 (100.0)	
Staph simulans (1)	0 (0.0)	1 (100.0)		1 (100.0)	1 (100.0)	
Staph succinus (1)	0 (0.0)	1 (100.0)		1 (100.0)	1 (100.0)	
Strep mitis/oralis (4)	3 (75.0)	4 (100.0)	0.285	4 (100.0)	4 (100.0)	
Strep pneumoniae (3)	3 (100.0)	3 (100.0)		3 (100.0)	3 (100.0)	
Enterococcus (22)	16 (72.7)	22 (100.0)	0.008	21 (95.5)	22 (100.0)	
Enterococcus avium (1)	1 (100.0)	1 (100.0)		1 (100.0)	1 (100.0)	
Enterococcus faecalis (11)	9 (81.8)	11 (100.0)	0.138	10 (90.9)	11 (100.0)	0.306
Enterococcus faecium (7)	4 (57.1)	7 (100.0)	0.051	7 (100.0)	7 (100.0)	
Enterococcus gallinarum (1)	1 (100.0)	1 (100.0)		1 (100.0)	1 (100.0)	
Enterococcus hirae (2)	1 (50.0)	2 (100.0)	0.248	2 (100.0)	2 (100.0)	
Aerococcus viridans (1)	1 (100.0)	1 (100.0)		1 (100.0)	1 (100.0)	
Total (111)	93 (83.8)	111 (100.0)	0.000	110 (99.1)	111 (100.0)	0.316

Table 3 Comparison of the sensitivities of serotyping and MALDI-TOF MS in the identification of selected bacteria to the species and genus levels

Organism (n)	Number (%) of isolates with Correct identification to species level:			Number (%) of isolates with Correct identification to genus level:		
	Serotyping	MALDI-TOF MS	P value	Serotyping	MALDI-TOF MS	P value
Salmonella sp. (18)	-	-		18 (100.0)	18 (100.0)	
Shigella flexneri (6)	6 (100.0)	0	0.001	6 (100.0)	0	0.001
Shigella sonnei (2)	2 (100.0)	0	0.046	2 (100.0)	0	0.046
Shigella dysentriae (1)	1 (100.0)	0	0.157	1 (100.0)	0	0.157
Strep agalactiae (13)	12 (92.3)	13 (100.0)	0.308	13 (100.0)	13 (100.0)	
Strep pyogenes (7)	7 (100.0)	7 (100.0)		7 (100.0)	7 (100.0)	
Total (29/47)	28 (96.6)	20 (69.0)	0.005	47 (100.0)	38 (80.9)	0.002

(p = 0.000) minor errors and 7.7% (p = 0.000) major errors in the identification of Gram negative bacteria.

Discussion

MALDI TOF MS has been shown to reduce the turn-around time, hospital stays and costs as compared to biochemical based tests [11]. Thus far, the MALDI-TOF MS databases introduced into clinical laboratories have employed pattern recognition approaches in the identification of microorganisms. In an attempt to improve sensitivities, use of specific biomarkers rather than generic non-conserved markers in the databases led to the concept of biomarker based approach in identification [9]. The Putative Assigned Protein Masses for Identification Database (PAPMID™) (Mabritec AG, Switzerland) is a biomarker based database that comprises molecular masses of ribosomal proteins calculated from partial or whole bacterial genome sequences [8].

Our study included an assortment of bacteria and yeast isolated in a routine clinical laboratory using Vitek

2 and serotyping. Guo et al. showed an overall sensitivity of 99.6 and 93.37% in the identification of bacteria to the genus and species levels respectively [12]. The sensitivity of biomarker based MALDI-TOF MS in the identification of Gram negative bacteria to the genus and species level was 99 and 92% respectively. This was significantly better than that of the conventional methods including Vitek 2. Studies by Wang et al. show sensitivities of pattern recognition MALDI TOF MS in the identification of Gram negative bacteria ranging from 93.2 to 98.7% [13]. Shigella species that had been identified routinely using serotyping were all identified as E. coli by MALDI-TOF MS. Studies have shown the difficulty in discriminating these two species due to their close relationship [14–16]. All the Salmonella isolates were identified by MALDI-TOF MS as Salmonella enterica subsp. enterica except one that was identified as Salmonella species. No comparison was done at the species level as serotyping identified serogroups as per the Kauffmann-White Scheme while MALDI-TOF MS identified species

Table 4 Comparison of the sensitivities of Vitek 2 and MALDI-TOF MS in the identification of selected yeast to the species and genus levels

Organism (n)	Number (%) of isolates with Correct identification to the species level:			Number (%) of isolates with Correct identification to the genus level:		
	Vitek	MALDI-TOF MS	P value	Vitek	MALDI-TOF MS	P value
Candida albicans (5)	5	5		5	5	
Candida dubliniensis (2)	2	2		2	2	
Candida glabrata (4)	4	4		4	4	
Candida guillermondii (1)[a]	-	-		-	-	
Candida haemulonii (6)[a]	-	-		-	-	
Candida kefyr (1)	1	1		1	1	
Candida krusei (4)	4	4		4	4	
Candida parapsilosis (2)	2	2		2	2	
Candida tropicalis (4)	3	4	0.285	4	4	
Cryptococcus neoformans (1)	1	1		1	1	
Total (23)	22 (95.7)	23 (100.0)	0.312	23 (100.0)	23 (100.0)	

KEY: [a]6 isolates of Candida haemulonii and 1 isolate of Candida guillermondii have been excluded from analysis as they were not identified by MALDI-TOF MS.

and subspecies [17]. Serological and biochemical tests would still be recommended in confirmation of identification of *Salmonella* and *Shigella* species [18]. An isolate of *Acinetobacter* genomospecies 13BJ/14TU by PAPMID™ was identified by SARAMIS™ as *Acinetobacter* species and as *Acinetobacter baumannii* by Vitek 2™. The *Acinetobacter* genomospecies 13BJ/14TU have been shown to be intrinsically resistant to colistin hence the significance of correctly differentiating it from other *Acinetobacter* species [10, 19].

In the overall identification of Gram positive bacteria to the genus level, there was no significant difference in the sensitivities between MALDI TOF MS and conventional methods. However significant difference was observed in the identification of these bacteria to the species level. Coagulase negative *Staphylococci* (CoNS) were identified to the species level with a sensitivity of 100% by biomarker MALDI-TOF MS compared to 77.5% by Vitek 2™ (p = 0.000). This sensitivity was slightly better than that shown by Zhou et al. where pattern recognition MALDI-TOF MS identified 97.7% of CoNS correctly to the species level compared to 76.0% by an automated biochemical system [20]. The significance of correctly identifying CoNS especially lies in the fact that they are the commonest organisms found in positive blood cultures and the need to rule out surface contamination and reduce cost of unnecessary interventions [13, 20].

In the identification of *Enterococcus*, biomarker MALDI-TOF MS correctly identified all to the species level while Vitek 2™ identified only 72.7% correctly (P = 0.008). There was no clinically significant difference between the two methods in the identification of *Enterococcus* to the genus level. These findings mirror those of previous studies that show excellent identification of *Enterococcus* by MALDI-TOF MS especially in the discrimination between *E. faecalis* and *E. faecium* due to their significant differences in their resistance patterns [20, 21].

In the identification of *Streptococcus*, there was no significant difference between biomarker MALDI-TOF MS and Vitek 2™. However, 1 isolate of *S. mitis/oralis* was misidentified as *Streptococcus pneumoniae* by Vitek 2™. Studies had initially shown some difficulty in identification between these two species that was attributed to lack of extensive database [14, 22]. For *Streptococcus pyogenes* and *Streptococcus agalactiae*, biomarker MALDI-TOF MS was compared to serotyping which is conventionally used in identification in our laboratory. Biomarker MALDI-TOF MS correctly identified all to the genus and species level while the serotyping misidentified 1 to the species level.

In the identification of yeasts biomarker MALDI-TOF MS performed as well as Vitek 2™ with sensitivity of 100.0 and 95.7% (p = 0.312) respectively. Both *SARAMIS™ and PAPMID™* databases were unable to identify

one *Candida guillermondii* and six *Candida haemulonii* isolates. This could be attributed to lack of spectra or poor representation for these species in the databases. Previous studies have shown the sensitivity of MALDI TOF MS in the identification of yeasts to range from to 82.7 to 87.2%. In the cohort studied by Lohmann et al. there was a single isolate of *Candida haemulonii* that was not identified [6]. In various studies, *Candida auris*, an emerging multidrug resistant organism, has been misidentified as *Candida haemulonii* by Vitek 2 system [23–25]. This underscores the importance of incorporation of spectra in the database for adequate discrimination between these closely related species.

A limitation of the study is the few numbers in some of the groups of microorganisms like *Streptococcus pneumoniae*, *Cryptococcus* and *Candida* species. *Salmonella typhi* was not analysed since it did not meet the shipping criteria. A larger study including these microorganisms would yield more information on the performance of biomarker based MALDI-TOF MS. We also recommend studies on the performance of this biomarker based database on direct identification from blood culture broths as recent studies have shown improved clinical utility [26].

Conclusion

Our study has shown good performance of the biomarker based approach in a clinical laboratory setting with high sensitivities in the identification of clinically relevant microorganisms. We recommend the adoption of this approach in clinical laboratory settings to improve sensitivities and to reduce the need for molecular testing.

Abbreviations

ACN: Acetonitrile; AKUH, N: Aga Khan University Hospital, Nairobi, Kenya; CAMPY: Campylobacter agar; CHCA: Alpha–cyano-4-hydroxycinnamic acid; CLED: Cystine Lactose Electrolyte Deficient; CoNS: Coagulase negative Staphylococci; GBA: Gentamicin Blood Agar; IBM: International Business Machines Corporation; MALDI-TOF MS: Matrix Assisted Laser Desorption Ionization-Time of Flight Mass Spectrometry; PAPMID™: Putative Assigned Protein Masses for Identification Database; rRNA: Ribosomal Ribonucleic Acid; SARAMIS™: Spectral ARchive And Microbial Identifications System; SDA: Sabouraud Dextrose agar; SOPs: Standard operating procedures; SPSS: Statistical package for the social sciences; TFA: Trifluoroacetic acid; ul: Microliter

Acknowledgements

We would like to acknowledge Mr. James Orwa and Mr. Thaddeus Egondi of the Research Support Unit, AKUH, N, the staff at the Microbiology department, AKUH, N and the staff of Mabritec AG and Swiss Tropical and Public Health Institute, Switzerland for the statistical and technical support received throughout the study.

Funding

The study was funded by a grant from the Swiss-African kick-starting project. The funding body did not participate at any point in the design of the study, data collection and analysis or in the writing of the manuscript.

Authors' contributions

AK, CD, ZP and GR were involved in the conception and design of the study, data analysis and revision of the manuscript. AK and VP were involved in data collection, data analysis and drafting of the manuscript. All authors approved the final manuscript.

Competing interests

Valentin Pflüger is an employee of Mabritec AG, Riehen, Switzerland.

Author details

[1]Aga Khan University Hospital, Nairobi, Kenya. [2]Swiss Tropical and Public Health Institute, Basel, Switzerland. [3]Mabritec AG, Riehen, Switzerland.

References

1. Karas M, Bachmann D, Hillenkamp F. Influence of the wavelength in high-irradiance ultraviolet laser desorption mass spectrometry of organic molecules. Anal Chem [Internet]. 1985;57(14):2935–9. Available from: http://pubs.acs.org/doi/pdf/10.1021/ac00291a042. American Chemical Society; 1 [cited 2015 Jan 4]

2. Karas M, Bachmann D, Bahr U, Hillenkamp F. Matrix-assisted ultraviolet laser desorption of non-volatile compounds. Int J Mass Spectrom Ion Process [Internet]. 1987;78:53–68. Available from: http://www.sciencedirect.com/science/article/pii/0168117687870416. Sep [cited 2015 Jan 8]

3. Karas M, Hillenkamp F. Laser desorption ionization of proteins with molecular masses exceeding 10,000 daltons. Anal Chem [Internet]. 1988; 60(20):2299–301. Available from: http://dx.doi.org/10.1021/ac00171a028. American Chemical Society; Oct [cited 2015 Jan 8]

4. Hillenkamp F, Karas M, Beavis RC, Chait BT. Matrix-assisted laser desorption/ionization mass spectrometry of biopolymers. Anal Chem [Internet]. 1991; 63(24):1193A–203A. Available from: http://pubs.acs.org/doi/pdf/10.1021/ac00024a002. American Chemical Society; 22 [cited 2015 Jan 8]

5. Claydon MA. The rapid identification of intact microorganisms using mass spectrometry. Nat Biotechnol. 1996;14:303–8.

6. Lohmann C, Sabou M, Moussaoui W, Prévost G, Delarbre JM, Candolfi E, et al. Comparison between the biflex III-biotyper and the axima-saramis systems for yeast identification by matrix-assisted laser desorption ionization-time of flight mass spectrometry. J Clin Microbiol. 2013;51(4):1231–6.

7. Chean R, Kotsanas D, Francis MJ, Palombo E a, Jadhav SR, Awad MM, et al. Comparing the identification of Clostridium spp. by two Matrix-Assisted Laser Desorption Ionization-Time Of Flight (MALDI-TOF) mass spectrometry platforms to 16S rRNA PCR sequencing as a reference standard: a detailed analysis of age of culture and sample. Anaerobe [Internet]. 2014;30:85–9. Available from: http://linkinghub.elsevier.com/retrieve/pii/S1075996414001280. Elsevier Ltd

8. Ziegler D, Pothier JF, Ardley J, Fossou RK, Pflüger V, de Meyer S, et al. Ribosomal protein biomarkers provide root nodule bacterial identification by MALDI-TOF MS. Appl Microbiol Biotechnol [Internet]. 2015;99(13):5547–62. [cited 2017 May 24].

9. Suarez S, Ferroni a, Lotz a, Jolley K a, Guerin P, Leto J, et al. Ribosomal proteins as biomarkers for bacterial identification by mass spectrometry in the clinical microbiology laboratory. J Microbiol Methods. 2013;94(3):390–6.

10. Toh BEW, Zowawi HM, Krizova L, Paterson DL, Kamolvit W, Peleg AY, et al. Differentiation of Acinetobacter genomic species 13BJ/14TU from Acinetobacter haemolyticus by use of Matrix-Assisted Laser Desorption Ionization-Time Of Flight Mass Spectrometry (MALDI-TOF MS). J Clin Microbiol [Internet]. 2015;53(10):3384–6. Available from: http://www.pubmedcentral.nih.gov/articlerender.fcgi?artid=4572560&tool=pmcentrez&rendertype=abstract. [cited 2016 Jan 5]

11. Cherkaoui A, Hibbs J, Emonet S, Tangomo M, Girard M, Francois P, et al. Comparison of two matrix-assisted laser desorption ionization-time of flight mass spectrometry methods with conventional phenotypic identification for routine identification of bacteria to the species level. J Clin Microbiol [Internet]. 2010;48(4):1169–75. Available from: http://www.pubmedcentral.nih.gov/articlerender.fcgi?artid=2849558&tool=pmcentrez&rendertype=abstract. [cited 2014 Dec 9]

12. Guo L, Ye L, Zhao Q, Ma Y, Yang J, Luo Y. Comparative study of MALDI-TOF MS and VITEK 2 in bacteria identification. J Thorac Dis [Internet]. 2014;6(5):534–8. Available from: http://www.pubmedcentral.nih.gov/articlerender.fcgi?artid=4015025&tool=pmcentrez&rendertype=abstract. [cited 2014 Aug 12]

13. Wang W, Xi H, Huang M, Wang J, Fan M, Chen Y, et al. Performance of mass spectrometric identification of bacteria and yeasts routinely isolated in a clinical microbiology laboratory using MALDI-TOF MS. J Thorac Dis [Internet]. 2014;6(5):524–33. Available from: http://www.pubmedcentral.nih.gov/articlerender.fcgi?artid=4015010&tool=pmcentrez&rendertype=abstract. [cited 2014 Jul 13]

14. Seng P, Drancourt M, Gouriet F, La Scola B, Fournier P-E, Rolain JM, et al. Ongoing revolution in bacteriology: routine identification of bacteria by matrix-assisted laser desorption ionization time-of-flight mass spectrometry. Clin Infect Dis [Internet]. 2009;49(4):543–51. Available from: http://www.ncbi.nlm.nih.gov/pubmed/19583519

15. Martiny D, Busson L, Wybo I, Ait El Haj R, Dediste A, Vandenberg O. Comparison of the Microflex LT and Vitek MS systems for routine identification of bacteria by matrix-assisted laser desorption ionization - time of flight mass spectrometry. J Clin Microbiol. 2012;50:1313–25.

16. Khot PD, Fisher MA. Novel approach for differentiating shigella species and Escherichia coli by matrix-assisted laser desorption ionization-time of flight mass spectrometry. J Clin Microbiol. 2013;51(11):3711–6.

17. Brenner FW, Villar RG, Angulo FJ, Tauxe R, Swaminathan B. Salmonella nomenclature. J Clin Microbiol. 2000;38(7):2465–7.

18. Neville SA, LeCordier A, Ziochos H, Chater MJ, Gosbell IB, Maley MW, et al. Utility of matrix-assisted laser desorption ionization-time of flight mass spectrometry following introduction for routine laboratory bacterial identification. J Clin Microbiol. 2011;49(8):2980–4.

19. Lee SY, Shin JH, Park KH, Kim JH, Shin MG, Suh SP, et al. Identification, genotypic relation, and clinical features of colistin-resistant isolates of Acinetobacter genomic species 13BJ/14TU from bloodstreams of patients in a university hospital. J Clin Microbiol [Internet]. 2014;52(3):931–9. Available from: http://www.ncbi.nlm.nih.gov/pubmed/24403305. American Society for Microbiology (ASM). [cited 2017 Mar 10]

20. Zhou C, Hu B, Zhang X, Huang S, Shan Y, Ye X. The value of matrix-assisted laser desorption/ionization time-of-flight mass spectrometry in identifying clinically relevant bacteria: a comparison with automated microbiology system. J Thorac Dis [Internet]. 2014;6(5):545–52. Available from: http://www.pubmedcentral.nih.gov/articlerender.fcgi?artid=4015012&tool=pmcentrez&rendertype=abstract. [cited 2014 Aug 12]

21. Fang H, Ohlsson AK, Ullberg M, Özenci V. Evaluation of species-specific PCR, Bruker MS, VITEK MS and the VITEK 2 system for the identification of clinical Enterococcus isolates. Eur J Clin Microbiol Infect Dis. 2012;31(11):3073–7.

22. Clark AE, Kaleta EJ, Arora A, Wolk DM. Matrix-assisted laser desorption ionization-time of flight mass spectrometry: a fundamental shift in the routine practice of clinical microbiology. Clin Microbiol Rev [Internet]. 2013; 26(3):547–603. Available from: http://www.pubmedcentral.nih.gov/articlerender.fcgi?artid=3719498&tool=pmcentrez&rendertype=abstract. [cited 2014 Jul 21]

23. Kathuria S, Singh PK, Sharma C, Prakash A, Masih A, Kumar A, et al. Multidrug-resistant Candida Auris misidentified as Candida haemulonii: characterization by Matrix-Assisted Laser Desorption Ionization-Time Of Flight Mass Spectrometry and DNA sequencing and its antifungal susceptibility profile variability by Vitek 2, CLSI broth microdilution, and Etest method. J Clin Microbiol [Internet]. 2015;53(6):1823–30. Available from: http://www.ncbi.nlm.nih.gov/pubmed/25809970. American Society for Microbiology; [cited 2017 May 9]

24. Ben-Ami R, Berman J, Novikov A, Bash E, Shachor-Meyouhas Y, Zakin S, et al. Multidrug-resistant Candida haemulonii and C. auris, Tel Aviv, Israel. Emerg Infect Dis [Internet]. 2017;23:2. Available from: http://dx.doi.org/10.3201/eid2302.161486. @BULLET www.cdc.gov/eid @BULLET. Wolfson Medical Center; [cited 2017 May 9]

25. Okinda N, Kagotho E, Castanheira M, Njuguna A, Omuse G, Makau P, et al. Candidemia at a referral hospital in sub-Saharan Africa: emergence of Candida auris as a major pathogen. European. European Conference On Clinical Microbiology And Infectious Diseases; Barcelona, Spain. 2014.

26. French K, Evans J, Tanner H, Gossain S, Hussain A. The clinical impact of rapid, direct MALDI-ToF identification of bacteria from positive blood cultures. [cited 2017 May 8]; Available from: http://journals.plos.org/plosone/article/file?id=10.1371/journal.pone.0169332&type=printable.

6

Pyruvate oxidase of *Streptococcus pneumoniae* contributes to pneumolysin release

Joseph C. Bryant[1], Ridge C. Dabbs[1], Katie L. Oswalt[1], Lindsey R. Brown[1], Jason W. Rosch[2], Keun S. Seo[3], Janet R. Donaldson[4], Larry S. McDaniel[5] and Justin A. Thornton[1*]

Abstract

Background: *Streptococcus pneumoniae* is one of the leading causes of community acquired pneumonia and acute otitis media. Certain aspects of *S. pneumoniae*'s virulence are dependent upon expression and release of the protein toxin pneumolysin (PLY) and upon the activity of the peroxide-producing enzyme, pyruvate oxidase (SpxB). We investigated the possible synergy of these two proteins and identified that release of PLY is enhanced by expression of SpxB prior to stationary phase growth.

Results: Mutants lacking the *spxB* gene were defective in PLY release and complementation of *spxB* restored PLY release. This was demonstrated by cytotoxic effects of sterile filtered supernatants upon epithelial cells and red blood cells. Additionally, peroxide production appeared to contribute to the mechanism of PLY release since a significant correlation was found between peroxide production and PLY release among a panel of clinical isolates. Exogenous addition of H_2O_2 failed to induce PLY release and catalase supplementation prevented PLY release in some strains, indicating peroxide may exert its effect intracellularly or in a strain-dependent manner. SpxB expression did not trigger bacterial cell death or LytA-dependent autolysis, but did predispose cells to deoxycholate lysis.

Conclusions: Here we demonstrate a novel link between *spxB* expression and PLY release. These findings link liberation of PLY toxin to oxygen availability and pneumococcal metabolism.

Keywords: *Streptococcus pneumoniae*, Pneumococcus, Pneumolysin, Virulence, Toxin, Metabolism, Protein secretion, Cytotoxicity

Background

S. pneumoniae (pneumococcus) is a Gram-positive human pathogen identified as a cause of acute otitis media, bacteremia, septicemia, pneumonia, and meningitis [1] and is the leading cause of death in children under the age of five worldwide [2]. Diseases resulting from pneumococcal infection impart a major economic impact, with healthcare costs estimated to range from 3 billion to 6 billion dollars annually in the United States for otitis alone [3]. In spite of this major disease burden, it most often exists as a commensal organism of the nasopharynx, with carriage rates of up to 70 % depending on the demographic [4]. Pneumococcus produces and secretes a number of surface proteins which contribute to virulence including neuraminidase, hyaluronidase, and pneumococcal surface protein A (PspA) [5–7]. However, it produces relatively few protein exotoxins compared to other pathogenic species capable of such invasive disease.

The primary toxin expressed by *S. pneumoniae* is pneumolysin (PLY), which is a 53-kDa cholesterol dependent pore-forming cytolysin (CDC) [8]. PLY has been found to reduce ciliary beating within the lungs and deletion of the *ply* gene from the pneumococcal chromosome attenuates virulence in vivo [9]. Additionally, exposure to PLY activates differential gene expression within host cells [10].

* Correspondence: thornton@biology.msstate.edu
[1]Department of Biological Sciences, Mississippi State University, 295 E Lee Blvd., Harned Hall, Rm 219, Mississippi State, MS 39762, USA
Full list of author information is available at the end of the article

Unlike the other members of the CDC family, PLY lacks an N-terminal signal sequence for extracellular release via the Sec-dependent pathway [11]. Despite this, it is well established that PLY is released into the extracellular space [8, 9, 12].

Controversy exists as to the primary route PLY takes to exit the cytoplasm. The mechanism of release was long thought to be solely attributable to autolysis of the bacterial cell [13]. However, PLY release has been demonstrated in the absence of autolysis [14], suggesting that other mechanisms must contribute to liberation of PLY during pneumococcal growth. Other findings have demonstrated that PLY can actually traverse and associate with the pneumococcal cell wall [15]. Subsequent studies by Price et al. demonstrated that domain 2 of PLY is essential for the cell wall association and that the export pathway was conserved in *Bacillus subtilis* [12]. The composition of the pneumococcal peptidoglycan has been recently shown to restrict PLY release [16]. Interestingly, this study found that greater PLY release does not correlate with enhanced virulence and that rather a controlled release of PLY is important for pathogenesis. These findings underline the fact that, while capable of inflicting damage to the host, PLY is also a stimulator of host immune responses [17].

S. pneumoniae is unique among catalase-negative organisms due to the fact that it produces up to millimolar concentrations of hydrogen peroxide (H_2O_2), primarily through the activity of the enzyme pyruvate oxidase (SpxB) [18, 19]. In aerobic environments, pneumococcus utilizes SpxB to convert pyruvate to acetate, a reaction that produces acetyl phosphate, CO_2, and H_2O_2 [20]. Although SpxB-derived H_2O_2 is detrimental to survival of the pneumococcus at high concentrations, deletion of *spxB* has been shown to reduce virulence in vivo [21]. Pneumococcal H_2O_2 has also been shown to aid the pneumococcus in competing with other inhabitants of the upper respiratory tract [22] and possibly has a significant impact upon host cells and tissues. It is cytotoxic to numerous cell types including neuronal cells, neutrophils, and alveolar epithelial cells [23–26]. During stationary phase, pneumococcal H_2O_2 results in pneumococcal cell death resembling apoptosis of eukaryotic cells and this process does not require the major autolysin, *N*-acetylmuramoyl-l-alanine amidase (LytA) [27]. However, strains vary in their production of H_2O_2, with some strains producing significant concentrations prior to stationary phase. The impact of low and intermediate levels of H_2O_2 upon the physiology and structural integrity of *S. pneumoniae* has not been investigated in detail. Price et al. demonstrated PLY associating with the cell wall [15], indicating that by some mechanism PLY escapes the cell membrane. Additionally, pneumococcus is known to alter its membrane composition in response to endogenous reactive oxygen species [28] Based on these results, we hypothesized that pneumococcal H_2O_2 might have non-lethal effects on the physiology of the bacterium at early phases of growth, possibly affecting release of PLY. We investigated the impact of SpxB on PLY localization into bacterial supernatants and the effect of this upon host cell integrity.

Methods

Bacterial strains and culture conditions

S. pneumoniae strains TIGR4 [29], an unencapsulated mutant of TIGR4 (T4R) [30], WU2 [31], AW267, along with isogenic mutants of these strains were grown in Todd Hewitt media plus 0.5 % yeast extract (THY) to a mid-logarithmic phase of growth (OD_{600} of 0.5) at 37 °C. Clinical isolates were received from the Center for Disease Control and Prevention's Active Bacterial Core surveillance (ABCs) isolate bank (http://www.cdc.gov/abcs/index.html). For additional studies strains lacking the major cell wall amidase LytA ($\Delta lytA$) were created as described below. The $\Delta lytA$ isogenic mutants of T4R and WU2, along with $\Delta lytA\Delta spxB$ double mutants were cultured under the same conditions as parental strains. To neutralize pneumococcal H_2O_2 from supernatants, THY media was supplemented with catalase derived from *Aspergillus niger* (2 ng/mL; cat #C3515 Sigma Aldrich). A549 type 2 lung epithelial cells (ATCC) were cultured in F12-K medium (ATCC) supplemented with 10 % fetal bovine serum (FBS) at 37 °C in 5 % CO_2 atmosphere.

Mutant construction

Isogenic $\Delta spxB$ mutants were developed in the strains TIGR4, WU2, T4R, and AW267 by allelic replacement. Briefly, the *spxB* gene with 500 base pairs flanking each end was amplified from T4R chromosomal DNA using primers ALR1 and ALR2, designed to contain both KpnI and XbaI restriction sites. The product was digested with KpnI and XbaI restriction endonucleases (New England BioLabs, Ipswich, MA) and ligated into KpnI and XbaI digested pBluescript vector using T4 DNA ligase (Thermo Fisher Scientific). Inverse PCR was performed using primers ALR3 and ALR4 to amplify outward from just inside the coding sequence of *spxB* from a positive clone. Primers ALR3 and ALR4 were designed to include BamHI sites. The gene *ermB* was amplified using designed primers ALR5 and ALR6 including BamHI sites and both products were digested with BamHI (New England BioLabs, Ipswich, MA) overnight at 37 °C. The linearized vector that was amplified with primers ALR3 and ALR4 and the *ermB* insert amplified with primers ALR5 and ALR6 were ligated and transformed into DH5α *E. coli* cells. Positive colonies with bands of the appropriate size were grown in LB medium overnight and frozen at –80 °C. PCR products containing the knockout construct were amplified from positive clones and used to transform pneumococcal strains by standard methods and subsequent selection for transformants by plating on blood agar plates containing erythromycin (0.5 µg/mL). Mutants lacking the *lytA* gene were generated in strains T4R and WU2 by overlap extension PCR mutagenesis as previously described [32]. Briefly, 500–1000 base pair DNA sequences flanking each side of the *lytA* gene were amplified using primers LytA-KO1, LytASup LytASdn, and LytA-KO4 (Table 1) and fused by PCR to the spectinomycin resistance cassette amplified from the shuttle vector pNE-1. The *lytA* gene was replaced with a spectinomycin cassette following transformation of *S. pneumoniae* strains by standard methods and selected for by plating on blood agar plates supplemented with spectinomycin (500 µg/mL). The T4R Δ*ply* mutant was created using the same method (with primers PlyAF, PlyAR, PlyBF, PlyBR), but replacing the *ply* gene with an erythromycin cassette. Primer sequences for all mutants are listed in Table 1. Complemented mutants in strains TIGR4 and T4R were developed through cloning the *spxB* gene from T4R by

Table 1 Primer sequences

Primer name	Sequence (5'-3')
LytA-KO1	GCGGGTACCCAGTCCAGCTTTGGTTTCCT
LytASup	TAAAAATATCTCTTGCCAGTCCTTGCCTATATGGTTGCACG
LytASdn	GGTAATCAGATTTTAGAAAACAATAAACCCTCACAGTAGAGCCAGAT
LytA-KO4	CGCGGATCCTCACAGTAGAGCCAGATGGC
PlyAF	CTCAATCCAGCTACCTGTCGC
PlyAR	GTTTGCTTCTAAGTCTTATTTCCCTTCTACCTCCTAATAAG
PlyBF	GAGTCGCTTTTGTAAATTTGGGAGAGGAGAATGCTTGCG
PlyBR	GCTTGTTTAGCACGGTCG
ALR1	GCG GGTACCGCGTGCTATTGCAGATCAAA
ALR2	GCGTCTAGACATCGTTAATCGGAGATGGA
ALR3	CGCGGATCCATCTACGCCCCATGTTTTCAATACG
ALR4	CGCGGATCCACCATTCCGTCTCTTCTTGG
ALR5	CGCGGATCCGGAAATAAGACTTAGAAGCAAAC
ALR6	CGCGGATCCCCAAATTTACAAAAGCGACTC
SpxB-F	CGCGCCCGGGTGACAACACTTTCAAAACTG
SpxB-R	CGCGGAATTCTTATTTAATTGCGCGTGATTGC
ERM-F	GGAAATAAGACTTAGAAGCAAAC
ERM-R	CCAAATTTACAAAAGCGACTC
Spec-F	CGTGACTGGCAAGAGATATTTTTA
Spec-R	GGGTTTATTGTTTTCTAAAATCTGATTACC
PLY-F	CAGAGCGTCCTTTGGTCTATATT
PLY-R	CAGCCTCTACTTCATCACTCTTAC

amplifying the gene by PCR with primers SpxB-F and SpxB-R followed by digestion of the product with EcoRI and XmaI enzymes (New England Biolabs, Ipswich, MA). Following digestion of the product, the gene was ligated with the pNE-1 pneumococcal shuttle vector that was similarly digested. The ligation was transformed into *E. coli* strain DH5α. Purified plasmid was then used to transform Δ*spxB* strains by standard methods and complemented mutants were selected by plating on blood agar plates supplemented with erythromycin (0.5 µg/mL) and spectinomycin (500 µg/mL). Complementation of *spxB* was confirmed by PCR and by assaying supernatant from clones by a Pierce Quantitative Peroxide Assay (cat# 23280; Life Technologies).

H_2O_2 quantitation

Pneumococci were grown to mid-log phase (OD$_{600}$ of 0.5). Following centrifugation (16,000 × g for 5 min) of 1 mL of the culture, the supernatant was subsequently filtered through a 0.22 µm polyethersulfone (PES) membrane syringe filter (CellTreat) to remove remaining bacterial cells and analyzed for hydrogen peroxide production using a colorimetric hydrogen peroxide quantification assay per manufacturer instructions in a BioTek Synergy HT plate reader at an OD$_{540}$ (cat# 23280; Life Technologies). This analysis was performed in triplicate.

H_2O_2 Treatment

To ascertain the impact of H_2O_2 alone on the release of PLY, T4R and its isogenic Δ*spxB* mutant were grown to mid log phase (OD$_{600}$ of 0.5) in 5 mL of THY media and treated with either 0 µM or 500 µM H_2O_2. Samples were then placed on ice for 1 h, after which H_2O_2 was quantified following centrifugation of 1 mL of culture at 15,000 rpm for 5 min and filtration of the supernatant through PES (CellTreat) syringe filters as described above. Bacterial counts were determined before and after the H_2O_2 treatment by viable plate counts using blood agar plates.

Western blot

To determine the relative amount of PLY released between parental and mutant strains, sterile-filtered bacterial supernatants from cultures grown to mid-log phase (OD$_{600}$ of 0.5) were denatured by boiling for 5 min and separated on 10 % SDS-PAGE gels (Bio-Rad) prior to being transferred onto a 0.22 nm PVDF membrane (Millipore). The membranes were then blocked for 30 min (5 % milk) and probed with rabbit polyclonal anti-PLY antibody overnight. The blot was subsequently washed with and probed with goat anti-rabbit HRP conjugated secondary antibody (BioRad) for 1 h. Blots were incubated with Luminata Forte substrate (Millipore) for

1 min at 25 °C. The membranes were then developed following exposure to radiography film (GeneMate). Band density was calculated using ImageJ software after scanning film at 600 dpi (NCBI).

Dot blot

To quantitate PLY in the supernatants, recombinant PLY was serially diluted into THY media to create a standard of known concentration (20 ng to 5 μg). PLY standards (20 μL) and bacterial supernatants (100 μL) were applied onto PVDF membranes using a dot blotting vacuum manifold (Bio-Rad). Following 45 min of gravity filtration, light vacuum was applied to adhere the protein to the membrane and the wells were washed twice using 200 μL of phosphate buffered saline (PBS). The membrane was then blocked for 30 min in 5 % milk. The membrane was then probed with an anti-PLY rabbit polyclonal antibody at a 1:200 dilution overnight. Following exposure to the primary antibody, the blot was then probed with a goat anti-rabbit secondary for 1 h. The blot was developed using Luminata Forte substrate (Millipore) and exposed to X-ray film (GeneMate). Dot density was quantitated using ImageJ (NCBI) and then plotted against the known concentration of standards to determine quantity of PLY per 100 μL.

Real time PCR

Pneumolysin gene expression in T4R, WU2, AW267, and their respective isogenic Δ*spxB* mutants was quantitated by qRT-PCR. Bacterial strains were grown to an OD_{600} of 0.5 in THY. Bacterial RNA was extracted following sonication of the bacterial culture, and a total of 4 min of bead beating with 0.1 mm zircon beads. Bacterial RNA was purified using the Qiagen RNeasy kit, with the inclusion of an on column RNase-free DNase treatment for 1 h. Bacterial RNA was quantitated using a Qubit fluorometer and 50 ng was used to generate cDNA utilizing a Maxima cDNA synthesis kit for qRT-PCR (Life Technologies). Gene-specific primer sequences PLY-F and PLY-R are listed in Table 1. The fold-change in gene expression was calculated using the $\Delta\Delta C_T$ method utilizing *gyrA* as an internal control.

Flow cytometry

To assess the effect of bacterial supernatants upon human cells, 1×10^5 A549 epithelial cells were treated with filtered bacterial supernatant obtained from T4R, T4RΔ*spxB*, T4RΔ*ply*, and the complemented T4RΔ*spxB* (Δ*spxB*+). Briefly, epithelial cells were pelleted by centrifugation at 380xg for 4 min, suspended in 0.2 mL of filtered bacterial supernatant, and incubated at 37 °C for 30 min on a rotating platform. Cells were then pelleted at 380 × g for 4 min and suspended in 0.5 mL PBS with 3 μg/mL propidium iodide (PI) (Sigma Aldrich). The cells were assessed using an Attune Flow Cytometer (Life Technologies) and the fluorescence intensity shift in the BL3 channel was measured as an indicator of epithelial cell death.

Hemolysis assay

PLY release was quantitated by standard hemolytic assay. Briefly, bacterial supernatants from cultures grown to an OD_{600} of 0.3 were serially diluted (1:3) across the microtiter plate into phosphate-buffered saline (PBS) containing 0.1 % dithiothreitol (Sigma, St. Louis, MO). Washed 1 % sheep red blood cells in PBS were added to the wells and incubated at 37 °C for 30 min. After incubation, the sheep erythrocytes were pelleted and plates were imaged.

Extracellular DNA quantitation

Bacterial strains were grown in THY medium to an OD_{600} of 0.5 and extracellular DNA was quantitated from sterile filtered supernatant by a Qubit fluorometer using Qubit dsDNA HS Assay Kit (Thermo Fisher Scientific).

Autolysis

S. pneumoniae T4R and the isogenic Δ*spxB* mutant were grown in 12 mL of THY at 37 °C. Upon reaching OD_{600} 0.45, each 12 mL culture was split into three separate tubes. The separate tubes containing mid-log phase bacteria were incubated at room temperature until reaching an OD_{600} of 0.5. Upon reaching the desired OD, Triton X-100 (0.5 %) or sodium deoxycholate (0.05 %) was added to the respective culture tubes and the OD_{600} of the cultures was recorded every minute for a total of 10 min for each of the three culture tubes. Controls received no detergents. Additional control and experimental samples were grown in the presence of catalase (2 μg/mL).

Statistical analysis. All statistical analyses (Student's *t*-test (two-tailed) and linear best-fit regression) were performed using GraphPad Prism software (www.graphpad.com) with a *p*-value < 0.5 considered significant.

Results

Expression of SpxB enhances PLY-dependent cell death

To determine the relative contribution of H_2O_2 and PLY secreted by pneumococcus to cytolysis of airway epithelial cells, A549 human lung epithelial cells were exposed to filter-sterilized supernatants collected from mid-log phase cultures of *S. pneumoniae* and isogenic mutants lacking either *spxB* or *ply* genes and the loss of membrane integrity was assessed by uptake of PI. Compared to THY media alone, exposure to supernatant from the parental strain T4R led to a nearly complete loss of A549 cell viability (89 %), while exposure to supernatant from a strain lacking pyruvate oxidase (T4RΔ*spxB*) resulted in minimal loss in membrane integrity (1.65 %;

Fig. 1a). This effect was shown to be dependent upon release of PLY into the supernatant since supernatant from a strain containing a functional *spxB* gene but lacking the *ply* gene (T4RΔ*ply*) resulted in only 2.52 % of cells staining PI positive. Complementation of *spxB* (TR4Δ*spxB*+) restored the percentage of dead cells to nearly that of A549 cells treated with T4R supernatant (70 % vs. 89 %, respectively). Similar results were obtained using a standard hemolysis assay, with supernatant from strains expressing PLY and SpxB leading to

greater hemolysis than T4RΔ*spxB* (Fig. 1b) Strains lacking both SpxB and PLY had no hemolysis, comparable to T4RΔ*ply* (data not shown). These data demonstrate that expression of SpxB contributes to PLY-dependent cytotoxicity at mid-log phase growth.

SpxB is responsible for the production of H_2O_2 among S. pneumoniae isolates at mid-log phase

S. pneumoniae is known to produce up to millimolar concentrations of H_2O_2 during growth [18]. The enzyme

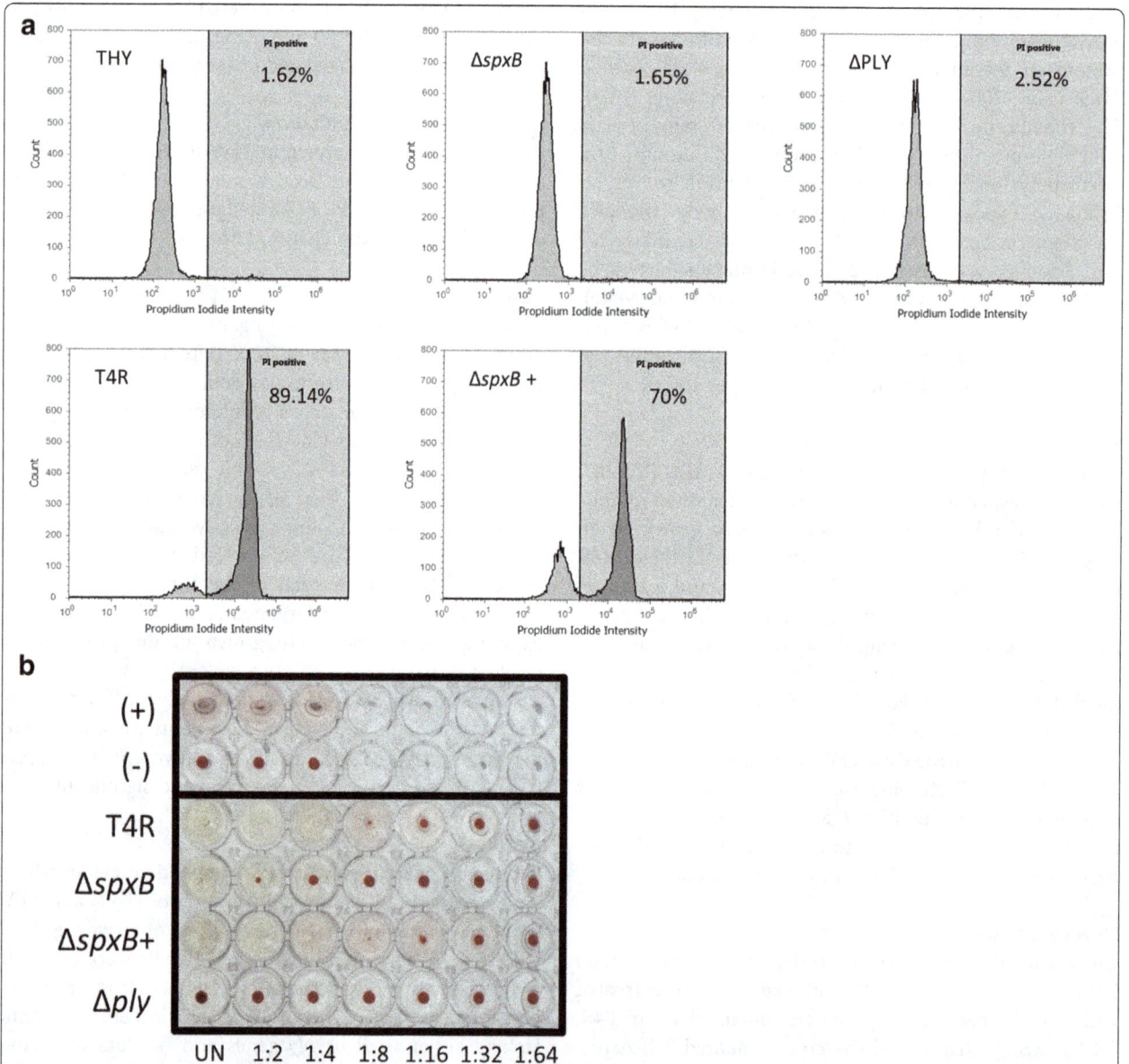

Fig. 1 Loss of SpxB reduces cytotoxic potential of culture supernatants. **a** A549 human epithelial cells were exposed to culture supernatants from parental strain T4R and mutant strains lacking SpxB or PLY. Cytotoxicity was assesed by measuring propidium iodide uptake using flow cytometry. **b** Hemolytic effect of PLY released in culture supernatant. Serially diluted culture supernatants from T4R and mutant strains lacking SpxB were incubated with washed sheep red blood cells for 30 min and subsequently centrifuged. The absence of a red pellet at the bottom of well indicates lysis of RBCs. (+) indicates distilled water positive control for complete hemolysis and (–) received no PLY. Results are representative of duplicate experiments

pyruvate oxidase (SpxB) is known to be responsible for the majority of peroxide production by pneumococcus grown aerobically [21]. However, the enzyme lactate oxidase is also capable of producing H_2O_2 [33]. We assessed the relative production of H_2O_2 among a panel of laboratory strains. Pneumococcal strains T4, T4R, WU2, and AW267 and their isogenic $\Delta spxB$ mutants were grown to mid-log phase (OD_{600} of 0.5) in THY and H_2O_2 was quantitated from filtered supernatants (Additional file 1: Figure S1 A-D). Peroxide concentrations were found to vary between strains. However, the deletion of $spxB$ in all strains examined resulted in a significant reduction of H_2O_2 ($P < 0.005$), which was comparable to supernatants from cultures grown in the presence of catalase. These results demonstrated that SpxB is primarily responsible for the production of the H_2O_2 by the strains examined in this study during mid-log phase growth.

SpxB enhances PLY release among S. pneumoniae isolates

To determine the contribution of SpxB upon PLY release, supernatants from TIGR4, T4R, WU2, and AW267, and their isogenic $\Delta spxB$ mutants, grown to OD_{600} of 0.5, were analyzed by SDS-PAGE and western blot. A significant reduction in the amount of PLY released into the supernatants was observed in the $\Delta spxB$ mutant of each strain examined (Fig. 2 a-d). Differences in PLY release between wild-type and $\Delta spxB$ strains were not due to differences in colony forming units in the mid-log cultures as determined by plate counts (Additional file 2: Figure S2). Interestingly, addition of catalase to the medium only attenuated PLY release in high-releasing strains AW267 and WU2. Since catalase cannot cross the cell membrane and therefore can only neutralize extracellular H_2O_2, it is possible that the strains have different sensitivity to endogenous versus exogenous H_2O_2. Complementation of $spxB$ restored the ability of strains to release significant concentrations of PLY (Fig. 2 e and f).

To determine if PLY release correlated with H_2O_2 production in additional strains, a panel of 15 clinical isolates was analyzed. Each strain was grown to mid-log phase (OD_{600} of 0.5), serial dilutions of the culture were plated, and supernatants were used for and PLY and H_2O_2 quantitation. A dot blot assay was used to quantitate PLY as plotted against known concentrations of recombinant PLY. PLY and H_2O_2 concentrations were normalized to 100,000 cells and a linear scatter plot was generated (Fig. 3). A significant correlation was observed between H_2O_2 production and PLY released among the isolates examined ($r^2 = 0.3167$; $p < 0.05$). This suggests that H_2O_2 production impacts PLY release in clinically-relevant strains.

Effects of SpxB on ply gene expression

It is possible that expression of SpxB affects ply gene expression by metabolically altering the intracellular

environment. To determine if deletion of $spxB$ has a deleterious effect on ply gene transcription, leading to reduced amounts of PLY in the supernatant, qRT-PCR was performed on RNA isolated from strains grown to mid-log phase (Fig. 4). Surprisingly, the transcription of ply was found to be increased in all $\Delta spxB$ mutants analyzed (AW267$\Delta spxB$: 2.83-fold, WU2$\Delta spxB$: 1.96-fold, T4R$\Delta spxB$: 2.3-fold) which was in contrast to PLY release results shown in Fig. 2. Therefore, the reduced PLY release seen in $\Delta spxB$ mutants is not due to reduced transcription of the ply gene.

Impact of exogenous H_2O_2 on PLY release

H_2O_2 is a by-product of SpxB activity and is known to have physiological effects upon pneumococcus. To determine the extent to which exogenous H_2O_2 can impact the release of PLY by inducing bactericidal lysis, T4R$\Delta spxB$ was treated with either 0 μM or 500 μM H_2O_2 on ice for 1 h and the concentration of PLY in the supernatant was determined via dot blot. Interestingly, no significant difference was found between the H_2O_2-treated bacteria and those not receiving H_2O_2 (Fig. 5a). To ensure exposure to H_2O_2 did not impact bacterial survival, and therefore PLY production, bacterial colony forming units were enumerated prior to and after H_2O_2 exposure. Concentrations up to 500 μM did not impact bacterial survival (Additional file 3: Figure S3). Additionally, no significant difference in DNA release was seen with strains lacking $spxB$ (Fig. 5b). These results, combined with the lack of catalase protection in the T4R strain (Fig. 2d), indicate that the contribution of SpxB to PLY release is not due to exogenous effects of H_2O_2 on bacterial viability and that H_2O_2 may exert its effects endogenously as it is being made.

SpxB-dependent PLY release is not due to autolysis

Another potential explanation for the difference in PLY release between the $\Delta spxB$ mutant and the parental strain is that loss of $spxB$ leads to decreased autolysis even prior to stationary phase. Assays for autolysis were performed to determine if this was a contributing factor. Autolysis of S. pneumoniae involves the activity of the cell wall amidase LytA [34]. To determine whether LytA affected SpxB-dependent PLY release, western blots were performed on mid-log (OD_{600} of 0.5) supernatants from parental strains T4R and WU2 as well as isogenic mutants of both strains lacking either $lytA$ or $spxB$, or both (Fig. 6a). Mutants lacking $spxB$, as expected, released less PLY, however loss of $lytA$ had no significant effect on PLY release. These results indicate that the major autolysin, LytA, is not contributing to SpxB-induced PLY release at mid-log phase growth, but does not rule out its contribution at later times, such as during stationary phase.

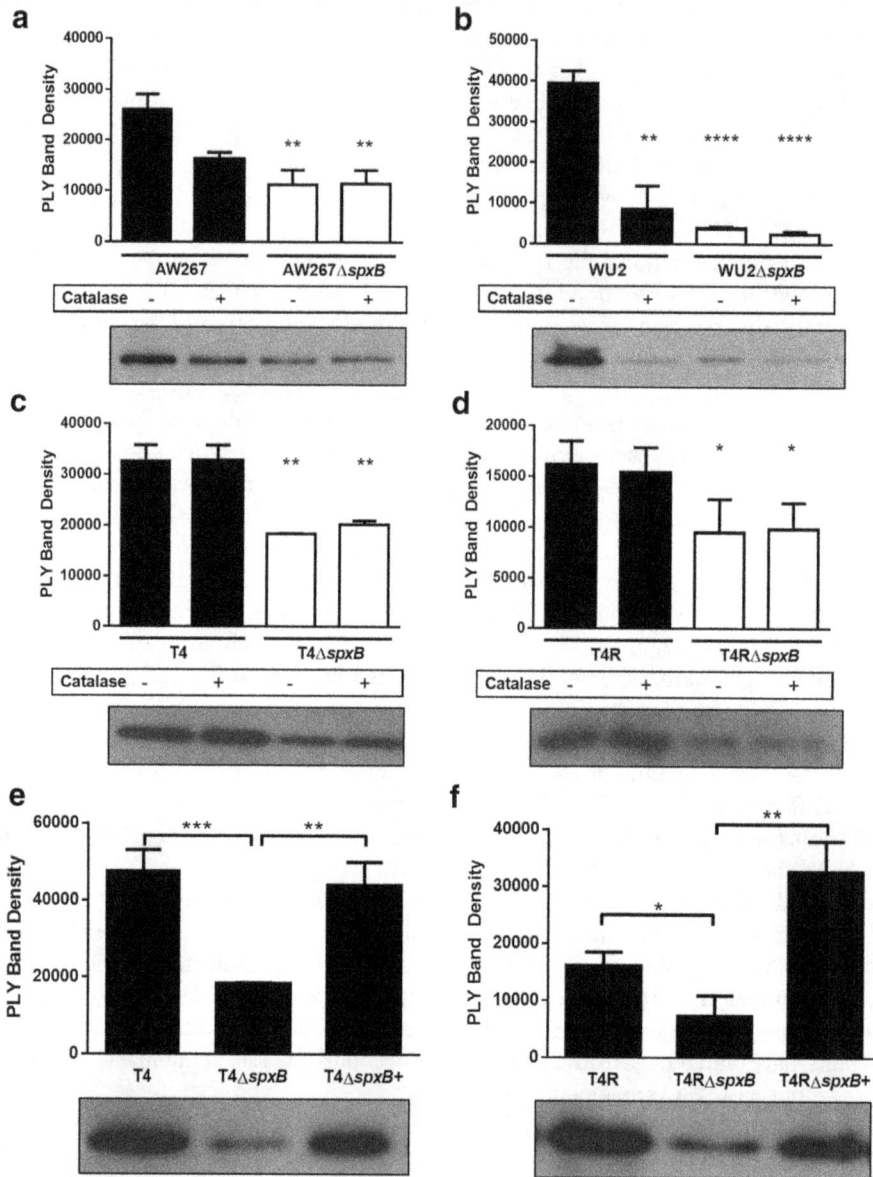

Fig. 2 Deletion of *spxB* greatly reduces PLY observed in the supernatant. **a-d** The amount of PLY released into culture supernatants from parental and mutant strains lacking SpxB (**a** AW267; **b** WU2, **c** T4, and **d** T4R) was measured by western blot. Band intensities were quantitated by densitometry. Results are representative of three independent experiments ± SD. (* $p < 0.05$, ** $p < 0.005$, **** $P < 0.00005$). Where indicated, the cultures were complemented with 10 ng of catalase. **e** and **f** The amount of PLY released in culture supernatants from parental, mutant strains lacking SpxB, and complemented strains (**e** T4; **f** T4R) was measured by western blot. Band intensities were quantitated by densitometry. Results are representative of three independent experiments ± SD. (* $p < 0.05$, ** $p < 0.005$, *** $P < 0.0005$). Representative images of western blots are shown

It is possible that H_2O_2 or another by-product of SpxB activity could weaken the bacterial cell membrane, thereby facilitating PLY release. To determine if SpxB expression affected detergent-induced lysis, T4R and its isogenic Δ*spxB* mutant were grown to an OD_{600} of 0.5 and then treated with the ionic detergent sodium deoxycholate (DOC). As expected, addition of DOC to midlog phase cultures initiated immediate lysis of both T4R and Δ*spxB* strains. However, wild type T4R lysed more

rapidly and to a greater extent than T4RΔ*spxB*. Supplementing the medium with catalase reduced lysis to levels comparable to T4RΔ*spxB*, indicating that H_2O_2 production may induce a condition that favors progression of lysis.

Discussion

Pneumolysin is a major virulence factor of pneumococcus and its release into the extracellular space has been shown to vary greatly between strains [9, 35]. However,

H₂O₂ vs PLY Released

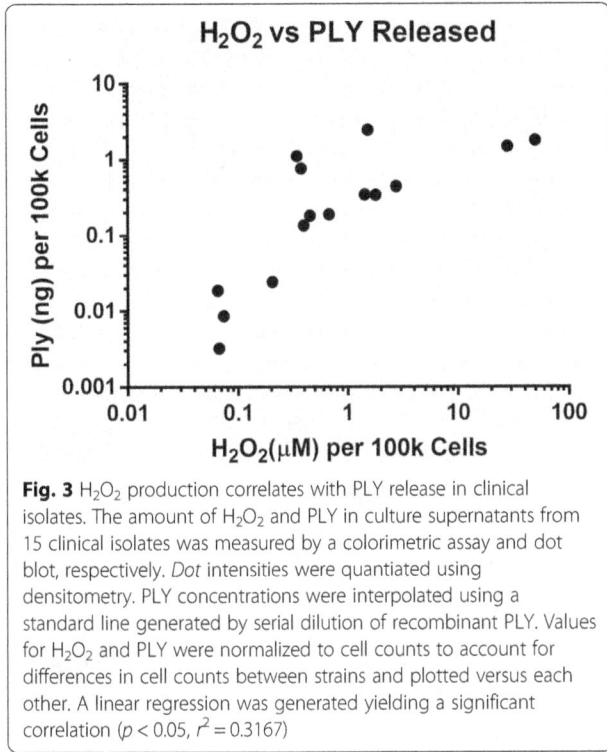

Fig. 3 H_2O_2 production correlates with PLY release in clinical isolates. The amount of H_2O_2 and PLY in culture supernatants from 15 clinical isolates was measured by a colorimetric assay and dot blot, respectively. *Dot* intensities were quantiated using densitometry. PLY concentrations were interpolated using a standard line generated by serial dilution of recombinant PLY. Values for H_2O_2 and PLY were normalized to cell counts to account for differences in cell counts between strains and plotted versus each other. A linear regression was generated yielding a significant correlation ($p < 0.05$, $r^2 = 0.3167$)

no single canonical mechanism of PLY release has been identified [8]. In this work, we have identified a novel link between expression of the metabolic enzyme pyruvate oxidase and release of PLY. Initially, we sought to determine the relative contribution of secreted H_2O_2 and PLY to host epithelial cytotoxicity. However, upon exposing A549 epithelial cells to pneumococcal supernatants, we found that supernatants from strains lacking

SpxB showed significantly decreased cytotoxicity, due to reduced extracellular PLY release. Since H_2O_2 has been shown to be the primary cause of pneumococcal autolysis during stationary phase [27], we initially suspected the loss of viability by H_2O_2 produced by SpxB might cause the enhanced release of PLY in culture supernatant. However, there was no significant difference in colony forming units between parental and *spxB* mutants. Additionally, there was no significant difference in extracellular DNA concentrations between parental and $\Delta spxB$ mutants. Together, these results suggest that enhanced release of PLY in parental strains during mid-log phase growth was not due to a bactericidal effect of H_2O_2 produced by SpxB. While pneumococcal H_2O_2 is known to have cytotoxic effects on host cells [24, 26], our results indicate that the loss of host cell integrity is primarily linked to PLY released into the supernatant, not to H_2O_2. However, it is possible that SpxB-dependent H_2O_2 induces stress within epithelial cells that may lead to alternative, non-cytolytic death processes. For instance, it has been shown that SpxB-induced H_2O_2 triggers genotoxicity and conserved stress responses within the epithelium [24, 36].

While H_2O_2-induced bactericidal effects do not appear to be the mechanism for PLY release, our findings do indicate that release may be linked to the production of H_2O_2, as indicated by the correlation between PLY release and H_2O_2 production in our clinical isolate panel. However, SpxB enzymatic activity produces other byproducts, including acetate and acetylphosphate that could potentially contribute to this mechanism of release [20]. Acetyl-phopshate contributes significantly to the ATP pool of pneumococcus [19]. We are currently

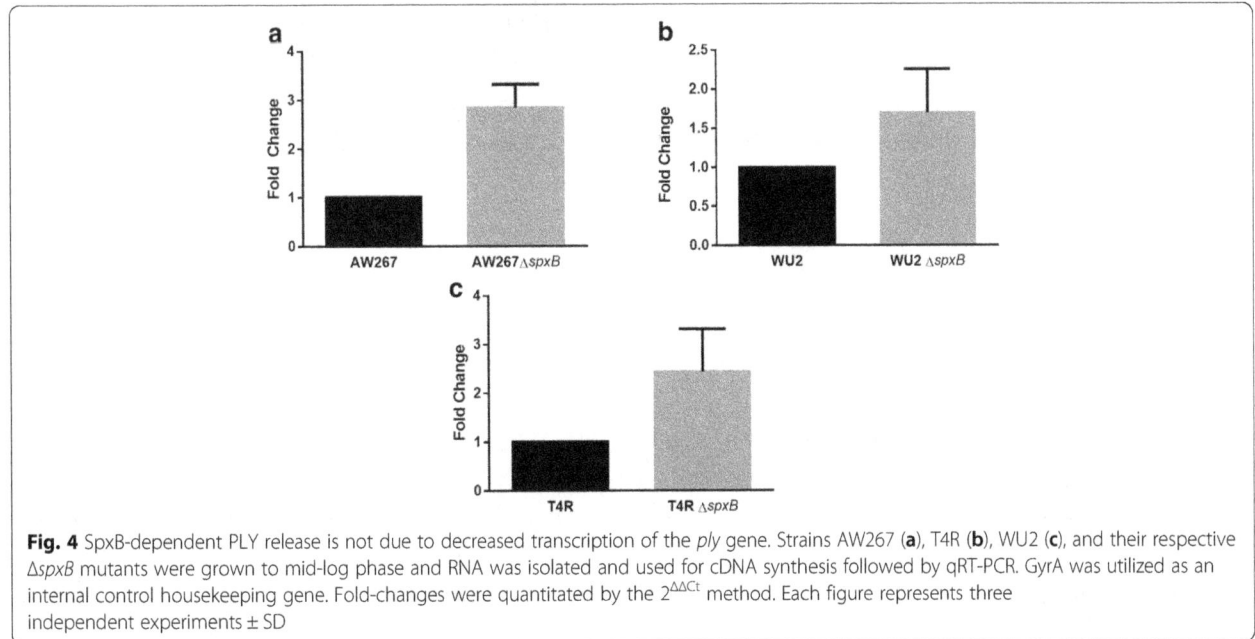

Fig. 4 SpxB-dependent PLY release is not due to decreased transcription of the *ply* gene. Strains AW267 (**a**), T4R (**b**), WU2 (**c**), and their respective $\Delta spxB$ mutants were grown to mid-log phase and RNA was isolated and used for cDNA synthesis followed by qRT-PCR. GyrA was utilized as an internal control housekeeping gene. Fold-changes were quantitated by the $2^{\Delta\Delta Ct}$ method. Each figure represents three independent experiments ± SD

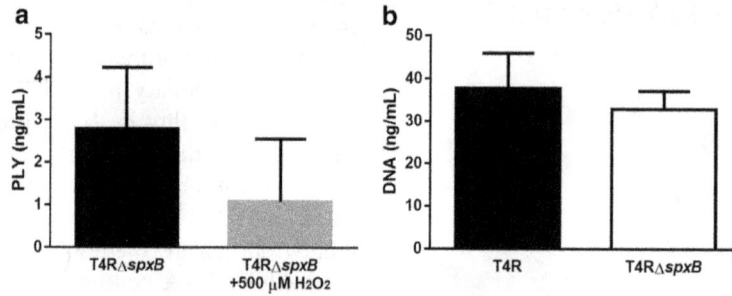

Fig. 5 Exogenous H_2O_2 does not induce PLY release. **a** The ΔSpxB mutant of T4R was grown to an OD of 0.5. The bacterial cells treated with 500 μM H_2O_2 for 1 h and PLY concentrations were quantitated via dot blot. Each figure represents three independent experiments ± SD. **b** T4R and its ΔspxB mutant were grown to an OD$_{600}$ of 0.5 and extracellular DNA was quantitated from sterile filtered supernatant by a Qbit fluorometer. Each figure represents three independent experiments ± SD

investigating whether PLY release is energy-dependent. Furthermore, the addition of catalase to the culture medium abrogated PLY release from some strains (WU2), while failing to prevent release from others (TIGR4). These results suggest that strain-specific differences in metabolism might impact the release of PLY. We are also currently investigating how the capsule genes possessed by different serotypes may alter their metabolic processes and thereby SpxB activity and subsequent PLY release. However, if H_2O_2 serves to release PLY, it appears that it may act from within the bacterial cell, as supplementation of exogenous H_2O_2, surprisingly, did not enhance PLY release, at least in the T4R strain. Also, while both *spxB* and *ply* are known to be differentially expressed in different biological niches during infection [37], *ply* was not found to be expressed greater in parental strains as compared to ΔspxB strains.

Our results indicate that the PLY release mechanism appears to be autolysis-independent, as mutants lacking the major autolytic cell wall amidase LytA were equally able to release PLY as parental strains. Furthermore, double mutants lacking both SpxB and LytA released

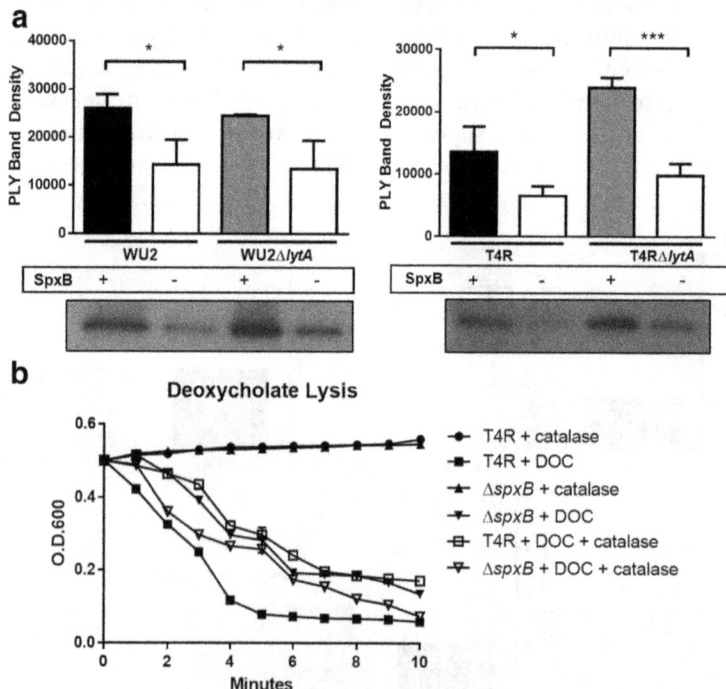

Fig. 6 Effects of SpxB upon detergent-induced lysis. **a** WU2 and T4R along with their ΔspxB, ΔlytA, and ΔspxB/ΔlytA mutants were grown to an OD of 0.5. Equal volumes of sterile supernatants were run on SDS-PAGE gels and probed for PLY via western blot. Each figure represents three independent experiments ± SD. (*$p < 0.05$). A representative western blot image for each strain is shown. **b** T4R and its ΔspxB mutant were grown to an OD of 0.5 prior to treatment with 0.05 % sodium deoxycholate (DOC). Spectrophotometric readings were taken every minute for 10 min to determine differences in the rate of induced autolysis

equal PLY to strains lacking only SpxB. However, it is possible that additional autolytic factors including the murein hydrolase LytB, competence induced bacteriocin (CibAB), or choline binding protein D (CbpD) could play a role in *spxB*-dependent PLY release [38–40]. Results from exposure of the parental strain T4R to deoxycholate indicate that *spxB* expression may predispose the bacterial cells to autolysis. This indicates that H_2O_2 or some other SpxB-induced byproduct could weaken the bacterial cell membrane allowing for release of PLY from the intracellular space, possibly in the absence of cell death. However, loss of cell membrane integrity would likely free intracellular stores of LytA which could trigger autolytic effects not seen in our studies [41, 42]. An intriguing possibility is that PLY possesses characteristics that allow it to preferentially traverse the SpxB-perturbed membrane. We are currently investigating whether other cryptically secreted proteins may escape the cytosol via the SpxB-dependent mechanism used by PLY.

We have yet to determine is whether or not H_2O_2 produced by SpxB impacts pneumococcal activators or repressors, thereby indirectly affecting expression of genes that could impact PLY release. In *Staphylococcus aureus*, exposure to exogenous H_2O_2 enhanced expression of multiple oxidative stress response genes controlled by the ferric uptake regulator (Fur) homolog PerR [43]. While pneumococcus doesn't possess PerR, the iron uptake regulator RitR regulates expression of a number of genes which are impacted by both iron availability and oxidative stress [44]. The MerR regulators of pneumococcus are another group of regulatory proteins that can be impacted by oxidative stress [45].

Since SpxB is an important metabolic enzyme for the organism during aerobic growth, our findings could link cytotoxic potential of strains to metabolic shifts that occur at different niches throughout the host during colonization or progression to invasive disease. This mechanism could constitute another reason for the attenuated virulence seen with *spxB*-negative mutants [21]. However, it was recently demonstrated that increased PLY release does not necessarily translate to enhanced virulence [16]. This underlines SpxB as an important virulence regulator, as our data indicate that SpxB expression contributes significantly to the release of PLY at early phases of growth, along with previous data indicating that it allows the organism to outcompete other common inhabitants of the nasopharynx [22]. These findings could represent a novel method for protein secretion that extends beyond pneumococcus to other bacterial species.

Conclusions

Expression of SpxB was shown to contribute to PLY release in various *S. pneumoniae* strains. While a panel of

clinical isolates demonstrated a correlation between H_2O_2 production and PLY release, catalase was able to prevent PLY release in a strain-dependent manner. This indicates endogenous/exogenous H_2O_2 may contribute in a strain-dependent fashion to SpxB-dependent PLY release. PLY release due to SpxB was not dependent upon cellular turnover based on plate counts and DNA release. Though SpxB-dependent PLY release was not dependent on the activity of the cell wall amidase LytA, deoxycholate-induced autolysis was greater in strains expressing SpxB, indicating possible weakening of the cell membrane when SpxB is expressed. These results identify a novel route of PLY release that may extend to secretion of other cytoplasmic proteins that lack signal sequences.

Additional files

Additional file 1: Figure S1. Deletion of *spxB* greatly reduces production of hydrogen peroxide. The amount of hydrogen peroxide in culture supernatants from wild type and mutant strains lacking SpxB (A. AW267; B. WU2; C. T4, D. T4R) was measured via a colorimetric peroxide assay. Asterisks indicate statistical significance (**** $p < 0.00005$, ** $p < 0.005$) and ND indicates a peroxide concentration below the detectable limits of the assay.

Additional file 2: Figure S2. Bacterial colony counts. Serial dilution plate counts were made at OD 0.5 of the indicated strains to determine if there was a difference in the bacterial counts when SpxB is removed, or when catalase is added. Each figure represents three independent experiments ± SD.

Additional file 3: Figure S3. Bacterial colony counts. Serial dilution plate counts were made from cultures grown to OD_{600} 0.5 prior to and 1 h post-exposure to various concentrations of H_2O_2. Results are shown as the average of 3 independent experiments ± SD.

Abbreviations
H_2O_2: Hydrogen peroxide; LytA: Autolysin; PLY: Pneumolysin; SpxB: Pyruvate oxidase

Acknowledgements
We would like to thank Mary Marquart for providing rabbit polyclonal anti-PLY antibody.

Funding
JAT's lab is supported by an Institutional Development Award (IDeA) from the NIGMS COBRE grant number (P20GM103646), the MSU Department of Biological Sciences, and the MSU Office of Research and Economic Development.

Authors' contributions
JB participated in design and completion of all experiments in the study and drafted the manuscript. RD performed autolysis assays and western blots. KO and LB performed molecular genetic studies and generated knockout strains. JR, KS, JD, and LM made substantial contributions to the conception and design of experiments. JT designed and coordinated all studies and contributed to drafting the manuscript. All authors read and approved the final manuscript.

Competing interests
The authors declare that they have no competing interests.

Author details
[1]Department of Biological Sciences, Mississippi State University, 295 E Lee Blvd., Harned Hall, Rm 219, Mississippi State, MS 39762, USA. [2]Department of Infectious Diseases, St. Jude Children's Research Hospital, Memphis, TN, USA. [3]Department of Basic Sciences, College of Veterinary Medicine, Mississippi State University, Mississippi State, MS, USA. [4]Department of Biological Sciences, University of Southern Mississippi, Hattiesburg, MS, USA. [5]Department of Microbiology and Immunology, University of Mississippi Medical Center, Jackson, MS, USA.

References
1. Bogaert D, De Groot R, Hermans PW. Streptococcus pneumoniae colonisation: the key to pneumococcal disease. Lancet Infect Dis. 2004;4(3):144–54.
2. O'Brien KL, Wolfson LJ, Watt JP, Henkle E, Deloria-Knoll M, McCall N, Lee E, Mulholland K, Levine OS, Cherian T. Burden of disease caused by Streptococcus pneumoniae in children younger than 5 years: global estimates. Lancet. 2009;374(9693):893–902.
3. Marcy M, Takata G, Chan LS, Shekelle P, Mason W, Wachsman L, Ernst R, Hay JW, Corley PM, Morphew T, et al. Management of acute otitis media. Evid Rep Technol Assess (Summ). 2000;15:1–4.
4. Le Polain de Waroux O, Flasche S, Prieto-Merino D, Edmunds WJ. Age-dependent prevalence of nasopharyngeal carriage of streptococcus pneumoniae before conjugate vaccine introduction: a prediction model based on a meta-analysis. PLoS One. 2014;9(1):e86136.
5. Berry AM, Lock RA, Thomas SM, Rajan DP, Hansman D, Paton JC. Cloning and nucleotide sequence of the Streptococcus pneumoniae hyaluronidase gene and purification of the enzyme from recombinant Escherichia coli. Infect Immun. 1994;62(3):1101–8.
6. Briles DE, Yother J, McDaniel LS. Role of pneumococcal surface protein A in the virulence of Streptococcus pneumoniae. Rev Infect Dis. 1988;10 Suppl 2:S372–4.
7. Camara M, Boulnois GJ, Andrew PW, Mitchell TJ. A neuraminidase from Streptococcus pneumoniae has the features of a surface protein. Infect Immun. 1994;62(9):3688–95.
8. Mitchell TJ, Dalziel CE. The biology of pneumolysin. Subcell Biochem. 2014;80:145–60.
9. Marriott HM, Jackson LE, Wilkinson TS, Simpson AJ, Mitchell TJ, Buttle DJ, Cross SS, Ince PG, Hellewell PG, Whyte MK, et al. Reactive oxygen species regulate neutrophil recruitment and survival in pneumococcal pneumonia. Am J Respir Crit Care Med. 2008;177(8):887–95.
10. Rogers PD, Thornton J, Barker KS, McDaniel DO, Sacks GS, Swiatlo E, McDaniel LS. Pneumolysin-dependent and -independent gene expression identified by cDNA microarray analysis of THP-1 human mononuclear cells stimulated by Streptococcus pneumoniae. Infect Immun. 2003;71(4):2087–94.
11. Walker JA, Allen RL, Falmagne P, Johnson MK, Boulnois GJ. Molecular cloning, characterization, and complete nucleotide sequence of the gene for pneumolysin, the sulfhydryl-activated toxin of Streptococcus pneumoniae. Infect Immun. 1987;55(5):1184–9.
12. Price KE, Greene NG, Camilli A. Export requirements of pneumolysin in Streptococcus pneumoniae. J Bacteriol. 2012;194(14):3651–60.
13. Mitchell TJ, Alexander JE, Morgan PJ, Andrew PW. Molecular analysis of virulence factors of Streptococcus pneumoniae. Soc Appl Bacteriol Symp Ser. 1997;26:62S–71.
14. Balachandran P, Hollingshead SK, Paton JC, Briles DE. The autolytic enzyme LytA of Streptococcus pneumoniae is not responsible for releasing pneumolysin. J Bacteriol. 2001;183(10):3108–16.
15. Price KE, Camilli A. Pneumolysin localizes to the cell wall of Streptococcus pneumoniae. J Bacteriol. 2009;191(7):2163–8.
16. Greene NG, Narciso AR, Filipe SR, Camilli A. Peptidoglycan Branched Stem Peptides Contribute to Streptococcus pneumoniae Virulence by Inhibiting Pneumolysin Release. PLoS Pathog. 2015;11(6):e1004996.
17. Marriott HM, Mitchell TJ, Dockrell DH. Pneumolysin: a double-edged sword during the host-pathogen interaction. Curr Mol Med. 2008;8(6):497–509.
18. McLeod JW, Gordon J. Production of Hydrogen Peroxide by Bacteria. Biochem J. 1922;16(4):499–506.
19. Pericone CD, Park S, Imlay JA, Weiser JN. Factors contributing to hydrogen peroxide resistance in Streptococcus pneumoniae include pyruvate oxidase (SpxB) and avoidance of the toxic effects of the fenton reaction. J Bacteriol. 2003;185(23):6815–25.
20. Taniai H, Iida K, Seki M, Saito M, Shiota S, Nakayama H, Yoshida S. Concerted action of lactate oxidase and pyruvate oxidase in aerobic growth of Streptococcus pneumoniae: role of lactate as an energy source. J Bacteriol. 2008;190(10):3572–9.
21. Spellerberg B, Cundell DR, Sandros J, Pearce BJ, Idanpaan-Heikkila I, Rosenow C, Masure HR. Pyruvate oxidase, as a determinant of virulence in Streptococcus pneumoniae. Mol Microbiol. 1996;19(4):803–13.
22. Pericone CD, Overweg K, Hermans PW, Weiser JN. Inhibitory and bactericidal effects of hydrogen peroxide production by Streptococcus pneumoniae on other inhabitants of the upper respiratory tract. Infect Immun. 2000;68(7):3990–7.
23. Braun JS, Sublett JE, Freyer D, Mitchell TJ, Cleveland JL, Tuomanen EI, Weber JR. Pneumococcal pneumolysin and H(2)O(2) mediate brain cell apoptosis during meningitis. J Clin Invest. 2002;109(1):19–27.
24. Rai P, Parrish M, Tay IJ, Li N, Ackerman S, He F, Kwang J, Chow VT, Engelward BP. Streptococcus pneumoniae secretes hydrogen peroxide leading to DNA damage and apoptosis in lung cells. Proc Natl Acad Sci U S A. 2015;112(26):E3421–30.
25. Rubins JB, Duane PG, Charboneau D, Janoff EN. Toxicity of pneumolysin to pulmonary endothelial cells in vitro. Infect Immun. 1992;60(5):1740–6.
26. Zysk G, Bejo L, Schneider-Wald BK, Nau R, Heinz H. Induction of necrosis and apoptosis of neutrophil granulocytes by Streptococcus pneumoniae. Clin Exp Immunol. 2000;122(1):61–6.
27. Regev-Yochay G, Trzcinski K, Thompson CM, Lipsitch M, Malley R. SpxB is a suicide gene of Streptococcus pneumoniae and confers a selective advantage in an in vivo competitive colonization model. J Bacteriol. 2007;189(18):6532–9.
28. Pesakhov S, Benisty R, Sikron N, Cohen Z, Gomelsky P, Khozin-Goldberg I, Dagan R, Porat N. Effect of hydrogen peroxide production and the Fenton reaction on membrane composition of Streptococcus pneumoniae. Biochim Biophys Acta. 2007;1768(3):590–7.
29. Tettelin H, Nelson KE, Paulsen IT, Eisen JA, Read TD, Peterson S, Heidelberg J, DeBoy RT, Haft DH, Dodson RJ, et al. Complete genome sequence of a virulent isolate of Streptococcus pneumoniae. Science. 2001;293(5529):498–506.
30. Fernebro J, Andersson I, Sublett J, Morfeldt E, Novak R, Tuomanen E, Normark S, Normark BH. Capsular expression in Streptococcus pneumoniae negatively affects spontaneous and antibiotic-induced lysis and contributes to antibiotic tolerance. J Infect Dis. 2004;189(2):328–38.
31. Briles DE, Nahm M, Schroer K, Davie J, Baker P, Kearney J, Barletta R. Antiphosphocholine antibodies found in normal mouse serum are protective against intravenous infection with type 3 streptococcus pneumoniae. J Exp Med. 1981;153(3):694–705.
32. Thornton JA. Splicing by Overlap Extension PCR to Obtain Hybrid DNA Products. Methods Mol Biol. 2016;1373:43–9.
33. Udaka S, Koukol J, Vennesland B. Lactic oxidase of Pneumococcus. J Bacteriol. 1959;78:714–25.
34. Ronda C, Garcia JL, Garcia E, Sanchez-Puelles JM, Lopez R. Biological role of the pneumococcal amidase. Cloning of the lytA gene in Streptococcus pneumoniae. Eur J Biochem. 1987;164(3):621–4.
35. Benton KA, Paton JC, Briles DE. Differences in virulence for mice among Streptococcus pneumoniae strains of capsular types 2, 3, 4, 5, and 6 are not attributable to differences in pneumolysin production. Infect Immun. 1997;65(4):1237–44.
36. Loose M, Hudel M, Zimmer KP, Garcia E, Hammerschmidt S, Lucas R, Chakraborty T, Pillich H. Pneumococcal hydrogen peroxide-induced stress signaling regulates inflammatory genes. J Infect Dis. 2015;211(2):306–16.
37. Orihuela CJ, Radin JN, Sublett JE, Gao G, Kaushal D, Tuomanen EI. Microarray analysis of pneumococcal gene expression during invasive disease. Infect Immun. 2004;72(10):5582–96.
38. Garcia P, Gonzalez MP, Garcia E, Lopez R, Garcia JL. LytB, a novel pneumococcal murein hydrolase essential for cell separation. Mol Microbiol. 1999;31(4):1275–81.
39. Kausmally L, Johnsborg O, Lunde M, Knutsen E, Havarstein LS. Choline-binding protein D (CbpD) in Streptococcus pneumoniae is essential for competence-induced cell lysis. J Bacteriol. 2005;187(13):4338–45.
40. Guiral S, Mitchell TJ, Martin B, Claverys JP. Competence-programmed predation of noncompetent cells in the human pathogen Streptococcus pneumoniae: genetic requirements. Proc Natl Acad Sci U S A. 2005;102(24):8710–5.

41. Mellroth P, Sandalova T, Kikhney A, Vilaplana F, Hesek D, Lee M, Mobashery S, Normark S, Svergun D, Henriques-Normark B, et al. Structural and functional insights into peptidoglycan access for the lytic amidase LytA of Streptococcus pneumoniae. MBio. 2014;5(1):e01120–01113.

42. Mellroth P, Daniels R, Eberhardt A, Ronnlund D, Blom H, Widengren J, Normark S, Henriques-Normark B. LytA, major autolysin of Streptococcus pneumoniae, requires access to nascent peptidoglycan. J Biol Chem. 2012;287(14):11018–29.

43. Horsburgh MJ, Clements MO, Crossley H, Ingham E, Foster SJ. PerR controls oxidative stress resistance and iron storage proteins and is required for virulence in Staphylococcus aureus. Infect Immun. 2001;69(6):3744–54.

44. Ulijasz AT, Andes DR, Glasner JD, Weisblum B. Regulation of iron transport in Streptococcus pneumoniae by RitR, an orphan response regulator. J Bacteriol. 2004;186(23):8123–36.

45. Brown NL, Stoyanov JV, Kidd SP, Hobman JL. The MerR family of transcriptional regulators. FEMS Microbiol Rev. 2003;27(2–3):145–63.

7

Trichoderma virens β-glucosidase I (*BGLI*) gene; expression in *Saccharomyces cerevisiae* including docking and molecular dynamics studies

Gammadde Hewa Ishan Maduka Wickramasinghe[1],
Pilimathalawe Panditharathna Attanayake Mudiyanselage Samith Indika Rathnayake[1],
Naduviladath Vishvanath Chandrasekharan[1], Mahindagoda Siril Samantha Weerasinghe[1],
Ravindra Lakshman Chundananda Wijesundera[2] and Wijepurage Sandhya Sulochana Wijesundera[3*]

Abstract

Background: Cellulose, a linear polymer of β 1–4, linked glucose, is the most abundant renewable fraction of plant biomass (lignocellulose). It is synergistically converted to glucose by endoglucanase (EG) cellobiohydrolase (CBH) and β-glucosidase (BGL) of the cellulase complex. BGL plays a major role in the conversion of randomly cleaved cellooligosaccharides into glucose. As it is well known, *Saccharomyces cerevisiae* can efficiently convert glucose into ethanol under anaerobic conditions. Therefore, *S.cerevisiae* was genetically modified with the objective of heterologous extracellular expression of the *BGLI* gene of *Trichoderma virens* making it capable of utilizing cellobiose to produce ethanol.

Results: The cDNA and a genomic sequence of the *BGLI* gene of *Trichoderma virens* was cloned in the yeast expression vector pGAPZα and separately transformed to *Saccharomyces cerevisiae*. The size of the *BGLI* cDNA clone was 1363 bp and the genomic DNA clone contained an additional 76 bp single intron following the first exon. The gene was 90% similar to the DNA sequence and 99% similar to the deduced amino acid sequence of 1,4-β-D-glucosidase of *T. atroviride* (AC237343.1). The BGLI activity expressed by the recombinant genomic clone was 3.4 times greater (1.7×10^{-3} IU ml^{-1}) than that observed for the cDNA clone (5×10^{-4} IU ml^{-1}). Furthermore, the activity was similar to the activity of locally isolated *Trichoderma virens* (1.5×10^{-3} IU ml^{-1}). The estimated size of the protein was 52 kDA. In fermentation studies, the maximum ethanol production by the genomic and the cDNA clones were 0.36 g and 0.06 g /g of cellobiose respectively. Molecular docking results indicated that the bare protein and cellobiose-protein complex behave in a similar manner with considerable stability in aqueous medium. The deduced binding site and the binding affinity of the constructed homology model appeared to be reasonable. Moreover, it was identified that the five hydrogen bonds formed between the amino acid residues of BGLI and cellobiose are mainly involved in the integrity of enzyme-substrate association.

(Continued on next page)

* Correspondence: sulochana@bmb.cmb.ac.lk
[3]Department of Biochemistry and Molecular Biology, Faculty of Medicine, University of Colombo, Kynsey Road, Colombo 08, Sri Lanka
Full list of author information is available at the end of the article

(Continued from previous page)

Conclusions: The BGLI activity was remarkably higher in the genomic DNA clone compared to the cDNA clone. Cellobiose was successfully fermented into ethanol by the recombinant *S.cerevisiae* genomic DNA clone. It has the potential to be used in the industrial production of ethanol as it is capable of simultaneous saccharification and fermentation of cellobiose. Homology modeling, docking studies and molecular dynamics simulation studies will provide a realistic model for further studies in the modification of active site residues which could be followed by mutation studies to improve the catalytic action of BGLI.

Keywords: Lignocellulose, β-glucosidase, Recombinant *S.cerevisiae*, Molecular docking, Molecular dynamics simulations, Homology modeling, Simultaneous saccharification and fermentation

Background

The conversion of cellulose and hemicellulose of lignocellulosic biomass into ethanol is a promising solution to the anticipated future fuel crisis. The sugar monomers of these two major components of plant biomass can be fermented to ethanol [1]. Therefore, second generation biofuel production based on enzymatic conversion of cellulosic biomass into ethanol was selected as a key area in the development of renewable energy technology in the 1980s [2, 3].

Cellulose consists mainly of long polymers of β 1–4, linked glucose units [4]. Cellulase is an enzyme complex consisting of endoglucanase (endo-1,4-β-D-glucanase, EGL, EC 3.2.1.4); cellobiohydrolase or exoglucanase (exo-1,4-β-D-glucanase, CBH, EC 3.2.1.91) and β-glucosidase (1,4-β-D-glucosidase, BGL, EC 3.2.1.21) that act synergistically to convert cellulose to glucose [5, 6]. The accumulation of cellooligosaccharides (cellobiose and cellotriose) inhibits the function of both CBH and EGL enzymes in simultaneous saccharification and therefore BGL plays a major role in the efficient conversion of the randomly cleaved inhibitory form of cellooligosaccharides into utilizable non-inhibitory glucose units [7].

Cellulases are produced by a variety of microorganisms including filamentous fungi and bacteria [8, 9]. Filamentous fungi are naturally excellent protein secretors and produce industrially important enzymes in feasible amounts [10]. Therefore, much attention has been given to fungal cellulolytic systems over those of bacteria [11]. It is reported that fungal strains secrete higher amounts of cellulases than bacterial species. The fungal species *Trichoderma* has been studied extensively as it is known to produce higher amounts of cellulases compared to other fungal species [12–14].

Saccharomyces cerevisiae can efficiently convert glucose into ethanol under anaerobic conditions [15]. It has the ability to propagate rapidly by budding and fission. This inherent phenomenon can be employed as a conclusive tool for successive utilization and subsequent fermentation of lignocelluloses into ethanol by expressing cellulase in *Saccharomyces cerevisiae* [16]. BGL has been expressed in *Saccharomyces cerevisiae* by many researchers with the objective of producing ethanol from lignocellulosic biomass [17, 18]. However, in many previous studies the expression levels have been reported to be considerably lower than the native host species, probably due to incompatibility of the signal peptide and the promoter of the yeast expression system [17, 19, 20]. The present study describes the characterization, cloning and expression of both genomic and cDNA clones of *BGLI* in *S. cerevisiae* from locally isolated *Trichoderma virens* using yeast integrative vector pGAPZα. The vector consists of the α-mating factor (MFα) signal sequence with the glyceraldehyde 3-phosphate dehydrogenase (GAP) promoter driven expression system [21–23]. The expression of both genomic and cDNA clones of *BGLI* by recombinant *S.cerevisiae* was compared with the objective of analyzing the effect of introns on expression in the eukaryotic system as the presence of introns have been known to enhance expression. Absence of a three-dimensional (3D) structure is a limitation in understanding the structural features and properties of BGLI. Therefore, a 3D structure of BGLI was built using homology modeling with quality assessments. Molecular dynamics (MD) simulations and protein docking studies were carried out to investigate the docking of substrates to the catalytic site of the enzyme and to study how enzyme dynamics are affected by cellulose. MD simulation is one of the most important tools for the computational study of bio-molecules. It provides the time-dependent behavior of the bio-molecule as well as comprehensive information of the conformational changes in the molecular system [24].

Methods

β-glucosidase (BGLI) activity assay of *Trichoderma virens*

A phenotypically characterized, local *Trichoderma* isolate was subjected to PCR based internal transcribed spacer (ITS) analysis and confirmed as *Trichoderma virens*. The cultures for the *BGLI* enzyme assay was prepared by placing a 5 mm diameter mycelium disc removed from a 7 day old culture and inoculated into conical flasks (100 mL) containing 25 mL of Mandel's medium (MM) [25] supplemented with 2% cellobiose as the sole carbon source [26]. Cultures were incubated at

30 °C, in a rotary shaker at 150 rpm and the enzyme extracts were harvested by filtration at 24 h, 48 h, 72 h, 96 h, 168 h, 192 h and 216 h intervals. Cell free enzyme extracts were obtained by centrifugation at 6200 g at 4 ° C for 10 min and then freeze dried. A 200 µL volume of the enzyme extract was mixed with cellobiose 2.8 mL (15 mM) as the substrate in a citrate buffer (50 mM, pH 4.8). Thereafter, the reaction mixture was incubated at 50 °C in a water bath for 15 min followed by addition of 3 mL of 3,5-dinitrosalicyclic acid (DNS) solution into each reaction mixture and placed them in a boiling water bath for 15 min to develop the colour. The colour intensity was measured at 540 nm [27]. The enzyme activity was calculated with reference to the glucose standard graph and all experiments were performed in triplicate. One unit of BGL activity is defined as 1 µmol of glucose produced from cellobiose in 1 mL of enzyme volume per second ($µmol\ ml^{-1}\ s^{-1}$) [28].

Cloning of *BGLI* in yeast expression vector pGAPZα

Gene specific primers were designed with reference to the sequence homology and the open reading frame of *BGL*1 was identified using the nucleotide blast search and ORF finder in NCBI. (https://www.ncbi.nlm.nih.gov/ orffinder/). Primers were designed with the objective of incorporating the α-mating factor signal sequence of pGAPZα vector (Invitrogen, USA) for extracellular expression of *BGL*1.

Genomic DNA was extracted from *T. virens* using a simple extraction method [29]. Genomic DNA was PCR amplified using *BGLI* gene specific primers containing restriction enzyme sites {BGLIFP: 5′-ATCGT<u>GAATTC</u> ATGTTGCCCAAGGACT-3′(*Eco*RI) and BGLIRP: 5′- TTGAT<u>TCTAGA</u>TCAAGCTCTTTGCGCT-3′ (*Xba*I)}. The PCR cycling conditions were as follows; initial denaturation at 94 °C for 2 min followed by; 94 °C /30 s, 53 °C / 30 s, 72 °C/ 30 s, for 35 cycles and then at 72 °C / 7 min. The PCR product was electrophoresed on a 0.8% agarose gel and was observed under UV illumination.

To construct the cDNA of *BGL*1, RNA was extracted from fungal mycelium harvested on the sixth day of its highest BGL activity as described below. The mycelia were washed in phosphate buffer solution (1X PBS, pH 7.5) to remove pigments and other components in the medium. The RNA was extracted using the guanidium thiocyanate method [30] followed by DNase (Promega,USA) treatment. The concentration of extracted total RNA was determined and its purity was examined by agarose gel electrophoresis. RNA was reverse transcribed using MMLV (Promega,USA) and oligodT primer (Promega,USA). The synthesized cDNA was then subjected to PCR amplification using gene specific primers as mentioned above. Amplified *BGL*1 gene from both genomic DNA and cDNA of *T.virens* were separately purified and

cloned into pGEM-T vector (Promega, USA). Thereafter, they were separately transformed into *E. coli* JM109 competent cells (Promega,USA) using the heat-shock method [31]. The transformants were selected on low salt Luria-Bertani (LB) agar medium (1% tryptone, 0.5% yeast extract, 0.5% NaCl, and 1.5% agar, pH 7.5) containing ampicillin (100 µg/mL), 0.2 mM,5-bromo-4-chloro-indolyl-D-galactoside (X-gal) and, 40 µg/mL isopropyl-thio-β-D-galactopyranoside (IPTG). Plasmid DNA extraction was carried out on selected white colonies of both genomic and cDNA amplified *BGL*1 clones and then custom sequenced (Macrogen, Korea). The sequence confirmed that recombinant *BGLI* clones were amplified from genomic DNA and cDNA of *T. virens* and were designated as pGEM-T/g*BGLI* and pGEM-T/c*BGLI* respectively. Both recombinant clones were digested with *Eco*RI and *Xba*I restriction enzymes, purified and ligated separately into pGAPZα vector. Ligated products were transformed into *E.coli* JM109 and the transformants were selected in zeocin (25 µg/mL) antibiotic containing low salt LB agar plates. The resulting clones were designated as pGAPZα/ g*BGLI* and pGAPZα/c*BGLI*.

Transformation into *Saccharomyces cerevisiae*

Both pGAPZα/g*BGLI* and pGAPZα/c*BGLI* recombinant vectors were linearized using *Bgl* II and purified. *Saccharomyces cerevisiae* (NCYC 87) was inoculated into 0.5 mL YPD broth (1% yeast extract, 2% peptone, 2% glucose) in a 1.5 mL microcentrifuge tube and incubated at 37 °C overnight in a rotary shaker. A volume of 500 µL from the above grown culture was inoculated into 50 mL of YPD broth in a 250 mL conical flask and incubated in a shaking water bath (150 rpm) at 37 °C until the OD_{600} reached 1.4. Yeast electro competent cells were prepared according to the procedure given in the pGAPZα vector manual (Invitrogen, USA). A volume of 80 µL *S.cerevisiae* competent cells was separately mixed with 5–10 µg of linearized pGAPZα/g*BGLI* and pGAPZα/c*BGLI* plasmid DNA. The mixture was subjected to electroporation under optimized conditions, (1.5 Kw, 200 mA and 25 uF and pulse time of 5 ms) in a 0.2 cm electroporation cuvette. The resulting transformation mixture was spread on to YPDS (1% yeast extract, 2% peptone, 2% glucose, 2% agar and 1 M sorbitol) plates with 100 µg/mL zeocin as the selection marker. The plates were incubated for 3 days at 37 °C to obtain positive transformants. Twenty yeast colonies were selected and streaked on fresh YPDS plates containing zeocin (100 µg/mL).

Screening for recombinant *Saccharomyces cerevisiae*

Colony PCR was performed on the above selected colonies to confirm the presence of the integrated *BGLI* in the *S.cerevisiae* genome. A non-recombinant *S.cerevisiae* and recombinant plasmids (pGAPZα/gBGLI and pGAPZα/cBGLI) were used as the negative and positive controls respectively. All PCR amplified products were subjected to agarose (0.8%) gel electrophoresis. Two putative clones designated as Y-pGAPZα/gBGLI and Y-pGAPZα/cBGLI were custom sequenced.

SDS-PAGE and expression analysis of BGLI containing recombinant S.cerevisiae

Recombinant S.cerevisiae, Y-pGAPZα/gBGLI and Y-pGAPZα/cBGLI were separately inoculated into YP (1% yeast extract, 2% peptone) medium containing 2% cellobiose broth. Cultures were incubated overnight at 37 °C in a rotary shaker at 200 rpm. The non-recombinant S.cerevisiae was used as the control. The enzyme was harvested by centrifugation of the culture medium at 12000 rpm for 2 min at 4 °C. The enzyme extract was concentrated by freeze drying. SDS-PAGE was conducted as described in Sambrook and Russel (2001).

Enzyme activity assay was carried out on the Y-pGAPZα/gBGLI and Y-pGAPZα/cBGLI recombinant S.cerevisiae using the non-recombinant S.cerevisiae as the control. The assay was carried out in triplicate with biologically independent clones. They were separately inoculated into 0.5 mL of YPD broth cultures in 1.5 mL microcentrifuge tubes and incubated for 3 days in a rotary shaker at 200 rpm at 37 °C. The enzyme harvests were freeze dried and dissolved in de-ionized water. The enzyme activity of the extracts was quantitatively determined as described above. Both enzyme and substrate controls were maintained throughout the assay procedure.

Fermentation studies of the recombinant Y-pGAPZα/gBGLI and Y-pGAPZα/cBGLI

Both Y-pGAPZα/gBGLI, Y-pGAPZα/cBGLI and non-recombinant S.cerevisiae were separately inoculated into YP broth (1% yeast extract, 2% peptone) with 5% cellobiose as the sole carbon source in 100 mL conical flasks. Anaerobic conditions were maintained by nitrogen gas supply at 25 °C at pH 4.5. During fermentation, samples were collected from day 1 to day 7 and centrifuged at 6200 g at 4 °C for 10 min to obtain cell free extracts and then it was stored at 4 °C. All experiments were performed in triplicate. The ethanol concentration in the extract was determined by a colourimetric ethanol assay (Megazyme, Ireland) using a standard graph.

Homology modeling and molecular dynamics simulation studies of BGLI

The tertiary structure of BGLI protein was constructed using the MODELLER (version 9.13) program [32, 33]. The Blast Protein tool [34] was used to search the RSCB PDB protein databank [35, 36] to find X-ray crystallographic structures with sequences similar to the target. Multiple sequence alignment was used for homology modeling and the generated model was based on the beta-glucosidase templates; 3AHY.pdb, 4MDP.pdb, 4MDO.pdb in the RSCB PDB protein databank.

Validity of the model was evaluated using various structure validation tools; VERIFY3D [37] was used to analyze the compatibility of the model with its amino acid sequence, PROCHECK [38] was applied to verify the geometrical and stereo-chemical constraints of the model, the overall quality factor was generated by ERRAT [39]. The binding site of the 3-D model generated above was identified using the COACH server (zhanglab.ccmb.med.umich.edu/COACH) [40, 41].

The 3D structure of the cellulose ligand consisting of five monomer units was constructed and geometrically optimized with 6-31 g** basis set using Gaussian 09 (linux version) software [42]. The optimized ligand structure was docked (Flexible Docking method) into the active site of the model structure of BGLI using DOCK6 software [43, 44]. The grid score energies were used to rank the binding strength of the protein and the ligand [45]. A low score is always a good indicator of high affinity.

The docked complex (protein + ligand) with lowest binding energy obtained from the docking process was selected as the starting configuration for molecular dynamics (MD) simulations using the GROMACS v4.6.5 [46]. The GROMOS54a7 all atom force field was employed for the model protein and the force field parameters for the ligand were obtained from PRODRG server [47]. The protein-ligand complex was placed in the center of a box of $9 \times 9 \times 9$ nm^3 volume. Na$^+$ ions were added to maintain electro neutrality and the system was solvated with SPC/E water molecules [48]. Electrostatic interactions were modeled by particle mesh Ewald (PME) with a short-range cutoff of 1.2 nm [49]. Temperature and pressure of the system were maintained at 300 K and 1 bar using Berendsen's weak coupling algorithm [50]. All bonds were constrained at their equilibrium distances using LINCS algorithm [51] while other internal motions (bending and torsion) were allowed during molecular dynamics simulation. The system was subjected to 2000 steps of energy minimization with steepest decent algorithm followed by 200 ps long MD simulation to equilibrate the simulation system. Following the equilibration step, 15 ns MD simulation was carried out using a desktop computer with Intel® Core™ i7−950 Processor. Configurations of the system at every 2 ps intervals were stored for further analysis. At the end of the simulation, the non-covalent interactions between ligand and the protein were analyzed by the LigPlot + v.14.5 software [52]. Further, the same simulation protocol mentioned above was employed for 15 ns

MD simulation for the model protein alone to study its stability in aqueous medium.

Results and discussion

The BGLI enzyme activity of *Trichoderma virens* was plotted against the time of broth extraction (Fig. 1). The maximum BGLI activity was determined as 1.5×10^{-3} IU ml^{-1} in the culture supernatant on the sixth day. Therefore, isolation of total RNA and the construction of the cDNA were carried out on day six old cultures grown under optimized growth conditions (6.5pH and 25 °C).

PCR amplification of both genomic DNA and cDNA using *BGLI* gene specific primes yielded approximately a 1.4 kb fragment and sequence analysis of their recombinant clones revealed pGEM-T/g*BGLI* to contain a 1439 bp insert while pGEM-T/c*BGLI* contained a 1363 bp insert (GenBank: KU535892.1). The genomic clone (GenBank: KM052276.1) consists of a single intron, 76 bp, (41 bp to 117 bp). Sequence similarity searching (NCBI blast searching and EMBL-EBI similarity searching) indicated that the amplified *BGLI* sequence had 90% similarity to the gene sequence of endo-1,4-beta-glucosidase of *Trichoderma atroviride* (clone JGIBTOG-13A22 (AC237343.1)) and only 83% similarity to *Trichoderma viride*. The similarity of translated open reading frame of the amplified *BGLI* fragment was 99% identical to the amino acid sequence of the endo-1,4-beta-glucosidase of *Trichoderma atroviride*. The InterProScan server protein domain analysis predicted the BGLI catalytic domain to belong to the glycoside hydrolase family 1 and BGLB super family.

The theoretical molecular weight of BGLI protein was calculated to be 52 kDa and isoelectric pH was 5.4. A signal peptide sequence was not identified by the signalP-4.1 server. This information shows that the BGL1 is an intracellular protein in *Trichoderma*. Two O -glycosylation sites at positions 306 and 307 and two N- glycosylation sites at positions 55 and 367 of the amino acid sequence were predicted using NetOGlyc 4.0 Server and NetNGlyc

1.0 Server. SDS-PAGE analysis confirmed the BGLI recombinant enzyme expressed by both genomic and cDNA clones to be ~52 kDA (Fig. 2).

The BGLI enzyme activities expressed by Y-pGAPZα/g*BGLI*) and Y-pGAPZα/c*BGLI S.cerevisiae* clones were 1.7×10^{-3} IU ml^{-1} and 5×10^{-4} IU ml^{-1}respectively. The BGLI activity observed for pGAPZα/g*BGLI* was comparable to the activity of locally isolated *Trichoderma virens* (1.5×10^{-3} IU ml^{-1}) denoting the successful approach of genetic engineering in the heterologous extracellular expression of BGLI using the (GAP) promoter driven expression system. Furthermore, the BGLI activity of pGAPZα/g*BGLI* clone was 3.4 times higher than that of the cDNA clone. This represents the direct and/or indirect intron mediated enhancement (IME) of eukaryotic gene expression [53–55]. Increased expression of genes have been observed in recombinant constructs containing an intron compared to its intronless cDNA clones [55, 56]. Present observations support this view. It is possible that the intronic region of the *BGLI* gene is positively influencing the transcription of the gene. The intronic region involved in IME should be within the transcribed region and located close to the start codon of the gene and should be in their natural orientation to increase the expression by enhancing RNA polymerase II initiation and processivity [55]. The intron (41 bp to 117 bp) in Y-pGAPZα/g*BGLI* clone is located following the first exon of the gene. Apart from the above, it has been reported that promoter proximal introns (5′- proximal introns) may be repositories for transcriptional regulatory elements such as enhancers, repressors and elements that modulate the function of the upstream promoter [57–59]. Furthermore,

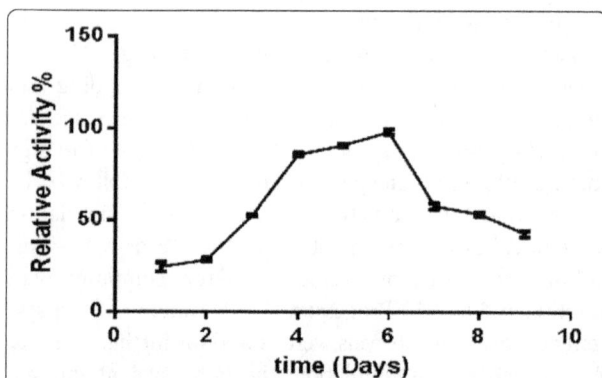

Fig. 1 Determination of β-glucosidase I (BGLI) enzyme activity against time of enzyme harvest from Mandel's medium (MM) containing cellobiose broth cultures at pH optimum (pH 6.5) at 25 °C

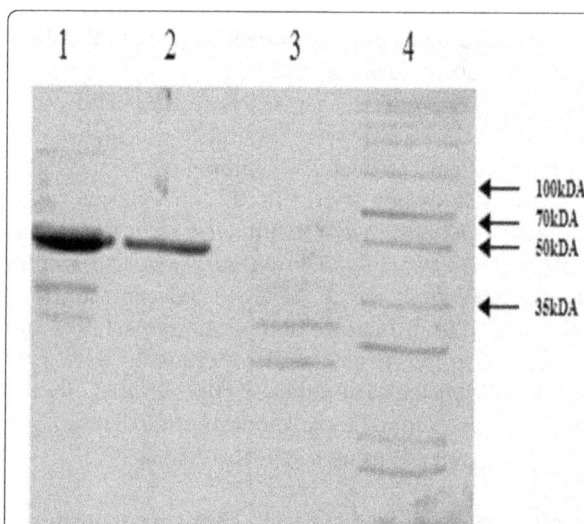

Fig. 2 Lanes 1 and 2: BGLI enzyme (52 kDA) secreted by recombinant *S.cerevisiae* clones Y-pGAPZα/g*BGLI* and Y-pGAPZα/c*BGLI* respectively. Lane 3: Enzyme extract of non-recombinant *S.cerevisiae*. Lane 4: Broad range protein molecular weight marker

studies on post splicing mechanisms have indicated that the Exon Junction Complex (EJC) consisting of several proteins play a major role in facilitating mRNA metabolism including pre mRNA splicing, mRNA export and association with spliced mRNA in the cytoplasm [60–62]. Ultimately this has been shown to facilitate translation, mRNA localization and protein folding efficiency [62–64].

The ethanol produced by recombinant Y-pGAPZα/gBGLI and Y-pGAPZα/cBGLI in the culture supernant of YP broth containing 5% cellobiose against time under anaerobic conditions at 30 °C is represented in Fig. 3. The non-recombinant S. cerevisiae was used as the control. The maximum production of ethanol was 18 g/L (0.36 g/1 g of cellobiose) by Y-pGAPZα/gBGLI and 2 g/L (0.06 g /1 g of cellobiose) by Y-pGAPZα/cBGLI from 5% of cellobiose in 50 ml of culture medium respectively on day four and day five of the fermentation.

Literature cites successful research on expression of β-glucosidase in S.cerevisiae from both eukaryotic and prokaryotic sources with the objective of simultaneous saccharification and direct fermentation (SSF) of cellobiose into ethanol [65–67]. Two recent studies report 0.47 g and 0.30 g of ethanol production per 1 g of cellobiose and pretreated cellulose respectively by intracellular expression of fungal BGL in S.cerevisiae [65, 68]. In both studies, cellodextrin transporter was introduced together with the BGL gene into S.cerevisiae. Therefore, the engineered S.cerevisiae was able to assimilate cellobiose and cello-oligosaccharides directly into the cell by cellodextrin transporter for intracellular hydrolysis. A similar result (0.36 g ethanol/1 g of cellobiose) was observed in the present study where BGLI expression was under the control of the glyceraldehyde 3-phosphate dehydrogenase (GAP) promoter driven yeast integrative vector (pGAPZα) and α-mating factor driven extracellular secretion. Although it is reported that the above vector has better tolerance to inhibitors, namely furan

derivatives, weak and phenolic compounds produced during the anaerobic fermentation of lignocellulosic biomass [69] the ethanol production was stationary after day four in Y-pGAPZα/gBGLI fermentation. However, this could also have been due to the limitation of growth factors of S.cerevisiae, retention of inhibitory un-hydrolyzed cellobiose and the lethal effect of ethanol on S.cerevisiae, thus limiting the enzymatic action of BGLI. One possible means of eliminating the inhibitory effect is by the continuous removal of ethanol from the medium during fermentation and it should be considered in further optimization studies to increase the ethanol yield.

In homology modeling, five probable models obtained from the MODELLER 9.13 software were ranked with respect to their normalized Discrete Optimized Protein Energy (zDOPE) and GA341 score and the model with the best scores was selected as the theoretical model for BGLI protein. This theoretical model contains 455 amino acids, 3654 atoms and 3762 bonds. Characterization of the BGLI with DSSP program [70] indicates that the secondary structure consists of 19 α-helices and 14 β-sheets as given below.

Alpha helices:
(14ALA-17ILE), (23LYS-25GLY), (30ILE-36ALA), (57THR-67LEU), (78TRP-81ILE), (93LYS-109ALA), (124GLU-130TYR), (132GLY-134LEU), (138GLU-153SER), (166PRO-176SER), (188GLU-215SER), (236PRO-258TYR), (264ALA-273ARG), (279ALA-285VAL), (343ALA-357TYR), (379LYS-383LEU), (386ASP-406ASP), (424TRP-429VAL), (449LYS-453SER)

β-sheets:
(7GLN-11ALA), (71SER-75SER), (112THR-116THR), (159ASN-161ILE), (218GLN-220GLY), (222VAL-224ASN), (227PHE-230PRO), (292TYR-295ASN), (299SER-304HIS), (318VAL-321LEU), (362ILE-366GLU), (410VAL-415ALA), (435THR-438ASP), (444GLN-447PRO)

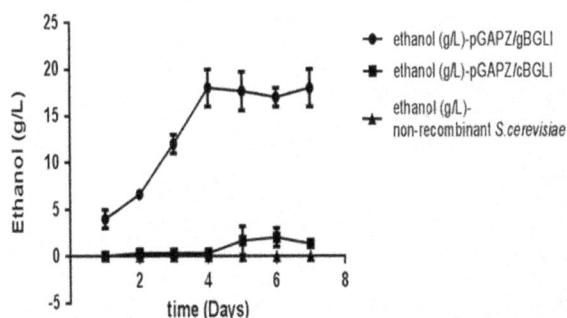

Fig. 3 Representation of ethanol produced in the fermented medium containing YP broth (1% yeast extract, 2% peptone) with 5% cellobiose from day 1 to day 7 by recombinant Y-pGAPZα/gBGLI and Y-pGAPZα/cBGLI against the control of non-recombinant S.cerevisiae under anaerobic conditions (pH 4.5 at 25 °C)

Table 1 Statistics of the 3D model of BGLI from the Ramachandran plot

Ramachandran plot statistics	BGLI	
Amino acids in most favored regions	358	91.6%
Amino acids in additional allowed regions	30	7.7%
Amino acids in generously allowed regions	2	0.5%
Amino acids in disallowed regions	1	0.3%
Number of non-glycine and non-proline residues	391	
Number of end residues	2	
Number of glycine residues	37	
Number of proline residues	25	
Total number of residues	455	

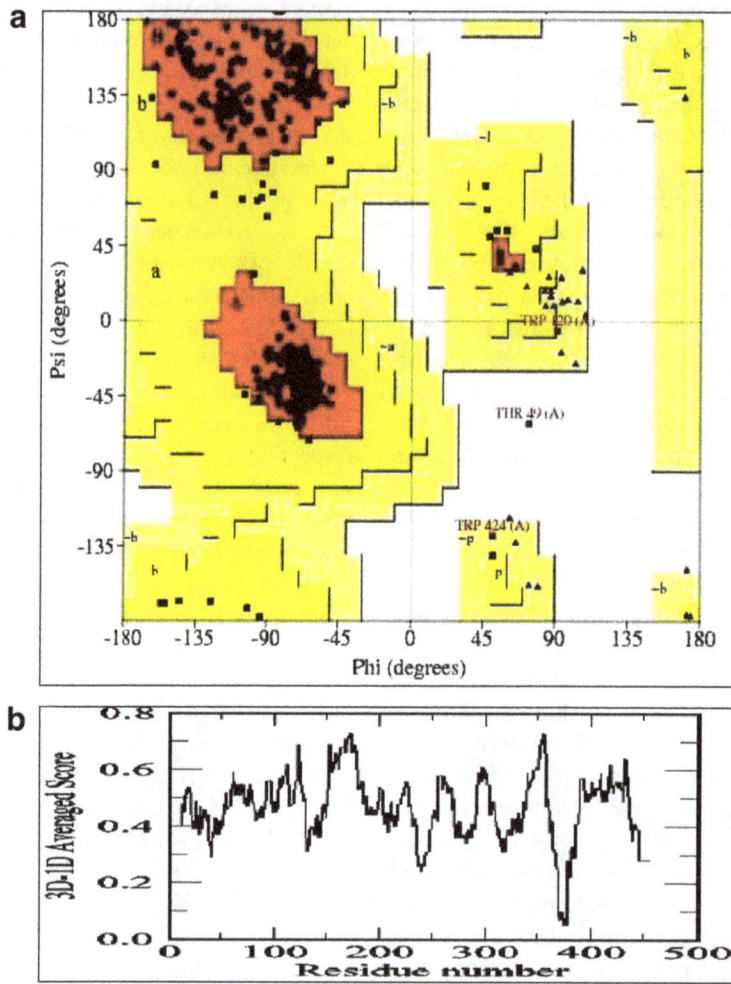

Fig. 4 a Ramachandran map of modeled BGLI protein. **b** Verify 3D score profile

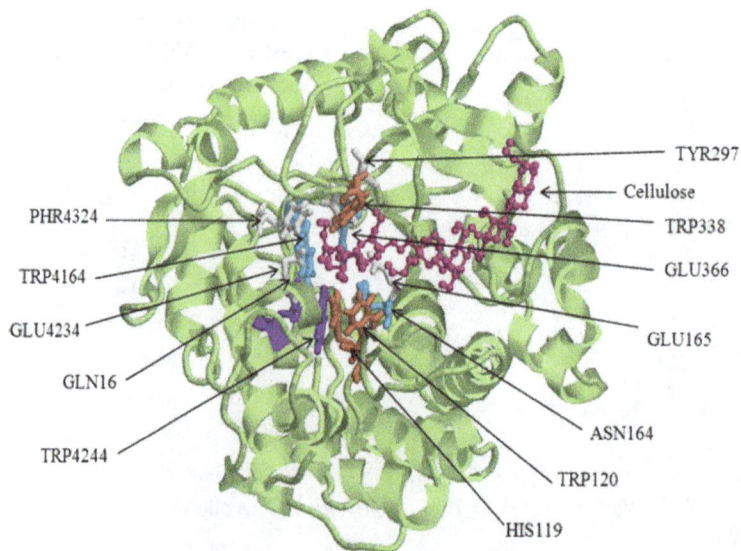

Fig. 5 Protein-ligand docked complex

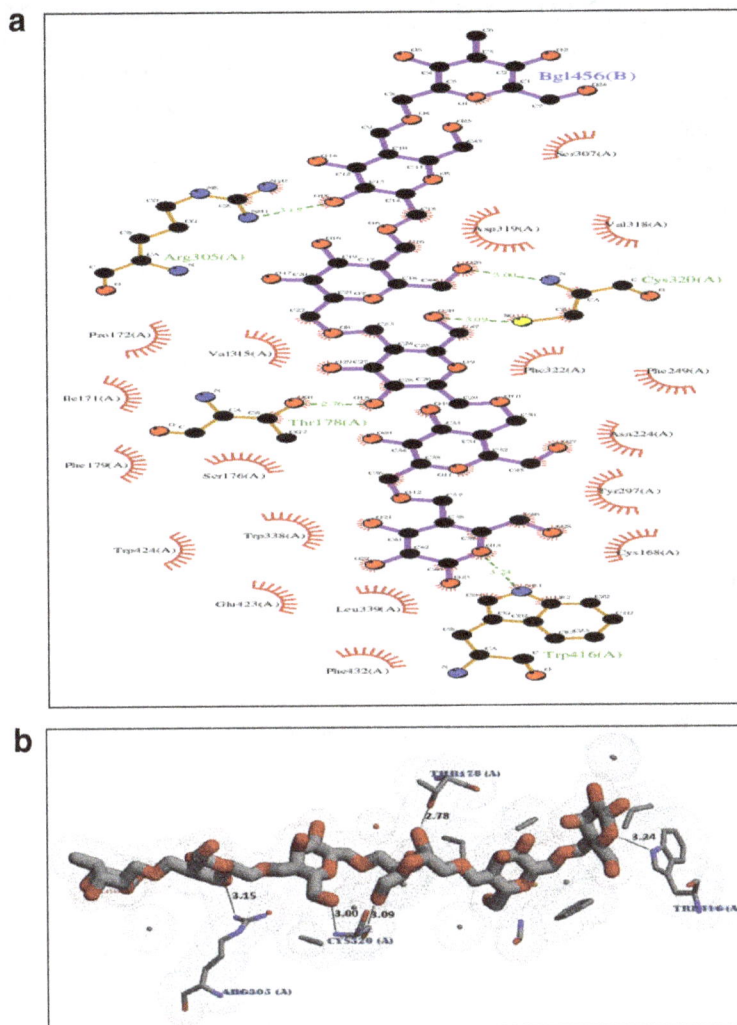

Fig. 6 a Three dimensional view of H bonds between ligand and the protein residues. **b** H bonds between ligand and the protein residues from LigPlot program

PROCHECK, VERIFY3D and ERRAT programs were used for the validation of the predicted model. PROCHECK analysis of BGLI is given in Table 1 and the Ramachandran plot generated by the same program is depicted in Fig. 4a. The statistical score of the Ramachandran plot shows that only 0.3% of amino acids are in the disallowed region.

VERIFY3D profile for BGLI protein shows that all the residues have an averaged 3D-1D score greater than zero (Fig. 4b) while 98.02% residues show more than 0.2 of averaged 3D-1D score. When 80% of residues show an average score of greater than 0.2%, the 3D structure of the protein is considered as reliable [37]. Further, the ERRAT program evaluated the overall quality factor as 88.4 for the modeled 3D structure of BGLI. Therefore considering the above results it can be concluded that the predicted 3D structure of BGLI is highly reliable.

For the molecular docking step, the BGLI-cellobiose complex with the lowest binding energy was selected. It was interesting to note that the selected ligand molecule successfully docked to the binding site which was previously identified by the I-TASSER-COACH Server. Residues; 16GLN, 119HIS, 120TRP, 164ASN, 165GLU, 297TYR, 338TRP, 366GLU, 416TRP, 423GLU, 424TRP, 432PHR are predicted by the COACH server as the consensus binding residues.

The recorded best grid score for the cellulose-protein was −121.07 kJ/mol and it indicates a fairly high binding affinity value and cellulose bind in a compatible binding pose. The best docked complex is presented in Fig. 5. This value represents the summation of van der Waals dispersive and electrostatic interaction energy, which approximately indicates the binding energy of the ligand.

The general catalytic mechanisms for glycoside hydrolases were proposed decades ago. Most glycoside hydrolases

Table 2 Detailed information of H bonds formed between ligand and the protein

Residue	Amino acid	Distance H-A	Distance D-A	Donor Angle	Protein donor	Side chain
178	THR	1.82	2.76	155.04	yes	yes
224	ASN	3.17	3.69	144.24	no	yes
305	ARG	2.58	3.15	116.58	yes	yes
305	ARG	3.33	3.65	100.43	yes	yes
320	CYS	2.08	3.00	152.99	yes	no
320	CYS	2.39	3.10	126.54	no	no
416	TRP	3.67	3.24	116.59	yes	yes

follow either retaining or inverting mechanisms [71]. These reactions typically occur with the assistance of acidic and a basic amino acid residues. The docking results revealed that the presence of glutamic and aspartic acids located in the active site may assist any of the above mechanisms.

Two MD simulations of 15 ns each were carried out for the protein-ligand complex and the bare protein in aqueous medium with 21,923 SPC water molecules. The non-covalent interaction (H bond) of the final configuration (after 15 ns) of protein-ligand complex identified from LigPlot + v.145 software is presented in Fig. 6a & b. The LigPlot analysis of the protein structure indicates that the ligand forms strong five hydrogen bonds with THR178 (2.76 Å), ARG305 (3.15 Å), CYS320 (3.00 Å) (3.09 Å), TRP416 (3.24 Å). Detailed information is presented in Table 2. The stability of all

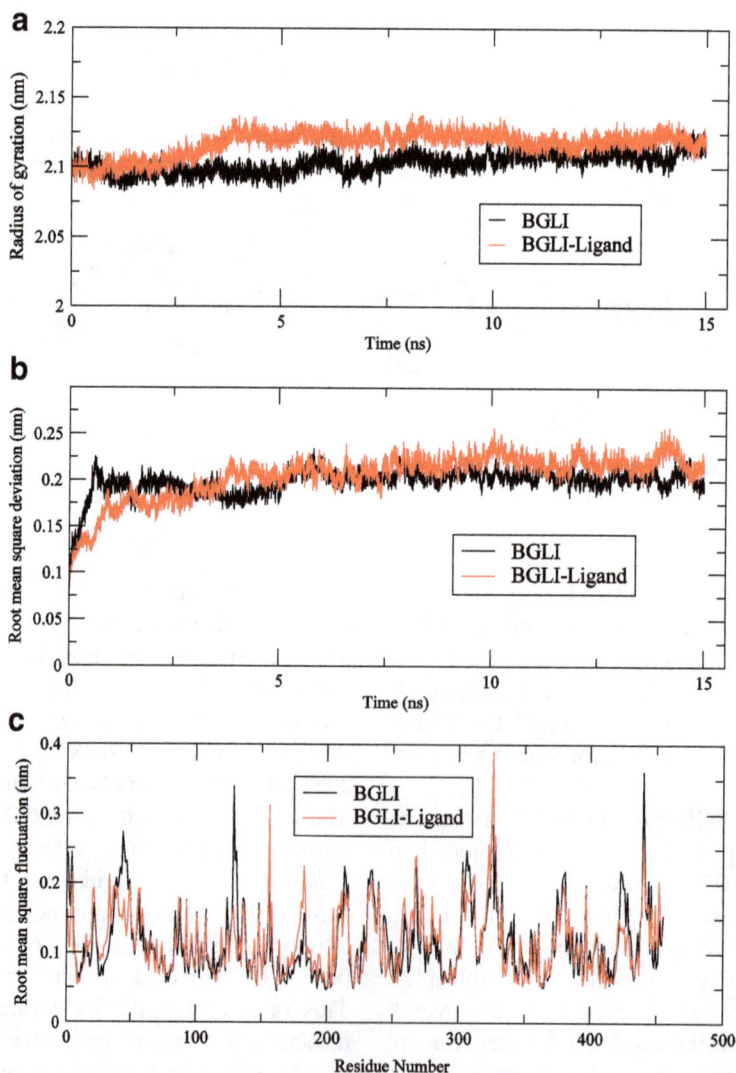

Fig. 7 a Root mean square deviations (RMSD) of the backbone. **b** radius of gyration (Rg) of the protein from 15 ns long MD trajectory. **c** Root mean square fluctuation (RMSF) of the residues in the protein over 15 ns long MD trajectory

these H bonds was studied using *g_dist* tool in the GROMACS program.

The analysis indicates that the distance between the centers of mass of the two groups of atoms which formed H bonds was nearly constant during the total simulation time. These results reveal the stability and effectiveness of the H bonding.

The stability of the protein after forming a complex with cellulose was studied by calculating root mean square deviations (RMSD), radius of gyration (Rg) and root mean square fluctuation (RMSF) of the protein. All three parameters of the protein of the complex were compared with that of the bare protein. Figure 7a & b compares RMSD of the backbone of the protein in two systems. Both systems indicate stable structures with RMSD of about 0.2 nm and there is no indication of increasing the RMSDs with time. Figure 7b gives the variation of radius of gyration (Rg) as a function of simulation time which indicates the compactness of the protein. As seen in the Fig. 7b, Rg of both systems were maintained approximately at the same value. Both these results suggest that the BGLI preserves its tertiary structure even after making a complex with the ligand.

Figure 7c represents the root mean square fluctuation (RMSF) of the protein residues in both simulation systems (protein alone and the protein with the substrate) indicating stable 3D structures for the bare protein and the protein in the complex. It is observed that most of the fluctuations are concentrated in the region of residues 323 and 328, residues 156 and 440 for both systems. Residues 44, 129–131 show relatively higher fluctuations in the free protein. Further, none of the high fluctuating residues of the protein of the complex were in the predicted active site and it can be safely postulated that the enzyme can perform well with the bound ligand via the proposed two mechanisms.

Conclusion

The genomic and cDNA of β-glucosidase 1 (BGL1) were isolated from *Trichoderma virens* and successfully characterized, cloned and expressed in *S.cerevisiae*. The expression of *S.cerevisiae* genomic DNA clone was determined to be higher than its cDNA clone. In the fermentation study a higher amount of ethanol (0.36 g/1 g of cellobiose) was obtained by *S. cerevisiae* genomic DNA clone than its cDNA (0.06 g/1 g of cellobiose). BGLI carrying *S.cerevisiae* will have the potential to be used in the industrial production of ethanol by the hydrolysis of the cellulose component in plant biomass by the combinatory simultaneous actions of endoglucanase and cellobiohydrolase, the other two enzymes of the cellulase complex.

The major ligand binding domain of the model enzyme was identified from the results of molecular docking studies. MD simulation results indicate an overall stable

confirmation of BGL-cellobiose complex that exhibits an almost similar structural flexibility shown by the free enzyme. Further, it has been found that mainly five hydrogen bonds are involved in maintaining the enzyme-substrate association. Thus these results lead to clear understanding of its binding site. The predicted model was a realistic stable model and the predicted active site residues would be a good starting point for the further efforts in the rational design of mutagenic experiments aimed at improving the catalytic activity of glycoside hydrolases.

Acknowledgements
National Science Foundation (NSF), Sri Lanka

Funding
This research was financially supported by National Science Foundation (NSF), Sri Lanka (grant no. RG/2012/BT/02).

Authors' contributions
GHIMW designed and performed all wet lab experiments, analyzed the data and drafted the manuscript. RLCW designed microbiological experiments and supervised the research. NVC supervised the research and was involved in gene cloning and expression studies. PPAMSIR carried out *in-silico* studies, analyzed the data and drafted the manuscript. MSSW designed and supervised the in-silico studies, analyzed the data and revised the manuscript. WSSW designed the overall research, supervised, involved in data analysis and revised the manuscript. All authors read and approved the final manuscript.

Competing interests
The authors declare that they have no competing interests.

Author details
[1]Department of Chemistry, Faculty of Science, University of Colombo, Colombo, Sri Lanka. [2]Department of Plant Sciences, Faculty of Science, University of Colombo, Colombo, Sri Lanka. [3]Department of Biochemistry and Molecular Biology, Faculty of Medicine, University of Colombo, Kynsey Road, Colombo 08, Sri Lanka.

References
1. Harmsen PFH. Huijgen WJJ. Bakker RRC. Literature review of Physical and Chemical Pretreatment Processes for Lignocellulosic Biomass: Bermúdez López LM; 2010. www.ecn.nl/docs/library/report/2010/e10013.pdf. Accessed 13 Dec 2013
2. Yang B, Dai Z, Ding SY, Wyman CE. Enzymatic hydrolysis of cellulosic biomass: a review. Biofuels. 2011;2(4):421–50.
3. Wright JD. Ethanol from lignocellulose: an overview. Energ Prog. 1988;8(2):71–8.

4. Ahmed S, Riaz S, Jamil A. Molecular cloning of fungal xylanases; an overview. Appl Microbial Biotechno. 2009;84:19–35.

5. Li XH, Yang HJ, Roy B, Wang D, Yue WF, Jiang LJ, et al. Miao1 YG. The most stirring technology in future: Cellulase enzyme and biomass utilization. Afr J Biotechnol. 2009;8(11):2418–22.

6. Gao JH. Weng D, Zhu M, yuan F, guan, Yu xi. Production and characterization of cellulolytic enzymes from the thermoacidophilic fungal Aspergillus terreus M11 under solidstate cultivation of corn stover. Bioresour Technol. 2008;99:7623–9.

7. Chauve M, Mathis H, Huc D, Casanave D, Monot F, Ferreira NL. Comparative kinetic analysis of two fungal β-glucosidases. Biotechnology for Biofuels. 2010;3:1–8.

8. Jayant M, Rashmi J, Shailendra M, Deepesh Y. Production of cellulase by different co-culture of Aspergillus niger and Penicillium chrysogenum from waste paper, cotton waste and baggas. Journal of Yeast and Fungal Research. 2011;2:24–7.

9. Lynd LR, Weimer PJ, Van Zyl WH, Pretorius IS. Microbial cellulose utilization: fundamentals and biotechnology. Microbiol Mol Biol Rev. 2002;66:506–77.

10. Bergquist PV, Teo O, Gibbs M. Expression of xylanase enzymes from thermophilic microorganisms in fungal host. Extermophiles. 2002;6:177–84.

11. Ljungdhal LG. Mechanism of cellulose hydrolysis by enzymes from anaerobic and aerobic bacteria. In: Coughlan MP (ed) enzyme systems for lignocellulose degradation. Elsevir. London; 1989. p. 5–16.

12. Amouri B, Gargouri A. Characterization of a novel β-glucosidase from a Stachybotrys strain. Biochem Eng. 2006;32:191–7.

13. Gautam SP, Bundela PS, Pandey AK, Awasthi MK, Sarsaiya S. Optimization for the production of Cellulase enzyme from municipal solid waste residue by two novel cellulolytic fungi. Biotechnol Res Int. 2011; doi:10.4061/2011/810425.

14. Pandey S, Srivastava M, Shahid M, Kumar V, Singh A, Trivedi A, Srivastava YK. Trichoderma species Cellulases Produced by Solid State Fermentation. J Data Mining Genomics Proteomics. 2015;doi:10.4172/2153–0602.1000170.

15. Ostergaard S, Olsson L, Nielsen J. Metabolic engineering of Saccharomyces cerevisiae. Microbiol Mol Biol Rev. 2000;64:34–50.

16. Kricka W, Fitzpatrick J, Bond U. Metabolic engineering of yeasts by heterologous enzyme production for degradation of cellulose and hemicellulose from biomass: a perspective. Front Microbiol. 2014;5:174.

17. Meko'o DJL, Xing Y, Shen LL, Bounda GA, WU J, Taiming LI, et al. Production of ethanol from cellobiose by recombinant β-glucosidase-expressing Pichia pastoris: submerged shake flask fermentation. Afr J Biotechnol. 2012;11(37):9108–17.

18. Yanase S1, Yamada R, Kaneko S, Noda H, Hasunuma T, Tanaka T, et al. Ethanol production from cellulosic materials using cellulase-expressing yeast. Biotechnol J. 2010;5(5):449–55.

19. Kotaka A, Bando H, Kaya M, Kato-Murai M, Kuroda K, Sahara H, et al. Direct ethanol production from barley beta-glucan by sake yeast displaying Aspergillus oryzae beta-glucosidase and endoglucanase. J Biosci Bioeng. 2008;105:622–7.

20. Jeon E, Hyeon JE, Eun LS, Park BS, Kim SW, Lee J, et al. Cellulosic alcoholic fermentation using recombinant Saccharomyces cerevisiae engineered for the production of clostridium cellulovorans endoglucanase and Saccharomycopsis fibuligera beta-glucosidase. FEMS Microbiol Lett. 2009;301:130–6.

21. Lin-Cereghino GP, Stark CM, Kim D, Chang J, Shaheen N, Poerwanto H, et al. The effect of α-mating factor secretion signal mutations on recombinant protein expression in Pichia pastoris. Gene. 2013;519:311–7.

22. Waterham HR, Digan ME, Koutz PJ, Lair SV, Cregg JM. Isolation of the Pichia pastoris glyceraldehyde-3-phosphate dehydrogenase gene and regulation and use of its promoter. Gene. 1997;186:37–44.

23. Cregg JM, Vedvick TS, Raschke WC. Recent advances in the expression of foreign genes in Pichia Pastoris. Bio/Technology. 1993;11:905–10.

24. Klepeis JL, Lindorff LK, Dror RO, Shaw DE. Long-timescale molecular dynamics simulations of protein structure and function. Curr Opin Struct Biol. 2009;19:120–7.

25. Mandels M, Sternburg D. Recent advances in cellular technology. J Ferment Technol. 1976;54:267–86.

26. Steiner J, Socha C, Eyzaguirre J. Culture conditions for enhanced cellulase production by a native strain of Penicillium purpurogenum. World J of Microbiol and Biotechnol. 1994;20:280–3.

27. Ghose TK. Measurement of cellulose activities. Journal of Pure & App Chem. 1987;59:257–68.

28. Zhang YP, Hong J, Ye X. Cellulase assays. Biofuels: methods and protocols. 2009:213–31.

29. Al-Samarrai TH, Schmid J. A simple method for extraction of fungal genomic DNA. The Society for Applied Microbiology. 1999;30:53–6.

30. Chomczynski P, Sacchi N. Single-step method of RNA isolation by acid guanidinium thiocyanate-phenol-chloroform extraction. Anal Biochem. 1987;162:156–9.

31. Sambrook J, Fritsch EF, Maniatis T. Molecular cloning: a laboratory manual. 2nd ed. New York: Cold Spring Harbor Laboratory Press; 1989.

32. Webb, B. and Sali, A. 2014. Comparative protein structure modeling using MODELLER. Current protocols in Bioinformatics. 47:5.6:5.6.1–5.6.32.

33. Renom MMA, Stuart A, Fiser A, Sánchez R, Melo F, Sali A. Comparative protein structure modeling of genes and genomes. Annu Rev Biophys Biomol Struct. 2000;29:291–325.

34. Madden TL, Tatusov RL, Zhang J. Applications of network BLAST server. Meth Enzymol. 1996;266:131–41.

35. Berman HM, Westbrook J, Feng Z, Gilliland G, Bhat TN, Weissig H, et al. The Protein Data Bank. Nucleic Acids Res. 2000;28:235–42. www.rcsb.org. Accessed 21 Aug 2016

36. Berman HM, Henrick K. Nakamura H. Announcing the worldwide Protein Data Bank Nature Structural Biology. 2003;10:980. www.wwpdb.org. Accessed 21 Aug 2016

37. Luthy R, Bowie JU, Eisenberg D. Assessment of protein models with three-dimensional profiles. Nature. 1992;356:83–5.

38. Laskowski RA, MacArthur MW, Moss DS, Thornton JM. PROCHECK - a program to check the stereochemical quality of protein structures. J App Cryst. 1993;26:283–91.

39. Colovos C, Yeates TO. ERRAT: an empirical atom-based method for validating protein structures. Protein Sci. 1993;2:1511–9.

40. Yang J, Roy A, Zhang Y. Protein-ligand binding site recognition using complementary binding-specific substructure comparison and sequence profile alignment. Bioinformatics. 2013;29:2588–95.

41. Yang J, Roy A, Zhang Y. BioLiP: a semi-manually curated database for biologically relevant ligand-protein interactions. Nucleic Acids Res. 2013;41:1096–103.

42. Frisch MJEA. Gaussian 09 Revision A 02. Gaussian Inc Wallingford CT. 2009;

43. Brozell, Scott R. Evaluation of DOCK 6 as a pose generation and database enrichment tool. Journal of computer-aided molecular design. 2012;26:749–773.

44. Allen, William J. DOCK 6: impact of new features and current docking performance. Journal of computational chemistry. 2015;36:1132–1156.

45. Wallace AC, Laskowski RA, Thornton JM. LIGPLOT: a program to generate schematic diagrams of protein-ligand interactions. Protein Eng. 1995;8:127–34.

46. Berendsen HJC, van der Spoel D, van Drunen R. GROMACS: a message-passing parallel molecular dynamics implementation. Comp Phys Commun. 1995;91:43–56. doi:10.1016/0010-4655(95)00042-e.

47. SchuÈttelkopf S, Alexander W, and Aalten DMV. PRODRG: a tool for high-throughput crystallography of protein–ligand complexes. Acta Crystallographica Section D: Biological Crystallography. 2004;60:1355–1363.

48. Berendsen H, Grigera J, Straatsma T. The missing term in effective pair potentials. J Phys Chem. 1987;91:6269–71.

49. Essmann U, Perera L, Berkowitz M, Darden T, Lee H, Pedersen L. A smooth particle mesh Ewald method. J Chem Phys. 1995;103:8577–93.

50. Berendsen H, Postma J, Gunsteren WV, DiNola A, Haak J. Molecular dynamics with coupling to an external bath. J Phys Chem. 1984;81:3684–90.

51. Hess B. P-LINCS: a parallel linear constraint solver for molecular simulation. J Chem Theory Comput. 2007;4:116–22.

52. Laskowski RA, Swindells MB. LigPlot+: multiple ligand-protein interaction diagrams for drug discovery. J Chem Inf Model. 2011;51:2778–86.

53. Mascarenhas D, Mettler IJ, Pierce DA, Lowe HW. Intron-mediated enhancement of heterologous gene expression in maize. Plant Mol Biol. 1990;15:913–20.

54. Akua T, Berezin I, Shaul O. The leader intron of AtMHX can elicit, in the absence of splicing, low- level intron-mediated enhancement that depends on the internal intron sequence.BMC Plant Biol. doi:10.1186/1471–2229-10-93.

55. Niu DK, Yang YF. Why eukaryotic cells use introns to enhance gene expression: splicing reduces transcription-associated mutagenesis by inhibiting topo isomerase I cutting activity. J Bio Med Central. 2011;6:24. doi:10.1186/1745-6150-6-24.

56. Rose AB, Beliakoff JA. Intron-mediated enhancement of gene expression independent of unique intron sequences and splicing. Plant Physiol. 2000; 122(2):535–42.

57. Kwek KY, Murphy S, Furger A, Thomas B, O'Gorman W, Kimura H, et al. U1 snRNA associates with TFIIH and regulates transcriptional initiation. Nat Struct Biol. 2002;9:800–5.

58. Fong YW, Zhou Q. Stimulatory effect of splicing factors on transcriptional elongation. Nature. 2001;414:929–33.

59. Furger A, Justin M. O'Sullivan, Binnie a, lee BA, and Proudfoot NJ. Promoter proximal splice sites enhance transcription. Genes Dev. 2002;16:2792–9.

60. Le Hir H, Gatfield D, Braun IC, Forler D, Izaurralde E. The protein Mago provides a link between splicing and mRNA localization. EMBO Rep. 2001;2:1119–24.

61. Kataoka N, Diem MD, Kim VN, Yong J, Dreyfuss G. Magoh, a human homolog of drosophila mago nashi protein, is a component of the splicing-dependent exon–exon junction complex. EMBO J. 2001;20:6424–33.

62. Le Hir H, Nott A, Moore MJ. How introns influence and enhance eukaryotic gene expression. Trends Biochem Sci. 2003;28(4):215–20.

63. Lykke-Andersen J, et al. Communication of the position of exon–exon junctions to the mRNA surveillance machinery by the protein RNPS1. Science. 2001;293:1836–9.

64. Dostie J, Dreyfuss G. Translation is required to remove Y14 from mRNAs in the cytoplasm. Curr Biol. 2002;12:1060–7.

65. Lee WH, Nan H, Kim HJ, Jin YS. Simultaneous saccharification and fermentation by engineered Saccharomyces cerevisiae without supplementing extracellular glucosidase. J Biotechnol. 2013;167:316–22.

66. Galazka JM, Tian C, Beeson WT, Martinez B, Glass NL, Cate JHD. Cellodextrin transport in yeast for improved biofuel production. Science. 2010;330:84–6.

67. Li S, Sun J, Galazka JM, Glass NL, Cate JHD, Yang X, et al. Overcoming glucose repression in mixed sugar fermentation by co-expressing a cellobiose transporter and glucosidase in Saccharomyces cerevisiae. Mol BioSyst. 2011;6:2129–32.

68. Ha SJ, Galazka JM, Kim SR, Choi JH, Yang X, Seo JH, et al. Engineered Saccharomyces cerevisiae capable of simultaneous cellobiose and xylose fermentation. Proc Natl Acad Sci U S A. 2011;108(2):504–9. doi:10.1073/pnas.1010456108.

69. Palmqvist E, Hagerdal BH. Fermentation of lignocellulosic hydrolysates. II: inhibitors and mechanisms of inhibition. Bioresour Technol. 2000;74:25–33.

70. Kabsch W, Sander C. DSSP: definition of secondary structure of proteins given a set of 3D coordinates. Biopolymers. 1983;22:2577–637.

71. Davies G. Henrissat B. Structures and mechanisms of glycosyl hydrolases. Structure. 1995;3:853–9.

8

Investigation of optimum ohmic heating conditions for inactivation of *Escherichia coli* O157:H7, *Salmonella enterica* serovar Typhimurium, and *Listeria monocytogenes* in apple juice

Il-Kyu Park[1†], Jae-Won Ha[2†] and Dong-Hyun Kang[1*] iD

Abstract

Background: Control of foodborne pathogens is an important issue for the fruit juice industry and ohmic heating treatment has been considered as one of the promising antimicrobial interventions. However, to date, evaluation of the relationship between inactivation of foodborne pathogens and system performance efficiency based on differing soluble solids content of apple juice during ohmic heating treatment has not been well studied. This study aims to investigate effective voltage gradients of an ohmic heating system and corresponding sugar concentrations (°Brix) of apple juice for inactivating major foodborne pathogens (*E. coli* O157:H7, *S.* Typhimurium, and *L. monocytogenes*) while maintaining higher system performance efficiency.

Results: Voltage gradients of 30, 40, 50, and 60 V/cm were applied to 72, 48, 36, 24, and 18 °Brix apple juices. At all voltage levels, the lowest heating rate was observed in 72 °Brix apple juice and a similar pattern of temperature increase was shown in 18–48 °Brix juice samples. System performance coefficients (SPC) under two treatment conditions (30 V/cm in 36 °Brix or 60 V/cm in 48 °Brix juice) were relatively greater than for other combinations. Meanwhile, 5-log reductions of the three foodborne pathogens were achieved after treatment for 60 s in 36 °Brix at 30 V/cm, but this same reduction was observed in 48 °Brix juice at 60 V/cm within 20 s without affecting product quality.

Conclusions: With respect to both bactericidal efficiency and SPC values, 60 V/cm in 48 °Brix was the most effective ohmic heating treatment combination for decontaminating apple juice concentrates.

Keywords: Ohmic heating, Apple juice, System performance efficiency, Foodborne pathogen, Inactivation

Background

The U.S. Food and Drug Administration stated that the possibility for contamination with foodborne pathogens is low in foods with pH below 4.6 [1]. However, acidic foods such as fruit juice have emerged as a novel substrate in which foodborne pathogens can maintain their viability since several illness outbreaks involving them have been documented [2]. Major foodborne pathogens implicated in fruit juice-borne outbreaks are *Escherichia coli* O157:H7 and *Salmonella enterica* serovar Typhimurium [3]. In the United States in 1996, a serious foodborne outbreak occurred in which one person died and 70 people were infected with *E. coli* O157:H7 traced to apple cider [4]. A multistate outbreak caused by *S.* Typhimurium was reported in the United States in 2005 which was associated with consumption of orange juice [5]. *Listeria monocytogenes* is a Gram positive bacterium

* Correspondence: kang7820@snu.ac.kr
†Equal contributors
[1]Department of Food and Animal Biotechnology, College of Agricultural Biotechnology, Center for Food and Bioconvergence, and Institute of GreenBio Science & Technology, Research Institute for Agricultural and Life Sciences, Seoul National University, Seoul 08826, Korea
Full list of author information is available at the end of the article

and has acid tolerance as do *E. coli* O157:H7 and *S.* Typhimurium [6]. Although outbreaks of foodborne illnesses linked to *L. monocytogenes* have not occurred in fruit juices, the National Advisory Committee on Microbiological Criteria for Foods suggested that *L. monocytogenes* should be categorized as a target bacterium even though no association has been identified between *L. monocytogenes* and fruit juices [7]. Apples used for producing juice can become contaminated with these pathogens from several sources, such as apples in orchards that have fallen onto the ground, contamination with manure, or those insufficiently washed [8, 9].

The U.S. Food and Drug Administration has regulated that facilities for pasteurization should ensure a minimum of 5-log pathogen reduction [10]. Thermal methods such as hot water or steam traditionally have been used to pasteurize apple juice. Although conventional heating guarantees food microbiological safety, it causes deterioration of overall quality involving nutritional degradation, color change, and flavor loss [11, 12]. Novel technologies such as radio frequency, microwave, and ohmic heating have emerged as alternatives in order to compensate for the drawbacks of traditional heating. Ohmic heating among innovative thermal technologies is an appropriate system to use for fruit juice pasteurization in that it is able to heat rapidly and uniformly with high temperature for a short time (HTST process) and is amenable to a continuous type design [13, 14]. Ohmic heating is a technology where heat is internally generated by the passage of alternating electric current in which foods act as a resistor [15], and the heating rate in ohmic heating is related to the electrical conductivity of liquid food products [14]. Because of this characteristic, many food engineers have studied ohmic heating associated with the electrical properties of foods. Castro et al. [16] studied the relationship between temperature and sugar content on the electrical conductivity of strawberry products during ohmic heating. Also, Icier and Ilicali [17] investigated the effect of orange juice concentration on system performance efficiency during ohmic heating. Therefore, not only the degree of antimicrobial effect but also several other factors such as the concentration of dissolved solids concerned with system performance efficiency should be considered in order to apply an ohmic heating pasteurization system practically by the fruit juice industry. To date, evaluation of the relationship between inactivation of foodborne pathogens and system performance efficiency based on differing soluble solids content of juices during ohmic heating treatment has not been well studied.

The purpose of this research was to investigate the optimum sugar concentration (°Brix) of apple juice and corresponding voltage gradient of an ohmic heating system for achieving both effective inactivation of foodborne pathogens including *E. coli* O157:H7, *S.* Typhimurium, and *L. monocytogenes* and higher system performance efficiency.

Methods

Bacterial strains and culture preparation

All bacterial strains, namely, *E. coli* O157:H7 (ATCC 35150, ATCC 43889, and ATCC 43890), *S.* Typhimurium (ATCC 19585, ATCC 43971, and DT 104) and *L. monocytogenes* (ATCC 19114, ATCC 19115, ATCC 15313) were obtained from the Bacterial Culture Collection at Seoul National University (Seoul, South Korea) and used for all experiments. All strains were stored at −80 °C in 0.7 ml of Tryptic Soy Broth (TSB; Difco Becton Dickinson, Sparks, MD, USA) and 0.3 ml of 50% glycerol (vol/vol). Working cultures were streaked onto Tryptic Soy Agar (TSA; Difco), incubated at 37 °C for 24 h, and stored at 4 °C. Each strain of *E. coli* O157:H7, *S.* Typhimurium, and *L. monocytogenes* was cultured in 5 ml TSB for 24 h at 37 °C, harvested by centrifugation at 4000 × *g* for 20 min at 4 °C, and washed three times with 0.2% peptone water (PW, Difco). The final pellets were resuspended in 0.2% PW, corresponding to approximately $10^8 \sim 10^9$ CFU/ml. Subsequently, suspended pellets of each strain of the three pathogens were mixed to produce a culture cocktail.

Sample preparation and inoculation

Pasteurized apple juice concentrate (pH 3.5, 72 °Brix), free of any preservatives, was purchased from a local grocery store (Incheon, Korea). Apple juice concentrate was diluted with sterile distilled water to 48, 36, 24, and 18 °Brix. Sugar concentration (°Brix) was measured by a digital refractometer (Atago co.,Ltd., Japan). Then, a 0.2-ml aliquot of the mixed culture cocktail (*E. coli* O157:H7, *S.* Typhimurium, and *L. monocytogenes*) was inoculated into each 25 ml sample of apple juice of different solids content. The final cell concentration was ca. $10^6 \sim 10^7$ CFU/ml.

Experimental apparatus

Ohmic heating treatments were conducted in a previously described apparatus [18]. The experimental device (Fig. 1) consisted of a two-channel digital storage oscilloscope (TDS2001C; Tektronix, Inc., Beaverton, CO), a precision power amplifier (4510; NF corp., Yokohama, Japan), a function generator (33210A; Agilent Technologies, Palo Alto, CA), a data acquisition instrument (34,790 A; Agilent Technologies), and an ohmic heating

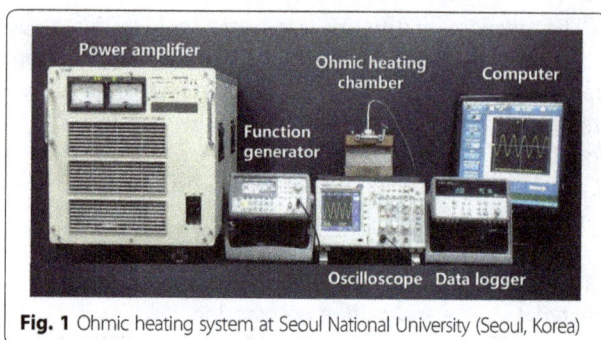

Fig. 1 Ohmic heating system at Seoul National University (Seoul, Korea)

chamber. In the middle of a rectangular container (an ohmic heating chamber, 2 by ×15 by ×6 cm) consisting of component Pyrex glass, two titanium electrodes and a K-type thermocouple coated with Teflon were located. The distance between the cross-sectional area and the two titanium electrodes was 2 cm and 60 cm², respectively. Multiple waveforms such as sine, square, ramp, pulse, triangle, noise, and custom waveforms could be produced by the function generator which permitted a frequency range of 1 MHz to 10 MHz and a maximum output signal of 5 V. These signals were expanded by the power amplifier from 45 to 20 kHz and a maximum output of 141 VAC. Each titanium electrode received signals amplified by the power amplifier. The signals, including waveform, frequency, voltage, and current, were measured using the two-channel digital storage oscilloscope. The data acquisition instrument was used to obtain temperature histories in this study.

Ohmic heating treatment
The ohmic heating chamber was filled with 25 ml of sample for treatment. A 20 kHz frequency and sine waveform were utilized in all experiments. Since electrochemical reactions can occur at standard line voltage frequency (60 Hz) during ohmic heating and it may affect inactivation of foodborne pathogens [18, 19], 20 kHz, a high frequency that does not cause electrochemical reactions, was chosen in this study. For obtaining temperature and electric current data, treatments were conducted at a fixed 30, 40, 50, and 60 V/cm setting in apple juice of 72, 48, 36, 24, and 18 °Brix for 90 s. Temperature and electric current were recorded every 1 s. For microbial inactivation experiments, inoculated samples were treated at a fixed 30 or 60 V/cm setting in 72, 48, 36, 24, and 18 °Brix apple juice for 0, 10, 20, 30, 40, 50, and 60 s.

Bacterial enumeration
For enumeration of bacteria, each treated 25 ml sample was immediately transferred into a sterile stomacher bag (Labplas Inc., Sainte-Julie, Quebec, Canada) containing 225 ml of iced 0.2% PW (maintained on crushed ice) and homogenized for 2 min with a stomacher (Easy Mix, AES Chemunex, Rennes, France). One ml aliquots of homogenized samples were tenfold serially diluted in 9 ml of 0.2% PW, and 0.1 ml of sample or diluent was spread-plated onto each selective medium. For the enumeration of *E. coli* O157:H7, *S.* Typhimurium and, *L. monocytogenes*, Sorbitol MacConkey agar (SMAC; Difco), Xylose Lysine Desoxycholate agar (XLD; Difco) and Oxford Agar Base (OAB; Difco) with antimicrobic supplement (Bacto™ Oxford Antimicrobic Supplement, Difco) were used as selective media, respectively. Where low numbers of surviving cells were anticipated, 250 µl of sample was spread-plated onto each of four plates to lower the detection limit (detection limit = 10 CFU/g). All agar media were incubated at 37 °C for 24–48 h before counting. To confirm the identity of the pathogens, colonies were selected randomly from the enumeration plates and subjected to serological or biochemical tests [*E. coli* O157:H7 latex agglutination assay (RIM, Remel, Lenexa, KS, USA), *Salmonella* latex agglutination assay (Oxoid, Ogdensberg, NY, USA), and API *Listeria* (bioMérieux, Inc. Hazelwood, MO, USA)].

System performance coefficient measurement
The system performance coefficient (SPC) of ohmic heating was determined from temperature, voltage, and current data [17] and calculated as follows (equation 1):

$$SPC = \frac{mCp\Delta T}{\sum \Delta VIt} \tag{1}$$

Where m is mass (g), Cp is specific heat capacity (J/g K), ΔT is difference between final temperature and initial temperature (K), ΔV is voltage applied (V), I is electric current (A), and t is time (s). $\sum \Delta VIt$ is the energy given to the system, $mCp\Delta T$ is energy given to the system minus energy loss during ohmic heating. The ratio of $mCp\Delta T$ to $\sum \Delta VIt$ indicates the system performance coefficient [17].

Color and pH measurement
To assess color changes of treated apple juice, a Minolta colorimeter (CR400; Minolta Co., Osaka, Japan) was used in this study. Color of apple juice were expressed by values of L*, a*, and b* (color lightness, redness, and yellowness, respectively) [20]. A pH meter (Seven Multi 8603; Mettler Toledo, Greifensee, Switzerland) was utilized to measure pH values.

Statistical analysis
All experiments were conducted three times with duplicate samples. Data were analyzed by the ANOVA procedure of SAS (Version 9.2. SAS Institute Inc., NC, USA),

and mean values were separated using Tukey-Kramer's multiple range test. A *P* value of <0.05 was used to indicate significant differences.

Results and discussion

Temperature profiles of different concentrations of apple juice

There are various factors affecting electrical conductivity of liquids. Electrical conductivity relies on chemical components, ion activity, and viscosity of liquids. Such an electrical characteristic, along with juice concentration, could have an influence on temperature rise and microbial inactivation [17, 21]. A study by Palaniappan and Sastry [22] stated that the relationship between electrical conductivity and temperature was linear but conductivity decreased with increasing soluble solids content in tomato and orange juices. The results of the present study were also consistent with previous reports. The heating rates of various concentrations of apple juice during ohmic heating at different voltage gradients are shown in Fig. 2. Temperature rise was more rapid in higher

concentrations than in lower concentrations of juice up to 36 °Brix. However, when approaching 48 °Brix, the rate of temperature increase began to decline. The slowest rate of temperature increase was observed at the maximum sugar concentration (72 °Brix) of apple juice since electric conductivity was suppressed as sugar concentration approached the maximum levels included in this study (data not shown).

System performance efficiency at different concentrations of apple juice and voltage gradients

The system performance coefficient (SPC), which affects processing cost, was considered as an important factor in this study. Icier and Ilicali [17] reported that SPC values of ohmic heating depended strongly on the voltage gradient applied to orange juice concentrates. For the 60 V/cm voltage gradients SPCs were approximately 0.52–0.59, which indicated that 41–48% of the electrical energy applied to the system was not used in heating orange juice concentrates. For low voltage gradients (20 V/cm),

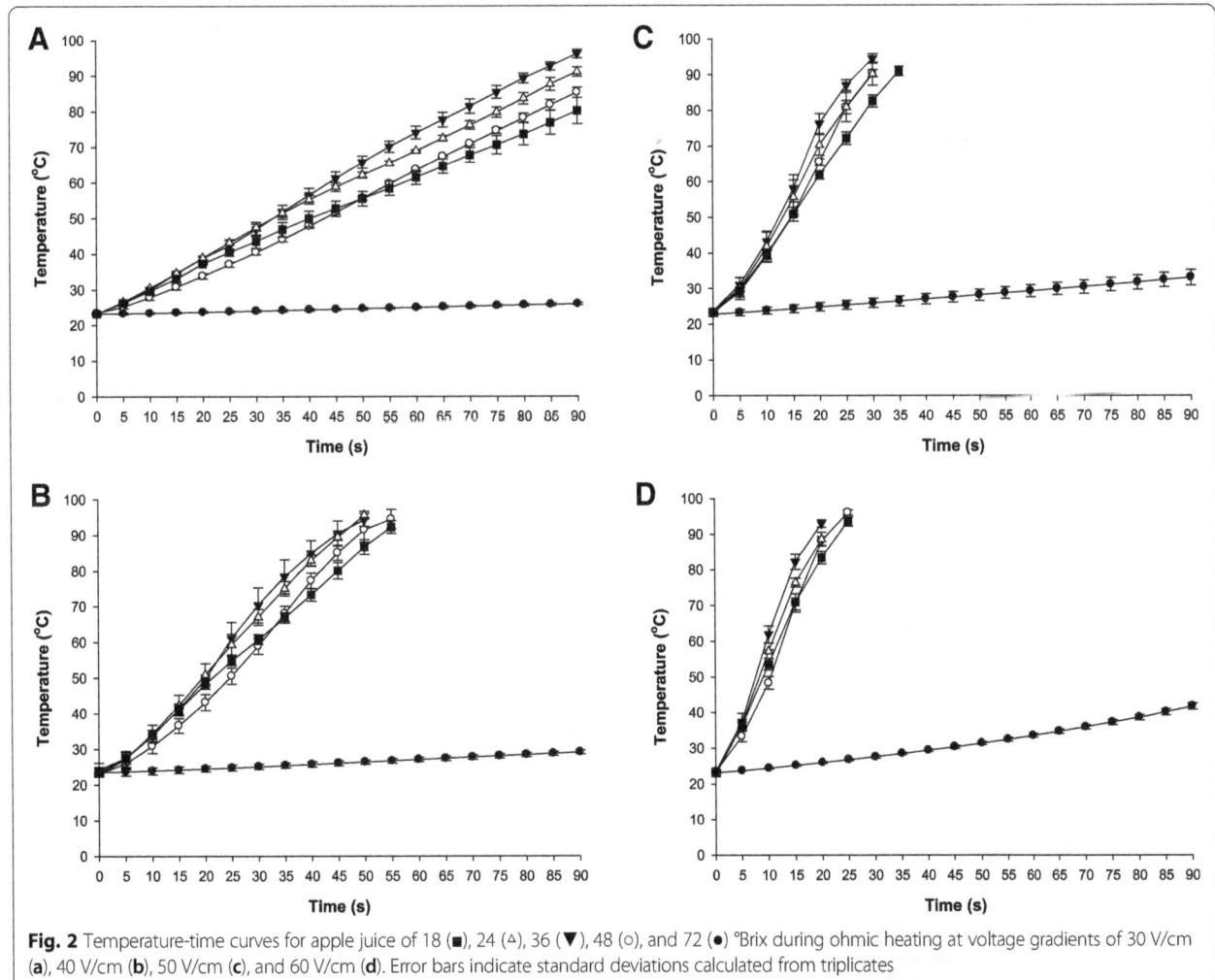

Fig. 2 Temperature-time curves for apple juice of 18 (■), 24 (△), 36 (▼), 48 (○), and 72 (●) °Brix during ohmic heating at voltage gradients of 30 V/cm (**a**), 40 V/cm (**b**), 50 V/cm (**c**), and 60 V/cm (**d**). Error bars indicate standard deviations calculated from triplicates

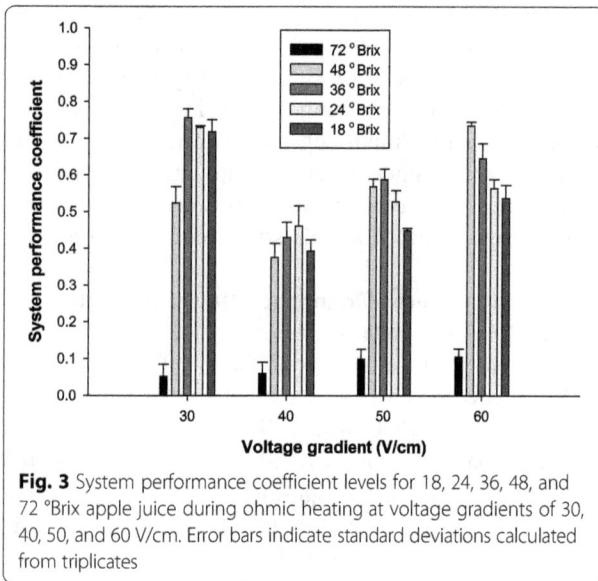

Fig. 3 System performance coefficient levels for 18, 24, 36, 48, and 72 °Brix apple juice during ohmic heating at voltage gradients of 30, 40, 50, and 60 V/cm. Error bars indicate standard deviations calculated from triplicates

the conversion of electrical energy into heat was greater. A similar tendency was also observed in the present study. Figure 3 shows system performance coefficients of ohmic heating at different sample concentrations and voltage gradients. Average SPC values at 40, 50, and 60 V/cm were not as high as that of 30 V/cm. The energy loss at a voltage gradient of 30 V/cm was the lowest when 36 °Brix apple juice was subjected to ohmic heating, which indicated that ca. 75% of the electrical energy applied to the system was utilized for heating (Fig. 3). When treated with 40 V/cm, the worst system performance efficiencies were detected at all sample concentrations. As applied voltage increased, overall SPC gradually increased from 40 to 60 V/cm. Following higher voltage gradients (60 V/cm), the peak system

efficiency was observed in 48 °Brix juice. The SPC value for 48 °Brix apple juice at 60 V/cm, which is the actual electrical energy used to heat the samples, was ca. 73%. In the case of 72 °Brix apple juice, SPC values were absolutely lower than in any other concentration of apple juice (Fig. 3). This can be correlated to electrical conductivity or resistance of juice at higher sugar concentrations.

Effect of ohmic heating for inactivation of foodborne pathogens at different voltage gradients

Control of foodborne pathogens is an important issue for the fruit juice industry and ohmic heating treatment has been considered as one of the promising antimicrobial interventions. In our previous study [18], reduction of *E. coli* O157:H7, *S.* Typhimurium, and *L. monocytogenes* resulting from ohmic heating was significantly higher ($P < 0.05$) than that resulting from conventional heating at equal temperatures of 55, 58, and 60 °C in apple juice. These results showed that electric field-induced ohmic heating led to additional bacterial inactivation due not only to thermal effect but also to electroporation-caused cell damage [18]. As the latest in a series of research studies on ohmic heating of apple juice, we attempted to optimize the processing conditions of ohmic heating based on system performance efficiency and inactivation level of pathogens to provide a practical methodology for the fruit juice industry.

Tables 1, 2 and 3 shows the reduction of *E. coli* O157:H7, *S.* Typhimurium, and *L. monocytogenes* in different apple juice concentrations during ohmic heating, respectively. At 30 V/cm, ohmic heating for 60 s achieved 0.95, 2.59, 6.78, 5.21, and 2.71 log reductions of *E. coli* O157:H7 in 72, 48, 36, 24, and 18

Table 1 Log reductions of *E. coli* O157:H7 in 72, 48, 36, 24, and 18 °Brix apple juice subjected to ohmic heating at 30 and 60 V/cm

Voltage gradient	°Brix	Log reduction [$\log_{10} (N_0/N)$][a] by treatment time (s)													
		0		10		20		30		40		50		60	
30 V/cm	72	0.00 ± 0.00	A	0.44 ± 0.22	B	0.42 ± 0.05	B	0.53 ± 0.16	B	0.56 ± 0.32	B	0.71 ± 0.22	BC	0.95 ± 0.03	C
	48	0.00 ± 0.00	A	0.26 ± 0.07	A	0.34 ± 0.02	A	0.41 ± 0.05	A	0.89 ± 0.22	B	1.38 ± 0.22	C	2.59 ± 0.63	D
	36	0.00 ± 0.00	A	0.28 ± 0.14	A	0.34 ± 0.04	A	0.91 ± 0.37	B	1.40 ± 0.30	C	3.33 ± 0.31	D	6.78 ± 0.11	E
	24	0.00 ± 0.00	A	0.24 ± 0.08	AB	0.22 ± 0.13	AB	0.71 ± 0.48	BC	1.26 ± 0.46	C	2.97 ± 0.47	D	5.21 ± 0.36	E
	18	0.00 ± 0.00	A	0.09 ± 0.08	A	0.17 ± 0.30	AB	0.17 ± 0.06	AB	0.67 ± 0.31	B	1.34 ± 0.50	C	2.71 ± 0.29	D
60 V/cm	72	0.00 ± 0.00	A	0.38 ± 0.22	AB	0.67 ± 0.22	B	0.62 ± 0.06	B	0.55 ± 0.29	AB	0.73 ± 0.32	B	0.73 ± 0.60	B
	48	0.00 ± 0.00	A	0.42 ± 0.13	B	6.33 ± 0.13	C	ND		ND		ND		ND	
	36	0.00 ± 0.00	A	0.82 ± 0.17	B	6.58 ± 0.22	C	ND		ND		ND		ND	
	24	0.00 ± 0.00	A	0.74 ± 0.58	B	6.88 ± 0.06	C	ND		ND		ND		ND	
	18	0.00 ± 0.00	A	0.50 ± 0.41	B	6.93 ± 0.11	C	ND		ND		ND		ND	

[a]The values are means ± standard deviations from three replications. Values in the same row followed by the same letter are not significantly different ($P > 0.05$).
ND not detected.

Table 2 Log reductions of *S.* Typhimurium in 72, 48, 36, 24, and 18 °Brix apple juice subjected to ohmic heating at 30 and 60 V/cm

Voltage gradient	°Brix	Log reduction [$\log_{10}$ (N$_0$/N)][a] by treatment time (s)												
		0		10		20		30		40		50		60	
30 V/cm	72	0.00 ± 0.00	A	0.64 ± 0.30	AB	0.72 ± 0.28	B	1.23 ± 0.28	B	0.90 ± 0.20	B	1.06 ± 0.77	B	1.40 ± 0.44	B
	48	0.00 ± 0.00	A	0.52 ± 0.21	B	0.33 ± 0.15	AB	0.69 ± 0.18	BC	1.08 ± 0.30	C	1.86 ± 0.54	D	2.88 ± 0.48	E
	36	0.00 ± 0.00	A	0.29 ± 0.06	AB	0.43 ± 0.15	B	0.80 ± 0.08	C	1.62 ± 0.25	D	3.99 ± 0.33	E	6.71 ± 0.13	F
	24	0.00 ± 0.00	A	0.22 ± 0.19	A	0.31 ± 0.14	A	0.93 ± 0.26	B	1.81 ± 0.25	C	3.42 ± 0.50	D	6.70 ± 0.16	E
	18	0.00 ± 0.00	A	0.09 ± 0.10	A	0.17 ± 0.11	AB	0.48 ± 0.23	B	0.87 ± 0.07	C	1.44 ± 0.22	D	3.27 ± 0.40	E
60 V/cm	72	0.00 ± 0.00	A	0.61 ± 0.12	AB	0.90 ± 0.30	B	1.00 ± 0.51	B	1.06 ± 0.54	B	1.26 ± 0.30	B	1.20 ± 0.66	B
	48	0.00 ± 0.00	A	0.42 ± 0.44	A	5.81 ± 0.06	B	ND		ND		ND		ND	
	36	0.00 ± 0.00	A	0.83 ± 0.32	B	6.10 ± 0.24	C	ND		ND		ND		ND	
	24	0.00 ± 0.00	A	0.79 ± 0.88	A	6.61 ± 0.13	B	ND		ND		ND		ND	
	18	0.00 ± 0.00	A	0.65 ± 0.45	B	6.68 ± 0.14	C	ND		ND		ND		ND	

[a]The values are means ± standard deviations from three replications. Values in the same row followed by the same letter are not significantly different ($P > 0.05$). *ND* not detected.

°Brix apple juice, respectively. Also, reductions of 1.40, 2.88, 6.71, 6.70, and 3.27 log CFU/ml in concentrations of 72, 48, 36, 24, and 18 °Brix, respectively, were observed in *S.* Typhimurium. In the case of *L. monocytogenes*, levels of log reduction following ohmic heating were 0.47, 1.74, 5.01, 3.91, and 1.13, respectively, in juice concentrations of 72, 48, 36, 24, and 18 °Brix. From these results at 30 V/cm, maximum log reductions of the three foodborne pathogens were observed in 36 °Brix apple juice. Dramatic levels of inactivation were achieved in 18–48 °Brix apple juice during ohmic heating at 60 V/cm. Reductions of *E. coli* O157:H7 were 6.32, 6.58, 6.88, and 6.93 log CFU/ml in 48, 36, 24, and 18 °Brix juice, respectively, after ohmic heating for 20 s. Similarly, ohmic heating for 20 s accomplished 5.80, 6.10, 6.60, and 6.68 log reductions of *S.* Typhimurium in 48, 36, 24, and 18 °Brix juice, respectively. Log reductions of

5.71, 5.70, 5.82, and 5.93 in 48, 36, 24, and 18 °Brix apple juice, respectively, were observed for *L. monocytogenes*. Thus, the time duration required for 5-log reduction at 30 V/cm in 36 °Brix apple juice was three times longer than for 60 V/cm at all apple juice concentrations with the exception of 72 °Brix. Also, commercial processing of higher concentration apple juice has the advantage of greater production yield (of 18 ° Brix juice). Therefore, with respect to bactericidal efficiency, SPC values, and treatment time, ohmic heating application of 60 V/cm in 48 °Brix apple juice could be more efficient than that of 30 V/cm in 36° Brix.

The influence of ohmic heating on quality of apple juice
Additionally, ohmic heating is a suitable technology for minimizing degradation of juice quality due to the fundamental property of ohmic heating, which

Table 3 Log reductions of *L. monocytogenes* in 72, 48, 36, 24, and 18 °Brix apple juice subjected to ohmic heating at 30 and 60 V/cm

Voltage gradient	°Brix	Log reduction [$\log_{10}$ (N$_0$/N)][a] by treatment time (s)												
		0		10		20		30		40		50		60	
30 V/cm	72	0.00 ± 0.00	A	0.34 ± 0.15	A	0.32 ± 0.16	A	0.34 ± 0.06	A	0.32 ± 0.22	A	0.37 ± 0.32	A	0.47 ± 0.49	A
	48	0.00 ± 0.00	A	0.21 ± 0.08	AB	0.36 ± 0.13	B	0.52 ± 0.14	BC	0.73 ± 0.30	C	1.34 ± 0.12	D	1.74 ± 0.24	E
	36	0.00 ± 0.00	A	0.43 ± 0.20	B	0.42 ± 0.14	B	0.67 ± 0.19	B	1.10 ± 0.18	C	1.90 ± 0.22	D	5.01 ± 0.35	E
	24	0.00 ± 0.00	A	0.04 ± 0.08	A	0.13 ± 0.12	A	0.21 ± 0.13	A	0.74 ± 0.20	B	1.18 ± 0.37	C	3.91 ± 0.26	D
	18	0.00 ± 0.00	A	0.07 ± 0.13	A	0.04 ± 0.10	A	0.27 ± 0.19	A	0.23 ± 0.28	A	0.42 ± 0.24	A	1.13 ± 0.41	B
60 V/cm	72	0.00 ± 0.00	A	0.31 ± 0.14	A	0.23 ± 0.12	A	0.33 ± 0.19	A	0.40 ± 0.31	A	0.40 ± 0.25	A	0.38 ± 0.40	A
	48	0.00 ± 0.00	A	0.57 ± 0.28	B	5.71 ± 0.27	C	ND		ND		ND		ND	
	36	0.00 ± 0.00	A	1.46 ± 0.09	B	5.71 ± 0.23	C	ND		ND		ND		ND	
	24	0.00 ± 0.00	A	0.50 ± 0.42	A	5.83 ± 0.13	B	ND		ND		ND		ND	
	18	0.00 ± 0.00	A	0.66 ± 0.11	B	5.94 ± 0.20	C	ND		ND		ND		ND	

[a]The values are means ± standard deviations from three replications. Values in the same row followed by the same letter are not significantly different ($P > 0.05$). *ND* not detected.

Table 4 Color values[b] and pH of treated and untreated apple juice of 18, 24, 36, 48, and 72 °Brix at 30 and 60 V/cm following ohmic heating

Voltage gradient	Mean ± SD[a]					
	Solids content (°Brix)	Treatment time (s)	pH	Color[b]		
				L*	a*	b*
30 V/cm	72	0	3.42 ± 0.00	26.47 ± 0.06	0.38 ± 0.01	4.11 ± 0.01
		60	3.42 ± 0.01	26.44 ± 0.08	0.38 ± 0.02	4.09 ± 0.05
	48	0	3.51 ± 0.00	25.43 ± 0.29	0.49 ± 0.03	4.66 ± 0.09
		60	3.51 ± 0.01	25.47 ± 0.72	0.47 ± 0.11	4.56 ± 0.22
	36	0	3.54 ± 0.00	24.85 ± 0.10	0.49 ± 0.04	5.16 ± 0.05
		60	3.54 ± 0.01	24.76 ± 0.19	0.54 ± 0.02	5.05 ± 0.19
	24	0	3.57 ± 0.01	24.72 ± 0.65	0.32 ± 0.04	5.02 ± 0.35
		60	3.57 ± 0.01	24.55 ± 0.08	0.38 ± 0.01	5.46 ± 0.09
	18	0	3.59 ± 0.00	24.23 ± 0.23	0.25 ± 0.02	5.30 ± 0.43
		60	3.60 ± 0.00	24.51 ± 0.19	0.27 ± 0.02	5.58 ± 0.12
60 V/cm	72	0	3.45 ± 0.01	26.01 ± 0.05	0.37 ± 0.03	4.10 ± 0.11
		60	3.44 ± 0.00	26.03 ± 0.02	0.36 ± 0.07	4.19 ± 0.08
	48	0	3.52 ± 0.01	25.36 ± 0.23	0.47 ± 0.02	4.26 ± 0.06
		20	3.53 ± 0.00	25.32 ± 0.39	0.46 ± 0.09	4.38 ± 0.15
	36	0	3.54 ± 0.01	24.56 ± 0.21	0.49 ± 0.01	5.28 ± 0.08
		20	3.54 ± 0.01	24.55 ± 0.11	0.52 ± 0.09	5.17 ± 0.02
	24	0	3.56 ± 0.00	24.45 ± 0.42	0.36 ± 0.01	5.39 ± 0.31
		20	3.55 ± 0.00	24.55 ± 0.18	0.37 ± 0.04	5.41 ± 0.19
	18	0	3.58 ± 0.01	24.43 ± 0.43	0.28 ± 0.07	5.35 ± 0.03
		20	3.57 ± 0.01	24.33 ± 0.12	0.27 ± 0.01	5.42 ± 0.10

[a]Results are expressed as means ± SD. Values in the same column are not significantly different ($P > 0.05$)
[b]Color values are L* (lightness), a* (redness), and b* (yellowness)

generates internal heat in food materials [14]. Color and pH values of 18, 24, 36, 48, and 72 °Brix apple juice following ohmic heating at 30 and 60 V/cm are shown in Table 4. All experiments were limited to a maximum treatment time of 60 s. In case of 60 V/cm, treatment time was restricted to 20 s in 18, 24, 36, and 48 °Brix apple juice because 20 s was a sufficient time interval for obtaining the target microbial reductions. L*, a*, and b* values of samples treated versus not treated with ohmic heating were not significantly ($P > 0.05$) different. The pH values of treated samples did not significantly differ from those of non-treated samples. Thus, the proposed parameters for optimal ohmic heating did not significantly affect the quality of apple juice product (Table 4).

Although ohmic heating is no longer regarded as a new technology, target microbe reductions have to be assessed in new application environments which include product type and production setting. In this study, optimized voltage gradient and juice concentration for ohmic heating gave a distinct advantage in terms of both bactericidal and economic aspects but also ensured minimal quality loss. However,

since ohmic heating was performed in a small-scale batch system, energy and performance criteria have limited significance relative to larger-scale processing units. Therefore, further research incorporating more sophisticated experimental conditions to industrial-scale continuous systems is needed.

Conclusions

Novel thermal processing interventions employed by the fruit juice industry for controlling foodborne pathogens involve the utilization of sophisticated systems, which enable reduced processing times and temperatures to prevent loss of nutritional and sensory quality while still securing outstanding bactericidal efficacy. Ohmic heating is one of the most promising thermal technologies for effectively inactivating foodborne pathogens in this respect. In the present study, the optimum processing parameters of ohmic heating treatment such as applied voltage gradients and °Brix of apple juice concentrates were investigated to provide benefits with regard to bactericidal, sensory, and economic aspects. These results can be utilized by the apple juice industry for effective application of ohmic heating.

Abbreviations
ATCC: American type culture collection; CFU: Colony forming unit; OAB: Oxford agar base; PW: Peptone water; SMAC: Sorbitol MacConkey agar; SPC: System performance coefficient; TSA: Tryptic soy agar; TSB: Tryptic soy broth; VAC: Volts alternating current; XLD: Xylose lysine desoxycholate agar

Funding
This research was supported by the Agriculture, Food and Rural Affairs Research Center Support Program, Ministry of Agriculture, Food and Rural Affairs, Republic of Korea. This research was also supported by the Public Welfare & Safety research program through the National Research Foundation of Korea (NRF) funded by the Ministry of Science, ICT and Future Planning (NRF-2012M3A2A1051679).

Authors' contributions
All authors contributed to the study design, interpretation of the data, intellectual discussion and/or revision of the manuscript. IKP planned and performed the laboratory work. DHK and JWH supervised the study and drafted the manuscript. All authors have read and approved the final version of the manuscript before submission.

Competing interests
The authors declare that they have no competing interest.

Author details
[1]Department of Food and Animal Biotechnology, College of Agricultural Biotechnology, Center for Food and Bioconvergence, and Institute of GreenBio Science & Technology, Research Institute for Agricultural and Life Sciences, Seoul National University, Seoul 08826, Korea. [2]Department of Food and Biotechnology, College of Engineering, Food & Bio-industry Research Center, Hankyong National University, Anseong-si 17579, Korea.

References
1. Feng P. *Escherichia coli* O157:H7: novel vehicles of infection and emergence of phenotypic variants. Emerg Infect Dis. 1995;1:47–52.
2. Sung HJ, Song WJ, Kim KP, Ryu S, Kang DH. Combination effect of ozone and heat treatments for the inactivation of *Escherichia coli* O157:H7, *Salmonella* Typhimurium, and *Listeria monocytogenes* in apple juice. Int J Food Microbiol. 2014;171:147–53.
3. Mihajlovic B, Dixon B, Couture H, Farber J. Qualitative microbiological risk assessment of unpasteurized fruit juice and cider. International Food Risk Analysis Journal. 2013;3:1–19.
4. Cody SH, Glynn MK, Farrar JA, Cairns KL, Griffin PM, Kobayashi J, Fyfe M, Hoffman R. An outbreak of *Escherichia coli* O157:H7 infection from unpasteurized commercial apple juice. Ann Intern Med. 1999;130:202–9.
5. Jain S, Bidol SA, Austin JL, Berl E, Elson F, Lemaile-Williams M, Deasy M, Moll ME. Multistate outbreak of *Salmonella* Typhimurium and Saint-paul infections associated with unpasteurized orange juice–United States, 2005. Clin Infect Dis. 2009;48:1065–71.
6. U.S. Food and Drug Administration. Hazard analysis and critical control points (HACCP); procedures for the safe and sanitary processing and importing of juice. Federal Register. 1998;63:20450–86.
7. Kroll RG, Patchett RA. Induced acid tolerance in *Listeria monocytogenes*. Lett Appl Microbiol. 1992;14:224–7.
8. Kenney SJ, Burnett SL, Beuchat LR. Location of *Escherichia coli* O157:H7 on and in apples as affected by bruising, washing, and rubbing. J Food Prot. 2001;64:1328–33.
9. Roering AM, Luchansky JB, Ihnot AM, Ansay SE, Kaspar CW, Ingham SC. Comparative survival of *Salmonella typhimurium* DT 104, *Listeria monocytogenes*, and *Escherichia coli* O157:H7 in preservative-free apple cider and simulated gastric fluid. Int J Food Microbiol. 1999;46:263–9.
10. U.S. Food and Drug Administration. Guidance for Industry: Juice HACCP Hazards and Controls Guidance First Edition; Final Guidance. 2004.
11. Aguilar-Rosas SF, Ballinas-Casarrubias ML, Nevarez-Moorillon GV, Martin-Belloso O, Ortega-Rivas E. Thermal and pulsed electric fields pasteurization of apple juice : Effects on physicochemical properties and flavour compounds. J Food Eng. 2007;83:41–6.
12. Choi LH, Nielsen SS. The effects of thermal and nonthermal processing methods on apple cider quality and consumer acceptability. J Food Qual. 2004;28:13–29.
13. Pereira RN, Vicente AA. Environmental impact of novel thermal and non-thermal technologies in food processing. Food Res Int. 2010;43:1936–43.
14. Lee SY, Sagong HG, Ryu S, Kang DH. Effect of continuous ohmic heating to inactivate *Escherichia coli* O157:H7, *Salmonella* Typhimurium, and *Listeria monocytogenes* in orange juice and tomato juice. J Appl Microbiol. 2012;112:723–31.
15. Sastry SK, Barach JT. Ohmic and inductive heating. J Food Sci. 2000;65:42–6.
16. Castro I, Teixeira JA, Salengke S, Sastry SK, Vicente AA. The influence of field strength, sugar and solid content on electrical conductivity of strawberry products. J Food Process Eng. 2003;26:17–30.
17. Icier F, Ilicali C. The effects of concentration on electrical conductivity of orange juice concentrates during ohmic heating. Eur Food Res Technol. 2005;220:406–14.
18. Park IK, Kang DH. Effect of electropermeabilization by ohmic heating for inactivation of *Escherichia coli* O157:H7, *Salmonella enterica* Serovar Typhimurium, and *Listeria monocytogenes* in buffered peptone water and apple juice. Appl Environ Microbiol. 2013;79:7122–9.
19. Lee SY, Ryu SR, Kang DH. Effect of frequency and waveform on inactivation of *Escherichia coli* O157:H7 and *Salmonella enterica* serovar Typhimurium in salsa by ohmic heating. Appl Environ Microbiol. 2013;79:10–7.
20. Chen Z, Zhu C, Zhang Y, Niu D, Du J. Effects of aqueous chlorine dioxide treatment on enzymatic browning and shelf-life of fresh-cut asparagus lettuce (*Lactuca sativa* L.). Postharvest Biol Technol. 2010;58:232–8.
21. Huixian S, Shuso K, Jun-ichi H, Kazuhiko I, Tatsuhiko W, Toshinori K. Effects of ohmic heating on microbial counts and denaturation of proteins in milk. Food Sci Technol Res. 2007;14:117–23.
22. Palaniappan S, Sastry SK. Electrical conductivity of selected juices : influences of temperature, solids content, applied voltage, and particle size. J Food Process Eng. 1991;14:247–60.

9

Environmentally triggered genomic plasticity and capsular polysaccharide formation are involved in increased ethanol and acetic acid tolerance in *Kozakia baliensis* NBRC 16680

Julia U. Brandt, Friederike-Leonie Born, Frank Jakob[*] and Rudi F. Vogel

Abstract

Background: *Kozakia baliensis* NBRC 16680 secretes a *gum*-cluster derived heteropolysaccharide and forms a surface pellicle composed of polysaccharides during static cultivation. Furthermore, this strain exhibits two colony types on agar plates; smooth wild-type (S) and rough mutant colonies (R). This switch is caused by a spontaneous transposon insertion into the *gumD* gene of the *gum*-cluster, resulting in a heteropolysaccharide secretion deficient, rough phenotype. To elucidate, whether this is a directed switch triggered by environmental factors, we checked the number of R and S colonies under different growth conditions including ethanol and acetic acid supplementation. Furthermore, we investigated the tolerance of R and S strains against ethanol and acetic acid in shaking and static growth experiments. To get new insights into the composition and function of the pellicle polysaccharide, the *polE* gene of the R strain was additionally deleted, as it was reported to be involved in pellicle formation in other acetic acid bacteria.

Results: The number of R colonies was significantly increased upon growth on acetic acid and especially ethanol. The morphological change from *K. baliensis* NBRC 16680 S to R strain was accompanied by changes in the sugar contents of the produced pellicle EPS. The R:Δ*polE* mutant strain was not able to form a regular pellicle anymore, but secreted an EPS into the medium, which exhibited a similar sugar monomer composition as the pellicle polysaccharide isolated from the R strain. The R strain had a markedly increased tolerance towards acetic acid and ethanol compared to the other NBRC 16680 strains (S, R:Δ*polE*). A relatively high intrinsic acetic acid tolerance was also observable for *K. baliensis* DSM 14400[T], which might indicate diverse adaptation mechanisms of different *K. baliensis* strains in altering natural habitats.

Conclusion: The results suggest that the genetically triggered R phenotype formation is directly related to increased acetic acid and ethanol tolerance. The *polE* gene turned out to be involved in the formation of a cell-associated, capsular polysaccharide, which seems to be essential for increased ethanol/acetic tolerance in contrast to the secreted *gum*-cluster derived heteropolysaccharide. The genetic and morphological switch could represent an adaptive evolutionary step during the development of *K. baliensis* NBRC 16680 in course of changing environmental conditions.

Keywords: *Kozakia baliensis*, Heteropolysaccharides, Pellicle, Ethanol/acetic acid tolerance, Adaptive evolution

* Correspondence: frank.jakob@wzw.um.de
Technische Universität München, Lehrstuhl für Technische Mikrobiologie, Gregor-Mendel-Straße 4, 85354 Freising, Germany

Background

Gram-negative bacteria produce extracellular hetero-(HePS) or homopolysaccharides (HoPS), which are attached to the bacterial cell as capsular polysaccharide (CPS) or secreted into the environment as extracellular polysaccharide (EPS). Bacterial polysaccharides are important for the survival of bacteria, for instance in bacteria–host interaction, biofilm formation [1] and stress adaptation [2].

Acetic acid bacteria (AAB) are obligate aerobes and belong to the class of α-Proteobacteria. They are oxidative bacteria that strongly oxidize ethanol to acetic acid. AAB are well known for their ability to produce large amounts of EPSs, either HoPS, like dextrans, levans [3–5] and cellulose, or different kinds of HePS, such as acetan [6] and gluconacetan [7]. Furthermore, a variety of AAB has the ability to grow floating on the surface of a static culture by producing a pellicle enabling a high aeration state. The pellicle consists of an accumulation of cells, which are tightly associated with each other by capsular polysaccharides as connecting element. The pellicle CPS can be a HoPS of cellulose, which is produced by many *Komagataeibacter* species, like *Komagataeibacter (K.) xylinus* [8] (formerly *Gluconobacter (G.) xylinum* [9]), or a HePS, such as produced by many *Acetobacter* strains [10–12]. This HePS can be composed of different sugar monomers, like the HePS of *A. tropicalis* SKU1100. which consists of glucose, galactose, and rhamnose [13] or of *A. aceti* IFO3284 that contains only glucose and rhamnose [10].

The genes involved in the cellulose pellicle biosynthesis are arranged in an operon structure, like the acs operon [13] or the bcs operon [14], and widely studied. It is assumed that the genes involved in the synthesis of other pellicle HePS in AAB are assigned to a particular cluster, called *pol*-cluster. Deeraksa et al. (2005) could show, that the pellicle HePS produced by *A. tropicalis* SKU1100 could be traced back to a gene cluster, *polABCDE*, which is required for pellicle formation. In this operon, the *polABCD* genes showed high similarity to *rfbBACD* genes, which are involved in dTDP-l-rhamnose biosynthesis. The downstream located *polE* gene showed only low similarity to known glycosyltransferases, whereas a transposon-induced disruption of the *polE* gene resulted in a non pellicle forming strain, due to the absence of CPS production. Instead of the CPS, however, the Pel⁻ strain showed a smooth-surfaced colony and the HePS was now secreted into the medium, which had the same composition as the capsular pellicle polysaccharide [11].

Acetobacter species are further known to exhibit high natural mutation frequencies [15, 16], also resulting in the formation of two or more different colony types. *A. pasteurianus* IFO3284 produces two altering types of colonies on agar medium that are inter-convertible by spontaneous mutation; a rough surface colony that can produce a pellicle (R strain) and smooth surface colony, which cannot produce a pellicle (S strain) [17]. Furthermore the R strains tolerate higher concentrations of acetic acid, whereas the pellicle formation is directly related to acetic acid resistance [18, 19]. It is further assumed that the pellicle CPS functions as a barrier-like biofilm against passive diffusion of acetic acid into the cells [18].

Microorganisms constantly face many difficult challenges, due to changing environmental conditions. The capacity to maintain functional homeostasis is essential for their survival. Recently, we have shown that the AAB *Kozakia (K.) baliensis* NBRC 16680 forms large amounts of a soluble unique HePS in the medium, as well as a pellicle during static cultivation [20]. Furthermore, *K. baliensis* NBRC 16680 forms a second type of colony form: a non-HePS producing rough-surfaced colony (R strain), caused by a transposon insertion into the *gumD* gene of the corresponding *gum*-like HePS cluster. The reason for this transposon insertion is unclear, whereby it was assumed that the transposon insertion represents a directed event, triggered by external factors.

Therefore, we investigated in this study, whether the morphology switch of *K. baliensis* NBRC 16680 is a random event, or triggered by environmental adaptations. In order to check if the morphological switch is connected to the formation of a CPS used in pellicle formation, we performed a ΔpolE deletion in *K. baliensis* NBRC 16680 R via a two step marker less gene deletion system. The different mutants were investigated regarding their growth and EPS production under different growth conditions including ethanol and acetic acid stress.

Methods

Bacterial strains, culture media, and culture conditions.

K. baliensis (NBRC 16680; National Institute of Technology and Evaluation (NITE) Biological Resource Center, Japan, DSM 14400; German Collection of Microorganisms and Cell Cultures(DSMZ)), as well as a mutant strain of *K. baliensis* NBRC 16680 R (ΔgumD) [20], with a rough phenotype, were used in this study. *K. baliensis* and its derivatives were grown at 30 °C in NaG media consisting of 20 g/L sodium gluconate, 3 g/L yeast extract, 2 g/L peptone, 3 g/L glycerol, 10 g/L mannitol and a pH adjusted to 6.0. *E. coli* strain TOP10 (Invitrogen, Karlsruhe, Germany), was grown at 37 °C and 180 rpm on a rotary shaker in LB medium consisting of 5 g yeast extract, 5 g NaCl and 10 g peptone. For selection of plasmids in *E. coli* TOP10. 50 μg/mL kanamycin was added to the LB medium. For selection of recombinant *K. baliensis* strains 50 μg/mL kanamycin or 60 μg/mL Fluorocytosin (FC) were used.

Convertibility of K. baliensis NBRC 16680 from wild-type (S) to rough strains (R)

The number of R strains was determined during/after cultivation of *K. baliensis* NBRC 16680 in standard NaG medium, NaG medium supplemented with 3% ethanol (NaG-EtOH) or 0.4% acetic acid (NaG-AA), respectively. *K. baliensis* NBRC 16680 was first cultivated overnight in 10 ml of unmodified NaG medium at 30 °C (200 rpm). About 1×10^8 CFU/mL seed culture was afterwards transferred to 10 ml of standard or modified NaG medium, respectively. The flasks were incubated at 30 °C with rotary shaking at 200 rpm for 48 h and samples adducted at 0, 24 and 48 h. Cell numbers of wild-type and rough colonies (*ΔgumD*) were counted on non-modified NaG agar plates. Each growth experiment in liquid culture was performed thrice in separate assays, while each assay contained further three technical plating replicates.

For targeting the transposon insertion side, random colony PCRs of 23 rough colonies were carried out with Phire Hot start DNA polymerase (Thermo Fisher scientific; Waltham, USA). A primer set of a genomic primer (G4F_Fw) and a primer, targeting the mobile element (TE_Rv) were used; primers are listed in Additional file 1. PCR products were subsequently sequenced via Sanger sequencing by GATC Biotech (Konstanz, Germany).

Deletion of the *polE* gene with a two step marker less gene deletion system

For plasmids preparation, the GeneJET Plasmid Miniprep Kit (Thermo Fisher scientific, Waltham, USA) was used. Genomic DNA from *K. baliensis* NBRC 16680 R was extracted with the E.Z.N.A. Bacterial DNA Kit (Omega Bio-tek, Norcross, USA) and DNA purification was done with the E.Z.N.A. Cycle-Pure Kit (Omega Bio-tek, Norcross, USA). Restriction enzymes, DNA ligase, and alkaline phosphatase (FastAP) were obtained from Fermentas (Waltham, USA). PCRs were performed according to the Phusion High-Fidelity DNA Polymerase manuals from New England Biolabs (Frankfurt, Germany). For construction of the deletion vector, a fusion PCR technique was used to ligate the PCR products of flanking regions according to a long flanking homology (LFH) protocol [21, 22]. The length of the homology sequences were 20 bp. Primers are listed in Additional file 1. An enzyme-free cloning technique [23] was used for the further construction of the deletion vector, with the pKOS6b plasmid as basis, including a multiple cloning site (MCS), a kanamycine resistance gene (KM^R) and the *codA* and *codB* gene (Additional file 2A) [24]. Flanking regions of the *polE* gene, covering approximately 950 bp of the upstream and downstream region of the *polE* gene were amplified via PCR. The upstream region of 957 bp was amplified with primer P1_*polE*_KpnI_Fw and P2-*polE*_Rv, and a second primer set amplified the downstream region (968 bp) of the *polE* gene, containing the primers P3_*polE*_Fw and P4-*polE*_XbaI_Rv (Additional file 1 & Additional file 2A). In the following step, a LFH PCR [21] was performed with P1_*polE*_KpnI_Fw and P4_*polE*_XbaI_Rv to merge the two

previously amplified fragments (1898 bp). The fused fragment, as well as the pKOS6b vector, were both digested (KpnI, XbaI) and finally ligated. The resulting deletion vector, pKOS6b*ΔpolE*, was verified via sanger sequencing (pK18MCS_Fw & pK18MCS_Rv), and further amplified in *E. coli* TOP 10. The transformation of pKOS6b*ΔpolE* into *K. baliensis* NBRC 16680 R, was carried out by electroporation [25–28] with Gene Pulser Xcell™ Electroporation Systems from Bio Rad (München, Germany). Therefore, cells were inoculated to an OD_{600} of 0.3 in NaG medium and finally grown to an OD_{600} of 0.9. The culture was centrifuged at 5000 g, at 4 °C for 10 min, and washed three times in 1 mM HEPES buffer (pH 7). Cells were resuspended in 1 mM HEPES buffer, supplemented with ¼ volume of glycerin and shock frozen in 50 µL aliquots. The electroporation took place in cuvettes with 2 mm electrode distance from Bio Rad (München, Germany). The electroporation was carried out under constant conditions: 2.5 kV, 25 µF, and 400 Ω. Fresh enriched NaG medium (450 mM mannitol, 15 g/L yeast extract, 15 mM CaCl$_2$, 10 mM MgSO$_4$ and 6 mM glycerin) was added immediately after the pulse. The treated cells were incubated on a rotary shaker over 14 h and subsequently plated on NaG plates containing 50 µg/mL kanamycin for the first selection step with an incubation time of 48 h. During the first recombination step on NaG-Kan plates, a random chromosomal integration of the plasmid took place, which was checked by colony PCR using a specific primer set of a plasmid (pK18MCS_Fw or pK18MCS_Rv) and a genome (CL_*polE*_Fw or CL_*polE*_Rv) specific primer. Phire Hot start DNA polymerase (Thermo Fisher scientific; Waltham, USA) was used for colony PCR reactions, to screen for mutants or to confirm integration of the deletion vector into the genome. PCR products were sequenced via sanger sequencing by GATC Biotech (Konstanz, Germany). The positive clones were further grown on NaG-plates with 60 µg/ml FC, to drive the directed loss of the plasmid and the final selection of *K. baliensis* NBRC 16680 R *ΔpolE* mutant colonies. After 3 days of incubation, the correct *ΔpolE* mutant colonies could be identified via colony PCR, with a genome specific primer set (CL_*polE*_Fw & CL_*polE*_Rv) resulting in a 1950 bp fragment for the *ΔpolE* mutant and a 3000 bp fragment for *K. baliensis* NBRC 16680 R (Additional file 2C).

Growth behavior of different *K. baliensis* strains in acetic acid and ethanol

*K. bali*ensis DSM 14400. NBRC 16680. the *ΔgumD* mutant [20] and the *ΔpolE* mutant strain (see 2.3) were grown on NaG agar plates, directly plated from the particular cryo stock, with either ethanol or acetic acid, in different concentrations. The ethanol supplemented plates contained 1% - 10% of ethanol (*v*/v, at intervals of 1%) and the acetic acid supplemented plates 0.1% - 1% of acetic acid (*v*/v, at intervals of 0.1%). Each strain was streaked onto the plates from cryo-stocks and incubated for 3 days at 30 °C.

Furthermore, a static cultivation in NaG medium with 3% ethanol and 0.6% acetic acid was carried out. The *K. baliensis* NBRC 16680 strain, R mutant [20] and the *ΔpolE* mutant (see below) were grown as seed cultures in unmodified NaG media, overnight. Cultures were inoculated with an OD_{600} 0.3 and cultivated up to 0.9, respectively. For static cultures, 300 μl of the seed culture were inoculated into 3 ml NaG medium and cultivated statically at 30 °C. Cells were harvested by centrifugation at 6000 g and cell pellets were dried overnight at 120 °C. The dry weight was measured each day, over a time span of 7 days.

Analysis of HePS composition

Main cultures of *K. baliensis* NBRC 16680, *ΔgumD* mutant [20] and *ΔpolE* mutant (see below) were performed in 500 mL Erlenmeyer flasks with 50 mL of modified NaG media, inoculated with 500 μl of the pre-cultures and kept at 30 °C in a rotary shaker (200 rpm) for 32 h. Afterwards, cells were removed and the EPS containing supernatants were precipitated with cold ethanol (2:1, *v/v*) and kept overnight at 4 °C. This step was repeated three times, followed by a dialysis step (MWCO 14 kDa) of the recovered (centrifugation) and in ddH₂O re-dissolved HePS. Finally, the purified HePSs were lyophilized and quantified by weighing. To obtain large amounts of pellicle EPS, *K. baliensis* NBRC 16680 R [20] and the *ΔpolE* mutant (see 2.3) were cultured in unmodified NaG medium in cell culture flasks (Greiner Bio-One, Austria), to ensure a large surface for oxygen supply. Briefly, 10% of the seed culture was inoculated to 30 mL NaG medium and incubated statically at 30 °C for 14 days. The pellicle EPS was purified from the culture and cells were separated from EPS via ultra-sonification (10 min) and mechanical disruption, related to the method of IAI Ali, Y Akakabe, S Moonmangmee, A Deeraksa, M Matsutani, T Yakushi, M Yamada and K Matsushita [11]. The culture was centrifuged (10 min, 10.000 g) and the supernatant was saved in another flask. The cell pellet was washed 2 times with 10 mM HEPES buffer (pH 7), and suspended in the same buffer. The suspension was again ultra-sonicated for 10 min, followed by a centrifugation step for 10 min, 13.000 g. The resulting supernatant was combined, with the present supernatant from the first centrifugation, and EPS was precipitated with cold ethanol (2:1, *v/v*) and kept overnight at 4 °C. This step was repeated three times, followed by a dialysis step (MWCO 14 kDa) of the recovered (centrifugation) and in ddH₂O re-dissolved EPS.

The monosaccharide composition of the isolated *K. baliensis* NBRC 16680 R pellicle or secreted HePS (NBRC 16680, *ΔpolE*) was investigated via high performance liquid chromatography (HPLC). For HPLC analysis the purified polysaccharide samples were hydrolyzed

with 10% of perchloric acid over 7 h at 100 °C, followed by a centrifugation step (4 °C, 10 min, 13,000 g) for removal of possible impurities, such as proteins. For the HPAEC analysis polysaccharide samples were hydrolyzed with 10% of perchloric acid over 2 h at 100 °C, as well followed by a centrifugation step (4 °C, 10 min, 13,000 g). The samples were further dissolved (1:10 or 1:100). The supernatant was analyzed using a Rezex RPM column (Phenomenex, Germany) coupled to a refractive index (RI) detector (Gynkotek, Germany) corresponding to the method of [29]. Sugar monomers were identified according to their retention time using suitable monosaccharide standards (D-glucose, D-galactose, D-mannose, D-rhamnose). The mobile phase was water, with a flow rate of 0.6 mL/min.

Results

Mutation of *K. baliensis* NBRC 16680 from S to R phenotype in dependence of different growth conditions

In *K. baliensis* NBRC 16680 spontaneous mutations occur, which cause a non-slimy phenotype, referred as rough (R) strain. In a previous work we have demonstrated, that this mutation can result from a transposon insertion in the *gumD* gene of the HePS forming cluster of *K. baliensis* NBRC 16680 [20]. The *gumD* gene encodes the first step of HePS formation, whereas a loss of the functional *gumD* gene leads to a total disruption of the HePS production and secretion in *K. baliensis* NBRC 16680. To clarify, if this mutation is a random event, or if it is a directed mutation possibly triggered by environmental factors, we performed an experimental series using different growth media. Different sugar combinations were tested (see 2.2) as well as stress inducing conditions like growth in ethanol (3%) or in acetic acid (0.6%) supplemented media. In addition, dilutions of the cryo-culture of *K. baliensis* NBRC 16680 were directly plated on NaG plates. A distinction was made between slimy glossy wild-type colonies (S) and rough dull mutant colonies (R), which were transparent, held against the light. The NaG plates with the directly plated dilutions of the cryo-cultures from *K. baliensis* NBRC 16680 showed only wild-type colonies. At 0 h, i. e. after inoculation with the overnight culture, solely wild-type colonies could be identified in standard NaG-medium, with cell numbers of about 1×10^8 CFU/mL (Fig. 1a). On the contrary, after inoculation of *K. baliensis* NBRC 16680 in NaG medium supplemented with ethanol or acetic acid, rough colonies could be detected at time point 0 in one of the three biological replicates, respectively (Fig. 1b*c*), which could have resulted from a too long contact to (and concomitant mutations in response to) acetic acid or ethanol. At 24 h the cell numbers of rough and wild-type strains were around $1,4 \times 10^9$ CFU/mL in the NaG medium and bacteria merged already into the starvation phase, with final cell counts of about 1×10^9 CFU/mL after 48 h (Fig. 1a). In

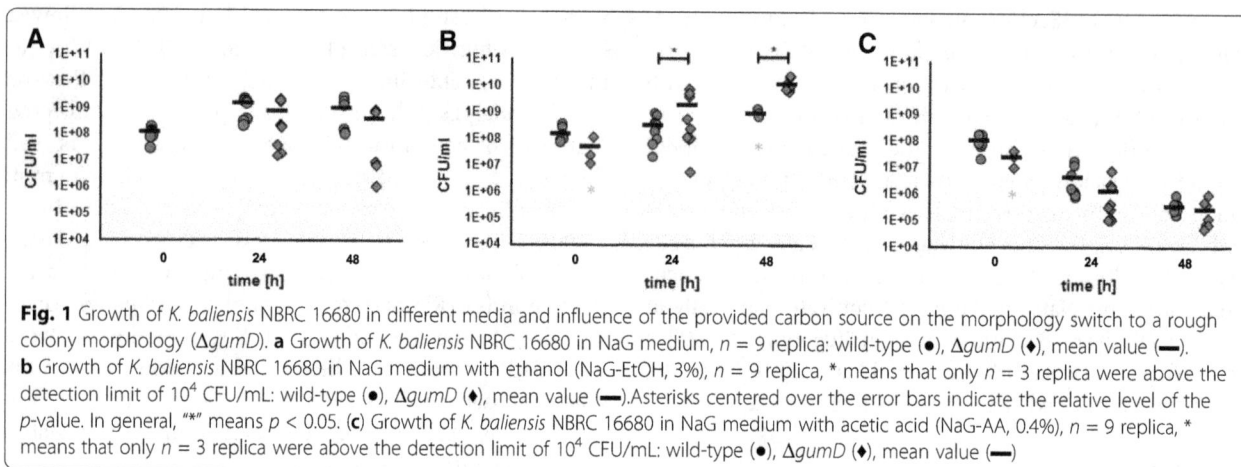

Fig. 1 Growth of *K. baliensis* NBRC 16680 in different media and influence of the provided carbon source on the morphology switch to a rough colony morphology (Δ*gumD*). **a** Growth of *K. baliensis* NBRC 16680 in NaG medium, *n* = 9 replica: wild-type (●), Δ*gumD* (◆), mean value (▬). **b** Growth of *K. baliensis* NBRC 16680 in NaG medium with ethanol (NaG-EtOH, 3%), *n* = 9 replica, * means that only *n* = 3 replica were above the detection limit of 10^4 CFU/mL: wild-type (●), Δ*gumD* (◆), mean value (▬). Asterisks centered over the error bars indicate the relative level of the *p*-value. In general, "*" means *p* < 0.05. **(c)** Growth of *K. baliensis* NBRC 16680 in NaG medium with acetic acid (NaG-AA, 0.4%), *n* = 9 replica, * means that only *n* = 3 replica were above the detection limit of 10^4 CFU/mL: wild-type (●), Δ*gumD* (◆), mean value (▬)

standard NaG medium the number of wild-type colonies exceled the number of mutant colonies or showed approximately equal numbers of wild-type and mutant colonies. During stress-inducing conditions (NaG-EtOH, NaG-AA), however, for NaG-EtOH an inverted picture emerged. At both survey marks (24 h, 48 h) significantly more mutant (R) than wild-type colonies were detectable, with around ten-power difference (Fig. 1b). After 48 h only in three of nine cases, wild-type colonies with a detection limit above of 10^4 CFU/mL could be detected (Fig. 1b*). In the NaG-AA medium, a continuous reduction of the cell numbers could be observed, in which the number of wild-type colonies always exceeded the number of mutated colonies (Fig. 1c).

In order to verify a possible integration of a mobile element in the *gumD* gene [20], or in front of the *gumD* gene, random colony PCR reactions were carried out, with a forward primer (G4F_Fw) targeting a location in front of the *gumD* gene (1.471.189–1.471.208 bp) and a reverse primer (TE-Rv) targeting the mobile element (Additional file 1) (Fig. 2b). Subsequently, the obtained PCR fragments were sequenced. For each mutated colony, a transposon insertion in the region of the *gumD* gene could be identified (Fig. 2c). These insertions, however, were not always located at the same site, but in a defined region of about 300 bp, around and in the *gumD* gene (Fig. 2c). Furthermore, it could be observed that in certain areas an integration of the mobile element occurred more often than in other areas, like up to nine times at 1471567 bp.

Influence of the *polE* gene on HePS formation and composition of *K. baliensis* NBRC 16680

In order to decode the role of the *polE* gene during pellicle formation of *K. baliensis* NBRC 16680 a *polE* deletion was carried out (see 2.3), with a markerless deletion system established by Kostner et al. [22, 24]. The pellicle forming ability of *K. baliensis* NBRC 16680 R Δ*polE* was

further analyzed under static conditions in NaG-medium for 5 days. The growth behavior, as well as the HePS production, was compared with the wild-type and the rough mutant strain of *K. baliensis* NBRC 16680. All three *K. baliensis* strains showed pellicle formation after three days (Fig. 3a). For the wild-type strain of *K. baliensis* NBRC 16680 as well as for the rough mutant strain, a distinct pellicle formation at the edge of the test-tube was visible, which spreads over the entire boundary surface after three days. *K. baliensis* NBRC 16680 R Δ*polE* showed only a slight pellicle production (Fig. 3a). The physiology of the pellicle was different from the other two *K. baliensis* strains, instead of a surface spanning layer, only a loose conglomerate was present. The time for pellicle formation, also varied between the three strains. In comparison to the other *K. baliensis* strains, the pellicle of *K. baliensis* NBRC 16680 R Δ*polE* was only scarcely visible after three days of static incubation. During cultivation of the three *K. baliensis* strains (NBRC 16680, NBRC 16680 R and RΔ*polE*) under shaking conditions, no significant aberration in the growth behavior could be observed (Fig. 3b). In case of the HePS formation in shaking cultures with NaG medium, after 48 h a slight HePS production of *K. baliensis* NBRC 16680 R Δ*polE* could be demonstrated, with HePS amounts of 180 mg/L (Fig. 3c). The rough mutant *K. baliensis* strain, which differs only in the presence of an intact *polE* gene from *K. baliensis* NBRC 16680 R Δ*polE*, showed no EPS production in shaking cultures (Fig. 3c). The wild-type strain of *K. baliensis* NBRC 16680 with an intact *gum*- and *pol*-cluster, showed the highest EPS production under shaking conditions, with 1,73 g/L EPS.

K. baliensis NBRC 16680 HePS was furthermore isolated from shaking cultures and the monomer compositions were determined. It was possible to obtain pellicle EPS from the rough mutant strain (*K. baliensis* NBRC 16680 R), as well as EPS from shaking cultures from the *K. baliensis* NBRC 16680 wild-type strain and the *polE*

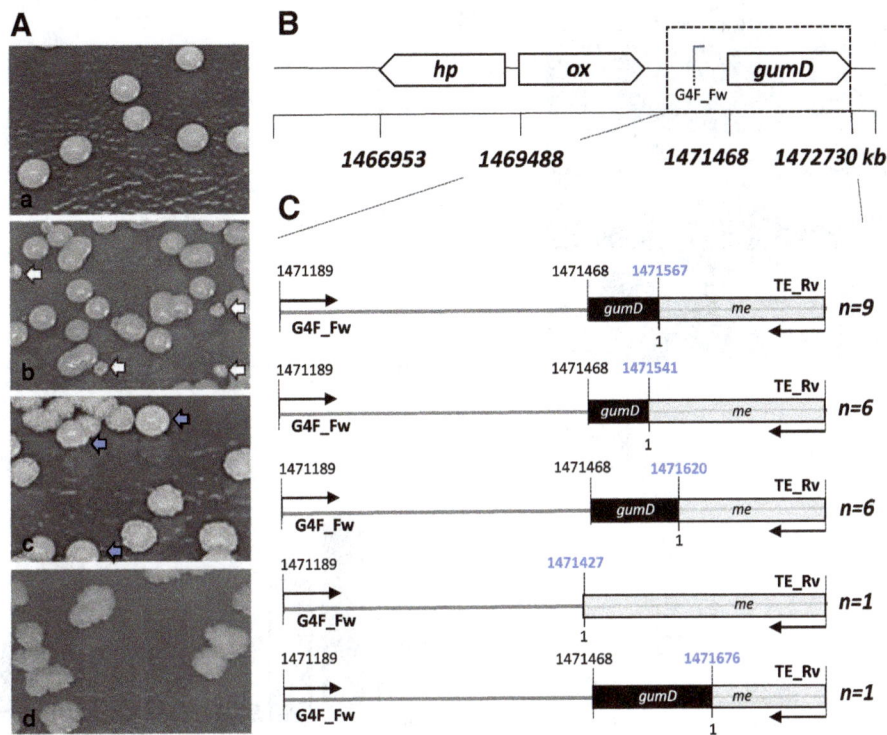

Fig. 2 Morphology and genetic switch of *K. baliensis* NBRC 16680 wild-type during cultivation in different media (*a*) In (*a*), growth of *K. baliensis* NBRC 16680 on NaG agar plates, plated at time point one after inoculation in NaG medium (0 h). Growth of *K. baliensis* NBRC 16680 on NaG agar plates, plated after 48 h of incubation in NaG medium (*b*), NaG-AA medium (*c*) and NaG-EtOH medium (*d*). Rough mutant colonies (Δ*gumD*) are indicated via a *white arrow*. Wild-type colonies are marked with a *grey arrow*. **b**) Shows a section of the *gum*-cluster of *K. baliensis* NBRC 16680 with the genomic location of the *gumD* gene (1471468–1,472,730 bp), oxidoreductase gene (*ox*, 1,469,488–1,470,627 bp) and a gene coding for a hypothetical protein (*hp*, 1,467,916–1,469,289 bp), based on JU Brandt, Jakob, F., Behr, J., Geissler, A.J., Vogel, R.F. [20]. Random colony PCRs of the respective rough colonies were carried out, targeting the transposon insertion side, using a genomic primer (G4F_Fw) and a primer, targeting the mobile element (TE_Rv). PCR products were subsequently sequenced. (**c**) Shows a schematic representation of the transposon insertion at the *gumD* locus in the rough colonys of *K. baliensis* NBRC 16680. The mobile element (*me*) is shown as grey bar with the corresponding insertion locus written in bold *blue*. The frequency of the found insertion site is marked on the left side, as n = x

deficient mutant, since for *K. baliensis* NBRC 16680 R, no EPS formation could be detected under shaking conditions. Because of the impaired pellicle formation of *K. baliensis* NBRC 16680 R:Δ*polE* (even during 14 days of cultivation in 30 mL cell culture flasks), no adequate amounts of EPS could be isolated for the monomer analysis. The monomer composition of wild-type *K. baliensis* NBRC 16680 HePS was composed of D-glucose, D-mannose and D-galactose (Fig. 3d). The statically cultivated rough mutant strain of *K. baliensis*, which is not able to produce HePS via the *gum*-cluster, showed a divergent monomer distribution, with generally higher amounts of D-mannose (46,2 ± 3,20%), and a consequently lower D-glucose (33,9 ± 5,03%) and D-galactose (19,9 ± 1,83%) level (Fig. 3d). The HePS derived from shaking cultures of the *polE* deficient *K. baliensis* NBRC 16680 R mutant showed a similar monomer distribution as the pellicle HePS of *K. baliensis* NBRC 16680 R, but with slightly variable percentages (Fig. 3d). Compared to *K. baliensis* NBRC 16680 R, a

smaller proportion of D-glucose (21,8 ± 1,03%) was observable, while the proportion of D-mannose (51,4 ± 1,20%) was still higher.

Physiological effects of Δ*gumD* and Δ*gumD* + Δ*polE* mutations

Since the insertion of a mobile element into the *gumD* gene of the *gum*-cluster of *K. baliensis* NBRC 16680 appears to be a mechanism affected by external factors, the question arises, whether this is a physiological adaptation of the bacterium to changes in its external environment. Therefore, the growth behavior of the different *K. baliensis* strains, under different cultivation conditions, was investigated. *K. baliensis* NBRC 16680 and DSM 14400, *K. baliensis* NBRC 16680 R and the Δ*polE* mutant of the rough *K. baliensis* strain were plated on NaG-medium plates, which were supplemented with different acetic acid (0.1%, 0.2%, 0.3%, 0.4%, 0.5%, 0.6%, 0.7%, 0.8%, 0.9% and 1%) and ethanol (1%, 2%, 3%, 4%, 5%, 6%, 7%, 8%, 9% and 10%) concentrations. Furthermore, the growth behavior of *K. baliensis* DSM 14400

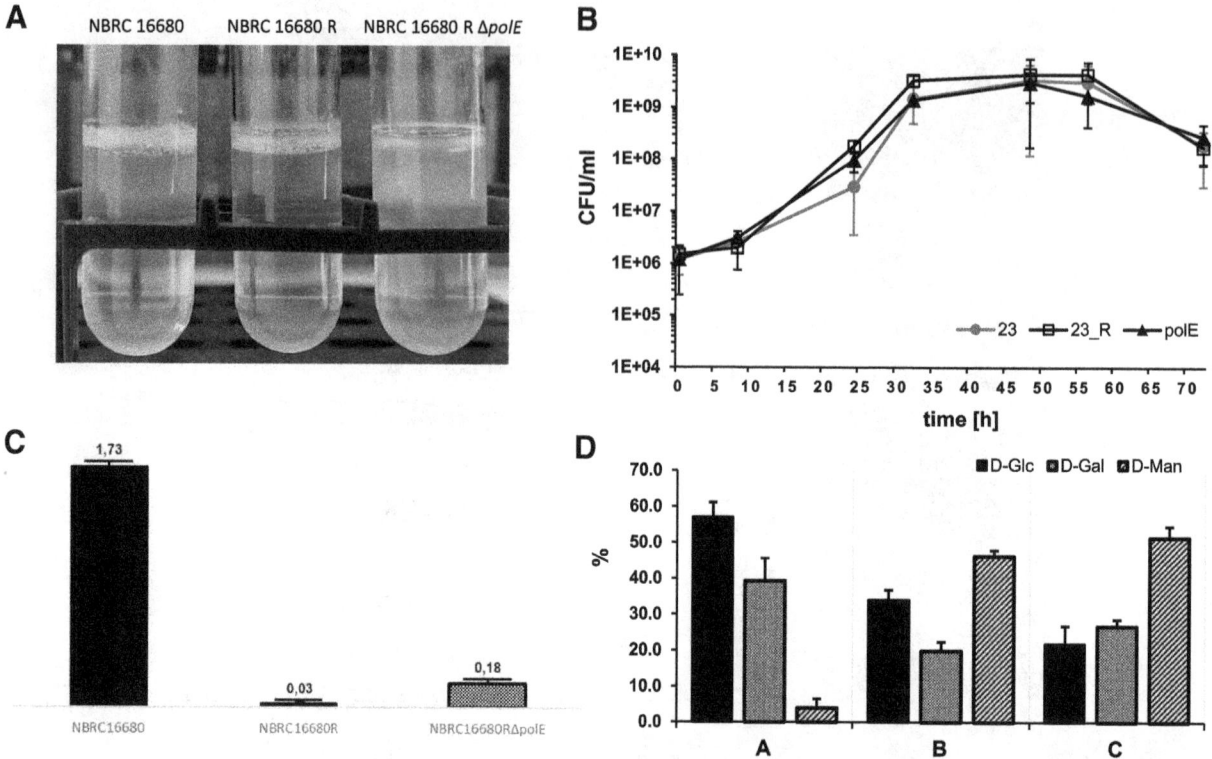

Fig. 3 Comparison of *K. baliensis* NBRC 16680 wild-type, the rough mutant strain and the *polE* deficient strain. **a** Phenotyps of *K. baliensis* NBRC 16680 during static cultivation of wild-type, rough mutant (*K. baliensis* NBRC 16680 R) and rough Δ*polE* mutant strain (*K. baliensis* NBRC 16680 R Δ*polE*). **b** Growth behavior of the different *K. baliensis* strains (NBRC 16680 (•), NBRC 16680 R (□), NBRC 16680 R Δ*polE*(▲)) in NaG medium. (**c**) Amount of precipitated HePS [g/L] from different *K. baliensis* strains (NBRC 16680. NBRC 16680 R, NBRC 16680 R Δ*polE*) during growth in NaG medium. (**d**) Neutral sugar composition of the isolated HePS for different *K. baliensis* strains during shaking or static cultivation. The proportions of the monomer occurring in the particular HePS are represented as percentage. (*a*) NBRC 16680 (shaking), (*b*) NBRC 16680 R (static), (*c*) NBRC 16680 R Δ*polE* (shaking)

was additionally tested, to investigate the variability in the growth behavior between two different wild-type strains of *K. baliensis*. *K. baliensis* DSM 14400 forms a HePS which is also composed of D-glucose, D-galactose and D-mannose, but with a deviating ratio compared to NBRC 16680 [20]. After a cultivation period of three days, distinct differences in the growth behavior of the different strains could be observed. Acetic acid had a significant influence on the growth of the different *K. baliensis* NBRC 16680 strains, while 0.7% acetic acid was the highest concentration, at which colony formation on agar plates was still possible for *Kozakia* strains NBRC 16680 R and DSM 14400 (Fig. 4a). The *K. baliensis* NBRC 16680 wild-type and the Δ*polE* mutant of the rough *K. baliensis* NBRC 16680 strain were not able to form colonies above 0.6% of acetic acid. In case of growth on NaG-EtOH plates, *K. baliensis* NBRC 16680 R and *K. baliensis* NBRC 16680 Δ*polE*, showed the highest EtOH tolerance, whereas *K. baliensis* NBRC 16680 R could even grow at 10% EtOH. Both wild-type strains of *K. baliensis* (NBRC 16680, DSM 14400) showed only poor growth behavior upon 7% EtOH and no growth at all, above 8% EtOH in the medium (Fig. 4b).

Furthermore, the effect of acetic acid and ethanol on the growth behavior of the different *K. baliensis* strains was monitored under static cultivation, by measuring the dry weight of the cells. It has to be noted, that the secreted, *gum*-cluster based HePS of the wildtype strain from *K. baliensis* NBRC 16680 was largely removed by centrifugation in advance. However, it can still slightly contribute to the dry weight as a small residue, which is still connected with the bacteria. In the NaG medium with 0.6% of acetic acid, the rough mutant strain displayed the fastest growth and reached the highest dry weight, after 7 days (1,77 ± 0.19 mg/ml). The wild-type strain of *K. baliensis* NBRC16680 showed slightly reduced final dry weight (1,64 ± 0.04 mg/ml), but an offset log phase resulting in a slower growth rate than the rough mutant strain (Fig. 5a). Similar results were obtained for the growth of the corresponding strains in NaG medium with 3% ethanol. *K. baliensis* NBRC16680 R exhibited the fastest growth, with a final dry weight of 1,17 ± 0.16 mg/ml. For *K. baliensis* NBRC16680 and the Δ*polE* mutant strain, a markedly reduced growth could be demonstrated (Fig. 5b).

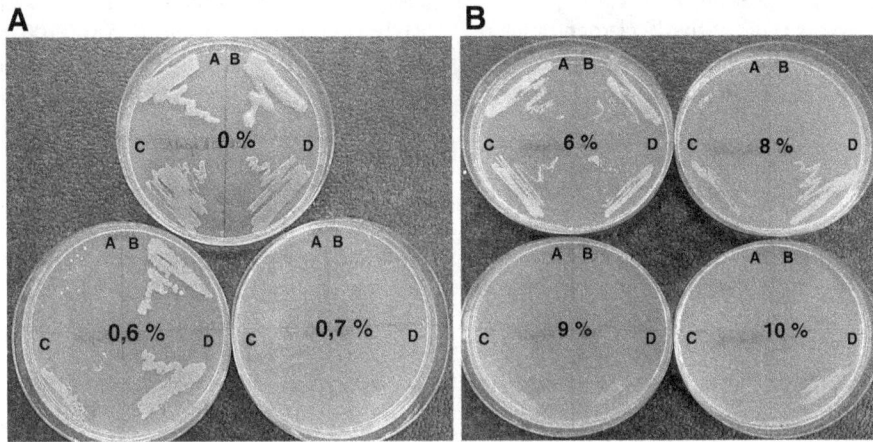

Fig. 4 Effect of acetic acid and ethanol on the growth behavior of different *K. baliensis* strains (**a**) Effect of acetic acid on growth of different *K. baliensis* strains (DSM 14400 (*a*), NBRC 16680 (*b*), NBRC 16680 R Δ*polE* (*c*), NBRC 16680 R (*d*)) at different acetic acid concentrations (0%, 0.6%, 0.7%). (**b**) Effect of ethanol on growth of different *K. baliensis* strains (DSM 14400 (*a*), NBRC 16680 (*b*), NBRC 16680 R Δ*polE* (*c*), NBRC 16680 R (*d*)) at different ethanol concentrations (6%, 8%,9%, 10%)

Discussion

K. baliensis NBRC 16680 produces and secretes large amounts of HePS via a *gum*-cluster encoded HePS biosynthesis. By insertion of a mobile element into the *gumD* locus of *K. baliensis* NBRC 16680 a mutant strain of *K. baliensis* is formed (*K. baliensis* NBRC 16680 R), which is unable to produce and secrete the *gum*-cluster derived HePS [20]. After cultivation of *K. baliensis* NBRC 16680 in standard or modified media (NaG, NaG-EtOH, NaG-AA), *K. baliensis* NBRC 16680 rough mutant strains were more frequently found in the presence of ethanol in the medium. For AAB an induced loss of various physiological properties has been observed, such as acetic acid resistance [30], ethanol oxidation, pellicle formation [31], and bacterial cellulose synthesis. The genetic mechanisms behind these

instabilities is often unclear. Takemura et al. [32] reported that the loss of the ethanol oxidation ability in *A. pasteurianus* NC11380 occurs by an insertion of a mobile element into the alcohol dehydrogenase-cytochrome c gene, resulting in the loss of alcohol dehydrogenase activity. This has also been demonstrated for *A. pasteurianus* NCI 1452, where an insertion sequence element (IS) is associated with the inactivation of the alcohol dehydrogenase gene [33]. Gene inactivation provoked by transposable elements could also be demonstrated for EPS forming clusters of AAB, like the cellulose synthase operon. In *A. xylinum* ATCC 23769, an IS element caused insertions 0.5 kb upstream of the cellulose synthase gene, associated with spontaneous cellulose deficiency [34]. In both cases, it was not possible to sustain revertants of the cellulose synthesis and ethanol oxidation

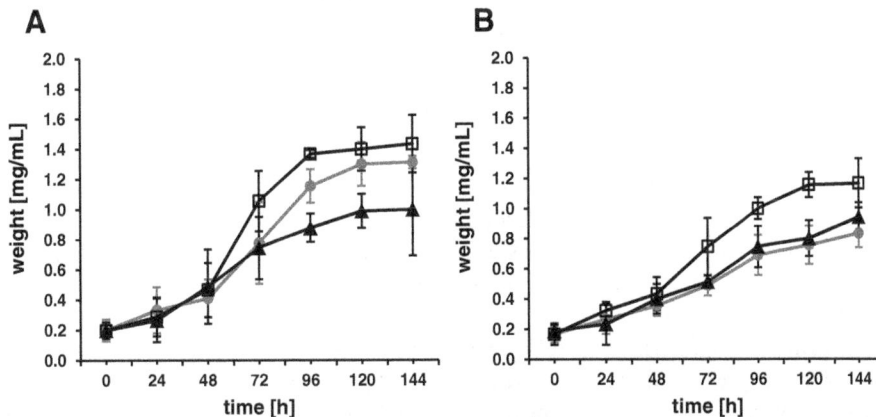

Fig. 5 Growth behavior of different *K. baliensis* strains during static cultivation in altered media (**a**) Growth behavior of the different *K. baliensis* strains (NBRC 16680 (•), NBRC 16680 R (□), NBRC 16680 R Δ*polE*(▲)) during static cultivation in NaG medium with acetic acid (0.6%). (**b**) Growth behavior of the different *K. baliensis* strains (NBRC 16680 (•), NBRC 16680 R (□), NBRC 16680 R Δ*polE*(▲)) during static cultivation in NaG medium with ethanol (3%) The growth was monitored by measuring the dry weight of cells

insufficient mutant strains, possibly as a consequence of remaining directed repeats (DR) of the IS element after relocation. Also in case of one previously investigated *K. baliensis* NBRC 16680 R mutant strain, the formation of directed repeats at the transposon insertion side could be observed [21]. Moreover, insertions of the mobile element in front of the *gumD* locus of the R strains as detected in the present study could result in a blocked transcription of *gumD* due to the presence of an energy-rich stem loop structure in the mobile element possibly causing *rho*-independent transcription termination [20]. It can actually not be ruled out that further mutations in the genome of *K. baliensis* NBRC 16680 simultaneously occurred. However, transposon insertion at the *gumD* locus of *K. baliensis* NBRC 16680 seems to be directly involved in R phenotype formation, since all of the 23 PCR checked R mutants exhibited the transposon insertion side at the *gumD* locus.

Morphotype variations, including the transition from a mucoid to a non-mucoid phenotype, are common events within the family of *Acetobacteraceae*, especially for *Acetobacter* species. This phenotypic change is often connected with the ability of the bacteria to form a pellicle on the medium surface [10, 11]. The pellicle is an assemblage of cells that permits them to float on the medium surface during static cultivation and ensures a high state of aeration. It was also shown that pellicle production could be associated with a change from a smooth phenotype to the rough, pellicle-forming strain. This change is accompanied by a transformation from secreted EPS to CPS, which could serve as a better barrier against ethanol [18]. Also the rough mutant of *K. baliensis* NBRC 16680 is still able to form a pellicle under static cultivation [20], suggesting that a second cluster, instead of the *gum*-cluster, is responsible for pellicle construction. For different *Acetobacter* strains, it has been shown, that the so-called *pol*-cluster is responsible for pellicle formation [35]. The *polABCDE* cluster shows a high level of homology to the *rfbBACD* genes of Gram-negative bacteria, which are involved in dTDP-rhamnose synthesis [36]. The *polE* gene has already undergone several assignments, since it has a relatively low homology level to glycosyltransferases, in general. A disruption of the *polE* gene in *A. tropicalis* SKU1100. however, leads to a defect in pellicle formation, thereby giving it a central role, either as rhamnosyl-transferase [35], or as galactosyl-transferase [11], which connects the CPS to the cell surface. Moreover, the mutant Pel⁻ cells secreted EPS into the culture medium. To investigate the relationship between *pol*-cluster and pellicle/CPS formation in *K. baliensis* NBRC 16680. *polE* knockouts were carried out in *K. baliensis* NBRC 16680 R, which does not form a *gum*-cluster dependent HePS [20]. Furthermore, we were interested to see, if a *polE* knockout has an effect on the ethanol tolerance of the respective *K. baliensis* strain. The Δ*polE* mutant of *K. baliensis* NBRC 16680 R showed the same rough colony

morphology as the rough strain (*K. baliensis* NBRC 16680 R). In contrast to *A. tropicalis* SKU1100. *K. baliensis* NBRC 16680 R Δ*polE* was still able to form a pellicle on the surface, which was, however, only a loose conglomerate of cells and required considerably more time for formation. Furthermore, the Δ*polE* mutant was able to secrete small amounts of EPS into the medium. The incoherent pellicle could therefore be formed by the secreted EPS, thus leading to the formation of a weak EPS/ cell-layer on the surface. The secreted EPS showed a similar composition as the *K. baliensis* NBRC 16680 R capsular HePS. This supports the hypothesis of Ali et al. [11] that PolE is responsible for CPS formation, via addition of some residue(s) that connect the HePS with the cell surface, e.g. β-d-galactopyranosyl residues. A *polE* gene deletion or interruption results in a switch from a rough CPS producing, to an EPS producing phenotype, connected with growth behavior variations [35].

It was shown that a change from a smooth phenotype to the rough phenotype is associated with an increased tolerance against acetic acid [35]. By static cultivation and in combination with growth experiments on NaG plates with different ethanol and acetic acid concentrations of the different *K. baliensis* strains (NBRC 16680, NBRC 16680 R, NBRC 16680 R Δ*polE*), variations in the growth behaviors could be observed. Additionally to *K. baliensis* NBRC 16680 a second *K. baliensis* strain (DSM 14400) was tested, to investigate the variability in the growth behavior between two wild-type strains of *K. baliensis*. The rough mutant strain could form colonies on NaG plates under both ethanol and acetic acid stress in contrast to the wild-type NBRC 16680 even at high acetic acid (0.7%) and ethanol (10%) concentrations. During static growth of *K. baliensis* NBRC 16680 and the rough mutant, a similar pattern compared to the NaG plates appeared, with higher growth rates of *K. baliensis* NBRC 16680 R in NaG medium with acetic acid and ethanol. The Δ*polE* strain, however, showed a diminished growth on acetic acid (up to 0.6%) and ethanol (up to 9%) agar plates, as well as during steady state growth. It is assumed, that the knockout of the *polE* gene results in a change from CPS to EPS, whereas CPS serves as a better barrier against acetic acid [37]. This indicates that the CPS of *K. baliensis* NBRC 16680 R should be involved (inter alia) in the protection against acetic acid and ethanol. The direct relationship between pellicle formation and acetic acid resistance could be proven for *A. pasteurianus* (IFO3283, SKU1108, MSU10), where the rough strains had clearly higher acetic acid resistance abilities than the smooth phenotypes, respectively [19]. The pellicle functions in this case as a biofilm-like barrier and prevents the passive diffusion of acetic acid into the cells [38]. Furthermore, Perumpuli et al. [18] could show, that ethanol in the medium significantly induced pellicle formation. This is in agreement with our observations for *K. baliensis* NBRC 16680, which mutated more frequently in NaG medium with ethanol, suggesting a directed mutation, which could at least partially be triggered under acetic acid fermenting conditions by ethanol or its oxidized product acetic acid.

K. baliensis DSM 14400 showed, similar to K. baliensis NBRC 16680 R, a high resistance against acetic acid, but was derogated by ethanol concentrations up to 8%. The high tolerance of K. baliensis DSM 14400 against acetic acid shows, that this strain has a different defense strategy in dealing with acetic acid, compared to K. baliensis NBRC 16680. Both strains were isolated from different environments (DSM 14400 isolated from palm brown sugar, NBRC 16680 isolated from ragi [39]). K. baliensis DSM 14400 might, therefore, have been commonly confronted with higher acetic acid concentrations, while they represent a new environmental factor for K. baliensis NBRC 16680. In the genome of K. baliensis DSM 14400. a cellulose synthase operon was identified on plasmid 3 (pKB14400_3) in contrast to K. baliensis NBRC 16680 including genes encoding the three cellulose synthase subunits A, B and C [20], suggesting that additionally formed cellulose could act as barrier against acetic acid. This is also the case for Komagataeibacter (Ko.) xylinus E25, isolated from vinegar, which generally deals with high acetic acid concentrations. Via comparison of diverse pol clusters from different acetic acid bacteria, a direct genetic connection between the pol- and gum-like clusters can be observed in Ko. xylinus (Fig. 6). Ko. xylinus E25 contains no polE gene in its genome, but is well known to produce a pellicle polysaccharide consisting of bacterial cellulose [13] and is also able to produce an extracellular HePS called acetan, composed of D-glucose, D-mannose, D-rhamnose, and D-glucuronic acid [6, 40]. In case of Ko. xylinus E25, the pol-cluster is flanking the acetan producing gum-like cluster, with the polAB gene upstream and the polCD gene downstream of the cluster (Fig. 6a). Acetan includes rhamnosyl residues, whose activated form is processed by polABCD [35], coding for the

enzyme of the TDP-rhamnose synthesis, connecting the acetan synthesis with the enzymes of the pol-cluster. The polE gene is missing in the Ko. xylinus E25 genome (GCA_000550765.1), possibly resulting in no HePS mediated pellicle formation, while a tight cellulose pellicle is commonly formed on the surface of the medium [41], which again contributes to increased acetic acid resistance of the bacterium [42]. This shows that the gum-like HePS biosynthesis and the pol-cluster are possibly related with each other and are most likely regulated according to the environmental requirements of the respective bacterium. A lack of oxygen, or an increasing ethanol concentration, could be a signal, which leads to an up-regulation of the polE gene in K. baliensis NBRC 16680. This can take place in association with an ethanol or acetic acid triggered stress response, probably resulting in an induced deactivation of the gum clusters, via a transposon insertion.

In the genome of Gluconacetobacter (Ga.) diazotrophicus PAl5, no polE gene is present as well (Fig. 6b). In contrast to Komagataeibacter xylinus, Ga. diazotrophicus is an obligate endophytic bacterium that lives symbiotically (N_2-fixing) in the intercellular space of roots, stem and leaves of sugarcane plants [30]. It produces a HePS composed of D-glucose, D-galactose and D-mannose, whose production is most likely attributable to its gum-like cluster similar to K. baliensis [31]. The produced HePS plays a key role during plant colonization, as shown for the molecular communication between the colonized plants and Ga. diazotrophicus [32]. The genetic switch of K. baliensis NBRC 16680 wild-type to its rough mutant could indicate an environmentally driven conversion/adaptation of a formerly more plant associated AAB (wild-type) to an increasingly acetic acid producing AAB (rough mutant). Furthermore, pol-cluster mediated capsular pellicle polysaccharide biosynthesis and concomitant inactivation of excess HePS production and secretion seems to increase the tolerance against

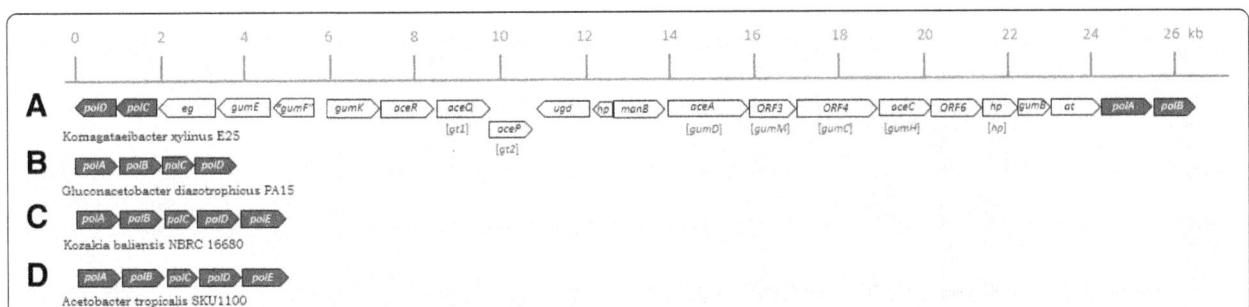

Fig. 6 Genetic organization of the pol-clusters from different acetic acid bacteria. The pol-clusters in combination with the gum-like cluster, responsible for the acetan synthesis, of Ko. xylinus E25 is depicted in (**a**). The so called acetan cluster has an overall size of ~26 kb and involves 22 genes, including glycosyltransferases (gt), mannose-phosphate-guanyltransferase (mpg), rhamnosyl transferase (aceR), hypothetical proteins (hp), and eight gum like genes marked as brackets under the particular genes (gumB, −C, −D, −E, −F, −H, −K, and −M). Furthermore, the cluster is flanked with the polAB gene upstream and the polCD gene downstream of the cluster. The nomenclature for the acetan cluster is based on AM Griffin, Morris, V. J., Gasson, M. J. [33]. In (**b**) is the pol-cluster of Ga. diazotrophicus Pal5, which consists of the polABCD genes. **c** and (**d**) show the related pol-clusters of K. baliensis NBRC 16680 [20] and A. tropicalis SKU1100 [35], including the polABCDE genes

acetic acid in certain AAB such as *K. baliensis* NBRC 16680, while specialized AAB additionally produce tight cellulose pellicles, which are typical for extremely acetic acid tolerant starter cultures used in vinegar production such as *Komagataeibacter xylinus*.

Conclusion

In summary the results obtained in this study show, that a switch from the *K. baliensis* NBRC 16680 wild-type to a rough mutant strain leads to a significantly increased acetic acid and ethanol resistance. This increased tolerance is probably accompanied by a morphologic switch, from secreted HePS in the wild type to capsular HePS in the rough mutant strain (Fig. 3d). The *polE* gene turned out to be involved for the formation of the resulting CPS. Although the exact role of PolE is still unknown, the results presented here show that *K. baliensis* NBRC 16680 R Δ*polE* displays a reduced tolerance against acetic acid and ethanol, most likely caused by a lack of cell-bound CPS and thus of pellicle formation. Since the morphological change is not reversible, this can also be understood as an adaptive evolutionary step of *K. baliensis* NBRC 16680 resulting in a shift of its ecological niche more towards an acetic acid-rich milieu.

Additional files

Additional file 1: Primers used for *polE* gene deletion in *K. baliensis* NBRC16680 RTable of primers used fort the *polE* gene deletion in *K. baliensis* NBRC 16680 R.

Additional file 2: Deletion of the *polE* gene (A0U90_11950).The deletion mechanism is depicted in (A) displayed with the particular basic vector pKos6b. The deletion of the *polE* gene (A0U90_11950) is shown in (B). In (C) Agarose gel of colony PCR verifying the deletion of the *polE* gene in the genome; lane 1 showing *K. baliensis* NBRC 16680 R with the *polE* gene (3000 bp), lane 2 showing *K. baliensis* NBRC 16680 R with a *polE* deletion (Δ*polE*), with a PCR product of 1950 bp.

Abbreviations

AA: Acetic acid; AAB: Acetic acid bacteria; *acs*: Cellulose synthesizing operon; *bcs*: Cellulose synthesizing operon; CPS: Capsular polysaccharide; DSM: Deutsche Sammlung von Mikroorganismen; EPS: Exopolysaccharide; EtOH: Ethanol; G.: *Gluconobacter*, *Ga.*: *Gluconacetobacter*, Gal: Galactose; Glc: Glucose; *gum*: Protein involved in HePS biosynthesis (derived from *X. campestris*); HEPES: 2-(4-(2-Hydroxyethyl)-1-piperazinyl)-ethansulfonsäure; HePS: Heteropolysaccharide; HoPS: Homopolysaccharide; HPLC: High-performance liquid chromatography; *K*: *Kozakia*; KM: Kanamycin; *Ko*: *Komagataeibacter*, Man: Mannose; MWCO: Molecular weight cut-off; NaG: Modified sodium-gluconate medium; NBRC: Biological Research Centre NITE, Japan; OD: Optical density; *pol*: Genes involved in pellicle formation; R: Rough strain; *rfb*: Genes involved in dTDP-rhamnose synthesis; *X.*: *Xanthomonas*

Acknowledgements

We want to thank Prof. Dr. Wolfgang Liebl, Dr. Armin Ehrenreich and Dr. David Kostner (Department of Microbiology, Technical University of Munich) for giving us the opportunity to construct a Δ*polE* deletion mutant in *K. baliensis* NBRC 16680 R with their developed gene deletion system.

Funding

This work was supported by the German Research Foundation (DFG) and the Technische Universität München within the funding program Open Access Publishing. Part of this work was funded by the German Federal Ministry for Economic Affairs and Energy via the German Federation of Industrial Research Associations (AiF) (FEI); project number AiF 18071 N. The funder had no role in the design, analysis, interpretation or writing of the manuscript.

Authors' contributions

JUB, FJ and RFV were involved in planning the experimental setup and writing the manuscript. JUB performed the main experimental work and FLB was involved in the implementation of the *polE* knockout system. JUB evaluated the data and wrote the main text of the manuscript. All authors read and approved the final manuscript.

Competing interests

The authors declare that they have no competing interests.

References

1. Roberts IS. The biochemistry and genetics of capsular polysaccharide production in bacteria. Annu Rev Microbiol. 1996;50:285–315.
2. Ferreira AS, Silva IN, Oliveira VH, Cunha R, Moreira LM. Insights into the role of extracellular polysaccharides in *Burkholderia* adaptation to different environments. Front Cell Infect Microbiol. 2011;16(1):1–9.
3. Jakob F. Novel fructans from acetic acid bacteria. Doctoral Dissertation, Freising, Germany: Technische Universität München; 2014.
4. Jakob F, Meißner D, Vogel RF. Comparison of novel GH 68 levansucrases of levan-overproducing *Gluconobacter* species. Acetic Acid Bacteria. 2012;1(1):2.
5. Jakob F, Pfaff A, Novoa-Carballal R, Rübsam H, Becker T, Vogel RF. Structural analysis of fructans produced by acetic acid bacteria reveals a relation to hydrocolloid function. Carbohydr Polym. 2013;92(2):1234–42.
6. Jansson P, Lindberg J, Wimalasiri KMS, Dankert MA. Structural studies of acetan, an exopolysaccharide elaborated by *Acetobacter xylinum*. Carbohydr Res. 1993;245:303–10.
7. Kornmann H, Duboc P, Marison I, Stockar U. Influence of nutritional factors on the nature , yield , and composition of exopolysaccharides produced by *Gluconacetobacter xylinus* I-2281. Appl Environ Microbiol. 2003;69(10):6091–8.
8. Brown RM, Willison JH, Richardson CL. Cellulose biosynthesis in *Acetobacter xylinum*: visualization of the site of synthesis and direct measurement of the in vivo process. PNAS. 1976;73(12):4565–9.
9. Yamada Y, Pattaraporn Y, Vu L, Thi H, Yuki M, Duangjai O, Somboon T, Yasuyoshi N. Description of *Komagataeibacter gen. nov.*, with proposals of new combinations (Acetobacteraceae). J Gen Appl Microbiol. 2012;58(5): 397–404.
10. Moonmangmee S, Kawabata K, Tanaka S, Toyama H, Adachi O, Matsushita K. A novel polysaccharide involved in the pellicle formation of *Acetobacter aceti*. J Biosci Bioeng. 2002;93(2):192–200.
11. Ali IAI, Akakabe Y, Moonmangmee S, Deeraksa A, Matsutani M, Yakushi T, Yamada M, Matsushita K. Structural characterization of pellicle polysaccharides of *Acetobacter tropicalis* SKU1100 wild type and mutant strains. Carbohydr Polym. 2011;86(2):1000–6.
12. Moonmangmee S, Toyama H, Adachi O, Teerakool G, Lotong N, Matsushita K. Purification and characterization of a novel polysaccharide involved in the pellicle produced by a thermotolerant *Acetobacter* strain. Biosci Biotechnol Biochem. 2001;66(4):777–83.
13. Saxena IM, Kudlicka K, Okuda K, Brown RM. Characterization of genes in the cellulose-synthesizing operon (*acs* operon) of *Acetobacter xylinum*: implications for cellulose crystallization. J Bacteriol. 1994;176(18):5735–52.
14. Wong HC, Fear AL, Calhoon RD, Eichinger GH, Mayer R, Amikam D, Benziman M, Gelfand DH, Meade JH. Emerick aW: genetic organization of the cellulose synthase operon in *Acetobacter xylinum*. Proc Natl Acad Sci U S A. 1990;87: 8130–4.
15. Azuma Y, Hosoyama A, Matsutani M, Furuya N, Horikawa H, Harada T, Hirakawa H, Kuhara S, Matsushita K, Fujita N, et al. Whole-genome analyses reveal genetic instability of *Acetobacter pasteurianus*. Nucleic Acids Res. 2009;37(17):5768–83.

16. Shimwell JLC, J. G. Mutant frequency in *Acetobacter*. Nature. 1964;201:1051–2.

17. Matsushita K, Ebisuya H, Ameyama M, Adachi O. Change of the terminal oxidase from cytochrome a1 in shaking cultures to cytochrome o in static cultures of *Acetobacter aceti*. J Bacteriol. 1992;174(1):122–9.

18. Perumpuli PABN, Watanabe T, Toyama H. Pellicle of thermotolerant *Acetobacter pasteurianus* strains: characterization of the polysaccharides and of the induction patterns. J Biosci Bioeng. 2014;118(2):134–8.

19. Kanchanarach W, Theeragool G, Inoue T, Yakushi T, Adachi O, Matsushita K. Acetic acid fermentation of *Acetobacter pasteurianus:* relationship between acetic acid resistance and pellicle polysaccharide formation. Biosci Biotechnol Biochem. 2014;74(8):1591–7.

20. Brandt JU, Jakob F, Behr J, Geissler AJ, Vogel RF. Dissection of exopolysaccharide biosynthesis in *Kozakia baliensis*. Microb Cell Factories. 2016;15(1):170.

21. Wach A. PCR-synthesis of marker cassettes with long flanking homology regions for gene disruptions in *S. cerevisiae*. Yeast. 1996;12(3):259–65.

22. Peters B, Junker A, Brauer K, Mühlthaler B, Kostner D, Mientus M, Liebl W, Ehrenreich A. Deletion of pyruvate decarboxylase by a new method for efficient markerless gene deletions in *Gluconobacter oxydans*. Appl Microbiol Biotechnol. 2013;97(6):2521–30.

23. Tillett D, Neilan B. Enzyme-free cloning: a rapidmethod to clone PCR products independent of vector restriction enzyme sites. Oxford university press. 1999;27(19):e26.

24. Kostner D, Peters B, Mientus M, Liebl W, Ehrenreich A. Importance of codB for new codA-based markerless gene deletion in *Gluconobacter* strains. Appl Microbiol Biotechnol. 2013;97(18):8341–9.

25. Hall PE, Anderson M, Johnston DM, Cannons RE. Transformation of *Acetobacter xylinus* with plasmid DNA by electroporation. Plasmid. 1992;28(3):194–200.

26. Creaven M, Fitzgerald RJ, O'Gara F. Transformation of *Gluconobacter oxydans subsp. suboxydans* by electroporation. Can J Microbiol. 1994;40(6):491–4.

27. Krajewski V, Simić P, Mouncey NJ, Bringer S, Sahm H, Bott M. Metabolic engineering of *Gluconobacter oxydans* for improved growth rate and growth yield on glucose by elimination of gluconate formation. Appl Environ Microbiol. 2010;76(13):4369–76.

28. Mostafa HE, Heller KJ, Geis A. Cloning of *Escherichia coli lacZ* and *lacY* genes and their expression in *Gluconobacter oxydans* and *Acetobacter liquefaciens*. Appl Environ Microbiol. 2002;68(5):2619–23.

29. Kaditzky SV, Rudi F. Optimization of exopolysaccharide yields in sourdoughs fermented by lactobacilli. Eur Food Res Technol. 2008;228:291–9.

30. Cavalcante A. V., Dobereiner, J.: a new acid-tolerant nitrogen-fixing bacterium associated with sugarcane. Plant Soil. 1988;108(1):23–31.

31. Bertalan M, Albano R, de Padua V, Rouws L, Rojas C, Hemerly A, Teixeira K, Schwab S, Araujo J, Oliveira A, et al. Complete genome sequence of the sugarcane nitrogen-fixing endophyte *Gluconacetobacter diazotrophicus* Pal5. BMC Genomics. 2009;10:450.

32. Meneses CHSG. Rouws, L. F. M., Simões-araújo, J. L., Vidal, M. S., Baldani, J. I.: exopolysaccharide production is required for biofilm formation and plant colonization by the nitrogen-fixing endophyte *Gluconacetobacter diazotrophicus*. Mol Plant-Microbe Interact. 2011;24(12):1448–58.

33. Griffin AM. Morris, V. J., Gasson, M. J.: genetic analysis of the acetan biosynthetic pathway in *Acetobacter xylinum*. Int J Biol Macromol. 1994;16(6):287–9.

34. Coucheron DH. An *Acetobacter xylinum* insertion sequence element associated with inactivation of cellulose production. J Bacteriol. 1991;173(18):5723–31.

35. Deeraksa A, Moonmangmee S, Toyama H, Yamada M, Adachi O, Matsushita K. Characterization and spontaneous mutation of a novel gene, *polE*,

involved in pellicle formation in *Acetobacter tropicalis* SKU1100. Microbiology. 2005;151(Pt 12):4111–20.

36. Marolda CLVM. Genetic analysis of the dTDP-rhamnose biosynthesis region of the *Escherichia coli* VW187 (O7:K1) rfb gene cluster: identification of functional homologs of rfbB and rfbA in the rff cluster and correct location of the rffE gene. J Bacteriol. 1995;177(19):5539–46.

37. Deeraksa A, Moonmangmee S, Toyama H, Adachi O, Matsushita K. Conversion of capsular polysaccharide, involved in pellicle formation, to extracellular polysaccharide by galE deletion in *Acetobacter tropicalis*. Biosci Biotechnol Biochem. 2006;70(10):2536–9.

38. Kanchanarach W, Theeragool G, Inoue T, Yakushi T, Adachi O, Matsushita K. Acetic acid fermentation of *Acetobacter pasteurianus* : relationship between acetic acid resistance and pellicle polysaccharide formation. Biosci Biotechnol Biochem. 2010;74(8):1591–7.

39. Lisdiyanti P, Kawasaki H, Widyastuti Y, Saono S, Seki T, Yamada Y, Uchimura T. *Kozakia baliensis* gen. Nov., sp. nov., a novel acetic acid bacterium in the α-Proteobacteria. Int J Syst Evol Microbiol. 2002;52:813–8.

40. Ishida T, Sugano Y, Shoda M. Effects of acetan on production of bacterial cellulose by *Acetobacter xylinum*. Biosci Biotechnol Biochem. 2002;66(8):1677–81.

41. Brown AJ. On an acetic ferment which forms cellulose. J Chem Soc. 1886; 49:432–9.

42. Toda K, Asakura T, Fukaya M, Entani E, Kawamura Y. Cellulose production by acetic acid-resistant *Acetobacter xylinum*. J Ferment Bioeng. 1997;84(3):228–31.

10

Antimicrobial activity and safety evaluation of peptides isolated from the hemoglobin of chickens

Fengjiao Hu[1], Qiaoxing Wu[1], Shuang Song[1], Ruiping She[1]*⊙, Yue Zhao[1], Yifei Yang[1], Meikun Zhang[2], Fang Du[1], Majid Hussain Soomro[1] and Ruihan Shi[1]

Abstract

Background: Hemoglobin is a rich source of biological peptides. As a byproduct and even wastewater of poultry-slaughtering facilities, chicken blood is one of the most abundant source of hemoglobin.

Results: In this study, the chicken hemoglobin antimicrobial peptides (CHAP) were isolated and the antimicrobial and bactericidal activities were tested by the agarose diffusion assay, minimum inhibitory concentration (MIC) analysis, minimal bactericidal concentration (MBC) analysis, and time-dependent inhibitory and bactericidal assays. The results demonstrated that CHAP had potent and rapid antimicrobial activity against 19 bacterial strains, including 9 multidrug-resistant bacterial strains. Bacterial biofilm and NaCl permeability assays, transmission electron microscopy (TEM) and scanning electron microscopy (SEM) were further performed to detect the mechanism of its antimicrobial effect. Additionally, CHAP showed low hemolytic activity, embryo toxicity, and high stability in different temperatures and animal plasma.

Conclusion: CHAP may have great potential for expanding production and development value in animal medication, the breeding industry and environment protection.

Keywords: Antimicrobial peptides, Hemoglobin, Hydrolysis, Antimicrobial activity, Bactericidal activity

Background

Due to the widespread use and even abuse of conventional antibiotics, antibiotic resistance is rampant all over the world, which limits the lifespan of commercial antibiotics and results in the urgent demand of new platforms for efficient antibiotic discovery [1, 2].

As an essential part of innate immunity, antimicrobial peptides (AMPs) have been receiving increasing attention because of their unique antimicrobial mechanism against both Gram-positive and Gram-negative bacteria, and even including some multidrug-resistant strains over recent decades [3, 4]. Antimicrobial peptides are ubiquitous in all living organisms. More than 5000 AMPs (http://www.camp.bicnirrh.res.in/index.php) have been identified and 2593 peptides have been derived naturally

(http://aps.unmc.edu/AP/main.php) since the discovery of the lysozyme by Alexander Fleming in 1922 [5]. Although substantial AMPs have been discovered over the past decades, only a small part of them have been used because of high costs and potential cytotoxicity [1, 6]. Hence, finding efficient, nontoxic and low-cost AMPs is urgent in promoting AMPs' practical applications.

The whole blood is a mixture of cells (erythrocytes, leucocytes and platelets) and plasma (colloids and crystalloids), which delivers nourishment and oxygen to and removes waste products from all parts of the body [7]. Components in blood, such as platelet concentrates [8], defensins [3], leukocyte extracts [9], also play important roles in antimicrobial host defense. Hemoglobin is the main component of the erythrocyte [10]. Aside from the basic function of transporting oxygen, hemoglobin has been found as a source of various biological peptides [11–13]. Many AMPs called hemocidins have been isolated from hemoglobin cleavage *in vivo* [14] or from

* Correspondence: sheruiping@126.com
[1]Department of Veterinary Pathology and Public Health, Key Laboratory of Zoonosis of Ministry of Agriculture College of Veterinary Medicine, China Agricultural University, Beijing 100193, China
Full list of author information is available at the end of the article

hemoglobin hydrolysis by chemical reagents, physical methods, or enzymes *in vitro* [15, 16]. To date, the hemocidins derived from human beings [17], bovines [12], rabbits [18], swine [15], crocodiles [19], fish [20], and shellfish [21] have been reported and most of them are made up of 2 to 60 residues, characterized by a common random coil structural and broad-spectrum antimicrobial activity [22–24]. As a byproduct and even one of the major dissolved pollutants in slaughter house wastewater [25], appropriate treatment of chicken blood is of great benefit to both environmental protection and economic development. However, hemocidins from poultry have not been documented yet. In this study, the hemocidins from chickens were isolated and their antimicrobial and bactericidal activities were further detected.

Methods

Materials and chemicals

All common chemical reagents and biological products were of analytical grade from commercial sources. Papain (2000 IU g^{-1}) was purchased from Sigma Chemical Co. (St. Louis, Mo, America).

Preparation of CHAP

The chicken hemoglobin antimicrobial peptides (CHAP) were prepared as modified method described before [26]. In brief, fresh chicken blood (Beijing Huadu Broiler Corporations, Beijing, China) was collected with heparin and then centrifuged with 2,000 × g at 4 °C for 10 min. The upper liquid and white cells were removed and washed with sterilized saline. The procedures described above were repeated 3 times. The cells were frozen, thawed, stirred and homogenized in deionized water (pH 7.0) with papain (1:1,000 *w/v*) proteolysis at 70 °C for 8 h. The digested suspensions were added with ice-cold aqueous 5% acetic acid solution (1:1 *v/v*) and extracted overnight at 4 °C. After being centrifuged at 8,000 × g for 30 min at 4 °C, the suspensions were collected as crude extracts, and the protein concentration was detected by NanoDrop 2000 UV–vis Spectrophotometer (Thermo Fisher Scientific, Massachusetts, America). The pH of the extracts was adjusted to 6.0 with sodium hydroxide. The crude extracts were loaded onto 10 × 300 mm Sephadex G-100 column and eluted by 0.2 mol L^{-1} sodium acetate buffer (pH 6.0) with the speed of 12.0 mL cm^{-2} h^{-1} Each elution was analyzed by agarose diffusion assay [27] with *Escherichia coli* ATCC 25922 as the indicator organism. The fractions with potent antibacterial activity were collected and detected with Tricine SDS-PAGE [28] and then subjected to mass spectrometry (Beijing Protein Innovation Co., Ltd., Beijing, China).

Bacterial strains and growth conditions

Staphylococcus aureus ATCC25923, *Staphylococcus aureus* ATCC 29213, *Staphylococcus albus* ATCC01331, *Escherichia coli* ATCC 25922, *Escherichia coli* O78, *Escherichia coli* C83922, *Escherichia coli* C83901, *Pseudomonas aeruginosa* ATCC27853, *Pasteurellae gallinarum* C48-3 were purchased from the China Veterinary Culture Collection Center (CVCC). *Aeromonas hydrophila*, *Bacillus cereus* and *Escherichia coli* were clinically isolated from crucian carps, pigeon and equines respectively by Laboratory of Veterinary Pathology and Public Health of the College of Veterinary Medicine, China Agricultural University. *Staphylococcus aureus* MR-L22, MR-QD-CD10, *Enterococcus faecalis* 53A, 52A, 37 N and *Pseudomonas aeruginosa* M140 and *Escherichia coli* T50 were all multi-resistance strains of clinics, and obtained from Beijing Key Laboratory of Detection Technology for Animal Food safety of the College of Veterinary Medicine. All the above Gram-negative strains were grown in Luria-Bertani (LB) agar and the Gram-positive bacteria were grown in brain heart infusion (BHI) agar.

Determination of antimicrobial and bactericidal activities

Agarose diffusion assay

The primary antibacterial activities of CHAP elution (100 µg mL^{-1}) were detected by modified agarose diffusion assay as described before [27]. Briefly, the single colony of each bacterial strain was grown in trypticase soy broth (TSB, 30 g L^{-1}) overnight at 37 °C under aerobic conditions. 2×10^8 CFU mL^{-1} bacteria culture of each strain was added to warm (50–55 °C) sterile agarose [1% agarose (low EEO, Sigma, St. Louis, MO), 0.03% nutrient broth, and 10 mM PBS buffer, pH 7.4] (1:100 *v/v*). 10 µL samples were added to 3 mm wells punched by agar punch (BioRad Laboratories, Hercules, Canda). 0.2 mol L^{-1} sodium acetate (solvent) and 20000 IU penicillin-streptomycin solution of the same volume were added as negative and positive control, respectively. After being incubated overnight at 37 °C, the diameter of the each clean zone of growth inhibition was measured as the antibacterial activity of CHAP against different strains.

Minimum inhibitory concentration (MIC) analysis

A micro dilution assay was employed to determine MIC according to the broth micro dilution guideline of Clinical and Laboratory Standards Institute (CLSI) [29]. Briefly, 50 µL of twofold serial dilutions of CHAP (25 to 0.20 µg mL^{-1}) was placed into wells of sterile 96-well cell culture plates. The 50 µL of bacterial suspensions (1×10^5 CFU mL^{-1}) were added to the peptides. The wells were added with 50 µL of Mueller- Hinto (MH) broth and 50 µL of bacterial culture was treated as positive and negative control, respectively. After 24 h incubation, the MICs were determined at 492 nm by spectrophotometer (Thermo Multiskan MK3, Thermo Fisher Scientific, Massachusetts, America).

Time-dependent inhibitory assay

A 500 μL aliquot of CHAP with 2 × MIC of the bacterium was added respectively to 500 μL bacterial suspensions (1×10^5 CFU mL^{-1}) in the sterilized 1.5 mL tubes as the treated groups. Bacteria treated with 500 μL solvent (0.2 mol L^{-1} sodium acetate) were set as the control groups. After being incubated for 30 min, 100 μL aliquot of the suspensions were pipetted into to a sterilized 1.5 mL tube. After centrifugation at $1,000 \times g$ for 5 min, the supernatant was removed, and the pellet was resuspended in 100 μL MHB medium. Tenfold serially diluted suspension was placed on agar plates and incubated at 37 °C until viable colonies could be seen and the numbers of colony-forming units (CFU) were counted. The inhibitory rate of each bacterium was calculated according to the following formula: the inhibitory rate = [(colonies of the treated group - colonies of the treated group)/colonies of the control group] × 100%.

In order to further detect the process and speed of the antimicrobial activity of CHAP, the time growth curves and inhibitory rates of *Escherichia coli* ATCC 25922, *Staphylococcus aureus* ATCC29213, *Staphylococcus aureus* MR-L22, *Enterococcus faecalis* 52A, *Pseudomonas aeruginoda* M140 and *Escherichia coli* T50 were achieved after the treated suspensions were incubated for 0, 5, 10, 30, 90 min respectively.

Minimal bactericidal concentration (MBC) analysis

The MBC values were determined in 96-well plates, which was similar to the method of MIC. MBC values were further confirmed by plating 100 μL samples of each well with no visible turbidity onto the MHB medium. The least concentration showing no visible growth on the plates was considered as the MBC value.

Time-dependent bactericidal assay

The time depending bactericidal curves of *Escherichia coli* ATCC 25922 were determined as the time-dependent inhibitory assay mentioned above by adding CHAP with concentration of its MBC value to the bacterial cultures grown to early and late exponential phase as the reference [2].

In order to detect the bacteriolysis against bacteria in stationary phase, 10 ml of bacterial culture (2×10^9 CFU mL^{-1}) was treated with 10 × MIC of CHAP. The culture treated with solvent (0.2 mol L^{-1} sodium acetate) was set as the control group. After 24 h incubation, 2 ml of each culture was added to a glass tube and was photographed [2].

Bacterial biofilms assay

Crystal violet staining method was applied to detect the effect of CHAP on the biofilm formation [30]. Briefly, *Staphylococcus aureus* ATCC29213 were cultured in TSB overnight. 100 μL bacterial suspensions (1×10^6 CFU mL^{-1}) with 2 × MIC, 1 × MIC, 1/2 × MIC, 1/4 × MIC, 1/8 × MIC, 1/16 × MIC of CHAP were added to 96-well plates and the bacterial suspensions with no CHAP and the sterilized TSB were treated as control groups. After static culture at 37 °C for 24 h or shake culture (50 rmp) at 37 °C for 72 h, the contents were aspirated and the wells were washed by 200 μL PBS for three times, methanol fixed for 1 h and stained with 200 μL crystal violet (5 g L^{-1}) for 30 min. The wells were washed by running water and air dried. The plates were determined at 600 nm by spectrophotometer.

NaCl permeability assay

The effect of CHAP on the NaCl permeability of bacteria was detected as modified protocol as follows [31]. 100 μL of bacterial suspensions (1×10^6 CFU mL^{-1}) with 1/2 × MIC of CHAP were added to 96-well plates and the bacterial suspensions with no CHAP as control. 100 μL of NaCl solutions with different concentrations (80, 100, 120, 140, 160, 180, 200 g L^{-1}) were added into each wells and incubated at 37 °C for 12 h. The bacterial concentration of each well was determined by measuring the optical density at 600 nm (OD_{600}).

Electron microscopy observations

Both transmission electron microscopy (TEM) and scanning electron microscopy (SEM) were conducted as previously described [32–34]. Briefly, *Escherichia coli* ATCC 25922 and *Staphylococcus aureus* ATCC29213 were cultured overnight, 10^7 CFU ml^{-1} bacteria were incubated with 1 × MIC of CHAP or diluents of the same volume at 37 °C for 30 min. All the samples were fixed and proceeded for the TEM and SEM respectively.

Hemolytic assay and embryotoxicity assay

The hemolytic activity was evaluated as previously described [35]. 4% (vol/vol) fresh chicken erythrocyte suspensions were added to a 96-well plate and incubated with CHAP at 360, 180, 90, 45, 22.5, 11.25 μg mL^{-1} individually at 37 °C for 1 h. Wells treated with PBS and 0.1% Triton X-100 of the same volume were taken as 0 and 100% hemolysis. The wells were determined by measuring the optical density at 492 nm (OD_{492}).

The embryotoxicity of CHAP was detected as the following measures. The 10-days-old-chicken embryos were randomly divided into 5 groups, each of 10 eggs, 0.2 mL of CHAP of 1 × MIC, 2 × MIC, 4 × MIC, 6 × MIC dose against *Escherichia coli* ATCC25922 were injected into the chorio-allantoic cavity, and embryos treated with the same volume of solvent (0.2 mol L^{-1} sodium acetate) were used as controls. The eggs were put in a hatching machine and hatchability and weight of the eggs were observed regularly until hatching.

Stability in different temperatures and in 50% plasma

CHAP (100 μg mL^{-1}) was treated with different temperatures varying from 30 °C, 40 °C, 50 °C, 60 °C, 70 °C, 80 °C, 90 °C, 100 °C, 121 °C for 30 min. The antimicrobial activities of these treated aliquots were determined with agarose diffusion assay and were compared with CHAP stored in 4 °C.

The stability of CHAP in 50% plasma was evaluated as previously described [29] with some modifications. Briefly, the plasma of chicken and rabbit was determined with no antimicrobial activity before the test. Then 640 μg ml^{-1} CHAP was diluted 1:1 in fresh chicken and rabbit plasma and pre-incubated at 37 °C for 0, 3, and 6 h respectively. After incubation, the antimicrobial activity of each sample was determined by agarose diffusion assay. The effect of CHAP diluted by its solvent was regarded as the 100%, and the effect of the treated samples was demonstrated as percentages.

Statistical Analysis

Experiments were conducted with biological replicates and experimental data were expressed as mean ± standard deviation of at least three determinations and analyzed by one-way ANOVA using SPSS 20.0 (SPSS Inc., Cary, NC, USA). Differences were considered to be statistically significant at $P < 0.05$ or $P < 0.01$.

Results

Preparation of CHAP

The crude extracts of CHAP were light yellow and the protein concentration was adjusted to 5 mg ml^{-1} before loading on the Sephadex G-100 column. There were two main peaks after the elution of Sephadex G-100 gelatin (Fig. 1a) and the tubes from 9 to 16 in the left half of the second peak showed potent antimicrobial activity (Fig. 1b). Detected by Tricine SDS –PAGE, the collected CHAP showed band around 3.3 KDa (Fig. 1c). This band was further analyzed and peptides of gallus hemoglobin subunit alpha were confirmed by mass spectrum (see Additional file 1).

Determination of antimicrobial and bactericidal activity
Antimicrobial effect of CHAP

The results of antibacterial activity of CHAP detected via agarose diffusion assay, MIC assay and inhibitory rate assay are shown in the columns 2–4 in Table 1.These results demonstrated that CHAP performed potent antimicrobial activities against both Gram-negative bacteria and Gram-positive bacteria, including 9 multidrug-resistant strains.

The time-dependent growth inhibitory activities of CHAP are shown in Fig. 2a-d. The results showed that CHAP not only significantly inhibited the growth of standard strains ($P < 0.05$), but also effectively inhibited the multi-resistant ones ($P < 0.05$) in 10 min (Fig. 2a and b). The inhibitory rates of all six strains reached 100% in 90 min (Fig. 2c and d). Although there were different growth inhibitory curves, CHAP showed more than 50% inhibitory rate against all six strains in 10 min.

Bactericidal effect of CHAP

The values of MBC are shown in column 5 of Table 1. By analyzing the values, most bacteria were killed by CHAP at concentrations ranging from 5 μg mL^{-1} to 80 μg mL^{-1}. However, 7 strains showed no obvious bactericidal effect with the maximum concentration of 160 μg mL^{-1}. The time-dependent bactericidal curves in Fig. 2e and f further revealed that the significant bactericidal effect of CHAP on both the bacteria grown to early and late exponential phases from 10–240 min ($P < 0.01$).

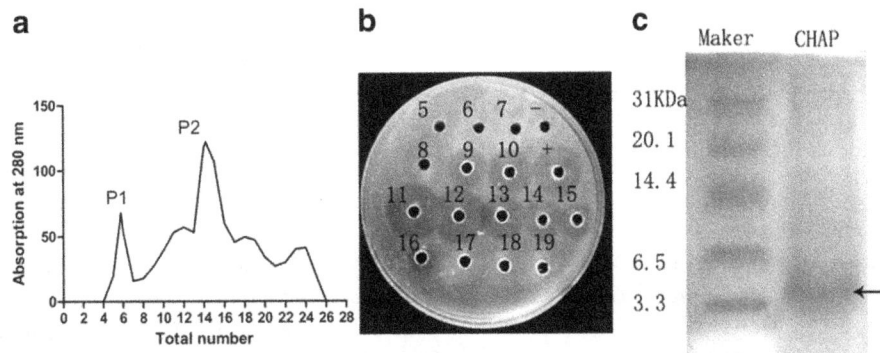

Fig. 1 Preparation of CHAP. **a** Sephadex G-100 gelatin separation of the extract from chicken blood. **b** Antibacterial activity detection of elution of Sephadex G-100 gelatin by agarose diffusion assay against *Escherichia coli* ATCC25922. **c** Tricine–SDS-PAGE of the interested elution and the band was around 3.3 KDa (*arrow*)

Table 1 Antibacterial activity and bactericidal activity of Chicken hemoglobin fragment peptides

Strains	D (mm)	MIC (μg mL^{-1})	IR (%)	MBC (μg mL^{-1})
Gram-negative bacteria				
Escherichia coli ATCC 25922	24.5	6.25	83.00	80
Escherichia coli C83901	18	12.5	66.53	80
Escherichia coli C83922,	20	6.25	55.24	80
Escherichia coli O78	18	6.25	47.83	80
Aeromonas hydrophila(crucian carp)	13	3.13	52.05	>160
Pseudomonas aeruginosa ATCC27853	15	3.13	52.94	80
Pasteurellae gallinarum C48-3	12	3.13	63.63	80
MR- Escherichia coli(equine)	33.5	6.25	62.73	80
MR- Pseudomonas aeruginosa M140	19	1.56	85.56	5
MR- Escherichia coli T50(swine)	11	6.25	89.92	80
Gram-positive bacteria				
Staphylococcus aureus ATCC25923	27	3.13	50.00	>160
Staphylococcus aureus ATCC 29213	21	3.13	94.26	5
Staphylococcus albus ATCC01331	14.5	1.56	73.36	40
MR-Bacillus cereus (pigeon)	13.5	3.13	73.36	20
MR-Staphylococcus aureus L22(swine)	15.5	6.25	88.6	>160
MR-Staphylococcus aureus QD-CD10 (swine)	15.5	6.25	56.93	>160
MR-Enterococcus faecalis 53A(pet)	22	1.56	91.10	>160
MR-Enterococcus faecalis 52A(pet)	13.5	6.25	35.00	>160
MR-Enterococcus faecalis 37 N(pet)	29	3.13	55.00	>160

MR multidrug resistance (in bold), D diameter of inhibition zone, MIC minimum inhibitory concentration, IR inhibitory rate in 30 min, MBC minimal bactericidal concentration

Especially the early exponential phase bacteria, they were killed completely in only 10 min. The bacteria in stationary phase resulted in lysis after being treated with CHAP for 24 h (Fig. 3a).

Bacterial biofilms assay

The formation of biofilms of *Staphylococcus aureus* ATCC29213 was decreased with the increase of the concentration of CHAP (Fig. 3b). The 2 × MIC, 1 × MIC, 1/2 × MIC, 1/4 × MIC, 1/8 × MIC of CHAP could inhibited the formation of biofilms ($P < 0.01$) in 24 h and 2 × MIC, 1 × MIC, 1/2 × MIC, 1/4 × MIC of CHAP could significantly decrease the formation of biofilms ($P < 0.01$) in 72 h. Remarkably, in the 2 × MIC and 1 × MIC groups, there were almost no biofilm formation.

NaCl permeability assay

As shown in Fig. 3c and d, the values of OD$_{600}$ of *Escherichia coli* ATCC 25922 and *Staphylococcus aureus* ATCC29213 cultures decreased with the increased concentration of NaCl solution and reached to their minimum values at concentration above 160 g L^{-1} and above 140 g L^{-1}, respectively. By adding CHAP, the value of OD$_{600}$ in both *Escherichia coli* ATCC 25922 and *Staphylococcus aureus*

ATCC29213 groups decreased to the lowest value at concentration of NaCl above 120 g L^{-1}.

Electron microscopy observations

The morphology of the *Escherichia coli* ATCC 25922 and *Staphylococcus aureus* ATCC29213 investigated by SEM is shown in the Fig. 3. Compared to the smooth, straight and unbroken surface of the control cells (Fig. 4a and c), the strains treated with CHAP for 30 min appeared severely damaged (Fig. 3b and d). The TEM images further demonstrated that the bacterial surfaces were damaged by the effect of CHAP (Fig. 5a and c) compared with the control group (Fig. 4b and d).

Hemolytic assay and embryotoxicity assay

The hemolysis of CHAP was 38.9% at the concentration of 360 μg mL^{-1} which was more than 50 times higher than the MIC values for all the detected bacteria (see Additional file 2). And the embryotoxicity assay showed that even CHAP of 6 × MIC dose against *Escherichia coli* ATCC25922 did not induce toxicity toward chicken embryos, that is, there was no dead or significant decrease of body weight compared to the control group ($P > 0.05$) (see Additional file 3).

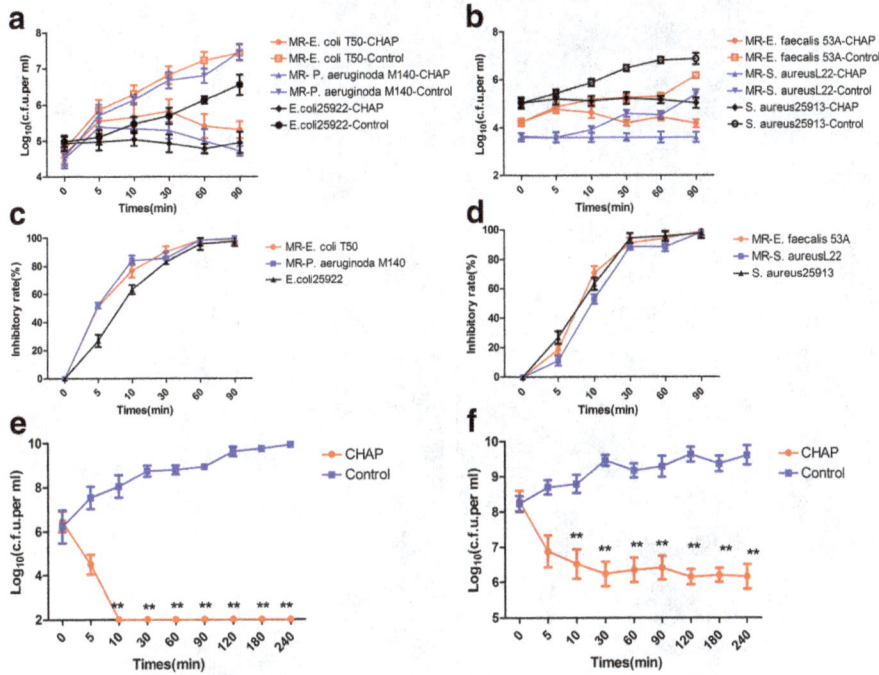

Fig. 2 a-b Time-dependent inhibitory curves of bacteria treated with CHAP and the bacteria treated with the solvent as control. **a** Gram-negative bacteria. **b** Gram-positive bacteria. **c-d** Time-dependent growth inhibitory rate curves of bacteria treated with CHAP and the bacteria treated with the solvent as control. **c** Gram-positive bacteria. **d** Gram-positive bacteria. **e-f** Time dependent bactericidal curves of *Escherichia coli* ATCC 25922 treated with CHAP and the bacteria treated with the solvent as control. **e** At early exponential phase. **f** At late exponential phase

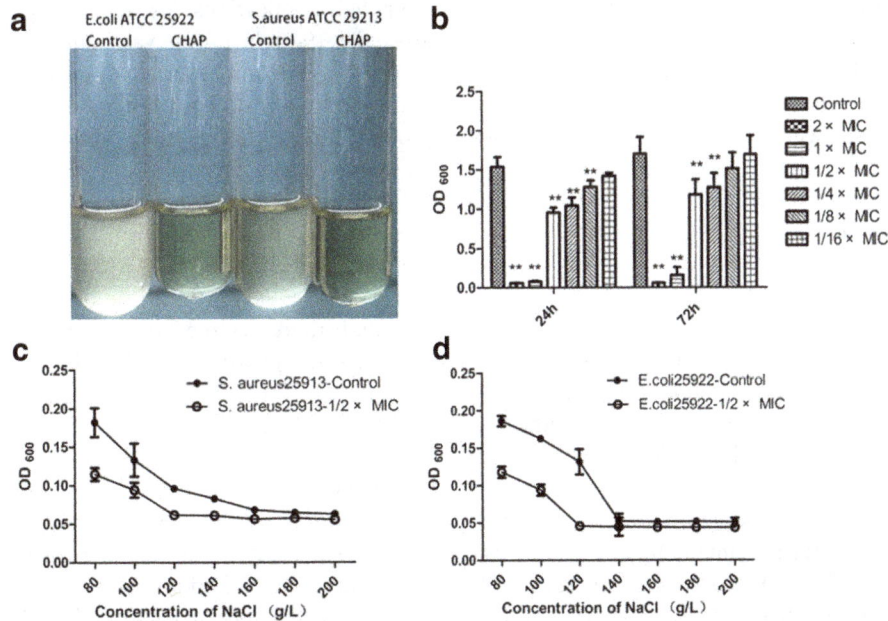

Fig. 3 a Bacteriolysis analysis. Bacteria in stationary phase treated with CHAP resulted in lysis with the solvent as control. **b** The inhibitory effect of CHAP on bacterial biofilm for 24 h and 72 h. **c-d** The effect of CHAP on NaCl permeability of *Escherichia coli* ATCC 25922 and *Staphylococcus aureus* ATCC29213

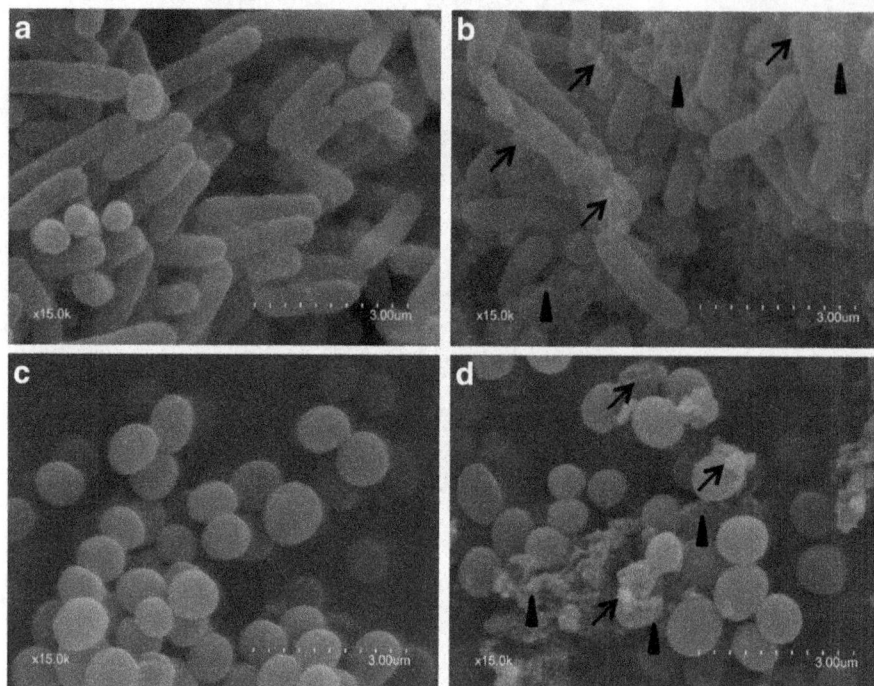

Fig. 4 The morphology of *Escherichia coli* ATCC 25922 and *Staphylococcus aureus* ATCC29213 were investigated by scanning electron microscopy. **a-b** The control group. **c-d** The *Escherichia coli* ATCC 25922 treated with CHAP. Viscous substances were adhering to almost all CHAP treated cells, which got large number of bacteria together (*arrowheads*); Some bacteria showed variable length, rough cell surfaces or globular protrusions on their surfaces, and even appeared to collapse (*arrows*)

Stability in different temperatures and in 50% plasma

The antimicrobial activity of CHAP did not decrease in different temperatures even when it was treated in 121 °C for 30 min compared with CHAP stored in 4 °C ($P > 0.05$) (Fig. 6a). It well proved that CHAP was capable of stability in various temperatures.

Compared with CHAP diluted in the solvent, the antimicrobial activity of CHAP showed no change in the treatment of chicken plasma and a slight but no significant decrease in treatment of rabbit plasma ($P > 0.05$) (Fig. 6b), which demonstrated that CHAP was of well stability in the plasma.

Discussion

Since the first anti-Streptococcus peptide was identified from the cow's milk [36], the enzyme strategy of isolating AMPs on a large scale has been a feasible method as Bolscher postulated [37]. Our lab isolated the peptides from the hemoglobin of chickens by using a simple and practical way and studied the antimicrobial activities against 19 bacterial strains, including 9 multidrug-resistant bacteria. At the same time, the properties such as hemolytic activity, embryotoxicity and stability in different temperatures and plasma were detected, which laid a foundation for its further employment in agricultural production, public health and medication.

Antibiotics have been helping humans to fight against hazardous infections since Alexander Fleming discovered the first antibiotic, penicillin, in 1928 [38]. However, resistance to most antibiotics was discovered shortly after their applications. For example, penicillin resistance arose in 1946 just one year after its introduction to clinics [1]. The main target of most AMPs is cell membrane and there are several models for explaining the process such as barrel-stave pore model, thoroidal pore model and carpet model [3, 29, 39], and it tends to be difficult for bacteria to totally change this basic structure to resist the effect of AMPs [3]. Given the unique antimicrobial mechanism of AMPs, it was not surprising that most AMPs induced little or no resistance [40, 41]. Although there are reports and doubts about the resistance of some peptides [41, 42], there is no report about the natural ones so far, which means the development of natural AMPs is high in potential.

In our study, both the antimicrobial and bactericide results showed that CHAP was capable of strong and rapid activities against various bacteria and even some multi-resistant strains, implying its wider utility in the prevention and treatment of infectious agents, which was similar with the hemosidins reported before [16, 17]. By specifically analyzing the results above, there was no obvious difference between the antibacterial activity against Gram-negative bacteria and Gram-positive bacteria, even between the

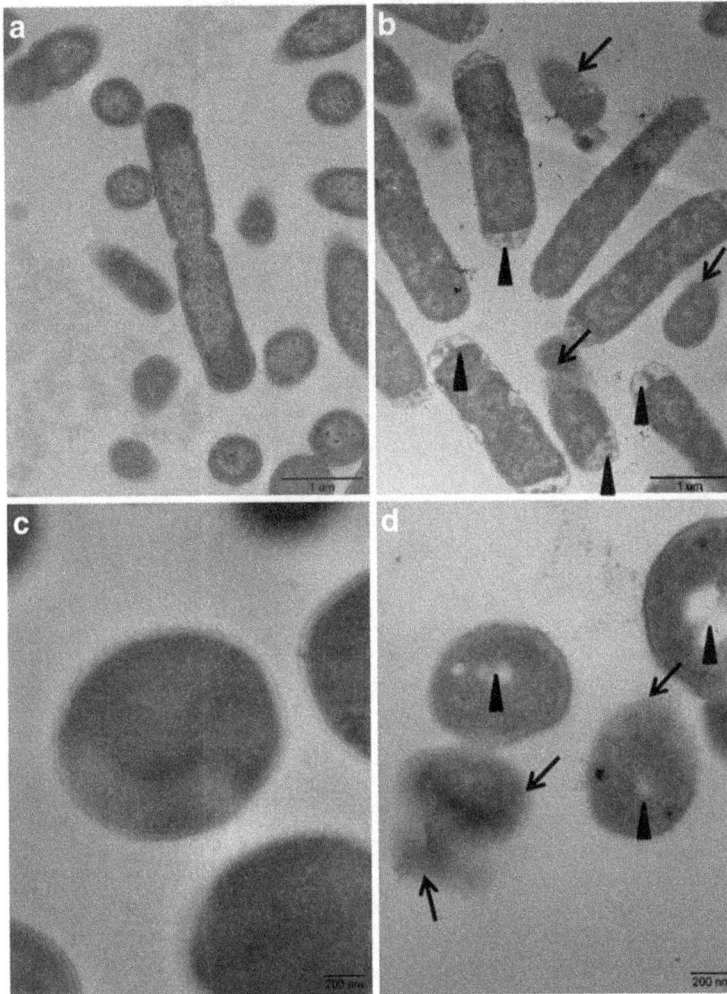

Fig. 5 The morphology of *Escherichia coli* ATCC 25922 *Staphylococcus aureus* ATCC29213 were investigated by transmission electron microscopy. **a-b** The control group. **c-d** The *Escherichia coli* ATCC 25922 treated with CHAP. Most of the bacteria were translucent and pores were evident on walls especially at the two terminals of each cell (*arrowheads*). There was some intracellular substance released from many bacteria (*arrows*).

standard strains and the multi-resistant strains, suggesting that the target of CHAP is the common component of bacteria such as the cell membrane like most AMPs reported before [29, 43]. The biofilm and the NaCl permeability results showed that CHAP could inhibit the formation of bacterial biofilms and change the permeability of some Gram-negative bacteria and Gram-positive bacteria to some extent. With the confirmation of the mechanism of most AMPs, the EM observations further revealed that CHAP could accumulate copious pathogens nearby and punch through their cell surfaces swiftly [29, 32].

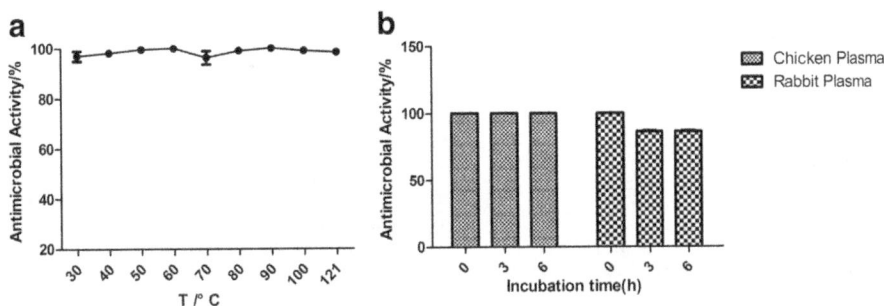

Fig. 6 The stability of CHAP. **a** The stability of the CHAP in different temperatures. **b** The stability of the CHAP in chicken and rabbit plasma

However the specific mechanism needs to be further investigated.

As a double-edged sword, the unique mechanism of targeting cell membranes could also lead to the low selection of some AMPs [3]. Hence, toxicity especially hemolysis and safety problems, are constantly an obstacle to their final applications [1, 40, 44]. According to our study, CHAP demonstrated low hemolysis and embryotoxicity even at rather high concentrations, which further implied that there was relative high selectivity of CHAP between eukaryote cells and prokaryote cells.

Good stability also plays an important role in the application of any biological product. As for the AMPs, the substances such as ions and proteolytic enzymes in the serum may reduce their biological ability to a large extent [29, 45]. In this study, CHAP kept high antimicrobial activity in two kinds of animal serum and different temperatures, suggesting its convenient application, transportation and storage.

Conclusions

In summary, this study firstly reported a practical method of isolating chicken hemosidins (CHAP) from the byproduct and even the pollutant of chicken-slaughtering industries. CHAP has an attractive antimicrobial and bactericidal ability with low hemolysis, low or none in toxicity and good temperature resistance and high stability in serum, which well accounts for their potential of expanding production and high development value in animal medication, breeding industry and environment protection.

Abbreviations

AMPs: Antimicrobial peptides; BHI: Brain heart infusion; CFU: Colony-forming units; CHAP: Chicken hemoglobin antimicrobial peptides; CLSI: Clinical and Laboratory Standards Institute; CVCC: China Veterinary Culture Collection Center; LB: Luria-Bertani; MBC: Minimal bactericidal concentration; MH: Mueller- Hinto; MIC: Minimum inhibitory concentration; OD: Optical density; SEM: Scanning electron microscopy; TEM: Transmission electron microscopy; TSB: Trypticase soy broth

Acknowledgements

We would like to thank the National Natural Science Foundation of China, Beijing Key Laboratory of Detection Technology for animal Food safety of the College of Veterinary Medicine, China Agricultural University, Beijing Protein Institute and Beijing Huadu CO.

Funding

This study was funded by the National Natural Science Foundation of China (31072110, 31272515).

Authors' contributions

RShe conceived and supervised the study; HF and RShe designed experiments; HF, WQ, SS, YZ, YY, MZ and FD performed experiments; HF, WQ, SS analyzed and interpreted the data; HF wrote the manuscript; RShe, MHS and RShi made manuscript revisions. All authors read and approved the final manuscript.

Competing interests

The authors declare that they have no competing interest.

Author details

[1]Department of Veterinary Pathology and Public Health, Key Laboratory of Zoonosis of Ministry of Agriculture College of Veterinary Medicine, China Agricultural University, Beijing 100193, China. [2]Beijing Huadu Broiler Corporations, Beijing 102211, China.

References

1. Lewis K. Platforms for antibiotic discovery. Nat Rev Drug Discov. 2013;12(5): 371–87.
2. Ling LL, Schneider T, Peoples AJ, Spoering AL, Engels I, Conlon BP, Mueller A, Schaberle TF, Hughes DE, Epstein S, et al. A new antibiotic kills pathogens without detectable resistance. Nature. 2015;517(7535):455–9.
3. Zasloff M. Antimicrobial peptides of multicellular organisms. Nature. 2002; 415(6870):389–95.
4. Nizet V. Antimicrobial peptide resistance mechanisms of human bacterial pathogens. Curr Issues Mol Biol. 2006;8(1):11–26.
5. Gallo RL. The birth of innate immunity. Exp Dermatol. 2013;22(8):517.
6. Ginsburg I, Koren E. Are cationic antimicrobial peptides also 'double-edged swords'? Expert Rev Anti Infect Ther. 2008;6(4):453–62.
7. Basu D, Kulkarni R. Overview of blood components and their preparation. Indian J Anaesth. 2014;58(5):529–37.
8. Drago L, Bortolin M, Vassena C, Taschieri S, Del Fabbro M. Antimicrobial activity of pure platelet-rich plasma against microorganisms isolated from oral cavity. BMC Microbiol. 2013;13:47.
9. Merchant ME, Leger N, Jerkins E, Mills K, Pallansch MB, Paulman RL, Ptak RG. Broad spectrum antimicrobial activity of leukocyte extracts from the American alligator (Alligator mississippiensis). Vet Immunol Immunopathol. 2006;110(3–4):221–8.
10. Hamidi M, Tajerzadeh H. Carrier erythrocytes: an overview. Drug Deliv. 2003; 10(1):9–20.
11. Brantl V, Gramsch C, Lottspeich F, Mertz R, Jaeger KH, Herz A. Novel opioid peptides derived from hemoglobin: hemorphins. Eur J Pharmacol. 1986; 125(2):309–10.
12. Zhao QY, Piot JM, Gautier V, Cottenceau G. Isolation and characterization of a bacterial growth-stimulating peptide from a peptic bovine hemoglobin hydrolysate. Appl Microbiol Biotechnol. 1996;45(6):778–84.
13. Karelin AA, Philippova MM, Ivanov VT. Proteolytic degradation of hemoglobin in erythrocytes leads to biologically active peptides. Peptides. 1995;16(4):693–7.
14. Fogaca AC, da Silva Jr PI, Miranda MT, Bianchi AG, Miranda A, Ribolla PE, Daffre S. Antimicrobial activity of a bovine hemoglobin fragment in the tick Boophilus microplus. J Biol Chem. 1999;274(36):25330–4.
15. Alvarez C, Rendueles M, Diaz M. Production of porcine hemoglobin peptides at moderate temperature and medium pressure under a nitrogen stream. Functional and antioxidant properties. J Agric Food Chem. 2012; 60(22):5636–43.
16. Mak P, Wojcik K, Silberring J, Dubin A. Antimicrobial peptides derived from heme-containing proteins: hemocidins. Antonie Van Leeuwenhoek. 2000;77(3):197–207.
17. Liepke C, Baxmann S, Heine C, Breithaupt N, Standker L, Forssmann WG. Human hemoglobin-derived peptides exhibit antimicrobial activity: a class of host defense peptides. J Chromatogr B Analyt Technol Biomed Life Sci. 2003;791(1–2):345–56.
18. Patgaonkar M, Aranha C, Bhonde G, Reddy KV. Identification and characterization of anti-microbial peptides from rabbit vaginal fluid. Vet Immunol Immunopathol. 2011;139(2–4):176–86.
19. Srihongthong S, Pakdeesuwan A, Daduang S, Araki T, Dhiravisit A, Thammasirirak S. Complete amino acid sequence of globin chains and biological activity of fragmented crocodile hemoglobin (Crocodylus siamensis). Protein J. 2012;31(6):466–76.
20. Fernandes JM, Smith VJ. Partial purification of antibacterial proteinaceous factors from erythrocytes of Oncorhynchus mykiss. Fish Shellfish Immunol. 2004;16(1):1–9.

21. Gambacurta A, Piro MC, Ascoli F. Cooperative homodimeric hemoglobin from Scapharca inaequivalvis. cDNA cloning and expression of the fully functional protein in E. coli. FEBS Lett. 1993;330(1):90–4.

22. Mak P, Wojcik K, Wicherek L, Suder P, Dubin A. Antibacterial hemoglobin peptides in human menstrual blood. Peptides. 2004;25(11):1839–47.

23. Nedjar-Arroume N, Dubois-Delval V, Adje EY, Traisnel J, Krier F, Mary P, Kouach M, Briand G, Guillochon D. Bovine hemoglobin: an attractive source of antibacterial peptides. Peptides. 2008;29(6):969–77.

24. Adje EY, Balti R, Kouach M, Dhulster P, Guillochon D, Nedjar-Arroume N. Obtaining antimicrobial peptides by controlled peptic hydrolysis of bovine hemoglobin. Int J Biol Macromol. 2011;49(2):143–53.

25. Kundu P, Debsarkar A, Mukherjee S. Treatment of slaughter house wastewater in a sequencing batch reactor: performance evaluation and biodegradation kinetics. Biomed Res Int. 2013;2013:134872.

26. Zhang Y, She R, Liu T, Wengui LI, Jia J. Studies on isolation, purification and antibacterial activities of antibacterial peptides in swine blood. Science & Technology Review (China). 2008;26(2):33–7.

27. Bao H, She R, Liu T, Zhang Y, Peng KS, Luo D, Yue Z, Ding Y, Hu Y, Liu W, et al. Effects of pig antibacterial peptides on growth performance and intestine mucosal immune of broiler chickens. Poult Sci. 2009;88(2):291–7.

28. Schagger H, von Jagow G. Tricine-sodium dodecyl sulfate-polyacrylamide gel electrophoresis for the separation of proteins in the range from 1 to 100 kDa. Anal Biochem. 1987;166(2):368–79.

29. Hou Z, Lu J, Fang C, Zhou Y, Bai H, Zhang X, Xue X, Chen Y, Luo X. Underlying mechanism of in vivo and in vitro activity of C-terminal-amidated thanatin against clinical isolates of extended-spectrum beta-lactamase-producing Escherichia coli. J Infect Dis. 2011;203(2):273–82.

30. Nair S, Desai S, Poonacha N, Vipra A, Sharma U. Antibiofilm activity and synergistic inhibition of S. aureus biofilms by bactericidal protein P128 in combination with antibiotics. Antimicrob Agents Chemother. 2016;60(12):7280–9.

31. De Oliveira MV, Intorne AC, Vespoli Lde S, Madureira HC, Leandro MR, Pereira TN, Olivares FL, Berbert-Molina MA, De Souza Filho GA. Differential effects of salinity and osmotic stress on the plant growth-promoting bacterium Gluconacetobacter diazotrophicus PAL5. Arch Microbiol. 2016;198(3):287–94.

32. Shi J, Ross CR, Chengappa MM, Sylte MJ, McVey DS, Blecha F. Antibacterial activity of a synthetic peptide (PR-26) derived from PR-39, a proline-arginine-rich neutrophil antimicrobial peptide. Antimicrob Agents Chemother. 1996;40(1):115–21.

33. Cao L, Dai C, Li Z, Fan Z, Song Y, Wu Y, Cao Z, Li W. Antibacterial activity and mechanism of a scorpion venom peptide derivative in vitro and in vivo. PLoS One. 2012;7(7):e40135.

34. Ding Y, Zou J, Li Z, Tian J, Abdelalim S, Du F, She R, Wang D, Tan C, Wang H, et al. Study of histopathological and molecular changes of rat kidney under simulated weightlessness and resistance training protective effect. PLoS One. 2011;6(5):e20008.

35. Stark M, Liu LP, Deber CM. Cationic hydrophobic peptides with antimicrobial activity. Antimicrob Agents Chemother. 2002;46(11):3585–90.

36. Jones FS, Simms HS. The bacterial growth inhibitor (lactenin) of milk : i. the preparation in concentrated form. J Exp Med. 1930;51(2):327–39.

37. Bolscher JG, van der Kraan MI, Nazmi K, Kalay H, Grun CH, Van't Hof W, Veerman EC, Nieuw Amerongen AV. A one-enzyme strategy to release an antimicrobial peptide from the LFampin-domain of bovine lactoferrin. Peptides. 2006;27(1):1–9.

38. Fleming AG. Responsibilities and Opportunities of the Private Practitioner in Preventive Medicine. Can Med Assoc J. 1929;20(1):11–3.

39. Pompilio A, Crocetta V, Scocchi M, Pomponio S, Di Vincenzo V, Mardirossian M, Gherardi G, Fiscarelli E, Dicuonzo G, Gennaro R, et al. Potential novel therapeutic strategies in cystic fibrosis: antimicrobial and anti-biofilm activity of natural and designed alpha-helical peptides against Staphylococcus aureus, Pseudomonas aeruginosa, and Stenotrophomonas maltophilia. BMC Microbiol. 2012;12:145.

40. Hancock RE, Sahl HG. Antimicrobial and host-defense peptides as new anti-infective therapeutic strategies. Nat Biotechnol. 2006;24(12):1551–7.

41. Bell G, Gouyon PH. Arming the enemy: the evolution of resistance to self-proteins. Microbiology. 2003;149(Pt 6):1367–75.

42. Habets MG, Brockhurst MA. Therapeutic antimicrobial peptides may compromise natural immunity. Biol Lett. 2012;8(3):416–8.

43. Melo MN, Ferre R, Castanho MA. Antimicrobial peptides: linking partition, activity and high membrane-bound concentrations. Nat Rev Microbiol. 2009;7(3):245–50.

44. Marr AK, Gooderham WJ, Hancock RE. Antibacterial peptides for therapeutic use: obstacles and realistic outlook. Curr Opin Pharmacol. 2006;6(5):468–72.

45. Bowdish DM, Davidson DJ, Lau YE, Lee K, Scott MG, Hancock RE. Impact of LL-37 on anti-infective immunity. J Leukoc Biol. 2005;77(4):451–9.

11

High-throughput sequencing technology to reveal the composition and function of cecal microbiota in Dagu chicken

Yunhe Xu[1], Huixin Yang[2], Lili Zhang[3], Yuhong Su[1], Donghui Shi[1], Haidi Xiao[1] and Yumin Tian[1*]

Abstract

Background: The chicken gut microbiota is an important and complicated ecosystem for the host. They play an important role in converting food into nutrient and energy. The coding capacity of microbiome vastly surpasses that of the host's genome, encoding biochemical pathways that the host has not developed. An optimal gut microbiota can increase agricultural productivity. This study aims to explore the composition and function of cecal microbiota in Dagu chicken under two feeding modes, free-range (outdoor, OD) and cage (indoor, ID) raising.

Results: Cecal samples were collected from 24 chickens across 4 groups (12-w OD, 12-w ID, 18-w OD, and 18-w ID). We performed high-throughput sequencing of the 16S rRNA genes V4 hypervariable regions to characterize the cecal microbiota of Dagu chicken and compare the difference of cecal microbiota between free-range and cage raising chickens. It was found that 34 special operational taxonomic units (OTUs) in OD groups and 4 special OTUs in ID groups. 24 phyla were shared by the 24 samples. Bacteroidetes was the most abundant phylum with the largest proportion, followed by Firmicutes and Proteobacteria. The OD groups showed a higher proportion of Bacteroidetes (>50 %) in cecum, but a lower Firmicutes/Bacteroidetes ratio in both 12-w old (0.42, 0.62) and 18-w old groups (0.37, 0.49) compared with the ID groups. Cecal microbiota in the OD groups have higher abundance of functions involved in amino acids and glycan metabolic pathway.

Conclusion: The composition and function of cecal microbiota in Dagu chicken under two feeding modes, free-range and cage raising are different. The cage raising mode showed a lower proportion of Bacteroidetes in cecum, but a higher Firmicutes/Bacteroidetes ratio compared with free-range mode. Cecal microbiota in free-range mode have higher abundance of functions involved in amino acids and glycan metabolic pathway.

Keywords: High-throughput sequencing technology, Feeding modes, Cecal microbiota, Composition and function

Background

Chickens have proportionally smaller intestines and shorter transit digestion times than mammals, but do not appear to any less efficient at digestion than their mammalian counterparts [1, 2]. Their digestive system is adapted to extract energy from difficult to digest food sources. This may be explained, in part, by the fact that the chicken gastrointestinal tract is home to a complex microbial community, the chicken gut microbiota, which underpins the links between diet and health [3, 4]. The host is unable to digest and utilize the complicated polysaccharide substance from the feedstuff in the absence of microbial fermentation [5]. Particularly relevant to the intensive farming of chickens is the cecum's role in digestion of non-starch polysaccaharides NSPs [6], which are found in the grains used in commercial chicken feed. The gut microbiota has one of the highest cell densities for any ecosystem and ranges from 10^7 to 10^{11} bacteria per g of gut content in poultry [7]. The most densely populated microbial community within the chicken gut is found in the ceca, a pair of blind-ended sacs that open off the large intestine [8]. This microbiota is also home to a rich collection of genes, the chicken gut microbiome, likely to include many sequences of scientific interest and biotechnological potential [4]. The coding capacity of microbiome vastly surpasses that of

* Correspondence: 13841607296@126.com
[1]Department of Animal Husbandry & Veterinary Medicine, Liaoning Medical University, Jinzhou, Liaoning 121000, China
Full list of author information is available at the end of the article

the host's genome, encoding biochemical pathways that the host has not developed.

An optimal gut microbiota can increase agricultural productivity, as evidenced by the ability of antibiotics to promote growth in chickens [9]. Studies on rumen microbe in ruminants have revealed that *Ruminococcus* and *Fibrobacter* species are important members of the rumen microbial community that enable the host to degrade and utilize fibrous plant materials efficiently as nutrients [10–12]. As a result, animal productivity has been improved through refining the animals' ability to degrade fiber by these microorganisms. Energy and nutrient extraction from feed requires interplay between the biochemical functions provided by the chicken and the microbiota present within the gastrointestinal tract (GIT). Highly productive chickens have been developed by selection for elite genetic traits; it is possible that in the future, gains in productivity and health outcomes could be influenced by selection of elite GIT microbiota [13]. Therefore, studies on the composition and functions of gut microbiota in animals raised in different feeding modes is significant for the improvement of feedstuff efficiency and animal productivity. At present, our ability to culture intestinal bacteria is limited, and hence, there is a need to profile and investigate this community using culture-independent techniques. Culture-independent analysis of the chicken cecal microbiota estimated 900 species of bacteria in 100 genera existing in the cecum of chickens, with most of them belonging to uncategorized genera [7, 14]. Previous studies have shown that the caeca microbial communities were more diverse in comparison to ilea [15]. Left and right ceca of chickens are harbouring similar bacterial communities [2]. But, the composition and function of cecal microbiota under different feeding mode are unknown.

Consumer interest in free-range and organic poultry is growing. The meat of the outdoor chickens had more protein than the indoor chickens [16]. Dagu chicken is a well-known local breed in China. Dagu chicken is native to Zhuanghe City, Liaoning and is free-range. This chicken has been called Cao Chicken (*cao* means grass in Chinese) because of the favorable living environment and fine feed resources of water and grass. Whether living habits influence the formation of gut microflora in Dagu chicken are unknown.

This study aims to explore the composition and function of cecal microbiota in Dagu chicken under two feeding modes, free-range and cage raising. Thus, providing base informations for designing high efficiency feed formula, developing applicable probiotics and regulating chicken meat quality.

Methods

Chicken farm and sampling

Zhuanghe City is located in the south end of the Liaodong peninsula. Its location, with coordinates N39.32′–40.5′, E122′–124.5′, indicates a typical mountainous hilly terrain. Dagu chicken in free-range farming is a traditional feeding in Zhuanghe and relies on abundant rivers and flourish pasture. Thus, this study was carried out in Zhuanghe Dagu chicken breeding center.

A total of 1000 1-day-old male Dagu chickens were selected. The chickens were raised in plastic mesh floors (80 cm above ground) for 6 weeks. The chickens were provided access to feed and water ad libitum. The house temperature was maintained at 35 °C during the first week, and it was reduced 2 °C per week until reaching the temperature of 23 °C. Six weeks later, 300 chickens with similar weights were randomly selected. Among them, 150 chickens were raised outside, which are in the outdoor group (OD group), while the other 150 chickens were raised inside their respective cages (50 cm × 50 cm × 50 cm,80 cm above ground) which are in the indoor group (ID group). The house temperature was maintained at 23 °C. The chickens were provided access to feed and water ad libitum. The difference of body weight between the two groups was not significant ($P >0.05$). Two groups were given the same compound feed (Additional file 1) as well as other environmental factors. The difference is for the OD group, each chicken in the OD group was let out every 5 am for self-help feeding in >30 m² area, where abundant water and grasses are found. The chickens were given supplementary feed at 1 pm, and kept indoors at 7 pm. When the chickens were 12 weeks old and 18 weeks old, weighed one by one, six of them with an average weight were randomly selected in each group,and then slaughtered. The cecum contents removed, preserved in liquid nitrogen, used for DNA extraction and PCR amplification. These samples were divided into four groups, namely, 12-w OD group, 12-w ID group, 18-w OD group, and 18-w ID group.

Gut microbes 16S rRNA sequencing

Microbial genomic DNA was extracted from cecal content samples by using the TIANGEN DNA stool mini kit (TIANGEN, cat#DP328) according to the producer's instructions (http://www.tiangen.com/asset/imsupload/up0921879001368428871.pdf). Variable region of 16S rRNA V4 was amplified using its universal primer sequence 520 F: AYTGGGYDTAAAGNG; 802R: TACNVGGGTATCTAATCC [17]. The PCR conditions were as follows: initial denaturation at 98 °C for 5 min; 98 °C denaturation for 30 s, 50 °C annealing for 30 s, and 72 °C extension for 30 s, which is repeated for 28 cycles; and a final extension at 72 °C for 5 min. PCR production was purified using QIAGEN Quick Gel Extraction Kit (QIAGEN, cat# 28706). PCR production from each sample was applied to construct a sequencing library by using Illumina TruSeq DNA Sample Preparation Kit (library was constructed using TruSeq Library Construction Kit). For each sample, barcoded V4 PCR amplicons were sequenced by the Illumina MiSeq PE250 platform.

Sequence reads were removed if sequence length was shorter than 150 bp, if average phred score was lower than 20, if containing ambiguous bases, if homopolymer run exceeded 6, or if there were mismatches in primers. Afterward, the sequences passed the quality filter that were assembled by Flash (http://www.genomics.jhu.edu), which required that the overlap of read 1 and read 2 ≥ 10 bp, and without any mismatches. The reads which could not be assembled were discarded. Chimera sequences were removed using UCHIME in mothur (version 1.31.2, http://www.mothur.org/). Amplification and sequencing of 16S rRNA v4 variable region was completed by Personal Biotechnology Co., Ltd. (Shanghai, China).

OTU clustering and statistical analysis

Sequences clustering was performed by uclust algorithm in QIIME (http://qiime.org/scripts/pick_otus.html), and clustered into operational taxonomic units (OTUs). The longest sequence in each cluster was selected as the representative. Taxonomy of each OTU was assigned by blasting the representative sequence against Greengenes reference database (Release 13.8, http://greengenes.secondgenome.com/). Unknown archaeal or eukaryotic sequences were filtered and removed. Ace, Chao, Simpson index were calculated using summary.single command in MOTHUR. A Venn diagram of between-group OTU was generated through R. The relative abundance of OTUs or taxa was compared between samples.

Diversity index data were analyzed statistically using analysis of variance (ANOVA) and significant differences between group means were determined using the least significant difference (LSD) test. Data of body weight and abundance at the phylum level between groups were analyzed statistically using T test. All values for diversity index and body weight are expressed as means ± standard errors (SE). Non-metric multidimensional scaling (NMDS) plots of sequence read abundance were generated with Vegan in R. All statistical analyses were performed using the SPSS 16.0 software.

Microbial function prediction

Functional genes were predicted through PICRUSt according to the abundance of OTU level [18]. The OTUs were mapped in gg13.5 database at 97 % similarity by QIIME's command "pick_closed_otus". The abundance of the OTUs was normalized automatically by using 16S

rRNA gene copy numbers from known bacterial genomes in the Integrated Microbial Genomes (IMG). The predicted genes and their function were aligned to the Kyoto Encyclopedia of Genes and Genomes (KEGG) database and the differences among groups were compared using STAMP (http://kiwi.cs.dal.ca/Software/STAMP) [19]. Two-side Welch's t-test and Benjamini–Hochberg FDR correction were employed in the two-group analysis. The relative abundance of KEGG metabolic pathways is referred to as a metabolic profile.

Results

OTU clustering and annotation

The trimmed and assembled sequences were clustered at 97 % similarity by calling uclust from Qiime. 1217 OTUs were obtained through database alignment by blast in Qiime. The total of OTUs obtained in each group were as follows: 1188 in the 12-w OD group, 1089 in the 12-w ID group, 1186 in the 18-w OD group, and 1158 in the 18-w ID group (Fig. 1). Figure 1 shows 34 special OTUs in OD groups (including 12-w OD and 18-w OD) and 4 special OTUs in ID groups (including 12-w OD and 18-w OD). The number of OTUs in each group slightly changed in the OD groups, whereas that

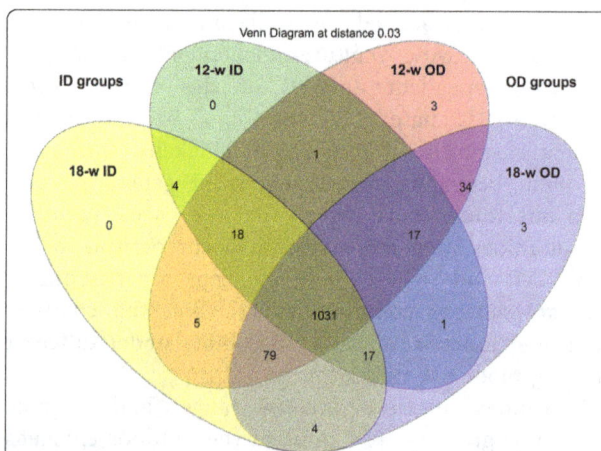

Fig. 1 Shared OUT analysis of the different groups. Numbers below groups indicate the number of OTUs within each sector. The number of species in 12-w OD is 1188; The number of species in 12-w ID is 1089; The number of species in 18-w OD is 1186; The number of species in 18-w ID is 1158; The number of species shared between 12-w OD and 12-w ID is 1067; The number of species shared between 12-w OD and 18-w OD is 1161; The number of species shared between 12-w OD and 18-w ID is 1133; The number of species shared between 12-w ID and 18-w OD is 1066; The number of species shared between 12-w ID and 18-w ID is 1070; The number of species shared between 18-w OD and 18-w ID is 1131; The number of species shared between 12-w OD, 12-w ID and 18-w OD is 1048; The number of species shared between 12-w OD, 12-w ID and 18-w ID is 1049; The number of species shared between 12-w OD, 18-w OD and 18-w ID is 1110; The number of species shared between 12-w ID, 18-w OD and 18-w ID is 1048; The total richness of all the groups is 1217

increased in the ID groups within days. The diversity of cecal microbiota in OD groups can be established earlier. The Chao and ACE in the 12-w OD group were significantly higher ($P < 0.05$) than those in the three other groups, but the Simpson in the OD groups was significantly lower ($P < 0.05$) than that in the ID groups. These results revealed that the richness of cecum microorganism in the 12-w OD group was higher than those in the three other groups, the evenness of cecum microorganism in the ID groups was higher than those in the OD groups (Table 1).

Differences of body weight and cecal microbiota in chickens raised in different feeding modes

In this study, chicken body weight in different feeding modes has obvious differences. Chicken body weight in the ID group was significantly higher than that in the OD group both 12-w or 18-w stage (Table 2).

A total of 24 phyla were shared by the 24 samples. Bacteroidetes was the most abundant phylum with the largest proportion, followed by Firmicutes and Proteobacteria (Fig. 2). Three significant differences ($P < 0.05$) in the 12-w groups and five significant ($P < 0.05$) differences in the 18-w groups were found (Table 3).

Spirochaetes had dynamic changes in the ID groups; its proportion was 5.73 % in the 12-w ID group, but it reduced to 1.4 % in the 18-w ID group. However, the proportion had a slight change in the OD groups (3.6 %, 3.05 %).

In the 12-w groups, Bacteroidetes, Firmicutes and Proteobacteria accounted for 83 % and 53 %, and Bacteroidetes for 52 % and 26.7 % in the OD and ID groups, respectively. In the 18-w groups, the three phyla accounted for 84.9 % and Bacteroidetes accounted for 53.66 % in the OD group, which exhibited a slight difference from that in the 12-w OD group. The proportion was 60.5 %, and Bacteroidetes had the largest share of 35 % in the ID group.

SAR406 mainly existed in the ID groups, accounting for 22.1 % in the 12-w group and 15.54 % in the 18-w group. SAR406 accounted for 0.17 % in the 12-w group and 0.37 % in the 18-w group in the OD groups, respectively.

At the genus level we detected 60 genera. 10 genera were significantly different ($P < 0.01$) between the 12-w

Table 2 Body weight

Group	Body weight g	
	OD	ID
12-W ($n = 150$)	1932.40 ± 13.24 [a]	2065.97 ± 11.36 [b]
18-W ($n = 144$)	2584.44 ± 18.39 [a]	2804.24 ± 15.76 [b]

Means with the different small letters within the same row are significantly; The means difference is significant at the 0.05 level

OD and 12-w ID groups (Additional file 2), 6 genera were significantly different ($P < 0.01$) between the 18-w OD and 18-w ID groups (Additional file 3).

NMDS results showed the difference in microorganism distributions in the four groups. The distribution was evidently different in the OD groups compared with that in the ID groups (Fig. 3). The microorganisms in the OD groups concentrated on one group whereas those in the ID groups concentrated on another. Numerical values in correlation analysis revealed that the cecal microbiota in the 12-w OD groups were quite different from those in the 12-w ID group (0.5729). However, the cecal microbiota in the 18-w OD group were remarkably similar to those in the 18-w ID group (0.9626) (Table 4). The results show that the richness and evenness of cecal microbiota in chickens raised in cages were noticeably different from those in chickens from free-range farming, especially at 12 weeks.

Microbial function analysis through PICTUSt was conducted to determine the differences in the functions of microbiota between the OD and ID groups. Numerous functions are involved in metabolic pathways. At KEGG level 2, cecal microbiota in the OD groups have higher abundance of functions involved in amino acids metabolic pathway (Fig. 4). At KEGG level 3, cecal microbiota in the 12-w OD group have higher abundance of functions involved in metabolic pathway such as metabolism of arginine, praline, histidine, glycine, serine, threonine, alanine, aspartate and glutamate, starch and sucrose, galactose, amino sugar and nucleotide sugar, and transcription machinery, DNA replication proteins than those in the 12-w ID group. Cecal microbiota in the 18-w OD group have higher abundance of functions involved in metabolic pathway such as metabolism of glycine, serine, threonine, arginine, praline, tryptophan, phenylalanine, tyrosine, and valine, leucine and isoleucine biosynthesis, amino acid related enzymes

Table 1 Diversity index

Group ($n = 6$)	12-w		18-w	
	OD	ID	OD	ID
Chao	4128 ± 733[b]	2273 ± 145[a]	2798 ± 223[a]	2814 ± 209[a]
ACE	5877 ± 1180[b]	3101 ± 218[a]	3884 ± 354[a]	3862 ± 323[a]
Simpson	0.037 ± 0.0087[ab]	0.088 ± 0.0231[c]	0.030 ± 0.0024[a]	0.064 ± 0.0141[bc]

Means with the same superscript within the same row are not significantly different,with the different small letters are significant; the means difference is significant at the 0.05 level

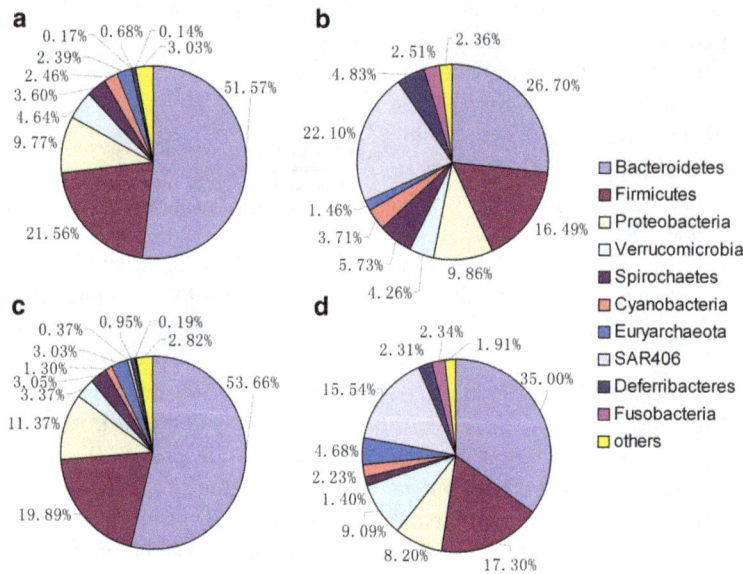

Fig. 2 Distribution of the cecum microbiota composition at the rank of phylum. **a** 12-w OD group. **b** 12-w ID group. **c** 18-w OD group. d, ID group. The proportions of each phylum in the 12-w OD and 12-w ID groups are as follows: Bacteroidetes: 51.57 %, 26.7 %; Firmicutes: 21.56 %, 16.49 %; Proteobacteria: 9.77 %, 9.86 %; Verrucomicrobia: 4.64 %, 4.26 %; Spirochaetes: 3.60 %, 5.73 %; Cyanobacteria: 2.46 %, 3.71 %; Euryarchaeota: 2.39 %, 1.46 %; SAR406: 0.17 %, 22.10 %; Deferribacteres: 0.68 %, 4.83 %; and Fusobacteria: 0.14 %, 2.51 %. The proportions of each phylum in the 18-w OD group and 18-w ID group are as follows: Bacteroidetes: 53.66 %, 35.00 %; Firmicutes:19.89 %, 17.30 %; Proteobacteria: 11.37 %, 8.20 %; Verrucomicrobia: 3.37 %, 9.09 %; Spirochaetes: 3.05 %, 1.40 %; Cyanobacteria: 1.30 %, 2.23 %; Euryarchaeota: 3.03 %, 4.68 %; SAR406: 0.37 %, 15.54 %; Deferribacteres: 0.95 %, 2.31 %; and Fusobacteria: 0.19 %, 2.34 %

than those in the 18-w ID group. In the OD groups, cecum contained more microbiota associated with glycosaminoglycan degradation and other glycan degradation (Additional file 4).

Discussion

Digestion and nutrient absorption are the basic function of the intestine, where gut microbiota play an important role. These microbiota have a significant influence on intestinal tract movement, growth and development, physiological functions, and non-specific immunity [20–25]. The diversity of gut microorganism is the foundation for animals' digestion and nutrient uptake, maintenance of biochemical functions and the intestine's physiological functions, and promotion of the immune system's development. Medical researches discovered that obesity is related to the changes of gut microbiota, diversity of gut microbiota

apparently decreases in obese patients [26]. The results of this study show that the body weight of caged chicken was significantly higher than that of free-range groups (Table 2). This is consistent with the results of other studies [27, 28]. Figure 1 shows 34 special OTUs in OD groups and 4 special OTUs in ID groups. The diversity of cecal microbiota in the OD groups was remarkably higher than that in the ID groups (Table 1). Bailey et al. discovered that long-term stress could reduce the diversity of gut microbiota in mice [29]. Chickens raised in OD and ID groups were exposed to distinct stresses and microbiota. Chickens raised in ID groups were exposed to more stresses, such as feeding density and space [30]. Chickens raised in OD groups may be due to the earlier contact to the natural environment; thus, the diversity can be established earlier.

Host and environmental factors influence the gut microbiota. The environmental factor is more important

Table 3 Comparisons for abundance at the phylum level

Phylum	12-w relative fold change (log$_2$ $^{OD/ID}$)	P value	18-w relative fold change (log$_2$ $^{OD/ID}$)	P value
Actinobacteria	1.3201	0.029	1.4615	0.069
Bacteroidetes	0.9498	0.000	0.6163	0.007
Elusimicrobia	1.0627	0.089	2.6878	0.033
Fusobacteria	−4.1746	0.054	−3.6384	0.043
SAR406	−7.0588	0.011	−5.4099	0.022
Tenericutes	0.4183	0.436	1.0385	0.045

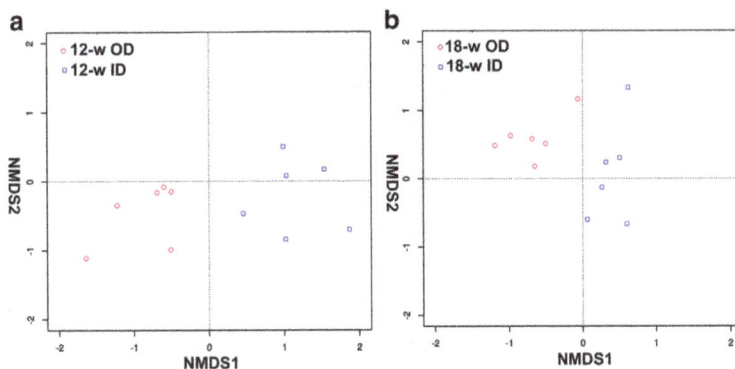

Fig. 3 NMDS ordination. **a** 12-w OD group and 12-w ID group. **b** 18-w OD group and 18-w ID group. NMDS plots demonstrate that free-range and cage ceca are harbouring different bacterial communities

than the host factor [31–34]. The phyla Firmicutes and Bacteroidetes dominate the intestine of mammals, followed by Fusobacteria, Proteobacteria, and Actinobacteria [35]. Bacteroidetes and Firmicutes have attracted considerable attention and are bounded to the host's metabolism. Numerous studies investigated the probiotic effect of Bacteroidetes; they found that Bacteroidetes help the host in polysaccharide decomposition to improve nutrient utilization [36], promote immune system development, improve host's immunity [37, 38], and maintain intestinal microecological balance [39, 40]. Results in this paper show that more (>50 %) Bacteroidetes existed in chickens in the OD groups (Fig. 2). and that the Firmicutes/Bacteroidetes ratio was smaller in the OD groups, with 0.42 and 0.62 in the 12-w groups and 0.37 and 0.49 in the 18-w groups. Research has shown that adding more dietary fiber can increase the amount of Bacteroidetes and lower the Firmicutes/Bacteroidetes ratio [41, 42]. The results show that compositions of cecal microbiota in chickens raised in two feeding modes were apparently different (Table 3, Fig. 3), especially at 12 weeks (Table 4). The difference may be attributed to the access of chickens from free-range farming to abundant microbiota in the outdoor environment; these chickens have abundance of food source and are able to intake more feedstuff containing fiber, which directly affects the composition of gut microbiota, increasing the Bacteroidetes content and lowering the Firmicutes/Bacteroidetes ratio.

Table 4 Correlation between groups for genus abundance

	12-w ID(N = 6)	18-w OD (N = 6)	18-w ID (N = 6)
12-w OD(N = 6)	0.5729	0.9936	0.9867
12-w ID (N = 6)		0.5767	0.6792
18-w OD (N = 6)			0.9626

Six samples from each group were used to calculate correlation

Obesity is related to the distribution of gut bacteria. High ratio of Firmicutes/Bacteroidetes causes obesity because more energy has been absorbed [43]. The small intestine is mainly involved in digestion and uptake of food, while a large amount of microorganisms related to microbial fermentation exists in the large intestine, especially the cecum [44]. Food rapidly passes the front of the intestinal tract but stays for several hours in the tail end of the tract [45]. Fat deposits mainly in the large intestine [46], which is closely related to the composition of microorganisms. In chicken production, bacteria related to productivity mainly include the phylum Firmicutes, along with Bacteroidetes and Proteobacteria [47]. Researchers suggested that fat pigs have more Firmicutes but fewer Bacteroidetes, especially fewer Bacteroides that are crucial in carbohydrate degradation [48, 49]. A study revealed that free-range farming can evidently reduce the growth performance and abdominal fat of chickens [27]. However, the efficiency of converting feedstuff to energy together with the chickens' productivity attracts increasing attention in the chickens production. In this paper, body weight of caged chickens was significantly higher than that of free-rage chickens (Table 2). We speculate that this may be due to that more Firmicutes and higher ratio of Firmicutes/Bacteroidetes in cecal microbiota improve the utilization efficiency of feed energy, of course, this needs further study.

Gut microbiota contains about 600,000 genes that are 25 times more compared with the genes in host's genome. Therefore, gut microbiota is usually regarded as one organ of the host and creates a gut microecosystem with the host's eucells [50, 51]. This microecosystem can execute numerous metabolic functions that alter with the change of microbiota' composition. In this paper, numerous functions are involved in metabolic pathways, such as metabolism of amino acid, carbohydrates, energy, lipid, replication and repair, nucleotides, and cofactors and vitamins. At KEGG level 2, there are 5

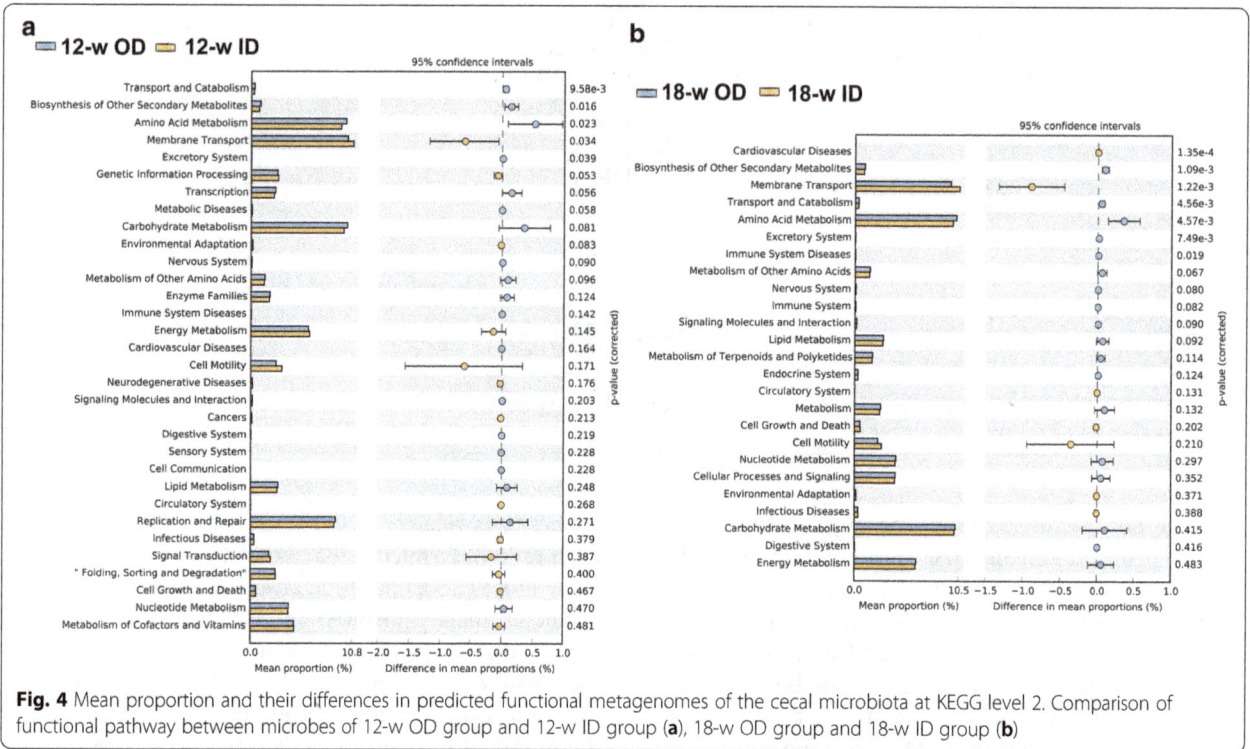

Fig. 4 Mean proportion and their differences in predicted functional metagenomes of the cecal microbiota at KEGG level 2. Comparison of functional pathway between microbes of 12-w OD group and 12-w ID group (**a**), 18-w OD group and 18-w ID group (**b**)

significant differences ($P < 0.05$) in abundance of functional categories between OD and ID group at 12-w, whereas 7 significant differences ($P < 0.05$) were found in between at 18-w (Fig. 2). At KEGG level 3, there are 42 significant differences ($P < 0.05$) in abundance of functional categories between OD and ID group at 12-w, among them 34 in OD group was significantly higher than that in ID group ($P < 0.05$). There are 72 significant differences ($P < 0.05$) in abundance of functional categories between the OD and ID group at 18-w, among them 44 in OD group was significantly higher than that in ID group ($P < 0.05$) (Additional file 4).

Cecal microbiota of OD group at 12-w and 18-w both has higher abundance of functions involved in metabolic pathway for certain amino acids, sugar compounds. Significant difference in amino sugar and nucleotide sugar metabolism pathways were observed in 12-w groups. Utilization of amino sugar and nucleotide sugar is important in chicken metabolism and growth. Amino sugar metabolism specifically is responsible for breaking down protein present in feed to amino acids or di- or tripeptides [52]. These were then transported from intestinal lumen to epithelial cell for energy. Nucleotide sugar metabolism on the other hand is important for purine and pyrimidine synthesis which is vital substrate for deoxyribonucleic acids derivatives. In addition, these components are also needed for producing high-energy nucleotides needed for cellular metabolism [53]. In this study, we observed that the genes responsible for amino

sugar and nucleotide sugar metabolism were up-regulated in 12-w OD group compared to 12-w ID group ($P < 0.05$) (Additional file 4). This may be the outdoor chickens needs more energy due to the large amount of movement. And movement promotes muscle development, and therefore the synthesis of more body protein. In contrast, the genes related to the metabolism of amino acids, amino sugars and nucleotide sugar were up-regulated in the cecum. Previous studies show that outdoor activities could make an improvement on the meat quality. The meat of chickens with outdoor access is darker, it has more protein contents and a better water-holding capacity [54, 55]. In addition, studies have revealed that feeding chickens with probiotics can improve meat quality and increase the output of breast and leg muscles [56]. All of these are likely to be related to the changes in compositions of gut microbiota. But, more scientific research is needed to confirm this.

Based on the research above, the many metabolic functions are involved in chickens' gut microbiota and these functions may vary because of the different compositions of gut microbiota. The compositions of chickens' cecal microbiota varied because the chickens were raised in different feeding modes. In-depth studies on the functions of dominant gut microbiota, such as Bacteroidetes and Firmicutes and their interaction, can help us develop a special probiotics and guide us to use the special probiotics to achieve the anticipated breed goals.

Conclusion

The composition and function of cecal microbiota in Dagu chicken under two feeding modes, free-range and cage raising are different. The cage raising mode showed a lower proportion of Bacteroidetes in cecum, but a higher Firmicutes/Bacteroidetes ratio compared with free-range mode. Cecal microbiota in free-range mode have higher abundance of functions involved in amino acids and glycan metabolic pathway. The results in this paper can provide relevant information for making strategies in raising Dagu chickens. This also provided valuable information for the study on microbiota in chicken gut.

Additional files

Additional file 1: Diet ingredients.

Additional file 2: Comparisons for abundance between the 12-w OD and 12-w ID groups at the genus level.

Additional file 3: Comparisons for abundance between the 18-w OD and 18-w ID groups at the genus level.

Additional file 4: Mean proportion and their differences in predicted functional metagenomes of the cecal microbiota at KEGG level 3. Comparison of functional pathway between microbes of 12-w OD group and 12-w ID group (a), 18-w OD group and 18-w ID group (b).

Acknowledgements
We thank T.-Y. Wang for help with sample collection. We also thank Y. Liu for technical assistance. All authors participated in the writing of the manuscript and agreed with the final format.

Funding
This study was funded by the National Natural Science Foundation of China (Grant no. 31172242; http://www.nsfc.gov.cn/) and Science and Technology Innovation Team Project of Liaoning Medical University (Grant no. 30420130104; http://www.lnmu.edu.cn/).

Authors' contributions
YT, YS and YX carried out the whole experiment and wrote the paper. LZ, HY, HX and DS participated in the sample collecting and analysis. All authors read and approved the final manuscript.

Competing interests
The authors declare that they have no competing interest.

Author details
[1]Department of Animal Husbandry & Veterinary Medicine, Liaoning Medical University, Jinzhou, Liaoning 121000, China. [2]Department of Veterinary Medicine, Nanjing Agricultural University, Nanjing, Jiangsu 210095, China. [3]Department of Food Science, Liaoning Medical University, Jinzhou, Liaoning, China.

References
1. McWhorter TJ, Caviedes-Vidal E, Karasov WH. The integration of digestion and osmoregulation in the avian gut. Biol Rev Camb Philos Soc. 2009;84(4):533–65.
2. Stanley D, Geier MS, Chen H, Hughes RJ, Moore RJ. Comparison of fecal and cecal microbiotas reveals qualitative similarities but quantitative differences. BMC Microbiol. 2015;15:51.
3. Björkstén B. The gut microbiota: a complex ecosystem. Clin Exp Allergy. 2006;36(10):1215–17.
4. Sergeant MJ, Constantinidou C, Cogan TA, Bedford MR, Penn CW, Pallen MJ. Extensive Microbial and Functional Diversity within the Chicken Cecal Microbiome. PLoS One. 2014;9(3):e91941. doi:10.1371/journal.pone.0091941.
5. Vrieze A, Holleman F, Zoetendal EG, de Vos WM, Hoekstra JB, Nieuwdorp M. The environment within: how gut microbiota may influence metabolism and body composition. Diabetologia. 2010;53(4):606–13. doi:10.1007/s00125-010-1662-7.
6. Jozefiak D, Rutkowski A, Martin SA. Carbohydrate fermentation in the avian ceca: a review. Anim Feed Sci Technol. 2004;113:1–15.
7. Apajalahti J, Kettunen A, Graham H. Characteristics of the gastrointestinal microbial communities, with special reference to the chicken. Worlds Poult Sci J. 2004;60:223–32. doi:10.1079/WPS200415.
8. Clench MH, Mathias JR. The avian cecum: a review. Wilson Bull. 1995; 107(l):93–121.
9. Huyghebaert G, Ducatelle R, Van Immerseel F. An update on alternatives to antimicrobial growth promoters for broilers. Vet J. 2011;187:182–8.
10. Mackie RI, White BA. Recent advances in rumen microbial ecology and metabolism: potential impact on nutrient output. J Dairy Sci. 1990;73(10): 2971–95. PMID:2178174.
11. Perumbakkam S, Mitchell EA, Craig AM. Changes to the rumen bacterial population of sheep with the addition of 2,4,6-trinitrotoluene to their diet. Antonie Van Leeuwenhoek. 2011;99(2):231–40. doi:10.1007/s10482-010-9481-x. PMID:20607404.
12. Weimer PJ. Redundancy, resilience, and host specificity of the ruminal microbiota: implications for engineering improved ruminal fermentations. Front Microbiol. 2015;6:296. doi:10.3389/fmicb.2015.00296. PMID:25914693.
13. Stanley D, Hughes RJ, Moore RJ. Microbiota of the chicken gastrointestinal tract: influence on health, productivity and disease. Appl Microbiol Biotechnol. 2014;98(10):4301–10.
14. Wei S, Morrison M, Yu Z. Bacterial census of poultry intestinal microbiome. Poult Sci. 2013;92:671–83.
15. Mohd Shaufi MA, Sieo CC, Chong CW, Gan HM, Ho YW. Deciphering chicken gut microbial dynamics based on high-throughput 16S rRNA metagenomics analyses. Gut Pathog. 2015;7:4. doi:10.1186/s13099-015-0051-7.
16. Fanatico AC, Pillai PB, Emmert JL, Owens CM. Meat quality of slow- and fast-growing chicken genotypes fed low-nutrient or standard diets and raised indoors or with outdoor access. Poult Sci. 2007;86(10):2245–55.
17. Blanton LV, Charbonneau MR, Salih T, Barratt MJ, Venkatesh S, Ilkaveya O, et al. Gut bacteria that prevent growth impairments transmitted by microbiota from malnourished children. Science. 2016; doi:10.1126/science.aad3311.
18. Langille MG, Zaneveld J, Caporaso JG, McDonald D, Knights D, Reyes JA, et al. Predictive functional profiling of microbial communities using 16S rRNA marker gene sequences. Nat Biotechnol. 2013;31(9):814–21. doi:10.1038/nbt.2676.
19. Parks DH, Tyson GW, Hugenholtz P, Beiko RG. STAMP: Statistical analysis of taxonomic and functional profiles. Bioinformatics. 2014;30(21):3123–4. doi:10.1093/bioinformatics/btu494.
20. De Angelis M, Siragusa S, Berloco M, Caputo L, Settanni L, Alfonsi G, et al. Selection of potential probiotic lactobacilli from pig feces to be used as additives in pelleted feeding. Res Microbiol. 2006;157(8):792–801.

21. Forder RE, Howarth GS, Tivey DR, Hughes RJ. Bacterial modulation of small intestinal goblet cells and mucin compositionduring early posthatch development of poultry. Poult Sci. 2007;86(11):2396–403.

22. Klasing KC. Nutrition and the immune system. Br Poult Sci. 2007;48(5):525–37.

23. Niba AT, Beal JD, Kudi AC, Brooks PH. Bacterial fermentation in the gastrointestinal tract of non-ruminants: influence of fermented feeds and fermentable carbohydrates. Trop Anim Health Prod. 2009;41(7):1393–407. doi:10.1007/s11250-009-9327-6.

24. Shira EB, Sklan D, Friedman A. Impaired immune responses in broiler hatchling hindgut following delayed access to feed. Vet Immunol Immunopathol. 2005;105(1-2):33–45.

25. Yin Y, Lei F, Zhu L, Li S, Wu Z, Zhang R, et al. Exposure of different bacterial inocula to newborn chicken affects gut microbiota development and ileum gene expression. ISME J. 2010;4(3):367–76. doi:10.1038/ismej.2009.128.

26. Turnbaugh PJ, Hamady M, Yatsunenko T, Cantarel BL, Duncan A, Ley RE, et al. A core gut microbiome in obese and lean twins. Nature. 2009;457(7228):480–4.

27. Wang KH, Shi SR, Dou TC, Sun HJ. Effect of a free-range raising system on growth performance, carcass yield, and meat quality of slow-growing chicken. Poult Sci. 2009;88(10):2219–23. doi:10.3382/ps.2008-00423.

28. Dou TC, Shi SR, Sun HJ, Wang KH. Growth rate, carcass traits and meat quality of slow-growing chicken grown according to three raising systems. Anim Sci Pap Rep. 2009;27:361–9.

29. Bailey MT, Dowd SE, Parry NM, Galley JD, Schauer DB, Lyte M. Stressor exposure disrupts commensal microbial populations in the intestines and leads to increased colonization by Citrobacter rodentium. Infect Immun. 2010;78(4):1509–19. doi:10.1128/IAI.00862-09.

30. Jones M, Millis AD. Divergent selection for social reinstatement and behaviors in Japanese quail: effects on sociality and social discrimination. Poult Avian Biol Rev. 1999;10(4):213–23.

31. Carmody RN, Gerber GK, Luevano Jr JM, Gatti DM, Somes L, Svenson KL. Diet dominates host genotype in shaping the murine gut microbiota. Cell Host Microbe. 2015;17(1):72–84. doi:10.1016/j.chom.2014.11.010.

32. Turnbaugh PJ, Ridaura VK, Faith JJ, Rey FE, Knight R, Gordon JI. The effect of diet on the human gut microbiome: a metagenomic analysis in humanized gnotobiotic mice. Sci Transl Med. 2009;1(6):6ra14. doi:10.1126/scitranslmed.3000322.

33. Tachon S, Zhou J, Keenan M, Martin R, Marco ML. The intestinal microbiota in aged mice is modulated by dietary resistant starch and correlated with improvements in host responses. FEMS Microbiol Ecol. 2013;83(2):299–309. doi:10.1111/J.1574-6941.2012.01475.x.

34. Fava F, Gitau R, Griffin BA, Gibson GR, Tuohy KM, Lovegrove JA. The type and quantity of dietary fat and carbohydrate alter faecal microbiome and short-chain fatty acid excretion in a metabolic syndrome 'at-risk' population. Int J Obes (Lond). 2013;37(2):216–23. doi:10.1038/ijo.2012.33.

35. Ley RE, Hamady M, Lozupone C, Turnbaugh PJ, Ramey RR, Bircher JS, et al. Evolution of mammals and their gut microbes. Science. 2008;320(5883):1647–51. doi:10.1126/science.1155725.

36. Bäckhed F, Ding H, Wang T, Hooper LV, Koh GY, Nagy A, et al. The gut microbiota as an environmental factor that regulates fat storage. Proc Natl Acad Sci. 2004;101(44):15718–23.

37. Stappenbeck TS, Hooper LV, Gordon JI. Developmental regulation of intestinal angiogenesis by indigenous microbes via Paneth cells. Proc Natl Acad Sci. 2002;99(24):15451–5.

38. Hooper LV. Bacterial contributions to mammalian gut development. Trends Microbiol. 2004;12(3):129–34.

39. Sears CL. A dynamic partnership: celebrating our gut flora. Anaerobe. 2005;11(5):247–51.

40. Hooper LV, Wong MH, Thelin A, Hansson L, Falk PG, Gordon JI. Molecular analysis of commensal host-microbial relationships in the intestine. Science. 2001;291(5505):881–4. PMID:11157169.

41. Trompette A, Gollwitzer ES, Yadava K, Sichelstiel AK, Sprenger N, Ngom-Bru C, et al. Gut microbiota metabolism of dietary fiber influences allergic airway disease and hematopoiesis. Nat Med. 2014;20(2):159–66.

42. Parnell JA, Reimer RA. Prebiotic fiber modulation of the gut microbiota improves risk factors for obesity and the metabolic syndrome. Gut Microbes. 2012;3(1):29–34.

43. Turnbaugh PJ, Ley RE, Mahowald MA, Magrini V, Mardis ER, Gordon JI. An obesity-associated gut microbiome with increased capacity for energy harvest. Nature. 2006;444(7122):1027–31.

44. DiBaise JK, Zhang H, Crowell MD, Krajmalnik-Brown R, Decker GA, Rittmann BE. Gut microbiota and its possible relationship with obesity. Mayo Clin Proc. 2008;83(4):460–9. doi:10.4065/83.4.460.

45. Kohl KD, Miller AW, Marvin JE, Mackie R, Dearing MD. Herbivorous rodents (Neotoma spp.) harbour abundant and active foregut microbiota. Environ Microbiol. 2014;16(9):2869–78. doi:10.1111/1462-2920.12376.

46. Zhao W, Wang Y, Liu S, Huang J, Zhai Z, He C, et al. The Dynamic distribution of porcine microbiota across different ages and gastrointestinal tract segments. PLoS One. 2015;10(2):e0117441. doi:10.1371/journal.pone.0117441.

47. Torok VA, Hughes RJ, Mikkelsen LL, Perez-Maldonado R, Balding K, MacAlpine R, et al. Identification and characterization of potential performance-related gut microbiotas in broiler chickens across various feeding trials. Appl Environ Microbiol. 2011;77(17):5868–78. doi:10.1128/AEM.00165-11.

48. Ley RE, Turnbaugh PJ, Klein S, Gordon JI. Microbial ecology: human gut microbes associated with obesity. Nature. 2006;444(7122):1022–3.

49. Arumugam M, Raes J, Pelletier E, Le Paslier D, Yamada T, Mende DR, et al. Enterotypes of the human gut microbiome. Nature. 2011;473(7346):174–80. doi:10.1038/nature09944.

50. Qin J, Li R, Raes J, Arumugam M, Burgdorf KS, Manichanh C, et al. A human gut microbial gene catalogue established by metagenomic sequencing. Nature. 2010;464(7285):59–65. doi:10.1038/nature08821.

51. Lederberg J. Infectious history. Science. 2000;288(5464):287–93.

52. Miska KB, Fetterer RH, Wong EA. The mRNA expression of amino acid transporters, aminopeptidase N, and the di- and tri-peptide transporter PepT1 in the embryo of the domesticated chicken (Gallus gallus) shows developmental regulation. Poult Sci. 2014;93:2262–70.

53. Rengaraj D, Lee BR, Jang HJ, Kim YM, Han JY. Comparative metabolic pathway analysis with special reference to nucleotide metabolism-related genes in chicken primordial germ cells. Theriogenology. 2013;79:28–39.

54. Mikulski D, Celej J, Jankowski J, Majewska T, Mikulska M. Growth performance, carcass traits and meat quality of slower-growing and fast-growing chickens raised with and without outdoor access. Asian Aust J Anim Sci. 2011;24(10):1407–16. doi:10.5713/ajas.2011.11038.

55. Chen X, Jiang W, Tan HZ, Xu GF, Zhang XB, Wei S, et al. Effects of outdoor access on growth performance, carcass composition, and meat characteristics of broiler chickens. Poult Sci. 2013;92(2):435–43. doi:10.3382/ps.2012-02360. PMID:23300311.

56. Kabir SML, Rahman MM, Rahman MB, Rahman MM, Ahmed SU. The dynamics of probiotics on growth performance and immune response in broilers. Int J Poult Sci. 2004;3(5):361–4.

12

Comparison of rumen bacterial communities in dairy herds of different production

Nagaraju Indugu[1], Bonnie Vecchiarelli[1], Linda D. Baker[1], James D. Ferguson[1], Jairam K. P. Vanamala[2,3] and Dipti W. Pitta[1]* ⓘ

Abstract

Background: The purpose of this study was to compare the rumen bacterial composition in high and low yielding dairy cows within and between two dairy herds. Eighty five Holstein dairy cows in mid-lactation (79–179 days in milk) were selected from two farms: Farm 12 (M305 = 12,300 kg; $n = 47$; 24 primiparous cows, 23 multiparous cows) and Farm 9 (M305 = 9700 kg; $n = 38$; 19 primiparous cows, 19 multiparous cows). Each study cow was sampled once using the stomach tube method and processed for 16S rRNA gene amplicon sequencing using the Ion Torrent (PGM) platform.

Results: Differences in bacterial communities between farms were greater (Adonis: $R^2 = 0.16$; $p < 0.001$) than within farm. Five bacterial lineages, namely *Prevotella* (48–52%), unclassified Bacteroidales (10–12%), unclassified bacteria (5–8%), unclassified Succinivibrionaceae (1–7%) and unclassified Prevotellaceae (4–5%) were observed to differentiate the community clustering patterns among the two farms. A notable finding is the greater ($p < 0.05$) contribution of Succinivibrionaceae lineages in Farm 12 compared to Farm 9. Furthermore, in Farm 12, Succinivibrionaceae lineages were higher ($p < 0.05$) in the high yielding cows compared to the low yielding cows in both primiparous and multiparous groups. *Prevotella*, S24-7 and Succinivibrionaceae lineages were found in greater abundance on Farm 12 and were positively correlated with milk yield.

Conclusions: Differences in rumen bacterial populations observed between the two farms can be attributed to dietary composition, particularly differences in forage type and proportion in the diets. A combination of corn silage and alfalfa silage may have contributed to the increased proportion of Proteobacteria in Farm 12. It was concluded that Farm 12 had a greater proportion of specialist bacteria that have the potential to enhance rumen fermentative digestion of feedstuffs to support higher milk yields.

Keywords: Dairy cows, Rumen microbiota, Dairy herds

Background

Nearly 70% of energy [1] and 60–85% of protein [2] requirements of the dairy cow are met from microbial fermentation, indicating a critical need for maximizing rumen function and describing rumen microbiota. However, it is still not known how diet and microbes interact to enhance milk yields in dairy cows. Typically, dairy herds are fed total mixed rations (TMR) and cows with greater milk production have greater dry matter intake (DMI) [3]. In TMRs for dairy cattle, carbohydrates constitute nearly 70% of the dietary dry matter (DM) and provide the major energy source for rumen microbes [4, 5]. Carbohydrates may be broadly classified into two distinct groups based on their solubility in neutral detergent [6, 7]. Neutral detergent fiber (NDF) is insoluble in neutral detergent solution and is composed of cellulose, hemicellulose and lignin. Cellulose and hemicellulose are fiber carbohydrates (FC) predominantly

* Correspondence: dpitta@vet.upenn.edu
[1]Department of Clinical Studies, School of Veterinary Medicine, New Bolton Center, University of Pennsylvania, Kennett Square, PA 19348, USA
Full list of author information is available at the end of the article

found in forages. FC ferment slowly and are important in regulating rumen function through formation of the rumen mat and influencing the rate of passage out of the rumen. Non-Fiber Carbohydrates (NFC) are soluble in neutral detergent solution and include sugars, starch, beta-glucans and pectins, which ferment more rapidly in the rumen [7]. To maintain a stable rumen environment and enhance microbial growth, a minimum NDF and a maximum NFC (% diet DM) are needed in high producing dairy cow diets. Excessive NFC can create acidosis, while excessive NDF and low NFC can constrain feed intake and milk production [8]. The proportion of NDF and NFC in the diet, the composition of NFC components, and the extent to which these carbohydrates ferment in the rumen can influence the ruminal microbiota [9, 10].

Corn silage represents a major feed resource and often comprises 50% to 70% of forage in diets in the Northeastern US [11, 12]. Dairy One (Ithaca, NY), a commercial laboratory, reports that NDF averaged 44.2% DM (SD 5.3) and starch 30.7% DM (SD 6.5) in 11,281 corn silage samples analyzed between May 1, 2014 and April 30, 2015 (http://dairyone.com/2016-fresh-corn-silage-results-ny-and-pa/). Corn silage is classified as forage, however contains varying amounts of grain, contributing both dietary NDF and starch [8]. Concentrations of NDF and starch in corn silage are dependent upon plant genetics, environmental conditions, and maturity and processing of the corn grain at harvest [8] and can have a major impact on ration fermentability, and most likely the rumen microbiota. Lettat et al., [12] found cows fed a corn silage based TMR had an increase in total bacteria, an increase in propionate and a decrease in methane production compared to cows fed an alfalfa silage based TMR. However, information concerning the influence of corn silage based diets on the microbial ecology in the rumen of dairy cows is limited.

Information on rumen microbial dynamics in dairy cows is emerging [13–16]. Recent reports include changes in the composition of ruminal microbiota in dairy cows in association with parity [15–17], diet [18, 19], breed [20, 21], feed efficiency [14, 15, 22], milk yield and composition [19, 23] and physiological status [16, 18, 24, 25]. The consensus of these reports indicates the preponderance of lineages from Bacteroidetes and Firmicutes among the rumen microbiota. However, it is not known which bacteria are relevant and in what proportions they are needed to enhance rumen fermentation of feedstuffs. In this study, our goal was to link how dietary components influence rumen bacterial populations, and how they may impact milk yield and composition in dairy cows. Here, we sampled rumen contents from primiparous and multiparous dairy cows selected from two dairy herds differing significantly in their average annual milk production. We analyzed the composition of rumen bacterial communities using Ion Torrent (PGM) sequencing and investigated their relationship with nutrition and production parameters.

Methods
Experimental details
This study included 85 animals of Holstein breed selected from two dairy herds in southeastern Pennsylvania, a higher producing farm (Farm 12; M305 = 12,300 kg; $n = 47$ including 24 primiparous, 23 multiparous cows (11 s, 4 third, 6 fourth, and 2 fifth lactation cows)) and a lower producing farm (Farm 9; M305 = 9700 kg; $n = 38$ including 19 primiparous, 19 multiparous cows (11 s, 6 third, and 2 fourth lactation cows). Production information including milk yield (kg/d), protein (%), and fat (%) for experimental cows were retrieved from Dairy Record Management Systems (DRMS). We differentiated cows into high and low milk production based on their previous milk test day results for each farm and parity group. As a result, we observed at least 9 kg difference in daily milk yield within each parity group. The selected cows were between 79 and 179 days in milk production (DIM) (Additional file 1: Table S1) with an average of 113 and 120 days for primiparous cows and 125 and 131 days for multiparous cows in Farm 12 and Farm 9, respectively. Dairy cows that were donors of rumen fluid were maintained according to the ethics committee and IACUC standards for the University of Pennsylvania (approval #805538).

TMR samples were collected immediately after presentation to the cows. Five samples were scooped from the dispersed feed at varying locations, combined into one sample and frozen at -20 °C. These feed samples were dried at 55 °C for 72 h and ground in a Wiley mill (Thomas Scientific, Swedesboro, NJ) using a 1-mm screen. DM, CP (Kjeldahl), ADF, sugars and ash were assayed according to AOAC [26]. The NDF was determined using sodium sulfite and a heat-stable α-amylase enzyme (A3306, Sigma-Aldrich, St. Louis, MO) according to the procedure of Van Soest et al. [27]. The starch content was determined using a commercial kit (Megazyme International Ireland Ltd., Bray, Ireland) based on the enzymatic method [23]. Mineral contents were analyzed by the atomic absorption spectroscopy method and protocol of AOAC [26].

Rumen sampling
Rumen contents were sampled once from animals on each farm 2 h post-feeding using the stomach tube method [28]. The initial volume collected, approximately

200 mL, was discarded due to possible contamination with saliva and the subsequent 250 ml sample (planktonic phase) was obtained, transferred into 15 ml falcon tubes and snap frozen in liquid nitrogen. The samples were then transported to the laboratory and archived at −80 °C.

DNA extraction, PCR and 16S rRNA gene sequencing

The genomic DNA was extracted from rumen samples using the PSP Spin Stool DNA Plus Kit (Invitek, Berlin, Germany). The DNA extraction method was adapted from Dollive et al. [29]. Briefly, the method involved taking 300 mg of sample in a Lysing Matrix E tube (MP BIomedicals, Solon, OH USA) and bead beaten in 1400 µl of stool stabilizer from the PSP kit to break open cell walls and release nucleic acid material. Samples were then heated at 95 °C for 15 min, placed on ice for 1 min, and spun down at 13,400 g for 1 min. The supernatant was then transferred to the PSP InviAdsorb tubes and the remainder of the protocol for the PSP Spin Stool DNA Plus was followed according to the manufacturer's instructions. The genomic DNA was amplified using specific primers 27F and BSR357, targeting the V1–V2 region of the 16S rRNA bacterial gene. The forward primer carried the Ion Torrent trP1 (5′-CCTCTCTATGGGC AGTCGGTGAT-3′) and the reverse primer carried the A adapter (5′-CCATCTCATCCCTGCGTGTCTCCGACTC AG-3′), followed by a 10–12 nucleotide (nt) sample-specific barcode sequence and a GAT barcode adapter. The PCR mix was prepared using the Accuprime Taq DNA polymerase System (Invitrogen, Carlsbad, CA).

The thermal cycling conditions involved an initial denaturing step at 95 °C for 5 min followed by 25 cycles (denaturing at 95 °C for 30 s, annealing at 56 °C for 30 s, extension at 72 °C for 90 s) and, finally, an extension step at 72 °C for 8 min as described in Pitta et al. [18]. Amplicons of 16S rRNA genes were purified using 1:1 volume of Agencourt AmPure XP beads (Beckman-Coulter, Brea, CA, USA). The purified PCR products from the rumen samples were pooled to achieve a concentration of 5–20 ng prior to sequencing in Ion Torrent (PGM) platform.

Data analysis

The 16S pyrosequence reads were analyzed using the QIIME pipeline [30], and in R 3.3.1 [31]. All sequences were quality filtered. Sequences shorter than 50 nt and longer than 480 nt, incorrect primer sequences, and those containing one or more ambiguous bases or homopolymers longer than 5 nt were discarded. Operational taxonomic units (OTUs) were formed at 97% similarity using UCLUST [32]. We randomly subsampled (rarified) the resulting OTUs to 3353 sequences per sample. Representative sequences from each OTU

were aligned to 16S reference sequences with PyNAST [33] which were used to infer a phylogenetic tree with FastTree [34]. Taxonomic assignments within the GreenGenes taxonomy [35] were generated using the RDP Classifier version 2.2 [36].

A non-parametric permutational multivariate ANOVA (PERMANOVA) test [36], implemented in the vegan package for R [37] was used to test the effects of milk production, parity and farm on overall community composition, as measured by weighted UniFrac distance. To test for differences in taxon abundance, a generalized linear model (GLM) was constructed with the statistical package for R. The model used a binomial link function and input data for the model consisted of a two-column matrix containing the number of reads assigned to the taxon (in column 1) and the number of reads assigned to other taxa (in column 2) and p $values$ were adjusted using a microbial taxonomy-wide detection rate.

The extent of relationship between bacterial communities was quantified using weighted pairwise UniFrac distances [38]. Communities with small weighted UniFrac distance are composed of phylogenetically similar organisms in similar proportions. To identify bacterial lineages that drive the clustering of microbial communities in each farm, we used the biplot function of the make_emperor.py script to plot the genus-level OTUs in PCoA (Principal Coordinate Analysis) space alongside each Farm. Spearman correlation was used to correlate physiological parameters with OTUs assigned to Succinivibrionaceae and with the abundant genera in Bacteroidetes, Firmicutes and Proteobacteria using R and visualized using the corrplot R package [39].

Results
Dietary information

The ingredient composition and the nutritive value of diets fed to both primiparous and multiparous dairy cows for both Farm 12 (M305 = 12,300 kg) and Farm 9 (M305 = 9700 kg) are presented in Table 1. Forage content was lower for cows on Farm 12 and averaged 49.8% of diet DM compared to an average of 56.0% forage on Farm 9. Both farms fed corn silage as their primary forage. Farm 12 fed 77.5% (primiparous) and 71.2% (multiparous) of the forage as corn silage, and 22.6% (primiparous) and 25.9% (multiparous) of the forage as alfalfa silage. Additionally, 2.9% of the forage was grass hay for the multiparous cows only on Farm 12. In contrast, Farm 9 fed 85.0% (primiparous) and 81.9% (multiparous) of the forage as corn silage and the remaining forage 15.0% (primiparous) and 18.1% (multiparous) was triticale silage. Carbohydrates fed as grains consisted of fine ground corn and wheat middlings on Farm 12 (combined, 26.2% DM), while fine

Table 1 Composition (% DM) of diets fed to primiparous and multiparous cows on Farm 12 and Farm 9

	Farm 12		Farm 9	
	Primiparous	Multiparous	Primiparous	Multiparous
Ingredient composition, %DM				
Corn Silage[a]	38.9	35.1	47.2	46.2
Alfalfa Silage[a]	11.3	12.8	—	—
Triticale[a]	—	—	8.3	10.2
Grass Hay[a]	—	1.4	—	—
Corn Ground Fine	18.2	20.6	15.2	14.9
Wheat Middlings	8.1	5.4	—	—
Soybean Hulls	—	—	5.5	5.4
Corn Distillers	—	—	3.7	3.6
Citrus Pulp	—	—	1.1	1.1
AminoPlus[b]	3.7	3.8	7.4	7.2
SoyPlus[c]	3.7	3.7	—	—
Canola Meal	9.7	8.9	—	—
Soybean Meal	0.4	1.7	5.5	5.4
Blood Meal	0.5	0.7	1.0	1.0
Urea	—	—	.0.4	0.4
Molasses	2.1	1.6	1.3	1.3
Energy Booster 100[d]	0.4	0.9	—	—
Megalac[e]	0.5	0.6	—	—
Pork-Vegetable Fat	—	—	0.5	0.5
Calcium Carbonate	0.99	1.01	1.09	1.11
Sodium Sesquicarbonate	0.87	0.92	0.74	0.72
Sodium Chloride	0.37	0.38	0.46	0.45
Magnesium Oxide	0.13	0.13	0.18	0.18
Methionine, MFP[f]	0.07	0.05	0.11	0.11
Methionine, SmartamineM[g]	—	0.02	—	—
Mineral-Vitamin Mix[h]	0.06	0.11	0.22	0.22
Rumensin 90[i,j], mg/kg	150.0	149.2	50.4	50.4
Chromium propionate 4%,[k,l] mg/kg	2.60	2.54	1.01	1.01
Chemical composition %DM				
Crude Protein	16.5	16.7	16.2	15.0
Soluble Protein, % CP	31.4	31.2	32.2	28.4
RDP[m] estimated, % CP	59.1	56.9	60.2	59.4
ADF	18.0	18.9	19.2	19.6
NDF	28.8	28.3	30.6	31.3
NFC[n]	44.0	44.6	42.6	43.1
Sugar	4.5	4.1	4.3	5.4
Starch	27.5	32.1	26.1	29.2
Fat	3.8	3.8	3.4	3.4
Ash	6.9	6.7	7.2	7.2
Ca	.92	.82	.96	.79
P	.48	.48	.34	.33

[a]Forage
[b]Heat treated soybean meal; Ag Processing Incorporated (Omaha, Nebraska)
[c]Extruded soybean meal; West Central Cooperative (Ralston, Iowa)
[d]Rumen inert fat; Milk Specialties (Eden Prairie, Minnesota)
[e]Rumen inert fat; Arm and Hammer Animal Nutrition (Princeton, New Jersey)
[f]Rumen available methionine source; Novus International (Saint Charles, Missouri)
[g]Rumen protected methionine; Adisseo (Alpharetta, Georgia)
[h]Concentration (DM basis); **Farm 12 Primiparous**: 39 mg of Fe/kg, 32,795 mg of Zn/kg, 7695 mg of Cu/kg, 18,091 mg of Mn/kg, 460 mg of Se/kg, 616 mg of Co/kg, 622 mg of I/kg, 3203 KIU of vitamin A/kg, 804 KIU vitamin D/kg, 22 KIU vitamin E/kg; **Farm 12 Multiparous**: 3300 mg of Fe/kg, 21,000 mg of Zn/kg, 5556 mg of Cu/kg, 11,118 mg of Mn/kg, 141 mg of Se/kg, 543 mg of Co/kg, 222 mg of I/kg, 1032 KIU vitamin A/kg, 1032 KIU vitamin D/kg, 10 KIU vitamin E/kg; **Farm 9 Primiparous and Multiparous** 4537 mg of Fe/kg, 19,638 mg of Zn/kg, 3724 mg of Cu/kg, 10,157 mg of Mn/kg, 130 mg of Se/kg, 102 mg of Co/kg, 305 mg of I/kg, 1067 KIU of vitamin A/kg, 267 KIU vitamin D/kg, 6 KIU vitamin E/kg;
[i]Elanco Animal Health (Greenfield, Indiana)
[j]Provided 14.6 and 15.0 mg monensin/kg diet DM to Farm 12 primiparous and multiparous cows, respectively, and provided 14.8 and 14.5 mg monensin/kg diet DM to Farm 9 primiparous and multiparous cows, respectively
[k]Kemin (Des Moines, Iowa)
[l]Provided 0.25 and 0.26 mg chromium/kg diet DM to Farm 12 primiparous and multiparous cows, respectively, and provided 0.30 and 0.29 mg chromium/kg diet DM to Farm 9 primiparous and multiparous cows, respectively
[m]Rumen degraded protein estimated using CPM-Dairy software
[n]Non-Fiber Carbohydrate calculated value (100-CP-NDF-Fat-Ash)

ground corn, soybean hulls, corn distillers and citrus pulp were included in the diets for Farm 9 (combined, 25.3% DM). However, the fine ground corn averaged 15.0% DM on Farm 9 and was higher on Farm 12 at 18.2% DM for primiparous cows and 20.6% DM for multiparous cows (Table 1). Analyzed TMR samples indicated the starch content of the multiparous cows on Farm 12 was the highest at 32.1%, compared to the range of the other production groups of 26.1% to 29.2% dietary starch. Subsequently, NDF was higher (31.0%) on Farm 9 than Farm 12 (28.5%) and NFC was lower (42.8%) on Farm 9 than Farm 12 (44.3%). Sugars were similar for all production and parity groups with the exception of the multiparous cows on Farm 9 which had the highest sugar content of 5.4% DM. Dietary fat content tended to be higher in Farm 12 due to dietary fat supplementation compared to Farm 9.

Production information

Test day records from DRMS, Raleigh, NC were extracted for cows sampled on each farm. Production parameters by farm and parity including milk yield (kg/d), milk fat (%) and milk protein (%) are presented in Table 2. A difference in milk yield of 9.5 and 18.1 kg/d was noted between the high and low producing cows for primiparous and multiparous groups, respectively for Farm 12. A difference of 11.7 and 9.7 kg/d was noted between high and low producing cows in the primiparous and multiparous groups, respectively for Farm 9. Milk fat (%) was the lowest (3.20%) in the high producing multiparous cows and was the highest (3.86%) in the low producing multiparous cows from both Farm 9 and Farm 12, most likely associated with the higher milk yields and dilution of milk components. Milk fat (%) was higher for the primiparous cows on Farm 9 (3.73%) as compared to the primiparous cows on Farm 12 (3.32%). Overall for both farms and parity groups, milk fat (%) and protein (%) was higher in low yielding cows compared to high yielding cows for both farms and parity groups.

Sequencing information

Approximately 660,913 reads were analyzed from 85 rumen bacterial communities, with an average of 7775 reads per sample and greater than 3353 reads per sample (Additional file 1: Table S1). Approximately 59,843 OTUs were produced by clustering at 97% sequence similarity. We randomly subsampled (rarified) OTUs to 3353 sequences per sample which produced approximately 42,223 OTUs. From these sequences, 18 bacterial phyla and 73 genera were identified.

Comparison of bacterial community composition within and between herds

The rumen bacterial communities in the planktonic phase for Farm 12 were significantly different (Adonis: $R^2 = 0.16$, $p < 0.001$; Additional file 2: Table S2; Fig. 1) from Farm 9. Bacterial communities in Farm 12 were distantly spaced and spread across the principal coordinate 1

(PC1) whereas, bacterial communities in Farm 9 formed tight clusters on PC2. PC1 was driven by Succinivibrionaceae lineages while the PC2 was driven by *Prevotella*, unclassified Bacteroidales, unclassified Prevotellaceae and an unclassified lineage. Within both herds, the rumen bacterial communities were significantly different by parity (PERMANOVA; $p < 0.05$; Additional file 2: Table S2). However, no such differences were evident by production level within parity group in either herd, with the exception of primiparous cows in Farm 12 (Additional file 2: Table S2).

Irrespective of the herd, Bacteroidetes (75%) and Firmicutes (10–11%) together comprised up to 86% of the bacterial abundance. However, the mean abundance of Proteobacteria (7% vs. 2%) was higher in Farm 12 compared to Farm 9 (Additional file 3: Table S3). Bacteroidetes was predominated by *Prevotella* (48.2%) followed by unclassified Bacteroidales and several other taxa in Farm 9. In Farm 12, *Prevotella* (51.8%) was higher, while unclassified Bacteroidales and RF16 were relatively reduced and S24-7 numerically increased compared to Farm 9 (Additional file 3: Table S3). Among Firmicutes, the majority of taxa decreased in Farm 12 compared to Farm 9, with the exception of *Coprococcus*, unclassified Mogibacteriaceae, *Pseudobutyrivibrio*, *Shuttleworthia* and unclassified Veillonellaceae. Noticeable differences were evident in the total abundance and genera that made up Proteobacteria between both herds. In Farm 12, more than 95% of Proteobacteria was composed of Succinivibrionaceae lineages, whereas in Farm 9, Proteobacteria was comprised of several genera in addition to Succinivibrionaceae lineages (Additional file 3: Table S3).

Differences in the individual bacterial populations are presented for the high and low yielding cows within each parity level in both herds (Table 3). In the high yielding primiparous cows of Farm 12, *Prevotella* from Bacteroidetes was significantly higher compared to the low yielding primiparous cows. The contribution from Firmicutes decreased while that of Proteobacteria

Table 2 Mean with SD of production parameters in high and low yielding cows by parity and farm

	Farm 12				Farm 9			
	Primiparous cows		Multiparous cows		Primiparous cows		Multiparous cows	
	High	Low	High	Low	High	Low	High	Low
	(n = 12)	(n = 12)	(n = 12)	(n = 11)	(n = 10)	(n = 9)	(n = 9)	(n = 10)
Days in milk	116	109	125	125	128	113	130	132
Milk yield (kg/d)	48.1	38.6	70.7	52.6	37.2	25.5	41.8	32.1
	± 1.3	± 1.2	± 4.3	± 3.3	± 7.1	± 2.7	± 3.4	± 3.4
Fat (%)	3.31	3.34	3.20	3.86	3.64	3.83	3.19	3.86
	± 0.57	± 0.61	± 1.25	± 0.90	± 0.36	± 0.62	± 0.48	± 0.67
Protein (%)	2.85	2.94	2.86	3.23	2.91	2.94	2.87	3.03
	± 0.22	± 0.20	± 0.27	± 0.28	± 0.22	± 0.30	± 0.23	± 0.23

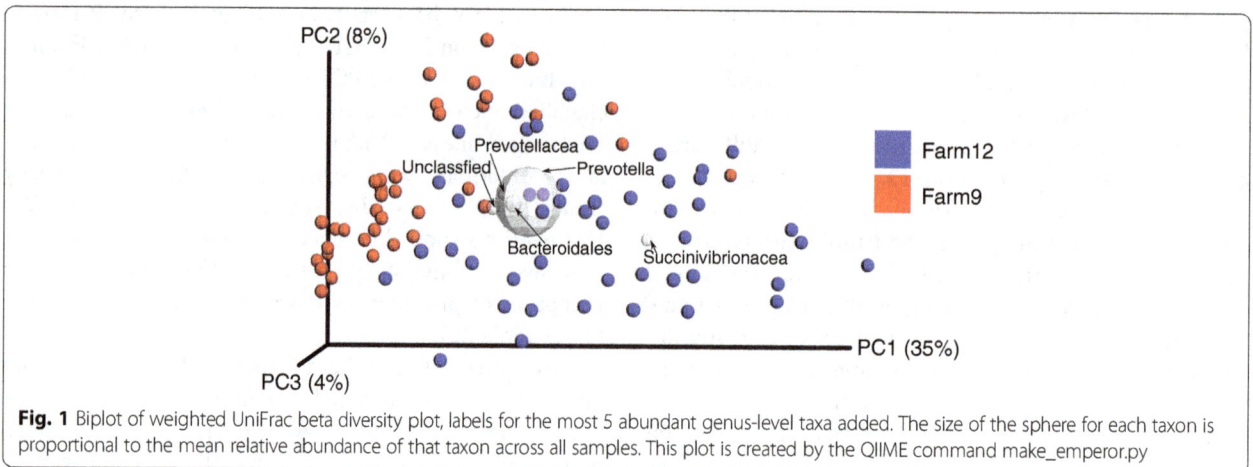

Fig. 1 Biplot of weighted UniFrac beta diversity plot, labels for the most 5 abundant genus-level taxa added. The size of the sphere for each taxon is proportional to the mean relative abundance of that taxon across all samples. This plot is created by the QIIME command make_emperor.py

increased in the high yielding primiparous cows compared to the low yielding primiparous cows. Members of Succinivibrionaceae lineages were almost doubled ($p < 0.05$) in high versus low yielders in the primiparous cows.

In the high yielding multiparous cows of Farm 12, the contribution from Bacteroidetes was lower compared to the low yielding multiparous cows, while several Firmicutes lineages increased, unlike what was observed in the primiparous cows. Distinct differences in Proteobacteria populations were not evident between high and low yielding multiparous cows, except for Succinivibrionaceae lineages which were increased in the high yielders ($p < 0.05$). No evident patterns were observed between high and low yielding cows in Farm 9.

Diversity of Succinivibrionaceae

As Succinivibrionaceae lineages were abundant in Farm 12 and were found to be higher ($p < 0.05$) in high yielding primiparous and multiparous cows, we performed additional analysis on the sequences that were annotated to Succinivibrionaceae lineages. Across all samples, we identified a total of 326 OTUs that were assigned to this bacterial family (Fig. 2). Of the 326 OTUs, 143 OTUs were common to both farms. About 141 were unique to Farm 12 and 42 OTUs were identified only in Farm 9. In Farm 12, OTU45762 was found in all samples and comprised 35% of Succinivibrionaceae abundance. In contrast, this OTU was not detected in a majority of samples in Farm 9. Instead OTU16670 was found to be the most abundant OTU in Farm 9, but was less than 5% in Farm 12. Similar patterns were observed for a majority of Succinivibrionaceae OTUs, where the OTUs that were abundant in Farm 12 contributed only a small portion in Farm 9 and vice versa. These results indicate the diversity among Succinivibrionaceae OTUs in Farm 12 is different compared to that of Farm 9.

Correlations between rumen bacterial populations and production traits

To investigate the relationship between bacterial populations and production traits, we performed a Spearman correlation test for the most abundant taxa within Bacteroidetes, Firmicutes and Proteobacteria (Fig. 3). In general, bacterial taxa that were positively correlated with milk yield, showed negative correlations with fat and protein content in milk. *Prevotella* and *S24-7* from Bacteroidetes, and Succinivibrionaceae lineages from Proteobacteria were positively correlated with milk yield. All taxa except Christensenellaceae, *Coprococcus*, Erysipelotrichaceae, Lachnospiraceae, *Shuttleworthia* and Veillonellaceae lineages from Firmicutes (Fig. 3) showed positive correlations with milk fat and protein content.

As the diversity of Succinivibrionaceae was different between farms, we performed correlations between the most abundant Succinivibrionaceae OTUs and production variables in this study (Additional file 4: Figure S1). The OTUs selected for this analysis were OTUs 45,762, 46,605, 22,052, 35,471, 14,464, 58,055, 7873, 20,365, and 11,472) from Farm 12 and the only abundant OTU (16670) from Farm 9. Interestingly, we found that all Succinivibrionaceae OTUs from Farm 12 were found to be positively correlated to milk yield and negatively correlated with fat and protein content, whereas OTU16670 from Farm 9 showed the opposite trend.

Discussion

In the Pennsylvania dairy sector, average annual milk production per cow (M305) is about 10,000 kg. In this study, Farm 12 (M305 = 12,300 kg) represented the top 5% of Pennsylvania dairy herds and Farm 9 (M305 = 9700 kg) represented the average dairy herd in Pennsylvania [40]. There are many sources to account for variation in milk yield between dairy farms. Genetics accounts for approximately 25% of the variation [41] and management factors,

Table 3: Mean abundance (%) and SEM of bacterial taxa (identified to the genus level) between low and high yielding cows within parity in Farm 12 and Farm 9

Taxa	Herd	Primiparous cows			Multiparous cows		
		Low	High	P-value[±]	Low	High	P-value[±]
Bacteroidetes (Phylum)							
Paraprevotellaceae (family)	Farm12	0.70 ± 0.05	0.57 ± 0.08	*	0.95 ± 0.06	0.82 ± 0.05	
	Farm9	0.85 ± 0.10	0.80 ± 0.05		0.69 ± 0.06	0.58 ± 0.05	
Bacteroidales (order)	Farm12	11.54 ± 0.29	10.32 ± 0.30	***	10.54 ± 0.59	9.87 ± 0.32	*
	Farm9	11.28 ± 0.52	12.03 ± 0.59	**	12.43 ± 0.43	11.89 ± 0.38	
BF311	Farm12	0.06 ± 0.01	0.08 ± 0.02		0.10 ± 0.04	0.04 ± 0.01	*
	Farm9	0.10 ± 0.03	0.18 ± 0.03	*	0.16 ± 0.03	0.17 ± 0.03	
CF231	Farm12	0.70 ± 0.07	0.61 ± 0.07		0.84 ± 0.06	0.73 ± 0.04	
	Farm9	1.16 ± 0.11	1.27 ± 0.12		1.24 ± 0.08	1.06 ± 0.09	
Prevotella	Farm12	50.35 ± 1.71	53.39 ± 1.29	***	51.51 ± 1.80	52.05 ± 1.08	
	Farm9	49.55 ± 2.72	45.05 ± 1.77	***	47.60 ± 2.13	50.59 ± 1.74	***
Prevotellaceae (family)	Farm12	4.16 ± 0.13	4.22 ± 0.27		3.45 ± 0.25	3.19 ± 0.16	
	Farm9	4.30 ± 0.30	4.36 ± 0.21		5.13 ± 0.27	4.96 ± 0.20	
RF16 (family)	Farm12	1.14 ± 0.20	0.93 ± 0.10	**	1.07 ± 0.12	1.09 ± 0.09	
	Farm9	2.03 ± 0.28	3.38 ± 0.47	***	1.51 ± 0.24	1.39 ± 0.43	
S24-7 (family)	Farm12	3.04 ± 0.34	2.34 ± 0.11	***	3.25 ± 0.36	2.99 ± 0.49	
	Farm9	2.21 ± 0.42	2.48 ± 0.27	*	1.70 ± 0.28	1.93 ± 0.31	
YRC22	Farm12	0.52 ± 0.06	0.54 ± 0.06		0.54 ± 0.05	0.54 ± 0.05	
	Farm9	0.63 ± 0.05	0.68 ± 0.06		0.85 ± 0.07	0.60 ± 0.06	***
Firmicutes (Phylum)							
Mogibacteriaceae (family)	Farm12	0.19 ± 0.03	0.14 ± 0.02		0.27 ± 0.04	0.18 ± 0.03	
	Farm9	0.16 ± 0.05	0.22 ± 0.03		0.11 ± 0.01	0.11 ± 0.02	
Anaerostipes	Farm12	NF	NF		0.04 ± 0.01	0.07 ± 0.01	
	Farm9	0.06 ± 0.02	0.09 ± 0.02		NF	NF	
Anaerovibrio	Farm12	NF	NF		0.06 ± 0.02	0.04 ± 0.01	
	Farm9	0.04 ± 0.01	0.06 ± 0.01		0.09 ± 0.02	0.08 ± 0.02	
Asteroleplasma	Farm12	0.10 ± 0.02	0.09 ± 0.03		0.07 ± 0.02	0.10 ± 0.02	
	Farm9	0.16 ± 0.04	0.10 ± 0.02		0.18 ± 0.03	0.19 ± 0.03	
Butyrivibrio	Farm12	0.24 ± 0.04	0.09 ± 0.02	***	0.26 ± 0.05	0.20 ± 0.03	
	Farm9	0.36 ± 0.06	0.24 ± 0.04		0.31 ± 0.05	0.36 ± 0.10	
Christensenellaceae (family)	Farm12	0.09 ± 0.02	0.06 ± 0.01		0.07 ± 0.02	0.05 ± 0.01	
	Farm9	0.08 ± 0.02	0.10 ± 0.02		0.08 ± 0.02	0.12 ± 0.02	
Clostridiales (order)	Farm12	2.70 ± 0.20	1.89 ± 0.14	***	3.33 ± 0.37	3.25 ± 0.38	
	Farm9	3.07 ± 0.35	3.05 ± 0.16		3.36 ± 0.39	2.98 ± 0.27	
Coprococcus	Farm12	0.07 ± 0.01	0.04 ± 0.01		0.18 ± 0.05	0.13 ± 0.02	
	Farm9	0.06 ± 0.01	0.07 ± 0.02		0.05 ± 0.01	0.05 ± 0.01	
Erysipelotrichaceae (family)	Farm12	NF	NF		0.03 ± 0.01	0.03 ± 0.01	
	Farm9	0.06 ± 0.01	0.05 ± 0.02		NF	NF	
Lachnospiraceae (family)	Farm12	1.31 ± 0.10	1.09 ± 0.12	*	1.53 ± 0.15	1.84 ± 0.15	**
	Farm9	1.48 ± 0.21	1.41 ± 0.10		1.56 ± 0.27	1.46 ± 0.15	
Moryella	Farm12	0.06 ± 0.02	0.05 ± 0.01		0.13 ± 0.02	0.10 ± 0.03	
	Farm9	0.13 ± 0.04	0.12 ± 0.02		NF	NF	

Table 3: Mean abundance (%) and SEM of bacterial taxa (identified to the genus level) between low and high yielding cows within parity in Farm 12 and Farm 9 *(Continued)*

p-75-a5	Farm12	NF	NF		NF	NF	
	Farm9	0.05 ± 0.01	0.06 ± 0.01		0.06 ± 0.01	0.07 ± 0.02	
Pseudobutyrivibrio	Farm12	0.07 ± 0.02	0.05 ± 0.01		0.10 ± 0.03	0.13 ± 0.03	
	Farm9	0.08 ± 0.02	0.05 ± 0.01		0.06 ± 0.02	0.07 ± 0.02	
RFN20	Farm12	0.53 ± 0.06	0.42 ± 0.05	*	0.49 ± 0.06	0.38 ± 0.04	
	Farm9	0.79 ± 0.07	0.87 ± 0.09		0.62 ± 0.07	0.65 ± 0.05	
Ruminococcaceae (family)	Farm12	1.40 ± 0.14	0.91 ± 0.13	***	1.43 ± 0.17	1.15 ± 0.12	**
	Farm9	1.51 ± 0.25	1.56 ± 0.17		1.24 ± 0.16	1.24 ± 0.12	
Ruminococcus	Farm12	0.89 ± 0.13	0.56 ± 0.10	***	0.93 ± 0.09	0.91 ± 0.11	
	Farm9	1.25 ± 0.23	1.17 ± 0.11		1.08 ± 0.26	0.99 ± 0.15	
Schwartzia	Farm12	0.06 ± 0.01	0.07 ± 0.02		0.07 ± 0.01	0.07 ± 0.01	
	Farm9	NF	NF		NF	NF	
Selenomonas	Farm12	0.14 ± 0.02	0.08 ± 0.02	*	0.17 ± 0.03	0.18 ± 0.03	
	Farm9	0.15 ± 0.03	0.09 ± 0.03		0.20 ± 0.06	0.18 ± 0.03	
Shuttleworthia	Farm12	0.08 ± 0.01	0.06 ± 0.01		0.08 ± 0.02	0.10 ± 0.02	
	Farm9	NF	NF		0.05 ± 0.01	0.04 ± 0.01	
Succiniclasticum	Farm12	0.17 ± 0.02	0.10 ± 0.02	*	0.20 ± 0.03	0.18 ± 0.03	
	Farm9	0.20 ± 0.02	0.13 ± 0.03		0.13 ± 0.02	0.13 ± 0.03	
Veillonellaceae (family)	Farm12	0.66 ± 0.08	0.63 ± 0.08		0.70 ± 0.10	0.70 ± 0.06	
	Farm9	0.55 ± 0.04	0.41 ± 0.05		0.55 ± 0.08	0.55 ± 0.08	
Proteobacteria (Phylum)							
Alphaproteobacteria (class)	Farm12	0.07 ± 0.01	0.06 ± 0.01		0.05 ± 0.02	0.08 ± 0.01	
	Farm9	0.14 ± 0.04	0.15 ± 0.03		0.08 ± 0.03	0.07 ± 0.02	
Deltaproteobacteria (class)	Farm12	NF	NF		NF	NF	
	Farm9	0.03 ± 0.01	0.05 ± 0.01		0.03 ± 0.01	0.05 ± 0.01	
RF32 (order)	Farm12	0.11 ± 0.02	0.11 ± 0.02		0.15 ± 0.03	0.09 ± 0.02	*
	Farm9	0.16 ± 0.05	0.18 ± 0.02		0.11 ± 0.03	0.11 ± 0.03	
Ruminobacter	Farm12	NF	NF		NF	NF	
	Farm9	0.05 ± 0.02	0.09 ± 0.02		NF	NF	
Succinivibrionaceae (family)	Farm12	4.48 ± 0.76	8.11 ± 2.05	***	5.92 ± 1.80	8.02 ± 1.45	***
	Farm9	1.41 ± 0.28	2.68 ± 0.92	***	0.85 ± 0.32	0.66 ± 0.30	*

*P-value comparing bacterial abundance of high and low milk production within parity
The magnitude of the P-value (*** $P<0.001$; ** $P<0.01$; * $P<0.05$)
NF Not found

including cow comfort, milking frequency, ration formulation and feeding management contribute to the remaining 75% [42]. While animal genetics, nutrition and management can greatly influence milk production differences among herds, our objective was to investigate how diet impacts microbiota and how rumen bacterial community compositions differ in dairy cows with varying milk production. To accomplish our goal and to obtain large numbers of cows varying in parity and production, we selected two commercial dairy herds, knowing that we would not be able to control the many aspects of herd variation that influence milk yield. We used the stomach tube method for rumen sample collection, realizing only the planktonic associated microbiota is typically retrieved, however this approach allowed us to sample large groups of cows across herds. To this end, we sampled rumen fluid from higher and lower yielding dairy cows across two farms in Pennsylvania and investigated the bacterial diversity using Ion Torrent (PGM) sequencing. The salient findings from this study are that each farm had a unique bacterial profile and the rumen bacterial composition differed substantially between the herds. The inter-herd differences were much greater than intra-herd differences in their rumen bacterial composition, which was not surprising due to

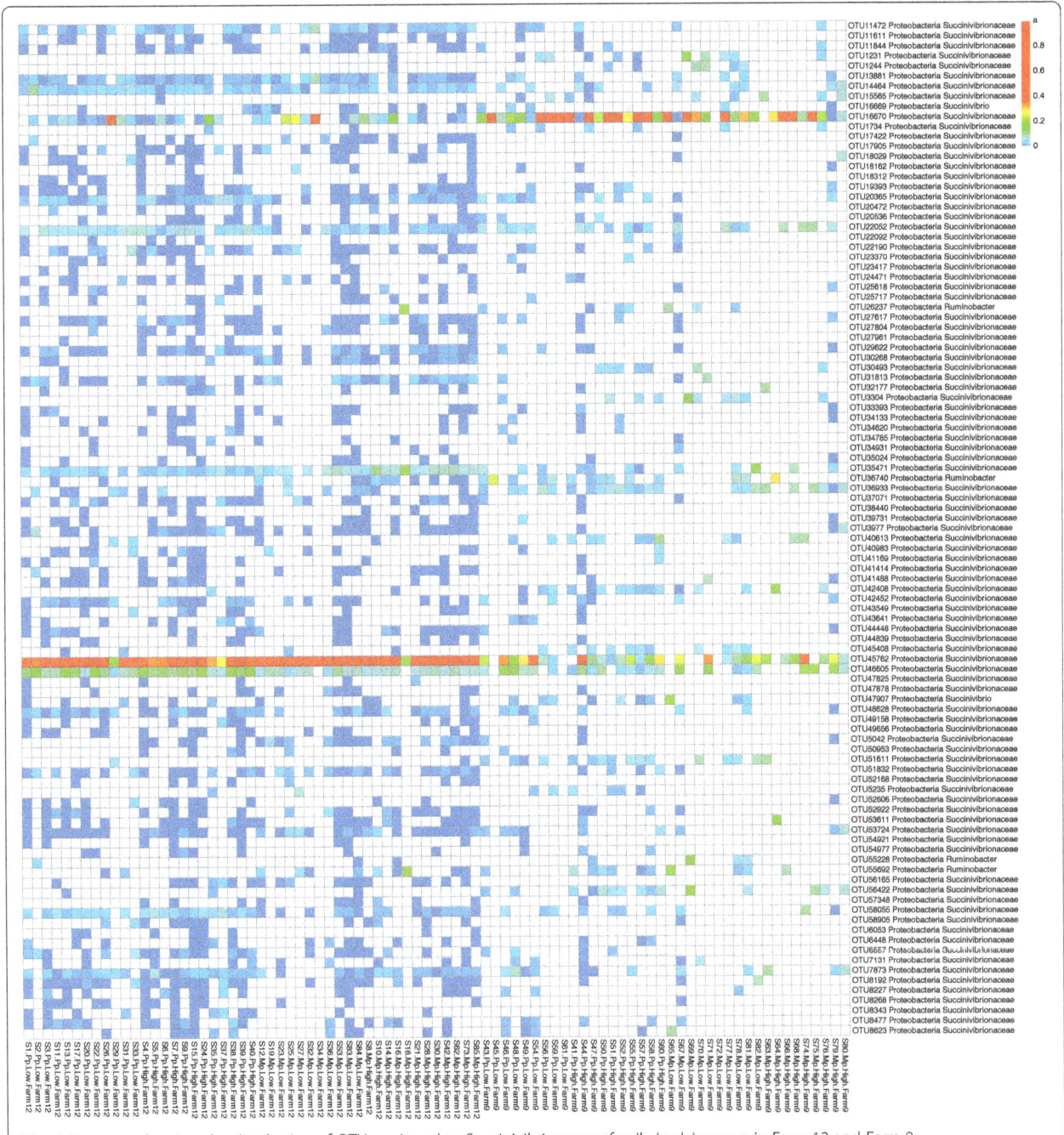

Fig. 2 Heatmap showing the distribution of OTUs assigned to Succinivibrionaceae family in dairy cows in Farm 12 and Farm 9

large differences in dietary composition between the herds. This study identified the presence of specialist rumen bacteria including Succinivibrionaceae lineages, *Coprococcus* and S24-7, which were associated with higher milk yields in Farm 12.

Rumen bacterial community composition

Across both farms in this study, Bacteroidetes, mostly comprised of *Prevotella*, Bacteroidales, and Prevotella-ceae lineage, and Firmicutes, mainly comprised of

Clostridiales, Lachnospiraceae, and Ruminococcaceae lineages constituted 75% and 10 to 11% of the rumen bacterial composition, respectively. The abundance of Bacteroidetes and Firmicutes in the rumen of dairy cows is a common finding reported by several authors [14–16]. However, the proportion of Bacteroidetes observed in this study was much higher compared to the above reports, but similar to our previous findings [17, 18]. Differences among reports can be attributed to differences in dietary composition, sampling times

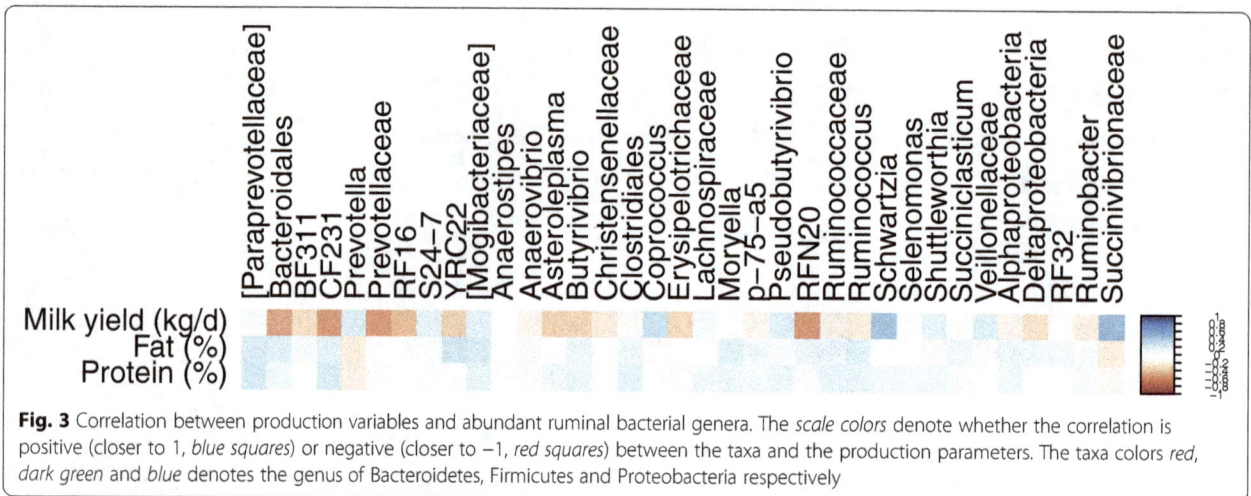

Fig. 3 Correlation between production variables and abundant ruminal bacterial genera. The *scale colors* denote whether the correlation is positive (closer to 1, *blue squares*) or negative (closer to −1, *red squares*) between the taxa and the production parameters. The taxa colors *red*, *dark green* and *blue* denotes the genus of Bacteroidetes, Firmicutes and Proteobacteria respectively

and methodologies employed in rumen bacterial diversity analysis [43].

Previous reports [15, 16] indicate Proteobacteria accounts for a very small proportion (1–5%) of the rumen bacterial population. Compared to these reports [15, 16], the contribution from Proteobacteria in Farm 12 was higher than expected at 7%, but on the lower end of reported ranges in Farm 9, where Proteobacteria comprised only 2%. Notably, in Farm 12, Succinivibrionaceae lineages represented 97% of the Proteobacteria. The proportions of Bacteroidetes, Firmicutes and Proteobacteria in dairy cows from Farm 9 were similar to our earlier reports from the same herd [17, 18]. This consistency in findings on Farm 9 suggests that the rumen bacterial profiles remain fairly stable for a dairy herd under similar management and dietary regimen.

Linking diets and rumen bacterial communities

The feed ingredients used in the TMR for Farm 9 and Farm 12 were different. Differences in dietary composition were well reflected in the inter-herd differences in rumen bacterial community composition, which may have contributed to differences in rumen digestion and influenced milk yields between herds. The rumen bacterial communities in Farm 12 were diverse and were driven by *Prevotella*, Bacteroidales, unclassified Prevotellaceae and Succinivibrionaceae, while in Farm 9 there was little variation within rumen bacterial communities, driven mostly by *Prevotella*, Bacteroidales and unclassified Prevotellaceae.

Rumen bacteria with a known function such as S24-7 for butyrate production [44], *Coprococcus* for propionate and butyrate production [22] and Succinivibrionaceae for succinate production [45] were found to be greater in abundance in Farm 12. Notably, *Schwartzia*, a genus from Firmicutes reported to utilize only succinic acid [46] was detected only in Farm 12. These bacteria with specific functions were in greater relative abundance on

Farm 12 and observed to be positively correlated with milk production. Recently it was shown that both *Coprococcus* and Succinivibrionaceae were positively correlated with gross feed efficiency in dairy cows [22]. Similarly, Succinivibrionaceae also compete with methanogens for hydrogen required to make succinate, a precursor for propionate [47]. Succinivibrionaceae has roles in mitigating methane production and in producing propionate to supply energy to the host for tissue metabolism [45]. One plausible mechanism for increased milk yield in Farm 12 could be Succinivibrionaceae converting succinate to propionate which is metabolized in the liver to glucose, a precursor for lactose synthesis regulating milk volume [48].

It is well known that the diet fed to a ruminant is a major driver in determining the composition of microbial communities in the rumen [19, 41, 42]. In this study we attribute the inter-herd differences in rumen bacterial composition to dietary factors, particularly forage %DM, forage type, and amount of corn grain and types of byproducts fed on the farms. Forage was fed at a higher %DM on Farm 9 (56%), however, corn silage was the major forage fed on both farms. The starch in properly ensiled corn silage is generally readily available in the rumen due to the moisture and softness of the kernel. However, starch availability in the rumen is also dependent upon maturity and processing of the corn kernel at harvest. Starch from corn silage on Farm 12 may have been more available to rumen microbes due to the use of more sophisticated self-propelled harvesting and kernel processing equipment, allowing kernels to be pulverized completely. Farm 9 used a pull type forage harvester and built-in processor producing kernels that were only knicked or broken into small visible pieces, possibly limiting rumen available starch. In addition to corn silage, Farm 12 fed alfalfa silage to complement their forage base. Alfalfa silage, when fed to steers, showed an increased abundance of Bacteroidetes and Proteobacteria as compared to Sainfoin silage [49]. Further, in experiments that involved feeding a combination of corn silage (33% DM) and alfalfa silage (25% DM), rumen Succinovibrionaceae was detected at 4% in the liquid and 6% in the solid fraction [15]. When corn silage comprised 60% DM and alfalfa silage 9% DM, rumen Succinivibrionaceae was reported at 4% in dairy cows [15]. Similar to these findings, the combination of corn silage and alfalfa silage fed on Farm 12 may have contributed to the higher abundance of rumen Succinivibrionaceae at 4–8%. Succinivibrionaceae OTUs were also more diverse and positively correlated with milk yield in Farm 12 compared to Farm 9, where Succinivibrionaceae OTUs associated with milk production were not detected. Triticale (a hybrid of rye and wheat) silage was fed with corn silage on Farm 9. Triticale is a grass forage and by nature has higher NDF values than legumes harvested at the same relative maturities. Triticale silage had higher NDF concentrations compared to barley and oat silages and resulted in higher molar proportions of acetate and lower molar proportions of butyrate in the rumen compared to barley silage [50]. Congruent to these findings, the NDF concentration on Farm 9 (31.0%) was relatively higher compared to Farm 12 (28.5%). Higher forage %DM and relatively higher NDF content may have contributed to the relative higher concentrations of Fibrobacter and Firmicutes [18, 51, 52] on farm 9 compared to farm 12. Further studies are required to investigate the extent of fermentability of corn silage alone and in combination with other legume and cereal silages on the rumen microbial populations, as these forages form a major component (50 to 70%) in lactating dairy cow diets [53]. Such studies will give insight into dietary-microbial interactions to enhance milk yields in dairy cows.

In addition to forages, non-forage fiber byproducts were fed on both farms. Wheat middlings were the only byproduct fed on Farm 12. In general, wheat middlings contain the highest concentration of starch (19%) compared to many other common byproducts fed to dairy cows. Farm 9 fed corn distillers (12% starch), soyhulls, and citrus pulp, both containing less than 5% starch. In addition, the mean proportion of ground corn (19.4%) in Farm 12 TMR was about 4.3% units greater than that of Farm 9 (15.1%), suggesting increased rumen starch availability on Farm 12. We speculate an increase in rumen Succinivibrionaceae may be associated with greater starch availability in the rumen of high yielding dairy cows in Farm 12. High-grain diets appear to favor the growth of these bacterial populations [45, 54], which agrees with the findings of this study, where corn grain was fed in higher amounts on Farm 12. In addition to higher concentrations of starch, processing of grain has

a major impact on the rumen availability of starch, and thus can influence the populations of Succinivibrionaceae in the rumen as revealed by the findings of (Shipp et al, Effects of corn processing method and dietary inclusion of wet distillers grains with soluble on rumen microbial dynamics in finishing steers Ginger, submitted), where the authors observed that Succinivibrionaceae was doubled when finishing beef cattle were fed steam flaked corn compared to dry rolled corn. Again, supporting the increase in rumen Succinivibrionaceae may be associated with greater starch availability in the rumen of high yielding dairy cows in Farm 12. The highest dietary starch concentration in this study (32.1%) was in the multiparous cows on Farm 12, which was also the group with the highest level of milk production. Additionally, DMI increases with the amount of milk produced [3, 53], suggesting cows on Farm 12 ate more TMR DM compared to Farm 9, thus consuming greater amounts of dietary starch to enhance the production of VFA and microbial protein for milk production.

Identification of specialized rumen bacteria is needed for improving productivity of dairy cows [55], and the findings of Pope et al. [45] have created renewed interest in Succinivibrionaceae among the scientific community. Several papers have been published recently concerning this bacterial population [47, 56, 57]. The abundance of Succinivibrionaceae lineages in Farm 12 is noteworthy, particularly the greater abundance of this bacterial population in the high yielding cows in both parity groups. Identification of dietary factors including forage varieties, starch concentration and processing methods can favor the growth of Succinivibrionaceae and other specialist bacteria in the rumen that may positively impact digestion and metabolism of feed substrates to ultimately improve milk production in dairy cows.

Conclusion

This study compared the rumen bacterial populations between two dairy herds differing by 2600 kg/cow in annual milk production. It can be concluded that within-herd differences are small compared to between-herd differences in the composition of rumen bacterial populations. In this study we attribute inter-herd differences in rumen bacterial composition to dietary factors, particularly forage %DM, forage type, and amount of corn grain and byproducts fed on the farms. The distinct abundance of Succinivibrionaceae lineages in Farm 12 was enlightening, particularly the greater abundance of this bacterial population in the high yielding cows in both parity groups. Our study suggests the growth of Succinivibrionaceae linages may have been associated with greater starch availability in the rumen, where corn grain was fed in higher amounts on Farm 12 and corn silage starch may have also been more readily available due to better kernel processing. Controlled studies are needed to more fully understand the impact of both NDF and starch components in corn silage and other forages on rumen microbial populations. Specific selection of NFC sources including starch, sugars, and non-forage fiber sources to compliment forage inputs will give additional insight into dietary microbial interactions important in improving milk yields on dairy farms. Identification of specialized rumen bacteria in dairy cows capable of improving nutrient utilization and feed conversion are needed to continually improve milk production and feed efficiency on dairy farms. Interactions between diet, rumen microbial and fermentation dynamics, and milk yield and components will expand our knowledge to benefit the dairy industry at large.

Additional files

Additional file 1: Table S1. Sequencing details and production data for samples from cows on Farms 12 and 9.

Additional file 2: Table S2. Multivariate permutational analysis (PERMANOVA) for differences in bacterial communities with regard to milk production and parity, both within and between Farm 12 and Farm 9.

Additional file 3: Table S3. Bacterial abundance (%) at phylum, family and genus level for low and high yielding cows within parity in Farm 12 and Farm 9.

Additional file 4: Figure S1. Spearman correlation between milk production parameters and most abundant Succinivibrionaceae OTUs across rumen samples. The scale colors denote whether the correlation is positive (closer to 1, blue squares) or negative (closer to −1, red squares) between the Succinivibrionaceae OTU and the milk production parameters.

Abbreviations
DIM: Days in milk production; DM: Dry matter; DMI: Dry matter intake; DRMS: Dairy Record Management Systems; FC: Fiber carbohydrates; GLM: generalized linear model; NDF: Neutral detergent fiber; NFC: Non-fiber carbohydrate; OTU: Operational taxonomic unit; PGM: Personnel genome machine; QIIME: Quantitative insights into microbial ecology; TMR: Total mixed rations

Acknowledgments
We are thankful to the Biomedical Research Core Facilities, University of Pennsylvania, for sequencing services.

Funding
This research received no specific grant from any funding agency in the public, commercial, or not-for-profit sectors.

Authors' contributions
NI analyzed the data, contributed reagents/materials/analysis tools, wrote the paper, prepared figures and/or tables, reviewed drafts of the paper. BV performed the experiments, contributed reagents/materials/analysis tools, wrote the paper. LDB, JDF and JKPV participated in the writing and drafting of the manuscript; DP conceived and designed the experiments, wrote the paper, reviewed drafts of the paper. All authors read and approved this final manuscript.

Competing interests
The authors declare that they have no competing interests.

Author details
[1]Department of Clinical Studies, School of Veterinary Medicine, New Bolton Center, University of Pennsylvania, Kennett Square, PA 19348, USA. [2]Department of Food Science, Pennsylvania State University, University Park, State College, PA 16802, USA. [3]Penn State Hershey Cancer Institute, Hershey, PA 17033, USA.

References
1. Bergman E. Energy contributions of volatile fatty acids from the gastrointestinal tract in various species. Physiol Rev. 1990;70(2):567–90.
2. Hackmann TJ, Firkins JL. Maximizing efficiency of rumen microbial protein production. Front Microbiol. 2015;6
3. Bargo F, Muller L, Delahoy J, Cassidy T. Performance of high producing dairy cows with three different feeding systems combining pasture and total mixed rations. J Dairy Sci. 2002;85(11):2948–63.
4. Dewhurst R, Davies D, Merry R. Microbial protein supply from the rumen. Anim Feed Sci Technol. 2000;85(1):1–21.
5. Hoover W, Stokes S. Balancing carbohydrates and proteins for optimum rumen microbial yield. J Dairy Sci. 1991;74(10):3630–44.
6. Goering HK, Van Soest PJ. Forage fiber analyses (apparatus, reagents, prcedures, and some applications). USDA Agr Handb. 1970;
7. Sniffen C, O'connor J, Van Soest P, Fox D, Russell J. A net carbohydrate and protein system for evaluating cattle diets: II. Carbohydrate and protein availability. J Anim Sci. 1992;70(11):3562–77.
8. Allen M. Maximizing digestible intake of corn silage-based diets. Part 2. In: Mich Dairy Rev. vol. 14. Michigan: Michigan State University; 2009. p. 4–6.
9. Khafipour E, Li S, Plaizier JC, Krause DO. Rumen microbiome composition determined using two nutritional models of subacute ruminal acidosis. Appl Environ Microbiol. 2009;75(22):7115–24.
10. Martin SA. Manipulation of ruminal fermentation with organic acids: a review. J Anim Sci. 1998;76(12):3123–32.
11. Hassanat F, Gervais R, Julien C, Massé D, Lettat A, Chouinard P, Petit H, Benchaar C. Replacing alfalfa silage with corn silage in dairy cow diets: Effects on enteric methane production, ruminal fermentation, digestion, N balance, and milk production. J Dairy Sci. 2013;96(7):4553–67.
12. Lettat A, Hassanat F, Benchaar C. Corn silage in dairy cow diets to reduce ruminal methanogenesis: Effects on the rumen metabolically active microbial communities. J Dairy Sci. 2013;96(8):5237–48.
13. Jami E, Mizrahi I. Composition and similarity of bovine rumen microbiota across individual animals. PLoS One. 2012;7(3):e33306.
14. Jami E, White BA, Mizrahi I. Potential role of the bovine rumen microbiome in modulating milk composition and feed efficiency. PLoS One. 2014;9(1):e85423.
15. Jewell KA, McCormick CA, Odt CL, Weimer PJ, Suen G. Ruminal bacterial community composition in dairy cows is dynamic over the course of two lactations and correlates with feed efficiency. Appl Environ Microbiol. 2015; 81(14):4697–710.
16. Lima FS, Oikonomou G, Lima SF, Bicalho ML, Ganda EK, de Oliveira Filho JC, Lorenzo G, Trojacanec P, Bicalho RC. Prepartum and postpartum rumen fluid microbiomes: characterization and correlation with production traits in dairy cows. Appl Environ Microbiol. 2015;81(4):1327–37.
17. Pitta DW, Indugu N, Kumar S, Vecchiarelli B, Sinha R, Baker LD, Bhukya B, Ferguson JD. Metagenomic assessment of the functional potential of the rumen microbiome in Holstein dairy cows. Anaerobe. 2016;38:50–60.
18. Pitta D, Kumar S, Vecchiarelli B, Shirley D, Bittinger K, Baker L, Ferguson J, Thomsen N. Temporal dynamics in the ruminal microbiome of dairy cows during the transition period. J Anim Sci. 2014a;92(9):4014–22.
19. Kumar S, Indugu N, Vecchiarelli B, Pitta DW. Associative patterns among anaerobic fungi, methanogenic archaea, and bacterial communities in response to changes in diet and age in the rumen of dairy cows. Front Microbiol. 2015;6:781.
20. Cersosimo LM, Bainbridge ML, Kraft J, Wright A-DG. Influence of

periparturient and postpartum diets on rumen methanogen communities in three breeds of primiparous dairy cows. BMC Microbiol. 2016;16(1):78.

21. Bainbridge ML, Cersosimo LM, Wright A-DG, Kraft J. Rumen bacterial communities shift across a lactation in Holstein, Jersey and Holstein× Jersey dairy cows and correlate to rumen function, bacterial fatty acid composition and production parameters. FEMS Microbiol Ecol. 2016;**92**(5).

22. Shabat SKB, Sasson G, Doron-Faigenboim A, Durman T, Yaacoby S, Miller MEB, White BA, Shterzer N, Mizrahi I. Specific microbiome-dependent mechanisms underlie the energy harvest efficiency of ruminants. ISME J. 2016;

23. Li F, Li Z, Lei S, d Ferguson J, Cao Y, Yao J, Sun F, Wang X, Yang T. Effect of dietary physically effective fiber on ruminal fermentation and the fatty acid profile of milk in dairy goats. J Dairy Sci. 2014;97(4):2281–90.

24. Minuti A, Palladino A, Khan M, Alqarni S, Agrawal A, Piccioli-Capelli F, Hidalgo F, Cardoso F, Trevisi E, Loor J. Abundance of ruminal bacteria, epithelial gene expression, and systemic biomarkers of metabolism and inflammation are altered during the peripartal period in dairy cows. J Dairy Sci. 2015;98(12):8940–51.

25. Wang X, Li X, Zhao C, Hu P, Chen H, Liu Z, Liu G, Wang Z. Correlation between composition of the bacterial community and concentration of volatile fatty acids in the rumen during the transition period and ketosis in dairy cows. Appl Environ Microbiol. 2012;78(7):2386–92.

26. AOAC. Official methods of analysis, vol. I. 15th ed. Arlington: AOAC; 1990.

27. VanSoest P, Robertson J, Lewis B, et al. J Dairy Sci. 1991;74(10):3583–97.

28. Lodge-Ivey S, Browne-Silva J, Horvath M. Technical note: bacterial diversity and fermentation end products in rumen fluid samples collected via oral lavage or rumen cannula. J Anim Sci. 2009;87(7):2333–7.

29. Dollive S, Peterfreund GL, Sherrill-Mix S, Bittinger K, Sinha R, Hoffmann C, Nabel CS, Hill DA, Artis D, Bachman MA. A tool kit for quantifying eukaryotic rRNA gene sequences from human microbiome samples. Genome Biol. 2012;13:R60.

30. Caporaso JG, Kuczynski J, Stombaugh J, Bittinger K, Bushman FD, Costello EK, Fierer N, Pena AG, Goodrich JK, Gordon JI. QIIME allows analysis of high-throughput community sequencing data. Nat Methods. 2010a;7(5):335–6.

31. Team RC. R: A language and environment for statistical computing. Vienna: R Foundation for Statistical Computing; 2013. ISBN 3-900051-07-0; 2016

32. Edgar RC. Search and clustering orders of magnitude faster than BLAST. Bioinformatics. 2010;26(19):2460–1.

33. Caporaso JG, Bittinger K, Bushman FD, DeSantis TZ, Andersen GL, Knight R. PyNAST: a flexible tool for aligning sequences to a template alignment. Bioinformatics. 2010b;26(2):266–7.

34. Price MN, Dehal PS, Arkin AP. FastTree 2–approximately maximum-likelihood trees for large alignments. PLoS One. 2010;5(3):e9490.

35. McDonald D, Price MN, Goodrich J, Nawrocki EP, DeSantis TZ, Probst A, Andersen GL, Knight R, Hugenholtz P. An improved Greengenes taxonomy with explicit ranks for ecological and evolutionary analyses of bacteria and archaea. ISME J. 2012;6(3):610–8.

36. Wang Q, Garrity GM, Tiedje JM, Cole JR. Naive Bayesian classifier for rapid assignment of rRNA sequences into the new bacterial taxonomy. Appl Environ Microbiol. 2007;73(16):5261–7.

37. Anderson MJ. A new method for non-parametric multivariate analysis of variance. Austral Ecol. 2001;26(1):32–46.

38. Lozupone C, Knight R. UniFrac: a new phylogenetic method for comparing microbial communities. Appl Environ Microbiol. 2005;71(12):8228–35.

39. Wei T, Simko V: Visualization of a Correlation Matrix. R package version 0.77. 2016.

40. Dairy Research Management Services (DRMS) Records [http://www.drms.org/].

41. Kiddy CA. A review of research on genetic variation in physiological characteristics related to performance in dairy cattle. J Dairy Sci. 1979;62(5):818–24.

42. Grant R, Albright J. Feeding behavior and management factors during the transition period in dairy cattle. J Anim Sci. 1995;73(9):2791–803.

43. McCann JC, Wickersham TA, Loor JJ. High-throughput methods redefine the rumen microbiome and its relationship with nutrition and metabolism. Bioinformatics and biology insights. 2014;8

44. Evans CC, LePard KJ, Kwak JW, Stancukas MC, Laskowski S, Dougherty J, Moulton L, Glawe A, Wang Y, Leone V. Exercise prevents weight gain and alters the gut microbiota in a mouse model of high fat diet-induced obesity. PLoS One. 2014;9(3):e92193.

45. Pope P, Smith W, Denman S, Tringe S, Barry K, Hugenholtz P, McSweeney C, McHardy A, Morrison M. Isolation of Succinivibrionaceae implicated in low methane emissions from Tammar wallabies. Science. 2011;333(6042):646–8.

46. Van Gylswyk N, Hippe H, Rainey F. Schwartzia succinivorans gen. nov., sp. nov., another ruminal bacterium utilizing succinate as the sole energy source. Int J Syst Evol Microbiol. 1997;47(1):155–9.

47. McCabe MS, Cormican P, Keogh K, O'Connor A, O'Hara E, Palladino RA, Kenny DA, Waters SM. Illumina MiSeq phylogenetic amplicon sequencing shows a large reduction of an uncharacterised succinivibrionaceae and an increase of the Methanobrevibacter gottschalkii Clade in feed restricted cattle. PLoS One. 2015;10(7):e0133234.

48. Liu H, Zhao K, Liu J. Effects of glucose availability on expression of the key genes involved in synthesis of milk fat, lactose and glucose metabolism in bovine mammary epithelial cells. PLoS One. 2013;8(6):e66092.

49. Romero-Pérez GA, Ominski KH, McAllister TA, Krause DO. Effect of environmental factors and influence of rumen and hindgut biogeography on bacterial communities in steers. Appl Environ Microbiol. 2011;77(1):258–68.

50. McCartney D, Vaage A. Comparative yield and feeding value of barley, oat and triticale silages. Can J Anim Sci. 1994;74(1):91–6.

51. Pitta D, Pinchak W, Dowd S, Dorton K, Yoon I, Min B, Fulford J, Wickersham T, Malinowski D. Longitudinal shifts in bacterial diversity and fermentation pattern in the rumen of steers grazing wheat pasture. Anaerobe. 2014b;30:11–7.

52. Pitta DW, Parmar N, Patel AK, Indugu N, Kumar S, Prajapathi KB, Patel AB, Reddy B, Joshi C. Bacterial diversity dynamics associated with different diets and different primer pairs in the rumen of Kankrej cattle. PLoS One. 2014c; 9(11):e111710.

53. NRC. Nutrient requirements of dairy cattle. Washington, DC: National Academy of Sciences; 2001. p. 381.

54. Petri RM, Schwaiger T, Penner GB, Beauchemin KA, Forster RJ, McKinnon JJ, McAllister TA. Characterization of the core rumen microbiome in cattle during transition from forage to concentrate as well as during and after an acidotic challenge. PLoS One. 2013;8(12):e83424.

55. Weimer PJ. Redundancy, resilience, and host specificity of the ruminal microbiota: implications for engineering improved ruminal fermentations. Front Microbiol. 2015;6:296.

56. Edwards JE, N.R. McEwan, A.J. Travis, Wallace RJ: 16S rDNA library-based analysis of ruminal bacterial diversity. Antonie Van Leeuwenhoek 2004, 86: 263-281.

57. Wallace RJ, Rooke JA, McKain N, Duthie C-A, Hyslop JJ, Ross DW, Waterhouse A, Watson M, Roehe R. The rumen microbial metagenome associated with high methane production in cattle. BMC Genomics. 2015;16(1):1.

13

Comparative analysis of solar pasteurization versus solar disinfection for the treatment of harvested rainwater

André Strauss, Penelope Heather Dobrowsky, Thando Ndlovu, Brandon Reyneke and Wesaal Khan[*]

Abstract

Background: Numerous pathogens and opportunistic pathogens have been detected in harvested rainwater. Developing countries, in particular, require time- and cost-effective treatment strategies to improve the quality of this water source. The primary aim of the current study was thus to compare solar pasteurization (SOPAS; 70 to 79 °C; 80 to 89 °C; and ≥90 °C) to solar disinfection (SODIS; 6 and 8 hrs) for their efficiency in reducing the level of microbial contamination in harvested rainwater. The chemical quality (anions and cations) of the SOPAS and SODIS treated and untreated rainwater samples were also monitored.

Results: While the anion concentrations in all the samples were within drinking water guidelines, the concentrations of lead (Pb) and nickel (Ni) exceeded the guidelines in all the SOPAS samples. Additionally, the iron (Fe) concentrations in both the SODIS 6 and 8 hr samples were above the drinking water guidelines. A >99% reduction in *Escherichia coli* and heterotrophic bacteria counts was then obtained in the SOPAS and SODIS samples. Ethidium monoazide bromide quantitative polymerase chain reaction (EMA-qPCR) analysis revealed a 94.70% reduction in viable *Legionella* copy numbers in the SOPAS samples, while SODIS after 6 and 8 hrs yielded a 50.60% and 75.22% decrease, respectively. Similarly, a 99.61% reduction in viable *Pseudomonas* copy numbers was observed after SOPAS treatment, while SODIS after 6 and 8 hrs yielded a 47.27% and 58.31% decrease, respectively.

Conclusion: While both the SOPAS and SODIS systems reduced the indicator counts to below the detection limit, EMA-qPCR analysis indicated that SOPAS treatment yielded a 2- and 3-log reduction in viable *Legionella* and *Pseudomonas* copy numbers, respectively. Additionally, SODIS after 8 hrs yielded a 2-log and 1-log reduction in *Legionella* and *Pseudomonas* copy numbers, respectively and could be considered as an alternative, cost-effective treatment method for harvested rainwater.

Keywords: Solar Pasteurization, Solar Disinfection, Microbial Indicators, *Legionella* spp., *Pseudomonas* spp., EMA-qPCR

Background

Several countries around the world utilise alternative water sources, such as rainwater harvesting (RWH) and surface water, to meet the increasing water demand and augment available water supplies. Rainwater harvesting in particular has been identified by the South African government as an alternative and sustainable water source that could provide water directly to households [1, 2]. Rainwater is considered a pure water source, however, during the harvesting process, it can become polluted with microorganisms and atmospheric particles such as, organic and inorganic matter (e.g. heavy metals and dust) [2–4]. Depending on the roof maintenance, leaves, animal faecal matter (which may contain chemicals such as phosphorous, nitrogen and trace elements) [4] and other debris particles, may also wash into the rainwater storage tank after a rain event and negatively affect the microbial quality of the tank water [4–6].

It has thus been concluded that stored harvested rainwater is not suitable for potable purposes due to the microbial quality in particular not complying with drinking water standards as established by the Department of Water Affairs and Forestry (DWAF) [7] and World Health Organization (WHO) [8] and it was recommended

---
[*] Correspondence: wesaal@sun.ac.za
Department of Microbiology, Faculty of Science, Stellenbosch University, Private Bag X1, Stellenbosch 7602, South Africa

that harvested rainwater should be treated before utilisation as a primary water source [5, 9]. In developing countries, particularly, researchers seek cost- and time-effective treatment methods in order to improve the quality of harvested rainwater, for utilisation as a potable water source and for other domestic activities [6]. Solar disinfection (SODIS) and solar pasteurization (SOPAS) systems have been considered as efficient and cost-effective treatment methods for harvested rainwater [1, 6].

A SODIS system is based on the effect of ultra-violet (UV) light and heat from the sun, which inactivates microorganisms [6, 10]. A very simple example of a SODIS system is outlined by Amin and Han [1] and Amin et al. [6] where a transparent container is filled with harvested rainwater, placed onto a reflective surface and is exposed to direct sunlight for at least 6 to 8 hrs. Advantages of this system include cost-effectiveness and due to its simplicity it can be implemented worldwide [11]. Recent studies have also shown that SODIS improves the microbial quality of harvested rainwater [1, 6], although certain microorganisms and endospores may persist. Furthermore, the turbidity of the water may decrease the efficiency of the system due to the systems' dependence on direct UV radiation penetration. Although the SODIS system is easier to implement than the SOPAS system, the efficiency of both systems decreases with cloudy weather conditions [1, 6, 10] and both systems may not improve the chemical quality of the harvested rainwater [10, 12].

A SOPAS system relies on the thermal effect (at least 70 °C), without UV radiation to inactivate microbes [13]. An example of a simple SOPAS system is the contemporary solar geyser, where water fills the borosilicate glass tubes, which is exposed to solar radiation. The energy which is obtained from solar radiation is transferred to the water which effectively heats up [14]. In addition, the time needed to treat water will decrease with an increase in temperature. Thus, the time required to treat water will decrease with a factor of 10 for every 10 °C increase in temperature above 50 °C [15]. This system is considered a cost-effective treatment method that is not influenced by the turbidity of the water [16, 17]. Research has also indicated that microbes will be inactivated when the water reaches a temperature of 55 °C or higher [6, 18, 19]. In a study conducted by Dobrowsky et al. [17], an Apollo™ SOPAS system (manufactured in China) successfully reduced the bacterial indicator counts in the rainwater samples pasteurized at the temperature ranges of 72 to 74 °C, 78 to 81 °C, and 90 to 91 °C, to below the detection limit (≥99.9%). Furthermore, Legionella spp. and Pseudomonas spp. were detected at the higher pasteurization temperatures (>78 °C), using the Polymerase Chain Reaction (PCR), however the viability of these organisms at temperatures higher than 72 °C was not confirmed. In a follow up study, Reyneke et

al. [20] then utilised ethidium monoazide bromide quantitative polymerase chain reaction (EMA-qPCR) to verify that viable Legionella spp. were detected in solar pasteurized rainwater samples (>70 °C).

Legionellosis is a lung infection caused by Legionella spp. where the bacterium enters the lungs by inhalation of aerosolized contaminated water. It is well known that Legionella can proliferate at high temperatures [17, 21], however the growth temperature for Legionella is between 25 °C and 45 °C with an optimum temperature of 36 °C [22]. In a recent study conducted by Reyneke et al. [20] the research group showed that Legionella spp. are viable at temperatures higher than 70 °C. Numerous Pseudomonas spp. are associated with water environments as well as heated water sources such as hot tubs, physiotherapy and hydrotherapy pools and whirlpools [23, 24]. This is one of the most common opportunistic pathogens associated with nosocomial infections in individuals with a vulnerable immune system [23]. It normally enters the human body through a skin wound or during surgery where it is then taken up into the blood stream leading to bacteraemia that could cause pneumonia, endocarditis, osteomyelitis, gastrointestinal infections, urinary tract infections and is a leading cause of septicaemia [24, 25]. Pseudomonas is generally spread through contaminated water that comes into contact with a human host, or surgical equipment and the hands of hospital personnel that transfer it to a patient in the case of nosocomial infections [23].

Results obtained by Dobrowsky et al. [17] and Reyneke et al. [20] however, also indicated that significant concentrations of iron (Fe), aluminium (Al), lead (Pb) and nickel (Ni) may have been leaching from the 100 L stainless steel storage tank of the Apollo™ SOPAS system, which may have negatively affected the chemical quality of the treated rainwater. In the current study a new Phungamanzi™ SOPAS system, which was designed and manufactured in South Africa and which consists of a 125 L high grade polyethylene storage tank, was utilised for the solar pasteurization of rainwater. The primary aim of the current study was to conduct a comparative analysis of the new SOPAS system versus SODIS for the treatment of rainwater. The treatment times of the SODIS systems included 6 and 8 hrs, while the treated rainwater for the SOPAS system was collected at different temperature ranges (70 to 79 °C; 80 to 89 °C; and 90 °C and above). To monitor the general microbial quality of the rainwater, indicator bacterial counts, including, Escherichia coli (E. coli), enterococci and faecal coliforms as well as the heterotrophic plate count (HPC), were determined using culture based methods. Chemical analysis was also performed (monitoring the concentration of cations and anions) in order to determine whether the treatment methods utilised alter the chemical quality of the rainwater. Finally, the efficiency of the two treatment methods in reducing the level of viable *Legionella* spp. and *Pseudomonas* spp. in roof harvested rainwater was analysed utilising EMA-qPCR. Ethidium monoazide bromide is a nucleic acid binding dye that can be used to bind to the deoxyribonucleic acid (DNA) of cells (after photoactivation) with damaged and permeable membranes (non-viable cells). The binding of the dye to the DNA prevents PCR amplification of the DNA and thereby leads to a strong signal reduction during qPCR as only the DNA from intact (viable) cells will be amplified [20, 26].

Methods

Description of the sampling site

A RWH system was installed on Welgevallen Experimental farm, Stellenbosch University (GPS co-ordinates: 33° 56′ 36.19″S, 18° 52′ 6.08″E), South Africa. The roof used as the catchment area was constructed from asbestos, while the gutter system leading to the polyethylene rainwater tank (2 000 L tank installed on a metal stand) was constructed from Chrysotile (white asbestos) (Fig. 1a). Furthermore, the sampling site is surrounded by trees and is located next to a dairy farm. However, no tree branches obstructed the catchment area.

Solar pasteurization system

The Phungamanzi™ solar pasteurization system (manufactured in South Africa) was donated to Stellenbosch University by Crest Organization, Stellenbosch. This SOPAS system was connected to the 2 000 L polyethylene

RWH tank, which was installed on a metal stand so that rainwater was able to flow from the rainwater storage tank into the SOPAS system in a passive manner (Fig. 1a). The water from the RWH tank flowed through the system components (Fig. 1a) as follows; water flowed from the RWH tank (A) through a pipe (B) into the high grade polyethylene tank (C) of the solar system, which has a 125 L storage capacity. The water then moved through the high borosilicate glass cylinders (D) in order to capture heat. Due to the thermo-siphoning effect, as the water was heated, the water moved into the main storage tank. The pasteurized water was then collected from the outlet tap (E).

Solar disinfection system

Two SUNSTOVE 2000™ solar oven systems (Sunstove Organization, South Africa), were placed on the rooftop of the JC Smuts building (33° 55′ 51.7″S 18° 51′ 55.3″E) at Stellenbosch University, South Africa, for the solar disinfection of the rainwater samples. As indicated in Fig. 1b, the solar oven has a very simplistic design, with the inside of the system constructed from a reflective aluminium plate and a black polyethylene material enclosing the system. In addition, in order to trap solar radiation, the inner section of the system was covered with a transparent Perspex lid.

Sample collection

For both the SOPAS and SODIS systems, water samples were collected from July 2015 till October 2015, with a

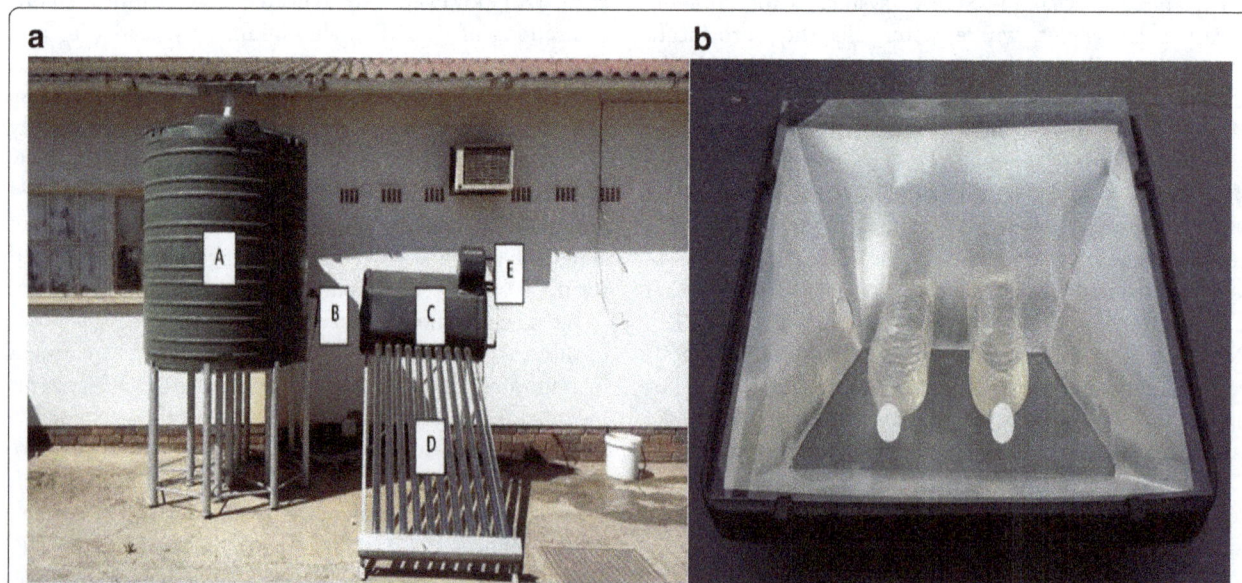

Fig. 1 a The SOPAS system utilised in the current study was connected to a RWH tank installed on a metal stand. A: Untreated RWH tank (capacity: 2 000 L), B: Inlet pipe leading into the SOPAS tank, C: High grade polyethylene tank (capacity: 125 L), D: 10 × High borosilicate glass collector tubes, E: Outlet pipe and water collection point. **b** The SODIS system with two polyethylene terephthalate (PET) bottles containing harvested rainwater. The SODIS system was constructed from a black polyethylene material that was lined with a reflective aluminium surface. The system was covered with a transparent Perspex lid to increase insulation

sampling event conducted one to four days after a rain event. Throughout the sampling period, for the SOPAS system, untreated rainwater (collected directly from RWH tank A) and solar pasteurized rainwater samples were collected in sterile 5 L polypropylene containers, respectively. Solar pasteurized samples were collected at the temperature ranges of 70 to 79 °C; 80 to 89 °C; and 90 °C and above. A MadgeTech TC101A thermocouple temperature Data Logger (MadgeTech, Inc.) was installed inside the SOPAS system in order to monitor the temperature of the treated rainwater for one month (01/08/2015 to 31/08/2015). The temperature data was obtained from the log tagger and analysed using the Data Logger Software Ver. 4.1.5 (Madge Tech, Inc.).

The SODIS treatment of rainwater was performed five times and for each sampling occasion, four sterile transparent 2 L polyethylene terephthalate (PET) bottles were filled to three-quarter capacity with roof harvested rainwater, obtained from the RWH tank A (Fig. 1a). Space was left in each bottle for aeration purposes and directly after collection each bottle was shaken for approximately 10 s in order to oxygenate the water [6, 27]. Two PET bottles were placed on the base of each respective SODIS system (Fig. 1b) and the one SODIS system was exposed to direct sunlight for 6 hrs, while the second SODIS system was exposed to direct sunlight for 8 hrs [28]. Furthermore, for each sampling occasion an untreated rainwater sample was also collected from tank A in a 5 L PET bottle.

The pH and temperature of each water sample was measured on site, using a hand-held pH meter (Milwaukee Instruments, Inc., USA) and mercury thermometer (ALLA® France, France), respectively. The daily temperature and rainfall data were obtained from the South African Weather Services (personal communication) and the solar irradiation data was obtained from the Stellenbosch Weather Services, Stellenbosch University, Faculty of Engineering (http://weather.sun.ac.za/).

Chemical analysis

The chemical quality, including cation and anion concentrations of untreated and pasteurized (SOPAS) rainwater samples, collected for the various temperatures (cations: 71 °C, 86 °C and 93 °C) was determined. In addition, the chemical quality of untreated and SODIS rainwater samples collected after 6 hrs of treatment (cations: 70 °C and 89 °C) and 8 hrs of treatment (cations: 63 °C and 86 °C), were also analysed. For the determination of cation and metal ion concentrations, Falcon™ 50 mL high-clarity polypropylene tubes (Corning Life Sciences, USA) containing polyethylene caps were pre-treated with 1% nitric acid before sampling. The cation and metal ion concentrations [aluminium (Al), chromium (Cr), copper (Cu), iron (Fe), manganese

(Mn), vanadium (V), and zinc (Zn), amongst others] were then determined using inductively coupled plasma atomic emission spectrometry (ICP-AES) [29]. This analysis was completed by the Central Analytical Facility (CAF), Stellenbosch University.

Furthermore, the anion analyses [SOPAS: untreated and 71 °C; SODIS untreated and treated at 6 hrs (52 °C; 70 °C and 89 °C) and 8 hrs (63 °C and 86 °C)] of the samples were performed by PathCare Reference Laboratory (PathCare Park, Cape Town, South Africa). All anions including, chloride, fluoride, nitrate and nitrite, phosphate and sulphates were measured utilising a Thermo Scientific Gallery™ Automated Photometric Analyser. The turbidity [Nephelometric Turbidity Units (NTU)] of selected (untreated and treated) water samples was also determined by PathCare Reference Laboratory (PathCare Park, Cape Town, South Africa).

Microbial analysis of treated and untreated rainwater samples

Enumeration of traditional indicator bacteria in rainwater samples

A serial dilution was prepared (10^{-1}–10^{-3}) for each rainwater sample collected during the sampling period [SOPAS (untreated and pasteurized samples) and SODIS (untreated and treated samples)] and using the spread plate method, 100 µL of the undiluted rainwater sample and each dilution (10^{-1}–10^{-3}) was cultured in duplicate onto Slanetz and Bartley Agar (Oxoid, Hampshire, England) that was incubated for 44 - 48 hrs at 36 ± 2 °C, m-FC Agar (Merck, Darmstadt, Germany) that was incubated for 22 – 24 hrs at 35 ± 2 °C and R2A Agar (Oxoid, Hampshire, England) that was incubated for 72 – 96 hrs at 35 ± 2 °C, to enumerate enterococci, faecal coliforms and HPC, respectively.

For each sample, E. coli was enumerated by filtering a total volume of 100 mL (undiluted) through a sterile GN-6 Metricel® S-Pack Membrane Disc Filter (Pall Life Sciences, Michigan, USA) with a pore size of 0.45 µm and a diameter of 47 mm, at a filtration flow rate of approximately ≥ 65 mL/min/cm^2 at 0.7 bar (70 kPa), in duplicate. The membrane filters were then incubated on Membrane Lactose Glucuronide Agar (MLGA) (Oxoid, Hampshire, England) at 35 ± 2 °C for 18 - 24 hrs.

Rainwater concentration, EMA treatment and DNA extraction

For each sampling event, 1 L rainwater sample [SOPAS (untreated and pasteurized samples) and SODIS (untreated and treated samples)] was concentrated as outlined in Reyneke et al. [20]. The concentrated rainwater samples utilised for Legionella spp. detection were treated with 2.5 µg/mL ethidium monoazide bromide (EMA) as previously described by Delgado-Viscogliosi et al. [30]. The same parameters were then utilised for the detection of *Pseudomonas* spp. in the concentrated rainwater samples. Following the addition of EMA, the samples were incubated on ice for 10 min followed by a 15 min halogen light exposure (keeping the samples on ice to avoid over-heating during the photoactivation step). The EMA treated samples were then washed with 1 mL NaCl (0.85%) followed by centrifugation (16 000 × g for 5 min). The DNA extractions were completed using the Soil Microbe DNA MiniPrep™ Kit (Zymo Research, USA) as per manufacturer's instructions by first re-suspending the obtained pellet in the lysis solution and transferring the mixture to the ZR BashingBead™ Lysis Tubes.

Quantitative PCR for the detection of Legionella and Pseudomonas spp.

Following the EMA treatment and DNA extractions, EMA-qPCR was performed on a LightCycler®96 (Roche Applied Science, Mannheim, Germany) using the FastStart Essential DNA Green Master Mix (Roche Applied Science, Mannheim, Germany). To a final reaction volume of 20 μL, the following were added: 10 μL FastStart Essential DNA Green Master Mix (2x), 5 μL template DNA (diluted by 10 fold) and 0.4 μL of each primer (final concentration 200 nM) as previously described by Herpers et al. [31] for *Legionella* spp. and by Roosa et al. [32] for *Pseudomonas* spp.

For *Legionella* spp., the primers LegF (5′–CTAATT GGCTGATTGTCTTGAC–3′) and LegR (5′–CAATCG GAGTTCTTCGTG–3′) were utilised to amplify a 259 bp product of the 23S rRNA gene [31]. The amplification conditions for *Legionella* spp. were as follows: initial denaturation at 95 °C for 10 min, followed by 50 cycles of denaturation at 95 °C for 15 s, annealing at 60 °C for 15 s and extension at 72 °C for 11 s.

For *Pseudomonas* spp., the primers PS1 (5′-ATGAA CAACGTTCTGAAATTC-3′) and PS2 (5′-CTGCGGC TGGCTTTTTCCAG-3′) were utilised to amplify a 249 bp product of the *Pseudomonas* lipoprotein *oprI* gene [33]. The amplification conditions for *Pseudomonas* spp. were as follows: initial denaturation at 95 °C for 10 min, followed by 50 cycles of denaturation at 94 °C for 30 s, annealing at 58 °C for 30 s and extension at 72 °C for 30 s.

The standard curves for the *Legionella* spp. qPCR assays were produced by amplifying the 23S rRNA gene of *Legionella pneumophila* ATCC 33152, using primers LegF and LegR. In addition, the standard curves for the *Pseudomonas* spp. qPCR assays were produced by amplifying the lipoprotein *oprI* gene of *P. aeruginosa* ATCC 27853, using primers PS1 and PS2. The PCR products were then purified using the DNA Clean & Concentrator™-5 Kit (Zymo Research) and were verified by DNA sequencing followed by quantifying the DNA in triplicate using the NanoDrop® ND-1000

(Nanodrop Technologies Inc., Wilmington, Delaware, USA). A serial 10-fold dilution (*Legionella* spp.: 10^8 to 10^1; *Pseudomonas* spp.: 10^9 to 10^0) of the PCR products was prepared in order to generate the standard curves, where the regression coefficient (R^2) was kept higher than 0.98 and 1.00 for *Legionella* and *Pseudomonas* spp., for each experiment, respectively. For *Legionella* spp. and *Pseudomonas* spp. detection, a concentration of 1.00×10^8 and 1.00×10^9 gene copies/μL was prepared for the dilution with the highest copy number, respectively, while a concentration of 1.00×10^1 and 1.00×10^0 gene copies/μL was prepared for the dilution with the lowest copy number. The standard curves were generated by plotting quantitative cycle (C_q) values versus the log concentrations of standard DNA, as previously described by Chen and Chang [34], for determining the copy number of the 23S rRNA gene in *Legionella* spp. and the copy number of the lipoprotein *oprI* gene in *Pseudomonas* spp.in all samples analysed. Melt curve analysis was included for both *Legionella* and *Pseudomonas* spp. SYBR green real-time PCR assays in order to verify the specificity of the primer set by ramping the temperature from 65 to 97 °C at a rate of 0.2 °C/s with continuous fluorescent signal acquisition at 5 readings/°C.

The determination of bacterial removal efficiency of the treatment systems

The bacterial removal efficiency of each treatment system (SOPAS and SODIS) was obtained by comparing the bacterial counts obtained from the samples collected before treatment and the average bacterial counts obtained from samples collected after treatment. The percentage reduction was calculated using Eq. 1 [35].

$$\text{Percentage reduction} = 100 - (\text{Survivor count}/ \text{Initial count}) \times 100$$

(1)

Statistical analysis

The statistical software package Statistica™ Ver. 11.0 (Stat Soft Inc., Tulsa, USA) was used for the evaluation of the microbial analysis and the temperature of the collected rainwater samples (untreated, pasteurized and disinfected). To test the significance of the data set, an ANOVA analysis was performed for evenly distributed data while for non-evenly distributed data, a spearman rank order correlation was performed. A significant level of 5% was used as a standard in the hypothesis tests [36], while in all tests a *p*-value of <0.05 was considered statistically significant.

Results

Physico-chemical parameters for water samples collected from SOPAS and SODIS treatment systems

The temperature of the solar pasteurized water samples collected throughout the sampling period (n = 6) ranged from 71 °C (July 2015) to the highest temperature of 93 °C (October 2015). The temperature of the SODIS samples were also monitored after 6 hrs and 8 hrs of treatment, respectively, with the temperature of the 6 hr samples (n = 5) ranging from 52 °C (July 2015) to 89 °C (October 2015) and the temperature of the 8 hr SODIS samples (n = 5) ranging from 63 °C (August 2015) to 86 °C (October 2015). For both the SOPAS and the SODIS treatment, the highest total monthly rainfall over the sampling period was recorded in July 2015 (174.4 mm), which then decreased to 67.6 mm in August 2015, increased to 78.2 mm in September and then decreased to the lowest rainfall recorded in October 2015 (10.0 mm).

For the SODIS treatment, an overall average daily ambient temperature of 24.3 °C was recorded during the sampling period, with the lowest temperature of 17.2 °C recorded during July 2015 and the highest temperature of 29.7 °C recorded during October 2015. The temperature of the untreated water samples (collected directly from the RWH tank), averaged 20.2 °C, with the lowest temperature measured as 17.2 °C (July 2015) and the highest temperature measured as 25.2 °C (October 2015). In addition, an overall average pH of 8.0 was recorded for the untreated water samples, while an overall pH of 8.1 was recorded for the solar disinfected water samples after 6 hrs and 8 hrs of treatment, respectively.

For the SOPAS treatment, an overall average daily ambient temperature of 25.5 °C was recorded during the sampling period, with the lowest temperature of 17.2 °C recorded during July 2015 and the highest temperature of 30.6 °C recorded during October 2015. Similarly, the temperature of the untreated water samples (collected directly from the RWH tank), averaged 24.7 °C, with the lowest temperature measured as 19 °C (July 2015) and the highest temperature measured as 29.0 °C (October 2015). In addition, an overall average pH of 8.0 was recorded for the untreated water samples, while an overall pH of 7.6 was recorded for the solar pasteurized water samples.

Furthermore, a data logger probe was used to measure the water temperature inside the SOPAS system for a period of one month (01/08/2015 to 31/08/2015) (results not shown). An overall average ambient temperature of 21.1 °C was obtained with the lowest temperature recorded as 7.4 °C and the highest temperature recorded as 39.0 °C. In addition, the water temperature inside the SOPAS system had an overall average of 56.9 °C during the monitored month which ranged from 40.1 °C to 82.9 °C. Solar irradiation data was obtained from Stellenbosch Weather Service (Engineering Facility) and ranged from 0.01 W/m^2 to 881.37 W/m^2 with an overall average of 297.27 W/m^2. A direct positive correlation between the ambient temperature and solar irradiation (R = 0.69; $p < 0.05$) and the temperature of the water inside the system (R = 0.20; $p < 0.05$) was also obtained.

Chemical analysis of untreated and treated rainwater samples

Chemical analysis of the SOPAS rainwater samples

Untreated and solar pasteurized water samples (71 °C) collected during the first sampling event were analysed for their anion concentrations (results not shown). All anion concentrations of the untreated water sample and the solar pasteurized water sample were within the drinking water guidelines as stipulated by Australian Drinking Water Guidelines (ADWG) [37], DWAF [7] and South African National Standards (SANS) 241 [38]. A previous study conducted by Dobrowsky et al. [17] also indicated that there was no significant difference between the anion concentrations in the untreated and solar pasteurized water samples (55 to 91 °C). Anion analyses were thus not conducted on the untreated and solar pasteurized rainwater samples collected during the remainder of the sampling period. The turbidity of the untreated and pasteurized rainwater samples was also measured and according to DWAF [7], SANS 241 [38] and ADWG [37], the turbidity should not exceed 1.00 NTU. For both the untreated and solar pasteurized water sample, the turbidity was measured as 0.00 NTU, thus the turbidity complied with the respective drinking water guidelines.

The metal ions and cation concentrations were determined for pasteurized water samples collected at 71 °C, 86 °C and 93 °C and the corresponding unpasteurized samples (Table 1). The concentrations of the metal ions and cations in the untreated and SOPAS treated rainwater samples were below the recommended guidelines as stipulated by ADWG [37], DWAF [7] and SANS 241 [38], with the exception of Pb and Ni. However, while all the before and after SOPAS treatment samples were within the stipulated guidelines for Fe concentrations, the concentration of Fe in the before treatment sample (172.92 µg/L), collected with the corresponding 71 °C SOPAS sample, exceeded the DWAF [7] drinking water guideline of <100 µg/L. The Fe concentration in the SOPAS treatment sample collected at 71 °C then decreased significantly ($p < 0.05$) to 29.19 µg/L.

In addition, while Ni was within the SANS 241 [38] drinking water guideline in all the water samples analysed, it was detected above the drinking water guideline (<20 µg/L) according to ADWG [37] for all three samples collected after pasteurization (71 °C, 30.00 µg/L; 86 °C, 26.46 µg/L; and 93 °C, 25.59 µg/L). Furthermore,

Table 1 Cation and metal ion concentrations of the untreated water samples and the corresponding solar pasteurized water samples collected at various temperatures compared to the recommended drinking water guidelines

Metal	Before 71 °C	After 71 °C	Before 86 °C	After 86 °C	Before 93 °C	After 93 °C	SANS 241	DWAF	AWDG
Al (µg/L)	1.98	99.18	1.36	37.99	1.13	31.70	300	150	200
B (µg/L)	< 0.1	37.77	-	-	-	-	-	-	4000
V (µg/L)	0.07	1.45	0.05	0.62	0.04	0.60	200	1000	-
Mn (µg/L)	4.78	8.60	1.09	9.94	2.16	9.48	100	50	500
Fe (µg/L)	172.92	29.19	78.91	50.07	51.55	26.89	200	100	300
Co (µg/L)	0.05	0.22	0.05	0.24	0.05	0.23	500	-	-
Ni (µg/L)	1.60	30.00	5.21	26.46	0.55	25.59	150	-	20
Cu (µg/L)	2.65	525.42	2.96	549.63	3.94	495.44	1000	1000	2000
Zn (µg/L)	26.45	2529.53	17.37	2086.09	6.70	2003.86	5000	3000	3000
As (µg/L)	0.25	5.63	0.37	1.58	0.49	1.48	10	10	10
Mo (µg/L)	0.02	0.26	0.01	0.15	0.02	0.14	-	-	50
Cd (µg/L)	< 0.05	0.49	0.00	0.77	0.00	0.58	5	5	2
Ba (µg/L)	29.86	86.30	92.97	78.62	88.75	73.53	-	-	2000
Pb (µg/L)	0.59	74.12	<0.006	26.30	<0.006	19.67	20	10	10
Ca (mg/L)	3.05	7.37	4.87	5.42	4.74	5.49	150	32	-
K (mg/L)	0.50	1.60	0.47	0.80	0.50	1.04	50	50	-
Mg (mg/L)	0.31	0.83	0.45	0.57	0.44	0.58	70	30	-
Na (mg/L)	1.61	3.67	2.07	2.70	2.09	2.70	200	100	180
P (mg/L)	0.04	0.08	0.04	0.05	0.03	0.04	-	-	-
Si (mg/L)	0.31	1.37	0.64	1.77	0.65	1.78	-	-	-

Pb was detected above the drinking water guideline stipulated by ADWG [37], DWAF [7] and SANS 241 [38] for all three samples collected after pasteurization (71 °C, 86 °C and 93 °C) with a concentration of 74.12 µg/L, 26.30 µg/L and 19.67 µg/L recorded, respectively. It should however, be noted that both the Ni and Pb concentrations decreased with an increase in SOPAS temperature.

Chemical analysis of the SODIS rainwater samples

All anion concentrations of the SODIS rainwater samples [untreated and treated at 6 hrs (anions: 52 °C; 70 °C and 89 °C) and 8 hrs (anions: 63 °C and 86 °C)], were within the drinking water guidelines as stipulated by ADWG [37], DWAF [7] and SANS 241 [38] (results not shown). In addition, there was no significant ($p > 0.05$) increase in the anion concentrations after treatment. While the turbidity measurements of all the water samples before and after treatment, were within the 1.00 NTU recommended guideline [7, 37, 38], the turbidity of samples collected during the first sampling event in August 2015, were not within the drinking water guidelines. It should however be noted that the untreated water sample had a turbidity of 1.90 NTU, which already exceed the drinking water guidelines. After 6 hrs of treatment by SODIS (70 °C), the turbidity

increased to 2.14 NTU, while the sample treated for 8 hrs (63 °C) had a turbidity of 2.09 NTU.

The metal ions and cation concentrations were measured for representative SODIS sampling events [6 hrs (70 °C and 89 °C) and 8 hrs (63 °C and 86 °C) after treatment] and their corresponding untreated water sample (Table 2). Similar to the results obtained for the SOPAS treated water samples, the concentrations of all the metal ions and cations, in the untreated and SODIS rainwater samples were within the recommended guidelines as stipulated by ADWG [37], DWAF [7] and SANS 241 [38]. However, the concentrations of Fe in the untreated and treated (6 and 8 hrs) samples were significantly ($p < 0.05$) higher compared to the drinking water guidelines as stipulated by ADWG [37], DWAF [7] and SANS 241 [38]. The first untreated sample had an Fe concentration of 571.26 µg/L, which increased to 729.71 µg/L after 6 hrs of treatment (70 °C) and then decreased to a concentration of 645.39 µg/L after 8 hrs of treatment (63 °C). Similarly, an Fe concentration of 112.60 µg/L was recorded in the untreated sample corresponding to the temperature ranges of 89 °C (6 hrs of treatment) and 86 °C (8 hrs of treatment), with the Fe concentration increasing to 1015.32 µg/L (6 hrs) and decreasing to 505.35 µg/L after 8 hrs.

Table 2 Cation and metal ion concentrations of the untreated water samples and the corresponding SODIS treated water samples collected after 6 and 8 hrs compared to the recommended drinking water guidelines

Metal	Untreated	After 6 hrs (70 °C)	After 8 hrs (63 °C)	Untreated	After 6 hrs (89 °C)	After 8 hrs (86 °C)	SANS 241	DWAF	AWDG
Ti (µg/L)	0.51	0.53	0.32	0.09	0.12	0.13	-	-	-
Al (µg/L)	14.35	20.29	11.82	1.19	3.77	5.61	300	150	200
V (µg/L)	0.21	0.20	0.25	0.07	0.14	0.16	200	1000	-
Cr (µg/L)	0.06	0.04	0.06	0.12	0.12	0.13	100	50	50
Mn (µg/L)	3.31	3.24	3.09	1.09	15.30	13.59	100	50	500
Fe (µg/L)	571.26	729.71	645.39	112.60	1015.32	505.35	200	100	300
Co (µg/L)	0.03	0.04	0.03	0.05	0.14	0.13	500	-	-
Ni (µg/L)	0.26	0.25	0.15	0.54	0.46	0.44	150	-	20
Cu (µg/L)	3.09	2.35	9.72	1.00	1.10	1.50	1000	1000	2000
Zn (µg/L)	1.46	5.29	4.18	4.50	5.12	2.52	5000	3000	3000
As (µg/L)	0.45	0.54	0.50	0.38	0.64	0.52	10	10	10
Mo (µg/L)	<0.005	<0.005	<0.005	0.02	0.01	0.03	-	-	50
Cd (µg/L)	<0.004	<0.004	<0.004	0.00	0.00	0.00	5	5	2
Ba (µg/L)	3.79	3.75	2.93	98.05	99.33	89.68	-	-	2000
Pb (µg/L)	0.46	0.65	0.53	<0.006	0.10	0.16	20	10	10
Ca (mg/L)	2.86	2.84	2.88	4.83	4.83	4.83	150	32	-
K (mg/L)	0.35	0.38	0.40	0.49	0.49	0.47	50	50	-
Mg (mg/L)	0.35	0.36	0.37	0.45	0.44	0.46	70	30	-
Na (mg/L)	2.01	2.05	2.08	2.06	2.00	2.09	200	100	180
P (mg/L)	0.10	0.14	0.15	0.04	0.08	0.14	-	-	-

Indicator bacterial counts in untreated and treated rainwater samples
Indicator bacteria detected in untreated and SOPAS rainwater samples

For each untreated water sample and the corresponding pasteurized sample collected at various temperatures ranging from 71 °C to 93 °C, water samples were analysed for the presence of indicator bacteria including *E. coli*, HPC, enterococci and faecal coliforms (Table 3). Enterococci and faecal coliforms were not detected in any of the untreated as well as the pasteurized rainwater samples. However, the HPC for the untreated water

Table 3 Indicator counts for solar pasteurized water samples and the corresponding untreated water samples collected at various temperatures

Pasteurization Temperature	Indicator	Untreated Water Sample (Ave. CFU/100 mL)	Treated Water Sample (Ave. CFU/100 mL)	Reduction (%)
71 °C	*E. coli*	2	BDL	>99
	HPC	7.05×10^6	BDL	>99
77 °C	*E. coli*	2	BDL	>99
	HPC	6.62×10^7	BDL	>99
81 °C	*E. coli*	2	BDL	>99
	HPC	1.01×10^7	BDL	>99
86 °C	*E. coli*	2	BDL	>99
	HPC	1.46×10^7	BDL	>99
91 °C	*E. coli*	2	BDL	>99
	HPC	1.43×10^7	BDL	>99
93 °C	*E. coli*	3	BDL	>99
	HPC	7.4×10^7	BDL	>99

(Note: *BDL* below detection limit)

samples ranged from a minimum of 7.05×10^6 CFU/100 mL to a maximum of 7.4×10^7 CFU/100 mL and were reduced to below the detection limit (<1 CFU/mL) after pasteurization for all temperature ranges (71 °C to 93 °C).

Escherichia coli were also detected in all the untreated water samples with a minimum of 2 CFU/100 mL to a maximum of 3 CFU/100 mL recorded. Similarly, *E. coli* counts were reduced to below the detection limit after pasteurization (71 °C to 93 °C). For the untreated rainwater samples, both the HPC and the *E. coli* counts exceeded the drinking water guidelines as stipulated by the DWAF [7]. However, after pasteurization a >99% reduction in indicator counts was observed for all the pasteurized rainwater samples and the counts were within the DWAF [7] standards.

Indicator bacteria detected in untreated and SODIS rainwater samples

For each untreated water sample and the corresponding solar disinfected water sample, collected at various temperatures ranging from 52 °C to 89 °C and 63 °C to 86 °C treated for 6 hrs and 8 hrs, respectively, water samples were analysed for the presence of indicator bacteria including *E. coli*, HPC, enterococci and faecal coliforms. Similar to results obtained for the SOPAS samples, enterococci and faecal coliforms were not detected in any of the untreated as well as both the 6 hr and 8 hr disinfected water samples. However, the HPC in all the untreated water samples ranged from a minimum of 7.05×10^6 CFU/100 mL to a maximum of 9.95×10^7 CFU/100 mL and was reduced to below the detection limit (< 1 CFU/mL) after 6 hrs of disinfection (Table 4). *Escherichia coli* were also detected in all the untreated water samples (6 hrs of treatment) with counts ranging from a minimum of 2 CFU/100 mL to a maximum of 4 CFU/100 mL. The *E. coli* counts were

then also reduced to below the detection limit after 6 hrs of disinfection.

The HPC in all the untreated water samples corresponding to 8 hrs of SODIS treatment ranged from a minimum of 1.45×10^7 CFU/100 mL to a maximum of 9.95×10^7 CFU/100 mL and was reduced to below the detection limit (< 1 CFU/mL) after 8 hrs of disinfection (Table 5). *Escherichia coli* were also detected in all the untreated water samples (8 hrs of treatment) with counts ranging from a minimum of 2 CFU/100 mL to a maximum of 13 CFU/100 mL. The *E. coli* counts were then also reduced to below the detection limit (< 1 CFU/mL) after 8 hrs of disinfection.

For the untreated rainwater samples (6 hrs and 8 hrs), both the HPC and *E. coli* counts exceeded the drinking water guidelines stipulated by the DWAF [7]. However, after both 6 hrs and 8 hrs of SODIS treatment a significant ($p < 0.05$) reduction (>99%) in indicator counts was observed and all counts were within the DWAF [7] guidelines.

Quantitative PCR for the detection of *Legionella* spp.
Quantitative PCR for the detection of viable *Legionella* spp. in SOPAS samples

The presence of viable *Legionella* cells in the untreated and corresponding treated SOPAS samples were determined using qPCR assays in conjunction with the EMA pre-treatment. A standard curve was constructed with a linear range of quantification from 10^8 to 10^1 gene copies per µL using the LightCycler®96 software Ver. 1.1.0.1320 (Roche Diagnostics International Ltd). A qPCR efficiency of 1.86 (93%) was obtained, with a linear regression coefficient (R^2) value of 0.98. Using the standard curve, viable *Legionella* copy numbers were quantified in the untreated and corresponding solar pasteurized water samples collected at various temperatures

Table 4 Indicator counts for solar disinfected water samples collected after 6 hrs of treatment and the corresponding untreated water samples collected at various temperatures

Disinfected Temperature	Indicator	Untreated Water Sample (Ave. CFU/100 mL)	Treated Water Sample (Ave. CFU/100 mL) after 6 hrs	Reduction (%)
52 °C	*E. coli*	2	BDL	>99
	HPC	7.05×10^6	BDL	>99
68 °C	*E. coli*	4	BDL	>99
	HPC	7.95×10^7	BDL	>99
70 °C	*E. coli*	2	BDL	>99
	HPC	9.95×10^7	BDL	>99
75 °C	*E. coli*	2	BDL	>99
	HPC	8.7×10^7	BDL	>99
89 °C	*E. coli*	2	BDL	>99
	HPC	1.45×10^7	BDL	>99

(Note: *BDL* below detection limit)

Table 5 Indicator counts for solar disinfected water samples collected after 8 hrs of treatment and the corresponding untreated water samples collected at various temperatures

Disinfected Temperature	Indicator	Untreated Water Sample (Ave. CFU/100 mL)	Treated Water Sample (Ave. CFU/100 mL) 8 hrs	Reduction (%)
63 °C	E. coli	4	BDL	>99
	HPC	7.95×10^7	BDL	>99
67 °C	E. coli	2	BDL	>99
	HPC	9.95×10^7	BDL	>99
72 °C	E. coli	13	BDL	>99
	HPC	9.4×10^7	BDL	>99
76 °C	E. coli	2	BDL	>99
	HPC	8.7×10^7	BDL	>99
86 °C	E. coli	2	BDL	>99
	HPC	1.45×10^7	BDL	>99

(Note: *BDL* below detection limit)

and are represented as 23S rRNA gene copies per mL (Fig. 2a).

A significant reduction ($p < 0.05$) in viable *Legionella* copy numbers after solar pasteurization of the rainwater samples collected at all temperature ranges (70 to 79 °C, 80 to 89 °C and 90 °C and above) was obtained (Fig. 2a). For the temperature range of 70 to 79 °C, an average of 1.74×10^5 copies/mL was observed for the untreated water samples, which decreased to an average of 6.15×10^3 copies/mL for the pasteurized water samples. For the temperatures ranging from 80 to 89 °C, an average of 4.79×10^5 copies/mL was observed for the untreated

water samples, compared to an average of 4.57×10^4 copies/mL obtained for the pasteurized water. Lastly, for the temperatures 90 °C and above, an average of 6.49×10^5 copies/mL for the untreated water samples was obtained, which decreased to an average of 8.92×10^3 copies/mL for the pasteurized water samples.

At the lowest (70 to 79 °C) and highest (90 °C and above) pasteurization temperature ranges, a percentage reduction of 99.97% and 96.83% was observed (2-log reduction) in *Legionella* copy numbers, respectively, while the lowest percentage reduction (89.76%) in copy numbers was observed for the 80 to 89 °C temperature range (1-log reduction).

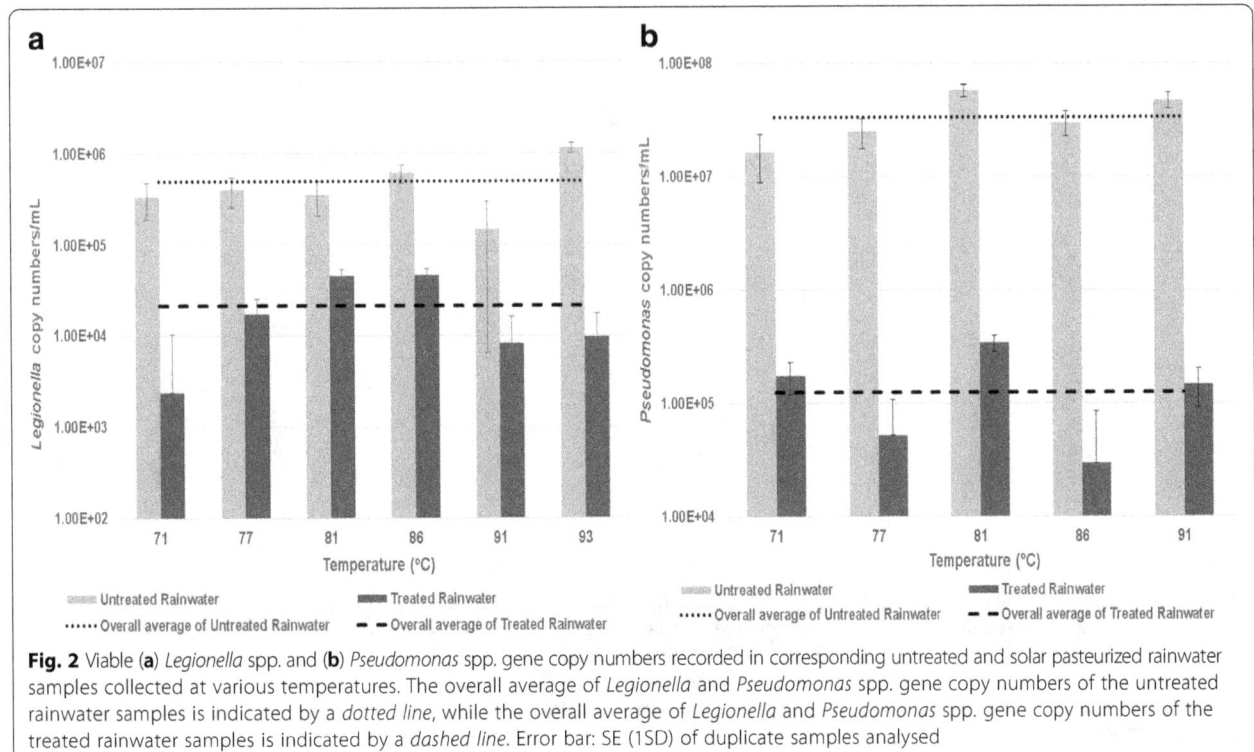

Fig. 2 Viable (**a**) *Legionella* spp. and (**b**) *Pseudomonas* spp. gene copy numbers recorded in corresponding untreated and solar pasteurized rainwater samples collected at various temperatures. The overall average of *Legionella* and *Pseudomonas* spp. gene copy numbers of the untreated rainwater samples is indicated by a *dotted line*, while the overall average of *Legionella* and *Pseudomonas* spp. gene copy numbers of the treated rainwater samples is indicated by a *dashed line*. Error bar: SE (1SD) of duplicate samples analysed

Quantitative PCR for the detection of viable *Legionella* spp. in SODIS samples

The same standard curve as described for the quantification of Pseudomonas copy numbers in th e untreated and SOPAS treated samples, was utilised to quantify viable *Legionella* copy numbers per mL for the untreated and corresponding solar disinfected water samples after 6 and 8 hrs (various temperatures recorded), respectively.

The results obtained for the qPCR assays showed that there was a reduction in viable *Legionella* copy numbers after SODIS treatment for 6 hrs (Fig. 3a). The lowest percentage reduction (24.46%) in *Legionella* copy numbers was observed for a solar disinfected sample with a temperature of 68 °C, where *Legionella* copy numbers decreased from 1.56×10^7 copies/mL for the untreated

sample to 1.18×10^7 copies/mL for the solar disinfected sample. The highest percentage reduction (74.09%) in copy numbers was observed for a solar disinfected sample with a temperature of 75 °C, where *Legionella* copy numbers decreased from 1.76×10^5 copies/mL for the untreated sample to 4.56×10^4 copies/mL for the solar disinfected sample. A significant ($p < 0.05$) reduction (72.6%) in *Legionella* copy numbers was also observed at 89 °C, where *Legionella* copy numbers of 4.13×10^4 copies/mL were observed for the untreated sample and then decreased to 1.13×10^4 copies/mL after SODIS at 6 hrs.

The results obtained for the qPCR assays, indicated that overall there was a 2-log reduction in viable *Legionella* copy numbers (except 63 °C sample) after SODIS of 8 hrs for the rainwater samples collected at temperatures

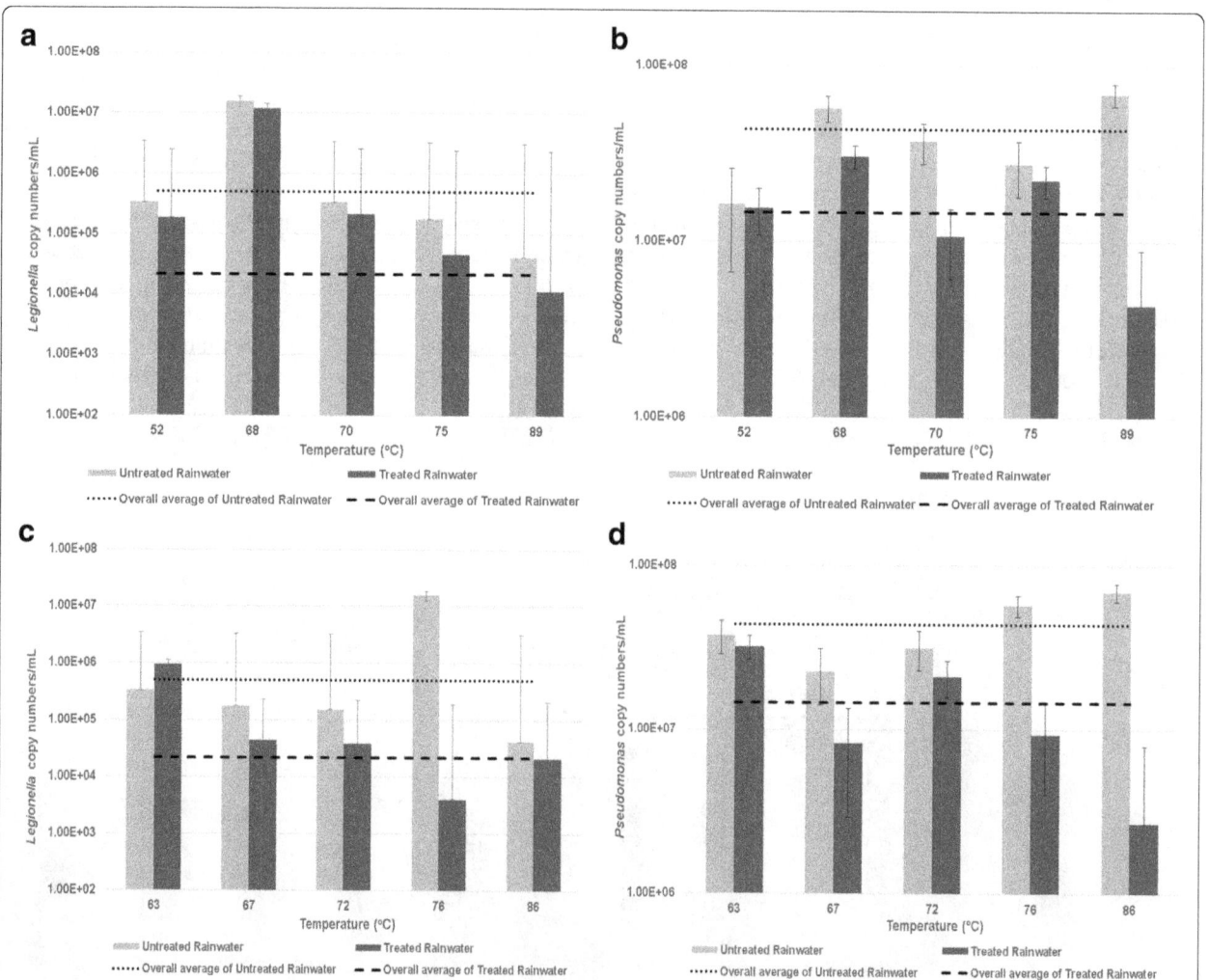

Fig. 3 Viable (**a**) *Legionella* spp. and (**b**) *Pseudomonas* spp. gene copy numbers recorded in corresponding untreated and solar disinfected (for 6 hrs) rainwater samples collected at various temperatures. Viable (**c**) *Legionella* spp. and (**d**) *Pseudomonas* spp. gene copy numbers recorded in corresponding untreated and solar disinfected (for 8 hrs) rainwater samples collected at various temperatures. The overall average of *Legionella* and *Pseudomonas* spp. gene copy numbers of the untreated rainwater samples is indicated by a *dotted line*, while the overall average of *Legionella* and *Pseudomonas* spp. gene copy numbers of the treated rainwater samples is indicated by a *dashed line*. Error bar: SE (1SD) of duplicate samples analysed

ranging from 67 to 86 °C (Fig. 3c). The lowest percentage reduction (50.07%) in copy numbers was observed for a SODIS temperature of 86 °C, where 4.12×10^4 copies/mL was observed in the untreated sample compared to the solar disinfected sample where 2.06×10^4 copies/mL was recorded. The highest percentage reduction (99.97%) in copy numbers was observed for a solar disinfected temperature of 76 °C, where 1.56×10^7 copies/mL was observed in the untreated sample compared to the solar disinfected sample where 4.03×10^3 copies/mL was recorded. For the temperatures of 67 °C and 72 °C, a percentage reduction in copy numbers of 75.43% and 75.41% was recorded, respectively. However, an increase in *Legionella* spp. copy numbers was observed for the solar disinfected sample with a temperature of 63 °C, where 3.32×10^5 copies/mL was observed in the untreated sample compared to 9.54×10^5 copies/mL recorded in the solar disinfected sample.

Quantitative PCR for the detection of *Pseudomonas* spp.

Quantitative PCR for the detection of viable *Pseudomonas* spp. in SOPAS samples

The quantification of viable *Pseudomonas* cells in the untreated and corresponding treated SOPAS samples was determined using qPCR assays in conjunction with the EMA pre-treatment. A standard curve was constructed with a linear range of quantification from 10^9 to 10^0 gene copies per μL using the software LightCycler®96 Version 1.1.0.1320 (Roche Diagnostics International Ltd). A qPCR efficiency of 1.83 (92%) was obtained, with a linear regression coefficient (R^2) value of 1.00. Using the standard curve, viable *Pseudomonas* copy numbers were quantified in the untreated and corresponding solar pasteurized (treated) water samples collected at various temperatures and are represented as *Pseudomonas* lipoprotein *oprI* gene copies per mL (Fig. 2b).

A significant reduction ($p < 0.05$) in viable *Pseudomonas* copy numbers after solar pasteurization of the rainwater samples collected at all temperature ranges (70 to 79 °C, 80 to 89 °C and 90 °C and above) was obtained (Fig. 2b). For the temperature range of 70 to 79 °C, an average of 2.07×10^7 copies/mL was observed for the untreated water samples, which decreased to an average of 1.13×10^5 copies/mL for the pasteurized water samples. For the temperatures ranging from 80 to 89 °C, an average of 4.37×10^7 copies/mL was observed for the untreated water samples, compared to an average of 1.84×10^5 copies/mL obtained for the pasteurized water. Lastly, for the temperatures ranging from 90 °C and above, an average of 3.57×10^7 copies/mL for the untreated water samples was obtained, which decreased to an average of 7.31×10^4 copies/mL for the pasteurized water samples. It should however be noted that while an average of 2.45×10^7 *Pseudomonas* copies/mL was observed in the untreated water sample (collected with the 93 °C

SOPAS sample), no amplification of the *oprI* gene was recorded in the 93 °C pasteurized water sample resulting in a C_q value below detection limit obtained (not presented on Fig. 2b).

For the pasteurization temperature ranges of 70 to 79 °C and 80 to 89 °C a reduction of 99.45% and 99.58% was observed in *Pseudomonas* copy numbers, respectively, thus a 2-log reduction was observed for both these temperature ranges. In addition, the greatest percentage reduction of 99.80% (3-log reduction) in copy numbers was observed for the 90 °C and above temperature range.

Quantitative PCR for the detection of viable *Pseudomonas* spp. in SODIS samples

The same standard curve as described for the quantification of Pseudomonas copy numbers in th e untreated and SOPAS treated samples, was utilised to quantify viable *Pseudomonas* copy numbers per mL for the untreated and corresponding solar disinfected water samples after 6 and 8 hrs (various temperatures recorded), respectively.

The results obtained for the qPCR assays showed that there was a reduction in viable *Pseudomonas* copy numbers after SODIS treatment for 6 hrs (Fig. 3b). The lowest percentage reduction (5.53%) in *Pseudomonas* copy numbers was observed for a solar disinfected sample with a temperature of 52 °C, where *Pseudomonas* copy numbers decreased from 1.63×10^7 copies/mL for the untreated sample to 1.54×10^7 copies/mL for the solar disinfected sample. The highest percentage reduction (93.73%) in copy numbers was observed for a solar disinfected sample with a temperature of 89 °C, where *Pseudomonas* copy numbers decreased from 6.90×10^7 copies/mL for the untreated sample to 4.33×10^6 copies/mL for the solar disinfected sample yielding a 1-log reduction.

The results obtained for the qPCR assays, indicated that there was an overall 1-log reduction in viable *Pseudomonas* copy numbers after SODIS of 8 hrs (Fig. 3d) for the rainwater samples collected at temperatures ranging from 63 to 86 °C. The lowest percentage reduction (14.37%) in copy numbers was observed for a SODIS temperature of 63 °C, where 3.73×10^7 copies/mL was observed in the untreated sample compared to the solar disinfected sample where 3.19×10^7 copies/mL was recorded. The highest percentage reduction (96.12%) in copy numbers was observed for a solar disinfected temperature of 86 °C, where 6.90×10^7 copies/mL was observed in the untreated sample compared to the solar disinfected sample where 2.68×10^6 copies/mL was recorded.

Discussion

The efficiency of two solar based treatment systems (SOPAS and SODIS) were evaluated for the treatment of roof harvested rainwater. Numerous chemical and microbial parameters were investigated in order to determine which system effectively improved the overall quality of the harvested rainwater to within drinking water guidelines. Chemical analysis of the solar pasteurized and corresponding untreated rainwater samples then indicated that all cation (with the exception of Pb and Ni) and anion concentrations were within the drinking water guidelines as stipulated by the ADWG [37], DWAF [7] and SANS 241 [38]. Nickel and Pb were detected in all three pasteurization water samples (71 °C, 86 °C and 93 °C) analysed at concentrations exceeding the drinking water guidelines. Although the SOPAS system has a storage tank constructed from high grade polyethylene, it contains SABS approved Ni coated dezincification resistant (DZR) brass connector points utilised for mounting purposes. Nickel could have thus leached from the Ni coated brass metal during exposure to high temperatures in the SOPAS system. However, only long term exposure to Ni at high concentrations may be toxic to humans as the concentration of beta-microglobulin increases in the kidneys [37]. In addition, the Pb detected could have leached from the surface of the polyethylene storage tank into the water, as the high grade polyethylene storage tank is treated with Pb (personal communication, Crest Organization) which acts as a stabilizer and is often used to treat polyethylene surfaces exposed to high temperature [39]. Significantly high concentrations of Pb have a severe effect on the human central nervous system and results in the interference with calcium metabolism (bone formation), red blood cell production and contributes to kidney failure [37].

For the SODIS system, chemical analysis revealed that the cation (with the exception of Fe) and anion concentrations, were also within the drinking water guidelines as stipulated by the ADWG [37], DWAF [7] and SANS 241 [38]. It should however, be noted that the untreated water samples had iron concentrations which exceeded the drinking water guidelines. These concentrations then increased in the SODIS samples treated for 6 and 8 hrs, respectively. Suib [40] indicated that the synergistic effect of solar photons and hydrogen peroxide generates hydroxide inside microbial cells by Fenton's reaction, causing Fe and hydrogen peroxide to flow through the cell membrane. Furthermore, when cells are irradiated with near UV photons, an increase in ferrous (Fe^{2+}) iron occurs due to increased membrane permeability, resulting in an increased Fe concentration in the surrounding environment. As SODIS uses both heat and UV to treat the water samples, this phenomenon could have been observed in the treated water samples.

Numerous studies have indicated that the microbial quality of harvested rainwater does not comply with drinking water guidelines [18, 41, 42]. The untreated rainwater, SOPAS and SODIS rainwater samples were thus analysed for the presence of the indicator bacteria E. coli, HPC, enterococci and faecal coliforms. Escherichia coli

and HPC were detected in all the untreated water samples collected for SOPAS analysis, and were effectively reduced (>99%) to below the detection limit in all the samples collected at the various temperature ranges (71 °C to 93 °C). These results correlate with a study conducted by Dobrowsky et al. [17], where the research group showed that indicator counts in solar pasteurized water were reduced to below the detection level at temperatures of 72 °C and above. Similar to the results obtained for the SOPAS system, the E. coli and HPC counts recorded in the untreated water samples were also above the drinking water guidelines as stipulated by DWAF [7] and were reduced to below the detection limit after 6 and 8 hrs of SODIS treatment, with a minimum final temperature of 52 °C and 63 °C recorded, respectively. A study conducted by Berney et al. [43] showed that SODIS with strong irradiation conditions of up to 6 hrs disrupts a sequence of basic cellular functions in E. coli that leads to cell death. Overall the results thus indicate that the SOPAS system and SODIS systems (6 and 8 hrs of treatment), successfully reduced indicator bacteria numbers by >99%, at a minimum temperature of 71 °C for the SOPAS system and 52 °C for the SODIS system. These results correlate to a study conducted by Spinks et al. [18] where the research group suggested that a minimum temperature of 55 °C was sufficient to eliminate enteric pathogenic bacteria in water samples.

A poor correlation between indicator microorganisms and opportunistic bacteria has however, been reported [44–46] as previous studies have shown that opportunistic bacteria, such as Legionella and Pseudomonas spp. amongst others, persist in roof harvested rainwater when low indicator counts are recorded [17, 42]. Oliver [47] then indicated that opportunistic pathogenic bacteria such as Legionella spp. are able to enter a viable but non-culturable state and therefore in the current study, EMA-qPCR assays were utilised to test for the presence and viability of these organisms in solar pasteurized and solar disinfected treated rainwater samples. Although conventional PCR can effectively be utilised as a presence/absence indicator of a particular gene or organism, it cannot be used to indicate the viability of the organism detected. In contrast, EMA-qPCR can be used to analyse for the presence and the viability of an organism and is considered a beneficial method for the detection and quantification of intact microorganisms [20, 48].

The EMA-qPCR assays indicated that a significant ($p < 0.05$) reduction (94.70%) in viable Legionella copy numbers was obtained after SOPAS and yielded a 2-log reduction overall. For the SODIS system, Legionella copy numbers also decreased in samples treated for 6 and 8 hrs, respectively. In addition, treatment after 8 hrs yielded a greater decrease (75.22%) in copy numbers (2-log reduction) in comparison to treatment for 6 hrs (maximum of 1-log reduction for the various temperatures), where a 50.60% reduction was observed. However, an increase in copy numbers was obtained for one solar disinfection sample (63 °C) treated for 8 hrs. It is known that *Legionella* can form associations with protozoa where they exist as intracellular parasites and are able to proliferate at temperatures from 50 °C to 65 °C due to the presence of heat shock proteins [49, 50]. *Legionella* spp. are therefore able to out compete other organisms and survive at these high temperatures (>90 °C) [51]. Moreover, a study conducted by Vervaeren et al. [50] showed that *L. pneumophila* is able to proliferate in heat treated water (up to a temperature of 70 °C). According to a study conducted by Hussong et al. [52] viable but non-culturable *Legionella* spp. also regain culturability and remain pathogenic when favourable conditions arise.

The EMA-qPCR assays for *Pseudomonas* yielded similar results to those obtained for *Legionella*. A reduction of 99.61% (3-log reduction) in viable *Pseudomonas* copy numbers was obtained after SOPAS treatment. In addition, SODIS treatment after 8 hrs yielded a greater reduction of 58.31% in viable copy numbers of *Pseudomonas* spp. in comparison to treatment for 6 hrs (47.27%). It is hypothesized that samples treated for 8 hrs were exposed to UV irradiation for an extended time period resulting in a greater microbial reduction. Furthermore, it is well known that *Pseudomonas* can enter a viable but non-culturable state [53] and results obtained in the current study indicated that *Pseudomonas* spp. remain viable at a temperature of 89 °C after treatment by SODIS. Several studies [54–56] have thus utilised the addition of a photocatalytic material, to enhance the effect of microbial inactivation over a wide range of microorganisms and thereby increase the efficiency of a SODIS system. Titanium dioxide (TiO_2) is considered the most suitable photocatalyst due to the lack of toxicity and chemical and photochemical stability, however further research is needed to determine to potability of TiO_2 treated water [57].

Results obtained in the current study indicated that SOPAS treatment yielded a greater reduction in viable *Legionella* and *Pseudomonas* spp. (94.70% and 99.61%, respectively) copy numbers, compared to SODIS treatment after 6 (50.60% for *Legionella* spp. and 47.27% for *Pseudomonas* spp.) and 8 hrs (75.22% for *Legionella* spp. and 58.31% for *Pseudomonas* spp.). While not significant, treatment with SOPAS yielded a lower reduction in viable *Legionella* copy numbers compared to *Pseudomonas* copy numbers. It is hypothesized that *Legionella* spp. may have been able to persist due to: the presence of heat shock proteins to protect them from high temperatures; associations with amoebae species; and the formation of biofilms [58].

Conclusions and future research

Based on the indicator count analysis, treatment of harvested rainwater with both SOPAS and SODIS improved the microbial quality of rainwater and the water could be utilised for irrigation and domestic purposes such as cooking, laundry and washing. The SOPAS system can however, effectively treat larger volumes of rainwater in comparison to the SODIS system and based on the EMA-qPCR results obtained in the current study, SOPAS was the most effective for the reduction of viable *Legionella* and *Pseudomonas* spp. copy numbers in harvested rainwater. However, depending on the material utilised to construct the storage tank, metals and chemicals may leach into the water when temperatures higher than 71 °C are achieved inside the SOPAS system. In contrast, SODIS systems function as batch culture systems and are more cost-effective and easier to operate and maintain. Future research should however, focus on up-scaling SODIS systems to allow for the efficient treatment of larger volumes of rainwater.

Abbreviations
ANOVA: Analysis of variance; Ave.: Average; BDL: Below detection limit; bp: Base pairs; CAF: Central analytical facility; CFU: Colony forming units; C_q: Quantitative cycle; DNA: Deoxyribonucleic acid; DWAF: Department of Water Affairs and Forestry; DZR: Dezincification resistant; EMA: Ethidium monoazide; EMA-qPCR: Ethidium monoazide bromide quantitative polymerase chain reaction; HPC: Heterotrophic plate count; Hrs: Hours; ICP-AES: Inductively coupled plasma atomic emission spectrometry; m-FC: Membrane filter faecal coliform; MLGA: Membrane lactose glucuronide agar; NTU: Nephelometric turbidity unit; PCR: Polymerase chain reaction; PET: Polyethylene terephthalate; qPCR: Quantitative PCR; R^2: Linear regression coefficient; R2A: Reasoner's 2A agar; rRNA: Ribosomal ribonucleic acid; RWH: Rainwater harvesting; SANS: South African National Standards; SD: Standard deviation; SODIS: Solar disinfection; SOPAS: Solar pasteurization; USEPA: United States Environmental Protection Agency; UV: Ultra-violet; WHO: World Health Organization

Acknowledgements
The authors would like to thank Jacques de Villiers from the Crest Organization (Stellenbosch, Western Cape) for providing the Phungamanzi^{TM} solar pasteurization system. The South African Weather Services is thanked for providing the daily ambient temperature and rainfall data.

Funding
The Water Research Commission (WRC; Project K5/2368//3) and the National Research Foundation of South Africa (Grant number: 90320) funded this project. Opinions expressed and conclusions arrived at, are those of the authors and are not necessarily to be attributed to the National Research Foundation.

Authors' contributions
AS and WK conceived and designed the experiments. AS performed the experiments and analysed the data. PD, TN and BR co-supervised the experimental procedures. WK acquired funding for the study and contributed reagents/materials/analysis tools. AS and WK wrote the paper. All authors edited the drafts of the manuscript and approved the final version of the manuscript.

Competing interests
The authors declare that they have no competing interests.

References

1. Amin M, Han M. Roof-harvested rainwater for potable purposes: application of solar collector disinfection (SOCO-DIS). Water Res. 2009;43(20):5225–35.
2. De Kwaadsteniet M, Dobrowsky PH, Van Deventer A, Khan W, Cloete TE. Domestic rainwater harvesting: microbial and chemical water quality and point-of-use treatment systems. Water Air Soil Poll. 2013;224(7):1–19.
3. Helmreich B, Horn H. Opportunities in rainwater harvesting. Desalination. 2009;248(1):118–24.
4. Abbasi T, Abbasi SA. Sources of pollution in rooftop rainwater harvesting systems and their control. Crit Rev Env Sci Tec. 2011;41(23):2097–167.
5. Dobrowsky PH, Mannel D, De Kwaadsteniet M, Prozesky H, Khan W, Cloete TE. Quality assessment and primary uses of harvested rainwater in Kleinmond, South Africa. Water SA. 2014;40(3):401–6.
6. Amin MT, Nawaz M, Amin MN, Han M. Solar disinfection of *Pseudomonas aeruginosa* in harvested rainwater: a step towards potability of rainwater. PloS One. 2014;9(3):1–10.
7. Department of Water Affairs and Forestry (DWAF). South African Water Quality Guidelines, 2nd Edition Volume 1: Domestic Water Use. Pretoria: CSIR Environmental Services; 1996. http://www.iwa-network.org/filemanager-uploads/WQ_Compendium/Database/Selected_guidelines/041.pdf. Accessed 28 Sept 2015.
8. World Health Organization (WHO). Guidelines for Drinking-Water Quality, 4th Edition. Geneva: World Health Organization; 2011. http://apps.who.int/iris/bitstream/10665/44584/1/9789241548151_eng.pdf. Accessed 18 Sept 2015.
9. Dobrowsky PH, De Kwaadsteniet M, Cloete TE, Khan W. Distribution of indigenous bacterial pathogens and potential pathogens associated with roof-harvested rainwater. Appl Environ Microbiol. 2014;80(7):2307–16.
10. McGuigan KG, Conroy RM, Mosler H, Du Preez M, Ubomba-Jaswa E, Fernandez-Ibanez P. Solar water disinfection (SODIS): A review from bench-top to roof-top. J Hazard Mater. 2012;235:29–46.
11. Safapour N, Metcalf RH. Enhancement of solar water pasteurization with reflectors. Appl Environ Microbiol. 1999;65(2):859–61.
12. Islam MF, Johnston RB. Household Pasteurization of drinking-water: The *Chulli* Water-treatment System. J Health Popul Nutr. 2006;24(3):356–62.
13. Sommer B, Marinõ A, Solarte Y, Salas ML, Dierolf C, Valiente C, et al. SODIS - an emerging water treatment process. J Water SRT-Aqua. 1997;46:127–37.
14. Raveendhra D, Faruqui S, Saini P. Transformer less FPGA Controlled 2-Stage isolated grid connected PV system. PESTSE 2014. 2014;doi:10.1109/PESTSE. 2014.6805304.
15. Feachem RE, Bradley DJ, Garelick H, Mara DD. Sanitation and disease: Health Aspects of Excreta and Wastewater Management. Washington DC: John Wiley & Sons; 1983.
16. Burch JD, Thomas KE. Water disinfection for developing countries and potential for Solar Thermal Pasteurization. Sol Energy. 1998;64:87–97.
17. Dobrowsky PH, Carstens M, De Villiers J, Cloete TE, Khan W. Efficiency of a closed-coupled solar pasteurization system in treating roof harvested rainwater. Sci Total Environ. 2015;536:206–14.
18. Spinks AT, Dunstan RH, Harrison T, Coombes P, Kuczera G. Thermal inactivation of water-borne pathogenic and indicator bacteria at sub-boiling temperatures. Water Res. 2006;40(6):1326–32.
19. El Ghetany HH, Abdel Dayem A. Numerical simulation and experimental validation of a controlled flow solar water disinfection system. Desalin Water Treat. 2010;20(1-3):11–21.
20. Reyneke B, Dobrowsky PH, Ndlovu T, Khan S, Khan W. EMA-qPCR to monitor the efficiency of a closed-coupled solar pasteurization system in reducing *Legionella* contamination of roof-harvested rainwater. Sci Total Environ. 2016;553:662–70.
21. Bédard E, Fey S, Charron D, Lalancette C, Cantin P, Dolcé P, Laferrière C, Déziel E, Prévost M. Temperature diagnostic to identify high risk areas and optimize *Legionella pneumophila* surveillance in hot water distribution systems. Water Res. 2015;71:244–56.
22. Yee RB, Wadowsky RM. Multiplication of *Legionella pneumophila* in unsterilized tap water. Appl Environ Microbiol. 1982;43(6):1330–4.
23. Giamarellou H. Prescribing guidelines for severe *Pseudomonas* infections. J Antimicrob Chemother. 2002;49(2):229–33.
24. Mena KD, Gerba CP. Risk assessment of *Pseudomonas aeruginosa* in water. Rev Environ Contam T. 2009;201:71–115.
25. Lyczak JB, Cannon CL, Pier GB. Establishment of *Pseudomonas aeruginosa* infection: lessons from a versatile opportunist. Microb Infect. 2000;2(9):1051–60.
26. Fittipaldi M, Nocker A, Codony F. Progress in understanding preferential detection of live cells using viability dyes in combination with DNA amplification. J Microbiol Methods. 2012;91(2):276–89.
27. Reed RH. Solar inactivation of faecal bacteria in water: the critical role of oxygen. Lett Appl Microbiol. 1997;24(4):276–80.
28. Burton A. Purifying drinking water with Sun, Salt, and limes. Environ Health Perspect. 2012;120(8): 10.1289/ehp.120-a305.
29. Saleh MA, Ewane E, Jones J, Wilson BL. Monitoring Wadi El Raiyan lakes of the Egyptian desert for inorganic pollutants by ion-selective electrodes, ion chromatography and inductively coupled plasma spectroscopy. Ecotoxicol Environ Saf. 2000;45(3):310–6.
30. Delgado-Viscogliosi P, Solignac L, Delattre JM. Viability PCR, a Culture-Independent Method for Rapid and Selective Quantification of Viable *Legionella pneumophila* Cells in Environmental Water Samples. Appl Environ Microbiol. 2009;75(11):3502–12.
31. Herpers BL, De Jongh BM, Van Der Zwaluw K, Van Hannen EJ. Real-time PCR assay targets the 23S-5S spacer for direct detection and differentiation of *Legionella* spp. and *Legionella pneumophila*. Clin Microbiol Rev. 2003;41(10):4815–6.
32. Roosa S, Wauven CV, Billon G, Matthijs S, Wattiez R, Gillan DC. The *Pseudomonas* community in metal-contaminated sediments as revealed by quantitative PCR: a link with metal bioavailability. Res Microbiol. 2014;165:647–56.
33. Bergmark L, Poulsen PHB, Abu Al-Soud W, Norman A, Hansen LH, Sørensen SJ. Assessment of the specificity of *Burkholderia* and *Pseudomonas* qPCR assays for detection of these genera in soil using 454 pyrosequencing. FEMS Microbiology Letters. 2012;333:77–84.
34. Chen NT, Chang CW. Rapid quantification of viable Legionellae in water and biofilm using ethidium monoazide coupled with real-time quantitative PCR. J Appl Microbiol. 2010;109(2):623–34.
35. Brözel VS, Cloete TE. Resistance of bacteria from cooling waters to bactericides. J Ind Microbiol. 1991;8(4):273–6.
36. Dunn OJ, Clark VA. Applied statistics: analysis of variance and regression. New York: Wiley; 1974.
37. NHMRC, NRMMC. Australian Drinking Water Guidelines Paper 6, National Water Quality Management Strategy. Canberra: National Health and Medical Research Council, National Resource Management Ministerial Council Commonwealth of Australia; 2011.
38. South African Bureau of Standards (SABS). South African National Standards (SANS) 241: Drinking Water Quality Management Guide for Water Services Authorities. Annexure 1. ISBN 0-626-17752-9. 2005.
39. Pedersen K. Biofilm development on stainless steel and PVC surfaces in drinking water. Water Res. 1990;24(2):239–43.
40. Suib SL. New and future development in catalysis. Oxford: Elsevier. ISBN: 978—444-53872-2. 2013.
41. Ahmed W, Richardson K, Sidhu JPS, Jagals P, Toze S. Inactivation of fecal indicator bacteria in a roof-captured rainwater system under ambient meteorological conditions. J App Microbiol. 2014;116(1):199–207.
42. Ahmed W, Vieritz A, Goonetilleke A, Gardner T. Health risk from the use of roof-harvested rainwater in Southeast Queensland, Australia, as potable or non-potable water, determined using quantitative microbial risk assessment. Appl Environ Microbiol. 2010;76:7382–91.
43. Berney M, Weilenmann HU, Ihssen J, Bassin C, Egli T. Specific growth rate determines the sensitivity of *Escherichia coli* to thermal, UVA, and solar disinfection. Appl Environ Microbiol. 2006;72:2586–93.
44. Harwood VJ, Levine AD, Scott TM, Chivukula V, Lukasik J, Farrah SR, Rose JB. Validity of the indicator organism paradigm for pathogen reduction in reclaimed water and public health protection. Appl Environ Microbiol. 2005;71(6):3163–70.
45. Savichtcheva O, Okabe S. Alternative indicators of faecal pollution: relations with pathogens and conventional indicators, current methodologies for direct pathogen monitoring and future application perspectives. Water Res. 2006;40(13):2463–76.
46. Wilkes G, Edge T, Gannon V, Jokinen C, Lyautey E, Medeiros D, et al. Seasonal relationships among indicator bacteria, pathogenic bacteria, *Cryptosporidium* oocysts, *Giardia* cysts, and hydrological indices for surface waters within an agricultural landscape. Water Res. 2009;43(8):2209–23.
47. Oliver JD. Recent findings on the viable but non-culturable state in pathogenic bacteria. FEMS Microbiol Rev. 2010;34(4):415–25.
48. Postollec F, Falentin H, Pavan S, Combrisson J, Sohier D. Recent advances in quantitative PCR (qPCR) applications in food microbiology. Food Microbiol. 2011;28(5):848 61.

49. Fields BS, Benson RF, Besser RE. *Legionella* and Legionnaires' Disease: 25 Years of Investigation. Clin Microbiol Rev. 2002;15(3):506–26.
50. Vervaeren H, Temmerman R, Devos L, Boon N, Verstraete W. Introduction of a boost of *Legionella pneumophila* into a stagnant-water model by heat treatment. FEMS Microbiol Ecol. 2006;58(3):583–92.
51. United States Environmental Protection Agency (USEPA). *Legionella*: Drinking Water Health Advisory. Report number: EPA-822-B-01-005. 2001. https://www.epa.gov/sites/production/files/2015-10/documents/legionella-report.pdf Accessed 28 Aug 2015.
52. Hussong D, Colwell RR, O'Brien M, Weiss E, Pearson AD, Weiner RM, et al. Viable *Legionella pneumophila* not detectable by culture on agar media. Biotechnol. 1987;9(5):947–50.
53. Dwidjosiswojo Z, Richard J, Moritz MM, Dopp E, Flemming HC, Wingender J. Influence of copper ions on the viability and cytotoxicity of *Pseudomonas aeruginosa* under conditions relevant to drinking water environments. Int J Hyg Environ Health. 2011;214(6):485–92.
54. McCullagh C, Robertson JM, Bahnemann DW, Robertson PK. The application of TiO$_2$ photocatalysis for disinfection of water contaminated with pathogenic micro-organisms: a review. Res Chem Intermed. 2007;33(3-5):359–75.
55. Dalrymple OK, Stefanakos E, Trotz MA, Goswami DY. A review of the mechanisms and modelling of photocatalytic disinfection. Appl Catal B. 2010;98(1):27–38.
56. Malato S, Fernández-Ibáñez P, Maldonado MI, Blanco J, Gernjak W. Decontamination and disinfection of water by solar photocatalysis: recent overview and trends. Catal Today. 2009;147(1):1–59.
57. Byrne JA, Fernandez-Ibañez PA, Dunlop PS, Alrousan D, Hamilton JW. Photocatalytic enhancement for solar disinfection of water: a review. Int J Photoenergy. 2011;2011:1–12.
58. Rakić A, Štambuk-Giljanović N. Physical and chemical parameter correlations with technical and technological characteristics of heating systems and the presence of *Legionella* spp. in the hot water supply. Environ Monit Assess. 2016;188(2):1–12.

14

The highly variable microbiota associated to intestinal mucosa correlates with growth and hypoxia resistance of sea bass, *Dicentrarchus labrax*, submitted to different nutritional histories

François-Joël Gatesoupe[1,3]* , Christine Huelvan[2], Nicolas Le Bayon[2], Hervé Le Delliou[2], Lauriane Madec[2], Olivier Mouchel[2], Patrick Quazuguel[2], David Mazurais[2] and José-Luis Zambonino-Infante[2]

Abstract

Background: The better understanding of how intestinal microbiota interacts with fish health is one of the key to sustainable aquaculture development. The present experiment aimed at correlating active microbiota associated to intestinal mucosa with Specific Growth Rate (SGR) and Hypoxia Resistance Time (HRT) in European sea bass individuals submitted to different nutritional histories: the fish were fed either standard or unbalanced diets at first feeding, and then mixed before repeating the dietary challenge in a common garden approach at the juvenile stage.

Results: A diet deficient in essential fatty acids (LH) lowered both SGR and HRT in sea bass, especially when the deficiency was already applied at first feeding. A protein-deficient diet with high starch supply (HG) reduced SGR to a lesser extent than LH, but it did not affect HRT. In overall average, 94 % of pyrosequencing reads corresponded to Proteobacteria, and the differences in Operational Taxonomy Units (OTUs) composition were mildly significant between experimental groups, mainly due to high individual variability. The highest and the lowest Bray-Curtis indices of intra-group similarity were observed in the two groups fed standard starter diet, and then mixed before the final dietary challenge with fish already exposed to the nutritional deficiency at first feeding (0.60 and 0.42 with diets HG and LH, respectively). Most noticeably, the median percentage of *Escherichia-Shigella* OTU_1 was less in the group LH with standard starter diet. Disregarding the nutritional history of each individual, strong correlation appeared between (1) OTU richness and SGR, and (2) dominance index and HRT. The two physiological traits correlated also with the relative abundance of distinct OTUs (positive correlations: *Pseudomonas* sp. OTU_3 and *Herbaspirillum* sp. OTU_10 with SGR, *Paracoccus* sp. OTU_4 and *Vibrio* sp. OTU_7 with HRT; negative correlation: *Rhizobium* sp. OTU_9 with HRT).

Conclusions: In sea bass, gut microbiota characteristics and physiological traits of individuals are linked together, interfering with nutritional history, and resulting in high variability among individual microbiota. Many samples and tank replicates seem necessary to further investigate the effect of experimental treatments on gut microbiota composition, and to test the hypothesis whether microbiotypes may be delineated in fish.

Keywords: Host-microbe interaction, 16S rRNA, Pyrosequencing, Autochthonous bacteria, Alternative feed ingredients, Physiological status

* Correspondence: Joel.Gatesoupe@partenaire-exterieur.ifremer.fr
[1]NUMEA, INRA, Univ. Pau & Pays Adour, 64310 Saint Pée sur Nivelle, France
[3]PFOM/ARN, Ifremer, Centre de Bretagne, CS 10070, 29280 Plouzané, France
Full list of author information is available at the end of the article

Background

There is growing evidence that intestinal microbes play functional roles that are essential to health and nutrition, and a better understanding of the relationship between fish and their gut microbiota is crucial for sustainable aquaculture development [1]. Many factors have been suggested as influencing the origin and composition of gut microbiota of fish, including genetic background [2, 3], diet [4], stress [5], and many environmental factors (e.g., temperature [6]). Regarding European sea bass, *Dicentrarchus labrax*, one of the two main marine fish species produced by aquaculture in southern Europe [7], high similarity was noticed among the faecal microbial communities collected from individuals grouped in the same tank [8]. However, dominant and subordinate individuals of Arctic charr harboured distinct aerobic microbiota associated with intestinal mucosa [9], suggesting that social interaction may affect autochthonous gut microbiota in individuals cohabiting in the same tank. High-throughput sequencing methods have been recently applied to intestinal microbiota in fish. The pyrosequencing of 16S rRNA gene fragments allowed Roeselers et al. [10] to detect bacterial taxa that were shared by the intestinal communities in zebrafish of different origins, including specimens caught in the wild. The authors concluded that fish have a specific core intestinal microbiota, as is the case with higher vertebrates [11]. However, the life history and diet may deeply influence gut microbiome in fish [12].

During the last decade, the composition of fish feeds has considerably changed to find the way to expand aquaculture without depleting natural fish stocks. Fish meal can be almost completely replaced by plant protein sources in the diet of sea bass [13]. The replacement of fish oil by vegetable oils has been also achieved at least partially [14], or in short-term experiments [15], but the high requirement of sea bass for highly unsaturated fatty acids (HUFA) has hindered the simultaneous substitution of fish oil and fish meal by plant ingredients. Scarce information is available about the effect of dietary fatty acids on gut microbiota in fish, and most studies concerned salmonids [4]. Starch and other polysaccharides from vegetal protein sources are known to influence gut microbiota in fish, including sea bass [16].

The composition of starter diets impacted the bacterial community associated with sea bass larvae [17], but the long-term effects of early nutritional treatments on intestinal microbiota have not yet been tested in this species. In rainbow trout fry, a short hyperglucidic hypoproteic dietary stress had short-term and long-term effects on intestinal fungi in juveniles, while intestinal bacteria were not significantly affected [18].

The present experiment was aimed at investigating the short-term and long-lasting effects of dietary stress on gut microbiota in sea bass fed diets either deficient in HUFA, or hypoproteic with starch as substitute for energy supply. To this end, sea bass were challenged with the unbalanced diets, or fed a standard diet, at first feeding (phase 1). After a transition period of 5 months, juvenile fish originating from the different dietary groups were transferred into the challenge tanks for phase 2. This second phase corresponded to a common garden experiment, which mixed in the same tanks animals that were already challenged or not during phase 1. These fish were fed the unbalanced diets for two months, while other individuals fed in standard conditions from start feeding onwards were maintained in a control tank. The physiological status of the individuals was addressed through growth and resistance to hypoxia. These health criteria were compared between dietary groups, and correlated with the composition of the intestinal community of autochthonous bacteria in each individual.

Results

Individual variability and dissimilarity of intestinal microbiota

The microbial community was analysed in the intestinal mucosa of 54 fish after two days of fasting at the end of the experiment. The bacterial composition was computed from pyrosequencing data after RNA extraction and reverse transcription, with a view to compare the relative activity of every taxon. The number of valid reads per intestinal sample was highly variable, with significant differences between groups (Table 1). Normalization was thus critical before comparing the rarefaction curves and alpha-diversity. Within each experimental group, a wide variability appeared among samples. The rarefaction curves showed that the number of OTUs was still exponentially increasing even above 11,000 reads in some samples, whereas a rarefaction plateau appeared after much less reads in other ones [see Additional file 1]. Due to individual variability, there was no significant difference between the mean alpha-diversity indices of the experimental groups (Table 1).

The high variability could be further illustrated by comparing the numbers of OTUs shared among individuals and experimental groups. Over a total of 1111 OTUs, *Escherichia-Shigella* OTU_1 was the only taxon detected in every sample, with a relative abundance varying between 82.5 and 9.6 %. When the experimental groups were compared, 42 OTUs were shared by every group, while 135–159 OTUs were detected only in one group [see Additional file 2].

The intra-group similarity appeared relatively low in the three groups without transfer before the final dietary challenge, with a mean Bray-Curtis index of 0.52–0.53 (Table 1). Interestingly, the same index was differentially affected when the fish were transferred to the challenge tanks, depending on the diet. The individuals submitted to HUFA restriction only during phase 2 presented the

Table 1 Pyrosequencing yield and mean indices of bacterial diversity in the experimental groups of sea bass

	LH1-LH2	C1-LH2	C1-C2	C1-HG2	HG1-HG2	p-value
Valid reads per sample before normalization	$17634^{ab} \pm 1052$	$20894^{a} \pm 1006$	$16603^{b} \pm 867$	$18492^{ab} \pm 1135$	$20603^{a} \pm 900$	0.01
Diversity index after normalization:						
OTU richness	45.6 ± 3.7	50.7 ± 5.1	45.4 ± 3.6	47.3 ± 3.4	49.3 ± 2.9	0.83
Dominance	0.31 ± 0.03	0.38 ± 0.05	0.30 ± 0.03	0.41 ± 0.05	0.33 ± 0.03	0.21
Intra-group similarity (Bray-Curtis)	$0.53^{ab} \pm 0.02$	$0.42^{c} \pm 0.03$	$0.52^{ab} \pm 0.02$	$0.60^{a} \pm 0.02$	$0.52^{b} \pm 0.02$	≤0.001

The means ($\pm$ SE) were compared by ANOVA; in case of significant difference, superscript letters on the same line indicated the significant differences after post-hoc Tukey's test. The diversity indices were computed after normalizing the numbers of total reads per sample, based on the minimum observed (11,599 reads). Each group was named after its diets during the two challenge phases: HUFA-deficient diet at both phases (LH1-LH2), or only at phase 2 (C1-LH2); hypoproteic diet with high starch supply at both phases (HG1-HG2), or only at phase 2 (C1-HG2); control with standard diets (C1-C2)

lowest intra-group similarity in gut microbiota composition (group C1-LH2; mean Bray-Curtis index: 0.42; difference significant with every other group), while the highest similarity was observed among the fish transferred to the tank challenged with the starchy hypoproteic diet (group C1-HG2; mean Bray-Curtis: 0.60; difference significant only with groups HG1-HG2 and C1-LH2).

Despite individual variability, the non-parametric multivariate analysis of variance (PERMANOVA) revealed an overall significant difference between the mean bacterial profiles of the experimental groups, but none of the Bonferroni-corrected p values was less than 0.05 in the post-hoc pairwise comparison (Table 2). The overall significant difference was mainly due to the dissimilarity between the two groups submitted to HUFA deficiency, and between the control group reared in standard conditions (C1-C2) compared to each of the two groups transferred for phase 2.

Phylogenetic analysis of intestinal microbiota

The dominant phylum was Proteobacteria in every sample (94.4 ± 1.0 %), with mainly Gammaproteobacteria (56.5 ± 2.6 %) and Alphaproteobacteria (35.0 ± 2.9 %).

Table 2 Comparison of inter-groups similarities in gut microbiota composition

PERMANOVA between all groups (p-value = 0.02)

Group	LH1-LH2	C1-LH2	C1-C2	C1-HG2	HG1-HG2
LH1-LH2		*0.23*	*1*	*1*	*1*
C1-LH2	0.02		*0.07*	*1*	*1*
C1-C2	0.30	0.007		*0.20*	*0.64*
C1-HG2	0.21	0.15	0.02		*1*
HG1-HG2	0.47	0.18	0.06	0.65	

PERMANOVA analysis on Bray-Curtis distance (9999 permutations); post-hoc pairwise comparison: uncorrected (left) and Bonferroni-corrected p values (right, in italics); each group was named after its diets during the two challenge phases: HUFA-deficient diet at both phases (LH1-LH2), or only at phase 2 (C1-LH2); hypoproteic diet with high starch supply at both phases (HG1-HG2), or only at phase 2 (C1-HG2); control with standard diets (C1-C2)

Three other phyla were significantly represented: Bacteroidetes (2.3 ± 0.5 %), Actinobacteria (1.4 ± 0.3 %), and Firmicutes (1.1 ± 0.3 %). At the phylum level, the only significant difference after ANOVA corresponded to Actinobacteria, which were more abundant in group C1-C2, compared to C1-HG2, but some other differences were observed between groups after Linear discriminant Effective Size (LEfSe) pairwise analysis [see Additional file 3].

The proportion of 41 OTUs exceeded 5 % in at least one sample, with great variability among individuals [see Additional file 4]. At the OTU level, the most noticeable difference between groups concerned *Escherichia-Shigella* OTU_1, which was generally dominant, but significantly less prevalent in group C1-LH2 compared to group LH1-LH2 or C1-C2 by LEfSe, though there was no significant difference after ANOVA on every group, due to high individual variability (Fig. 1).

Pseudomonas OTU_3 was the only major OTU with a significant difference detected by ANOVA, with a mean relative prevalence of 14.7 % in the control group C1-C2, which represented more than the double of the mean proportions in the other groups. The hypoproteic diet with high starch supply had little effect on gut microbiota composition, which appeared rather similar between the two groups, with or without dietary challenge at first feeding. The most noticeable difference concerned *Alkanindiges* sp. OTU_14, which was not detected in group C1-HG2 [see Additional file 5].

There were other significant differences between groups at every phylogenetic level, as revealed by ANOVA, Kruskal-Wallis or LEfSe. It concerned some Gammaproteobacteria [see Additional file 5], Alpha- and Beta-Proteobacteria [see Additional file 6], Actinobacteria and Firmicutes [see Additional file 7], Bacteroidetes and Spirochaetae [see Additional file 8].

However, some of these differences appeared more visible between individuals than between groups. For example, three distinct OTUs of *Bacillus* sp. were relatively active, each one in three different samples from group C1-LH2 (OTU_178, OTU_196, and OTU_202, accounting for 1.2, 0.8, and 0.6 % of total reads in each of the three samples, respectively), while this genus was not detected

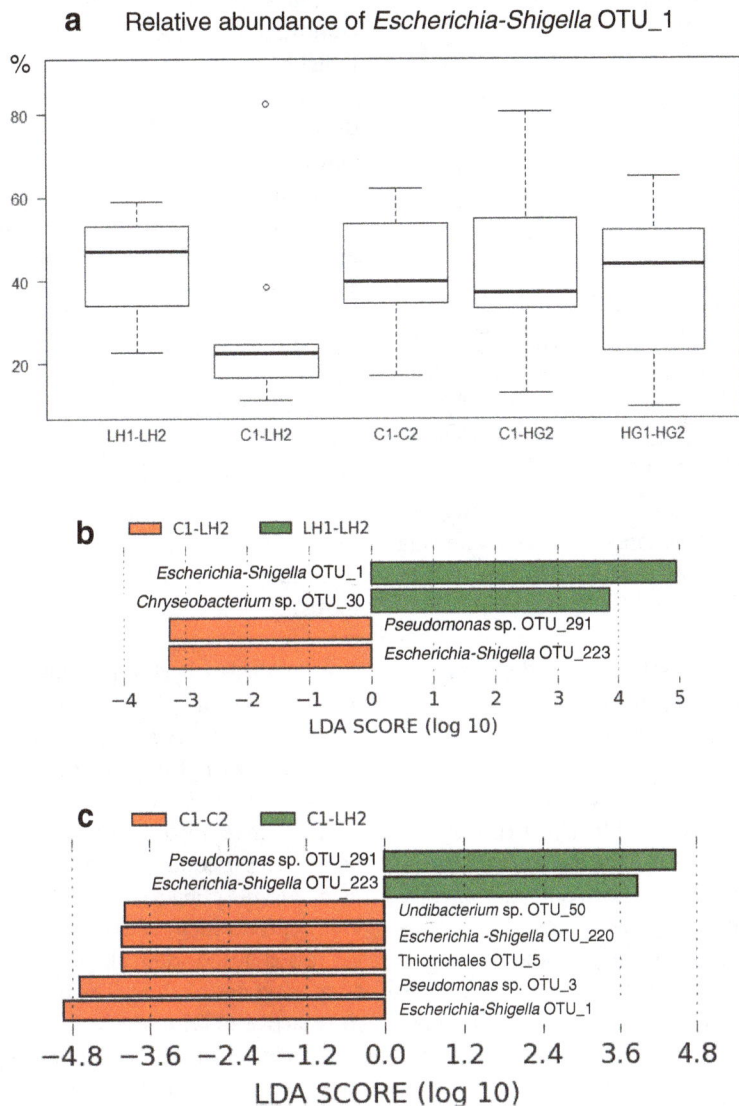

Fig. 1 OTU dissimilarity between the two groups submitted to HUFA deficiency, compared to the control group. *Legend:* When all the experimental groups were compared by ANOVA, the high individual variability masked the significance of the relatively low proportion of *Escherichia-Shigella* OTU_1 in group C1-LH2, fed the HUFA-deficient diet only during the final challenge (**a**). However, the LEfSe pairwise comparisons showed significant Linear Discriminant Analysis (LDA) scores when C1-LH2 was compared to the other HUFA-deficient group LH1-LH2 (**b**), and to the control group C1-C2 (**c**). The few other significant LEfSe differences between OTU abundances were also shown in diagrams **b** and **c**

elsewhere, except at a low level in one sample from group C1-HG2 (OTU_202, 0.05 % total reads).

Growth and resistance to hypoxia of sea bass

The highest mean weight corresponded to group HG1-HG2 at the three sampling dates, even at the beginning of phase 2 at 225 dph [see Additional file 9]. At this date, the mean weights of groups C1-C2, C1-LH2 and C1-HG2 were significantly lower than that of HG1-HG2. At 266 and 287 dph the mean weights of LH1-LH2 and C1-LH2 were significantly lower than that of HG1-HG2. The Specific Growth Rate (SGR) during this last three weeks was chosen as the best growth indicator, with a

view to reduce the incidence of the initial differences in mean weights at the beginning of phase 2. During the final three weeks, the highest mean SGR was observed in the control group C1-C2 (Table 3). The lowest SGR were observed in the groups fed the HUFA-deficient diet, especially when the dietary challenge was applied at both phases (group LH1-LH2). The hypoproteic diet with high starch supply resulted in mild SGR.

At the end of the experiment, the group fed HUFA-deficient diet at both periods was significantly less resistant to hypoxia than those with normal HUFA dietary supply. The two groups that were fed unbalanced diets only during the last two months presented also a significant

Table 3 Specific Growth Rate (SGR) and final Hypoxia Resistance Time (HRT) of sea bass

Experimental group	SGR	HRT (h)
LH1-LH2	$0.94^d \pm 0.03$	$6.63^c \pm 0.08$
C1-LH2	$1.05^c \pm 0.03$	$6.78^{bc} \pm 0.07$
C1-C2	$1.29^a \pm 0.02$	$6.96^{ab} \pm 0.07$
C1-HG2	$1.16^b \pm 0.03$	$7.20^a \pm 0.07$
HG1-HG2	$1.14^b \pm 0.02$	$7.03^{ab} \pm 0.07$

The individual specific growth rate (SGR) was computed during the last 3 weeks of the dietary challenge. The means ($\pm$ SE) without common superscript letter on the same column corresponded to significant differences according to the post-hoc pairwise comparisons after ANOVA ($p \leq 0.001$, Tukey's test) and Kruskal-Wallis test ($p \leq 0.001$, Dunn's test) for SGR and HRT, respectively. Each group was named after its diets during the two challenge phases: HUFA-deficient diet at both phases (LH1-LH2), or only at phase 2 (C1-LH2); hypoproteic diet with high starch supply at both phases (HG1-HG2), or only at phase 2 (C1-HG2); control with standard diets (C1-C2)

difference in mean Hypoxia Resistance Time (HRT), while the other pairwise comparisons did not indicate further significant differences (Table 3).

Correlation between gut microbiota composition and host's growth or resistance to hypoxia

Due to some missing data, the complete correlation analysis could be performed only with 51 fish [see Additional file 10]. The individual data of SGR and HRT were put in front of the two indices used to describe alpha-diversity in gut microbiota. The means of OTU richness and dominance were not significantly different between experimental groups (Table 1), but disregarding the groups, the individual scores appeared dependent on the physiological traits. The Partial Least Squares (PLS) analysis showed that SGR correlated strongly with OTU richness, whereas HRT correlated with dominance (Fig. 2a). The two physiological traits correlated also with the relative prevalence of two distinct sets of OTUs (Fig. 2b). SGR correlated with the prevalence of *Herbaspirillum* sp. OTU_10 and *Pseudomonas* sp. OTU_3 (correlation scores around 0.8 and 0.6, respectively), while HRT correlated positively with *Vibrio* sp. OTU_7 and *Paracoccus* sp. OTU_4, but negatively with *Rhizobium* sp. OTU_9 (correlation scores around 0.8 and 0.6, and −0.8, respectively). The correlation between SGR and *Pseudomonas* sp. OTU_3 could be partly explained by the higher prevalence of the bacterium in the fast-growing control group, but the four other correlations seemed independent from the experimental conditions.

Discussion

There is growing evidence indicating the impact of dietary components on fish gut microbiota, which seems essential for host health and well-being [4]. However, apparent contradiction may arise between some observations obtained under different conditions. The present experiment attempted to evaluate the short and long-term influence

of two kinds of nutritional deficiencies on the bacterial community associated with intestinal mucosa in sea bass individuals. When animals are subjected to nutritional stress, not all individuals may react in a similar way, partly depending on intestinal microbiota [19]. A common garden approach was used to test the effects of possible social interaction and inter-individual contamination.

High-throughput sequencing has notably changed the insight on intestinal microbiota in fish, including European sea bass, which has been recently studied with such methods. Carda-Diéguez et al. [20] identified around 78 bacterial families in autochthonous intestinal microbiota of sea bass from genomic DNA, which was in the same range as the 90 families detected in the present dataset obtained from cDNA. However, the phylogenetic profiles were quite different with large proportions of Bacteroidetes, Tenericutes, and Firmicutes in the fresh samples analysed in the previous study. Surprisingly, after storage at −80 °C, Carda-Diéguez et al. [20] reported also a shift of the bacterial profile towards an overwhelming dominance of Proteobacteria. The main genera were not the same as in the present dataset, also dominated by Proteobacteria, but where *Ralstonia* sp. and *Methylobacterium* sp. accounted only for 0.05 and 0.02 % of total reads, respectively, and where *Bradyrhizobium* sp. was not even detected. A large drift of bacterial profiles during cold storage was unlikely in the present study, as the samples were immediately soaked in RNAlater before deep-freezing. Another potential source of bias may be due to resorting to nested PCR for mucosal samples with low concentration of bacterial 16S rRNA. Yu et al. [21] expressed reservation about the detection of rare OTUs after using nested PCR. The present analysis focused on the most active OTUs, which were likely the most susceptible to dietary influence. For this reason, the dataset was limited to the 1111 OTUs that represented at least 0.0002 % of total reads. The most prevalent OTU was the only taxon shared by every individual, but this did not necessarily contradict the core theory applied to fish intestinal microbiota [10], as the rare or weakly active bacteria could not be detected with the present method.

A striking feature of the present data was the high variability of the bacterial profiles among individuals reared in the same conditions. This may also happen in wild fish living in the same environment. Star et al. [22] displayed heterogeneous pyrosequencing profiles among the intestinal contents of 11 specimens of Atlantic cod, which were caught in one location, and then kept in a common tank for seven to twelve days of fasting. These results confirmed that gut microbiota composition may keep original features in fish individuals, even after mix in the same environment, in the absence of feed supply, which could blur the analysis done from intestinal contents. Some individual variation was already observed in the faecal microbiota from European sea bass, but the

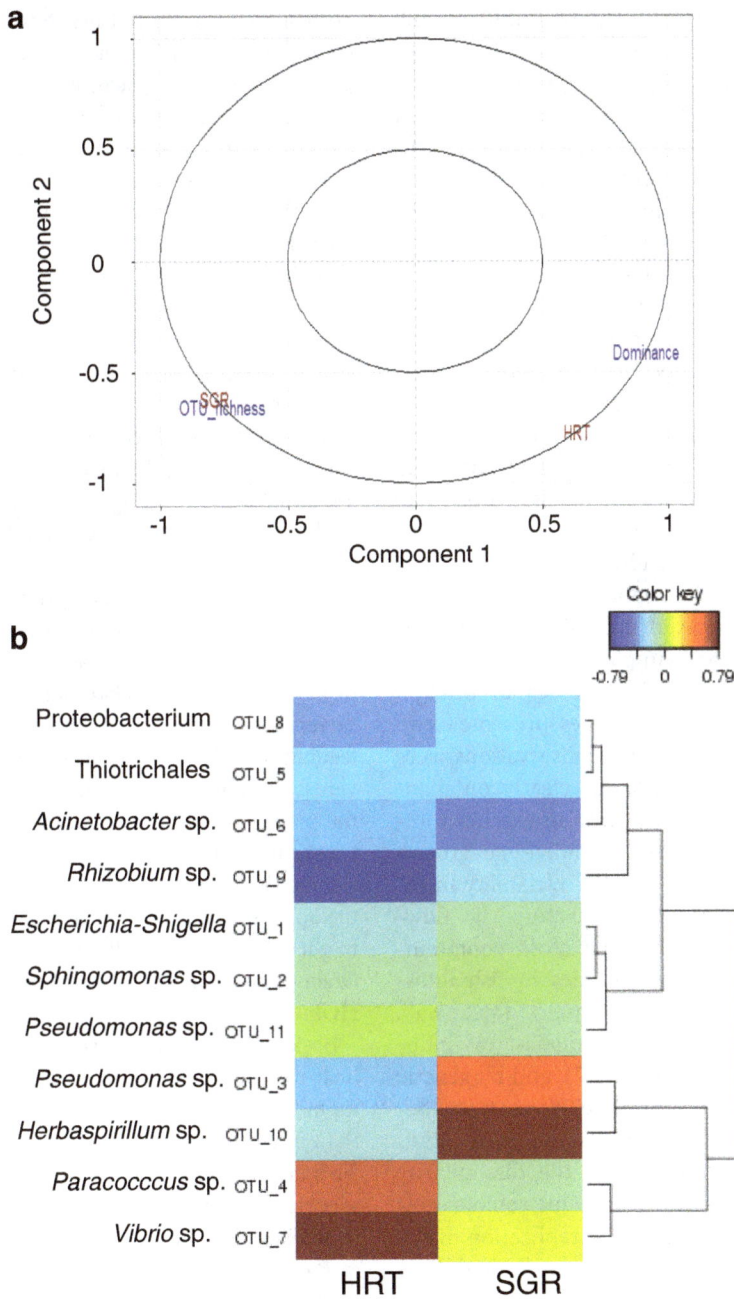

Fig. 2 Correlations between physiological traits and characteristics of gut microbiota, whatever the rearing history. *Legend:* **a** Partial Least Square (PLS) canonical correlation between SGR and HRT with alpha-diversity indices in gut microbiota (OTU richness and dominance). The projection vectors on the circle plot reached the maximum score for the four variables, which strongly correlated by pairs. Both pairs of vectors were almost perpendicular, showing the independence of the two variables in each matrix. **b** Heatmap of the sparse PLS canonical correlation between SGR and HRT with the 11 most abundant OTUs in gut microbiota, each one representing more than 0.5 % of total reads. The analysis was based on three components, taking into account all of the 11 abundance scores and the two physiological traits. The *red-brown* and *deep blue* colours indicate strong positive and negative correlations, respectively (correlation coefficient between the two variables ranging from 0.6 to 0.79 in absolute value, see the colour key scale)

intestinal bacterial profiles of fish confined in the same tank were much more similar than those reared in replicate tanks [8]. The study was based on the analysis of faeces that were emitted during the overnight isolation of individuals in separate aquaria, which led to conclude to a close intra-tank similarity between individual samples. After release, faecal microbiota evolved under the influence of nutrients and aquatic environment. Faeces collected after some delay cannot accurately reflect the intimate interaction between host intestinal mucosa and associated microbiota. Before comparing different microbial datasets, it is essential to distinguish the nature of the samples, which may come from faeces, digestive tract contents, or intestinal mucosa. The transient bacterial populations in the intestinal content ("allochthonous") are known to differ from the community adhering to mucus in fish intestine ("autochthonous") [23–25]. In the autochthonous community of sea bass intestine, Carda-Diéguez et al. [20] noted a higher diversity than in the samples of intestinal content, which appeared less reliable and less representative of the complexity of the bacterial consortium living in fish intestine. This later result confirmed the interest to focus on mucosal samples when studying the interaction between gut microbiota and the host, whereas the analysis of intestinal contents would be more dependent on the interaction with the diet.

Some relationship between individual characteristics and intestinal microbiota were reported in the literature. For example, the distribution of cultivable bacteria differed between the intestinal samples of dominant and subordinate Arctic charr [9]. Two classes of orange-spotted with slow and fast growth rates presented two distinct bacterial profiles [26]. Similar observations were done on Atlantic cod larvae, and less clearly on mangrove killifish larvae [3]. Sea bass can also exhibit dominant behaviour, leading to different individual growth rates, which may account partly for the variability in intestinal microbiota, besides genetic factors. In flow-through water systems, there may be cross contamination between the bacterial communities in fish intestine, fish skin, and those adhering to tank wall. Microbiota associated with cutaneous mucus was influenced by the diet in Atlantic salmon [27], and by genetic background in brook charr [28]. The interactions between fish individuals and bacterial communities within each tank may therefore cause partly the dissimilarity that was observed between intestinal microbiota collected from different tanks. The lack of replication of the experimental treatments in several tanks limited thus the conclusiveness about the effects of the diets in the present study. Some mean proportions of OTUs in the control group were significantly different from others in groups treated with a deficient diet, but it cannot be excluded that these differences might be due to tank cohabitation, rather than actually due to the diet (e.g., *Pseudomonas* sp. OTU_3).

The maintenance of discernible characteristics in gut microbiota composition after three months of common garden in fish individuals with two different nutritional histories was an innovative finding. The evidence came from the tank subjected to HUFA restriction during the final two months of dietary challenge, which was applied to fish either reared previously in standard conditions, or already deprived from normal HUFA supply at first feeding. Ringø et al. [29] "suggested that dietary fatty acids affect the attachment sites for the gastrointestinal microbiota, possibly by modifying the fatty acid composition of the intestine wall". Dietary HUFA are known to influence the immune status of sea bass [30], and that might also impact the association of bacteria to intestinal mucosa. Bacterial colonization of marine fish larvae may be affected by dietary fatty acids [31], but the hypothetical long-term effect of such initial colonization remains to be investigated in fish. However, the restriction of dietary HUFA at start feeding compromised durably the growth potential of sea bass, as indicated by the lower SGR observed in the individuals already challenged during the larval stages, compared to those submitted to similar deficiency only during the final period. The long-term effect of the initial deficiency might also undermine the development of the immune system and the cell-wall defences in the intestine, which could explain why gut microbiota remained dissimilar between the two groups in the same tank. The most visible differences between these two groups lied in the highest individual variability and in the lowest average proportion of *Escherichia-Shigella* OTU_1 that were observed in the group transferred from the control tank for feeding the HUFA-deficient diet in phase 2. The high individual variability can be perceived through the diversity of the prevalent OTUs that competed with OTU_1 in the latter group, though their average proportions did not result in significant differences between groups. In the other groups, the apparent similarity of the relative proportion of OTU_1 might mask possible differences in bacterial load, especially when comparing the control group to those submitted to HUFA deficiency.

Besides HUFA, other feed components could affect the host-microbe interaction. In particular, the proportion of lupin meal was much higher in the HUFA-deficient diet than in the two others diets of the final challenge (Table 4). This alternative protein source is rich in non-starch polysaccharides, and it influenced gut microbiota composition in sea bass [16]. It might induce the relatively high activity of *Bacillus* sp. in some samples, as these OTUs were close to strain DFEL3.4, previously isolated from the stomach of gilthead sea bream fed lupin meal in the same laboratory [32] [see Additional file 11].

The hypoproteic diet with high starch supply had little effect on gut microbiota composition. This was different from the results obtained previously in sea bass fed isoproteic diets with either lupin meal, or starch, or cellulose as carbohydrate source [16]. The impact of the diet on gut microbiota was likely moderated in the present experiment by introducing lupin meal in every diet during the final challenge.

Beyond the experimental treatments, the correlative study allowed to highlight the interaction between gut microbiota composition and individual characteristics of the host,

Table 4 Composition of the experimental diets

Ingredients (g kg⁻¹) (Dry matter basis)	Larval conditioning diets (phase 1)			Nutritional challenging diets (phase 2)		
	LH1	C1	HG1	LH2	C2	HG2
Fish meal	70	520	300	200	350	290
Defatted fish meal	400	–	–	–	–	–
Fish soluble	150	150	150	80	150	90
Soft white lupin	–	–	–	520	300	240
Marine phospholipids	–	–	20	–	–	–
Fish oil	–	–	–	–	–	10
Rapeseed oil	–	–	–	80	80	80
Soybean oil	40	–	–	–	–	–
Rapeseed lecithin				30	30	30
Soy lecithin	200	200	200			
Starch	–	–	210	70	70	240
Vitamin mix	80	80	80	10	10	10
Mineral mix	40	40	40	10	10	10
Cellulose	20	10	–	–	–	–
Proximate composition (%, dry matter basis)						
Crude protein (N × 6.25)	43.5	46.7	32.0	44.8	53.0	40.3
Crude lipids	28.4	23.1	22.1	15.0	13.7	13.2

probably depending on genetic background and/or traits of behaviour. Such relationship should be carefully interpreted, as the correlation network is likely intricate. However it seems possible to associate physiological traits such as growth potential and hypoxia resistance to some characteristics of gut microbiota in fish, as already attempted between quantitative trait loci and specific bacterial strains associated with the skin of brook charr [28]. Using DGGE, Forberg et al. [3] noted higher band richness in large killifish larvae compared to small individuals, but Shannon index and evenness were not significantly different, whereas the reverse observations were done with Atlantic cod larvae. Pyrosequencing allowed much deeper insight into bacterial diversity than DGGE, and the dominance index appeared clearly independent from OTU richness, which strongly correlated with SGR in sea bass. That suggested that the gut microbiota of fast-growing individuals might be more flexible, due to the presence of numerous taxa, which were not necessarily prevalent, but which could be activated in response to environmental changes, possibly benefiting to the host. The relationship between fish growth and gut microbiota may however depend on rearing conditions, especially on those aimed at managing the intestinal community. A synbiotic treatment with a probiotic strain of *Lactococcus lactis* and oligosaccharides increased growth in Siberian sturgeon, while decreasing gut microbial richness and Shannon index [33]. More surprising was the strong correlation between the dominance index in gut microbiota and the host's capacity to resist hypoxia. It is admitted in fish as in other vertebrates that the resistance to hypoxia depends on several

physiological traits including strong capacity for metabolic depression and high energy reserves to fuel anaerobic metabolism [34, 35]. A longer exposure to hypoxia might favour the relative prevalence of *Vibrio* sp. OTU_7 in some resistant individuals, possibly by stimulating specific metabolic reactions as in *Vibrio cholerae* [36–38]. However, the fish were exposed to hypoxia for 6–8 h, and this range of variation seemed rather short to affect differentially intestinal microbiota. The diet might interfere and, as contrary to lupin meal, dietary starch stimulated *Vibrio* spp. in autochthonous gut microbiota of sea bass [16]. In the present experiment, the HUFA-deficient diet, which depressed HRT, was rich in lupin meal with limited starch supply. Circumspection was nevertheless required to interpret partial least square correlation between HRT or SGR and the most prevalent OTUs, which were highly variable among individuals. For example, the negative correlation between HRT and *Rhizobium* sp. OTU_9 seemed mainly due to one individual, which showed the lowest HRT (6.04 h) and the highest percentage of OTU_9 (25.4 % total reads), whereas this OTU was detected only in seven individuals.

Conclusion

The main lesson is that active microbiota associated with intestinal mucosa may considerably vary among sea bass individuals, and large samples collected in several replicate tanks will be necessary to attempt at understanding the possible roles of gut microbiome. The severe HUFA restriction highlighted the interference between fish phenotype and gut microbiota, showing that its variability is not merely stochastic, but linked to life history or genetic background. This may suggest to investigate whether microbiotypes could be delineated in fish as in mammals [39], likely in terms of bacterial functions, rather than phylotypes.

Methods
Fish and rearing protocols
Newly-hatched sea bass larvae were provided by Aquastream (Ploemeur, France), allotted at 2 dph (day post hatch) in 15 tanks of 35 L, and then reared in the general conditions described elsewhere [40]. The 15 tanks were divided in three groups of five tanks, which were fed the compound diets from 7 dph onwards (Fig. 3). Two unconventional diets, LH1 and HG1, were tested in comparison with a control diet C1 (Table 4). Diet LH1 consisted in low HUFA content by using defatted fish meal and soybean oil (less than 0.3 % EPA + DHA, dry matter basis). This diet was administered for the 22 first days of feeding, till 28 dph. Starch replaced 40 % of fish meal in the hypoproteic and hyperglucidic diet HG1, which was used from 7 to 22 dph, and then the larvae of this group were fed diet C1 till 28 dph. At 29dph, all the larvae were fed *Artemia* nauplii, with a view to compensate for the growth deficit caused by the deficient diets. At 35 dph, the larvae of the control group, fed diet C1 from 7 to 28 dph, were big enough to be grouped together in one 450-L tank (10,014 fish in total from the five replicated tanks), and then progressively weaned onto standard diet from 35 to 52 dph. Totals of 3662 and 6269 larvae previously fed LH1 and HG1, respectively, were transferred to two other 450-L tanks at 44

dph, and then progressively weaned onto standard diet till 52 dph. The water temperature was maintained at 20 °C during all the rearing period, in an open system without recirculation. At 191 dph, as the juveniles grew up, 500 individuals were randomly selected from each of the three dietary groups C1, LH1 and HG1 (individual mean weight of 16.6, 16.7, and 15.4 g, respectively). At 202 dph, a PIT-tag (PIT: passive integrated transponder) was subcutaneously implanted in every fish that was selected for the second phase of the experiment. Five groups of 100 tagged individuals were named after the diets fed during the two experimental phases (Table 4). The groups were formed as follows: three lots were randomly selected from the initial group fed diet C1, one still fed control diet (final group C1-C2), while the two other groups were mixed either with fish from group LH1 or HG1 for the common garden test. At 227 dph, the second phase of nutritional challenge started by feeding either low HUFA diet LH2 or high-starch/low-protein diet HG2 in each of the two tanks where two groups cohabited, namely, groups C1-LH2 and LH1-LH2 on the one hand, and groups C1-HG2 and HG1-HG2 on the other hand. HUFA were restricted in diet LH2 by replacing 65 % of the protein sources by lupin meal (c. 0.5 % HUFA, dry matter basis), while 30 % of the protein sources were replaced by starch in diet HG2. The fish were fed these experimental diets for two months, until the end of the experiment. They were individually weighed at 225, 266 and 287 dph after light anaesthesia (2-phenoxyethanol, 200 µL L^{-1}). The individual specific growth rate (SGR) was computed between the last two weighing times (266 and 287 dph).

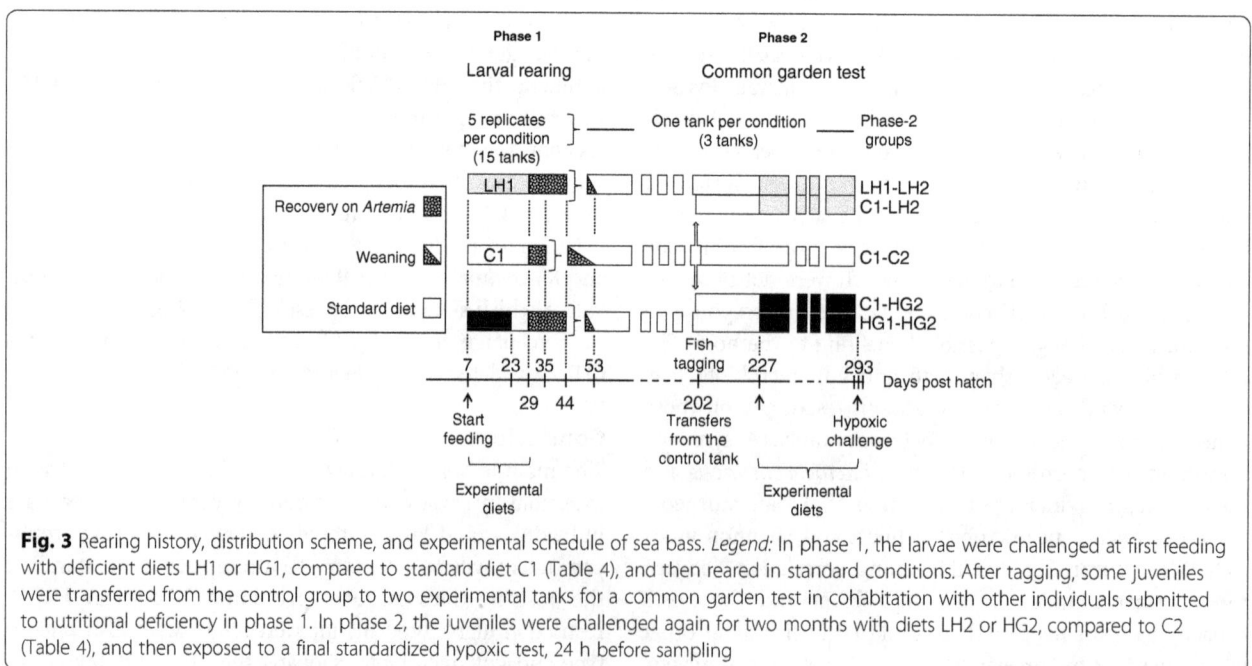

Fig. 3 Rearing history, distribution scheme, and experimental schedule of sea bass. *Legend:* In phase 1, the larvae were challenged at first feeding with deficient diets LH1 or HG1, compared to standard diet C1 (Table 4), and then reared in standard conditions. After tagging, some juveniles were transferred from the control group to two experimental tanks for a common garden test in cohabitation with other individuals submitted to nutritional deficiency in phase 1. In phase 2, the juveniles were challenged again for two months with diets LH2 or HG2, compared to C2 (Table 4), and then exposed to a final standardized hypoxic test, 24 h before sampling

Final hypoxic challenge test

At 292 dph, the fish were grouped in one tank, and fasted for one day before being challenged for resistance to hypoxia. The method of this challenge test was previously described [41, 42]. Briefly, the rearing water was deprived of oxygen by bubbling nitrogen gas in the tank, under constant monitoring of dissolved O_2. The oxygenation level was first dropped by 90 % from saturation within 1 h, and then further decreased by 1.2 % per hour, until reaching a minimum of c. 4 % air saturation. In the meantime, the fish were constantly observed, and each individual losing its maintenance of equilibrium was identified by reading its PIT-tag, and immediately placed in a fully aerated tank for recovery. The challenge lasted c. 8 h in total, and the interval of time from start to equilibrium lost was noted for every fish as its hypoxia resistance time (HRT).

Sampling for microbiological analysis

In each group, the fish were sorted based on HRT, and 12 individuals were selected per group for microbiological sampling (60 fish in total). Four individuals were selected among the most sensitive to hypoxia in each group, four others were among the mildly sensitive, and the four last were among the most resistant. As there were significant differences in hypoxia sensitivity between groups, the selection grid was not the same in every group [see Additional file 10].

At 294 dph, after two days of fasting, the 60 selected fish were euthanized with an overdose of anaesthetic (2-phenoxyethanol, 1 mL L^{-1}). The intestines were empty, and each one was dissected under sterile conditions, and separated from the perivisceral fat in a Petri dish on ice. The intestine was immediately plunged into a microtube with 1.5 mL RNAlater (Qiagen). After soaking for 24 h at c. 4 °C, the tubes were stored at −80 °C.

RNA extraction, RT-PCR, and pyrosequencing

The microbial profiles compared in the present experiment were based on the analysis of 16S rRNA after reverse transcription, which was preferred to the method based on genomic DNA. As explained previously [16], this is a way to focus on the relative ribosomal activity among bacteria, whereas the relative abundance of rDNA cannot provide information about bacterial activity, and the result is biased by the variable number of gene copies among species.

After thawing, the intestine was removed from RNAlater with sterile tweezers, cut open longitudinally along the entire length in a Petri dish, and plunged into Extract-All (Eurobio) chilled on ice. RNA was extracted according to the instructions of the manufacturer for biological tissues, with an additional step of bead-beating for 10 min after the initial step of homogenization with a dispersing aggregate unit. After purification, the RNA concentration was estimated by NanoDrop (Thermo Scientific), and aliquoted for reverse transcription (RT). The surplus was precipitated and stored at −80 °C. cDNA was transcribed with the Quanti-Tect® RT kit (Qiagen), and stored at −20 °C.

Due to the small proportion of 16S rRNA in the samples, nested PCR was required before pyrosequencing. The PCR mix contained Taq DNA polymerase (0.025 U μl^{-1}; MP Biomedicals), 0.2 mM of each dNTP (deoxyribonucleotide triphosphate; Eurogentec premix), and 0.4 μM of each primer (first round: EUB-8-f, 907-r; second round, V3-V4 region: PCR1F_460 and PCR1R_460 [see Additional file 12]). After initial denaturation at 94 °C for 2 min in T100 Thermal Cycler (Bio-Rad), 20 and 25 cycles were run in the first and second round, respectively, with 30 s denaturation at 94 °C, 30 s annealing at 55 °C (first round) or 62 °C (second round), and 1 min elongation at 72 °C. Both rounds ended with 1 min extension at 72 °C. Six of the sixty samples were ruled out because of insufficient PCR yield [see Additional file 10]. The 54 other PCR products were purified with GenElute PCR Clean-Up Kit (Sigma), further prepared by GenPhySE (INRA, UMR1388, Toulouse, France), and sequenced with Illumina MiSeq at GeT-PlaGe [43]. The pyrosequencing data were deposited in the National Center for Biotechnology Information-Short Reads Archive (NCBI-SRA) under the BioProject accession number PRJNA294963 [44].

Bioinformatics data processing

The raw sequence dataset was first treated with FROGS (Find Rapidly OTU with Galaxy Solution) [45]. Briefly, after merging the paired 250 bp reads, the software denoised the dataset, which was clustered with Swarm [46]. A first range of chimera was removed with vsearch [47], and then the dataset was further filtered using PhiX and removing the singletons. A second filtration level was obtained by keeping only the clusters that represented at least 0.0002 % of total reads. After double identification with RDP and Blast + the dataset was restricted to the bacterial kingdom. A total of 1160 different sequences were thus detected among the 1,017,693 remaining reads. Between 11,599 and 28,067 valid reads were counted in each sample, and the data were normalized on the basis of 11,599 reads per sample, before computing alpha-diversity and the rarefaction curves. The normalization resulted in 1111 operational taxonomic units (OTUs). The final identification was assigned at the lowest phylogenetic level of RDP-Blast concordance, after correcting some misleading affiliations.

Statistics

Fish growth and resistance to hypoxia, the diversity indices of intestinal microbiota, and the relative abundance of OTUs were compared between experimental groups by ANOVA or Kruskal-Wallis test, depending on normality and homoscedasticity. Post-hoc tests were used for multiple comparisons between dietary groups (Tukey's and

Dunn's tests after ANOVA and Kruskal-Wallis, respectively). It must be noticed that the most relevant pairwise comparisons in the present experiment were those between the groups (1) with or without the deficient diet at first feeding and reared in the same tank during the final phase (LH1-LH2 vs. C1-LH2; HG1-HG2 vs. C1-HG2) and (2) with standard diet at first feeding, but either transferred or not from the control tank before the final phase (C1-LH2, C1-HG2, and C1-C2). The Bray-Curtis index was used for comparing the similarity between bacterial profiles by PERMANOVA with PAST [48]. The bacterial profiles were further compared between two groups by Linear Discriminant Analysis (LDA) Effective Size (LEfSe) pairwise analysis under Galaxy environment [49, 50]. The canonical correlation between intestinal microbiota and fish growth or resistance to hypoxia was analysed after (sparse) Partial Least Squares, (s)PLS, classification with mixOmics [51, 52].

Additional files

Additional file 1: Rarefaction curves of OTU richness of each sample from the experimental groups, computed after normalization to 11599 reads per sample.

Additional file 2: Venn diagram showing the distribution of the 1111 OTUs among the intestinal samples of the five experimental groups.

Additional file 3: Histogram showing the overwhelming dominance of Alpha- and Gamma-Proteobacteria among the OTUs detected in every experimental group.

Additional file 4: Histogram showing the distribution of the most dominant OTUs (more than 5 % total reads in at least one sample).

Additional file 5: Mean relative abundance of phylogenetic clusters among Gammaproteobacteria with significant differences between experimental groups.

Additional file 6: Mean relative abundance of phylogenetic clusters among Alpha- and Beta-Proteobacteria with significant differences between experimental groups.

Additional file 7: Mean relative abundance of phylogenetic clusters among Actinobacteria and Firmicutes with significant differences between experimental groups.

Additional file 8: Mean relative abundance of phylogenetic clusters among Bacteroidetes and Spirochaetae with significant differences between experimental groups.

Additional file 9: Mean weights (± SE) of sea bass before, at midterm, and after the final dietary challenge of sea bass.

Additional file 10: Hypoxia resistance time (h) of the individuals selected for the microbiological analysis.

Additional file 11: Phylogenetic tree derived by neighbour joining of the three *Bacillus* sp. OTUs from the present intestinal samples, aligned with GenBank sequences.

Additional file 12: Sequences of the primers used in the nested PCR for pyrosequencing.

Abbreviations

ANOVA: Analysis Of VAriance; Blast: Basic local alignment search tool; C: Control diet (C1 and C2, also used to name experimental groups); cDNA: complementary DNA; DGGE: Denaturing Gradient Gel Electrophoresis; DNA: Deoxyribo-Nucleic Acid; dph: Day post hatch; FROGS: Find Rapidly OTU with Galaxy Solution; HG: Hyper-Glucidic diet (HG1 and HG2, also used to name experimental groups); HRT: Hypoxia Resistance Time; HUFA: Highly poly-Unsaturated Fatty Acids; LDA: Linear Discriminant Analysis; LEfSe: Linear discriminant Effective Size; LH: Low-HUFA diet (LH1 and LH2, also used to name experimental groups); OTU: Operational Taxonomy Unit; PAST: PAleontological STatistics software package; PCR: Polymerase Chain Reaction; PERMANOVA: PERmutational Multivariate Analysis Of Variance; PLS: Partial Least Squares; RDP: Ribosomal Database Project; rRNA: ribosomal Ribo-Nucleic Acid; RT-PCR: Reverse Transcription Polymerase Chain Reaction; SE: Standard error; SGR: Specific Growth Rate; sPLS: sparse Partial Least Squares

Acknowledgements

This work was carried out with financial support from the Commission of the European Communities, specific RTD programme of Framework Programme 7 (project FP7-KBBE-2011-5, ARRAINA project no. 288925, Advanced Research Initiatives for Nutrition and Aquaculture). We are grateful to the Genotoul bioinformatics platform Toulouse Midi-Pyrénées, France, and to Signae group for providing help in pyrosequencing, computing and storage resources thanks to Galaxy instance (http://signae-workbench.toulouse.inra.fr). Special thanks are also due to Drs G. Pascal and O. Zemb, INRA, UMR1388 GenPhySE, for their precious help about pyrosequencing and bioinformatics.

Funding

This work was funded by the Commission of the European Communities, specific RTD programme of Framework Programme 7 (project FP7-KBBE-2011-5, ARRAINA project no. 288925, Advanced Research Initiatives for Nutrition and Aquaculture).

Authors' contributions

JLZ and DM conceived and designed the experiments on sea bass, and revised critically the manuscript. PQ and NLB were particularly involved in the rearing management of the larval and juvenile stages, respectively. HLD, LM, and OM were particularly involved in the biochemical analyses concerning diets and physiological data. CH was particularly involved in the statistical analyses regarding rearing and physiological data. FJG conceived and conducted the microbiological tasks, from intestinal sampling to bioinformatics and statistics, before writing the manuscript. All authors participated in data acquisition processes, and read and approved the final manuscript.

Competing interests

The authors declare that they have no competing interests.

Author details

[1]NUMEA, INRA, Univ. Pau & Pays Adour, 64310 Saint Pée sur Nivelle, France. [2]Ifremer, UMR 6539 (LEMAR), PFOM/ARN, Centre de Bretagne, CS 10070, 29280 Plouzané, France. [3]PFOM/ARN, Ifremer, Centre de Bretagne, CS 10070, 29280 Plouzané, France.

References

1. Nayak SK. Role of gastrointestinal microbiota in fish. Aquac Res. 2010;41:1553–73.
2. Navarrete P, Magne F, Araneda C, Fuentes P, Barros L, Opazo R, et al. PCR-TTGE analysis of 16S rRNA from rainbow trout (*Oncorhynchus mykiss*) gut microbiota reveals host-specific communities of active bacteria. PLoS One. 2012;7:e31335.
3. Forberg T, Sjulstad EB, Bakke I, Olsen Y, Hagiwara A, Sakakura Y, et al. Correlation between microbiota and growth in Mangrove Killifish (*Kryptolebias marmoratus*) and Atlantic cod (*Gadus morhua*). Sci Rep. 2016;6:21192.

4. Ringø E, Zhou Z, Vecino JLG, Wadsworth S, Romero J, Krogdahl Å, et al. Effect of dietary components on the gut microbiota of aquatic animals. A never-ending story? Aquacult Nutr. 2016;22:219–82.

5. Olsen R, Sundell K, Hansen T, Hemre GI, Myklebust R, Mayhew T, et al. Acute stress alters the intestinal lining of Atlantic salmon, Salmo salar L.: an electron microscopical study. Fish Physiol Biochem. 2002;26:211–21.

6. Leamaster BR, Walsh WA, Brock JA, Fujioka RS. Cold stress-induced changes in the aerobic heterotrophic gastrointestinal tract bacterial flora of red hybrid tilapia. J Fish Biol. 1997;50:770–80.

7. Federation of European Aquaculture Producers. Annual Report. 2015. p. 20. http://www.feap.info/shortcut.asp?FILE=1361. Accessed 28 Oct 2016.

8. De Schryver P, Dierckens K, Quyen QBT, Amalia R, Marzorati M, Bossier P, et al. Convergent dynamics of the juvenile European sea bass gut microbiota induced by poly-beta-hydroxybutyrate. Environ Microbiol. 2011;13:1042–51.

9. Ringø E, Olsen RE, Øverli Ø, Løvik F. Effect of dominance hierarchy formation on aerobic microbiota associated with epithelial mucosa of subordinate and dominant individuals of Arctic charr, Salvelinus alpinus (L.). Aquac Res. 1997;28:901–4.

10. Roeselers G, Mittge EK, Stephens WZ, Parichy DM, Cavanaugh CM, Guillemin K, et al. Evidence for a core gut microbiota in the zebrafish. ISME J. 2011;5:1595–608.

11. Turnbaugh PJ, Gordon JI. The core gut microbiome, energy balance and obesity. J Physiol (London). 2009;587:4153–8.

12. Sullam KE, Rubin BER, Dalton CM, Kilham SS, Flecker AS, Russell JA. Divergence across diet, time and populations rules out parallel evolution in the gut microbiomes of Trinidadian guppies. ISME J. 2015;9:1508–22.

13. Kaushik SJ, Coves D, Dutto G, Blanc D. Almost total replacement of fish meal by plant protein sources in the diet of a marine teleost, the European seabass, Dicentrarchus labrax. Aquaculture. 2004;230:391–404.

14. Ozsahinoglu I, Eroldogan T, Mumogullarinda P, Dikel S, Engin K, Yilmaz AH, et al. Partial replacement of fish oil with vegetable oils in diets for European seabass (Dicentrarchus labrax): effects on growth performance and fatty acids profile. Turk J Fish Quat Sci. 2013;13:819–25.

15. Castro C, Corraze G, Panserat S, Oliva-Teles A. Effects of fish oil replacement by a vegetable oil blend on digestibility, postprandial serum metabolite profile, lipid and glucose metabolism of European sea bass (Dicentrarchus labrax) juveniles. Aquacult Nutr. 2015;21:592–603.

16. Gatesoupe FJ, Huelvan C, Le Bayon N, Severe A, Aasen IM, Degnes KF, et al. The effects of dietary carbohydrate sources and forms on metabolic response and intestinal microbiota in sea bass juveniles, Dicentrarchus labrax. Aquaculture. 2014;422:47–53.

17. Delcroix J, Gatesoupe FJ, Desbruyères E, Huelvan C, Le Delliou H, Le Gall MM, et al. The effects of dietary marine protein hydrolysates on the development of sea bass larvae, Dicentrarchus labrax, and associated microbiota. Aquacult Nutr. 2015;21:98–104.

18. Geurden I, Mennigen J, Plagnes-Juan E, Veron V, Cerezo T, Mazurais D, et al. High or low dietary carbohydrate: protein ratios during first-feeding affect glucose metabolism and intestinal microbiota in juvenile rainbow trout. J Exp Biol. 2014;217:3396–406.

19. Tsukumo DM, Carvalho BM, Carvalho Filho MA, Saad MJA. Translational research into gut microbiota: new horizons on obesity treatment: updated 2014. Arch Endocrinol Metab. 2015;59:154–60.

20. Carda-Dieguez M, Mira A, Fouz B. Pyrosequencing survey of intestinal microbiota diversity in cultured sea bass (Dicentrarchus labrax) fed functional diets. FEMS Microbiol Ecol. 2014;87:451–9.

21. Yu G, Fadrosh D, Goedert JJ, Ravel J, Goldstein AM. Nested PCR biases in interpreting microbial community structure in 16S rRNA gene sequence datasets. PLoS One. 2015;10:e0132253.

22. Star B, Haverkamp THA, Jentoft S, Jakobsen KS. Next generation sequencing shows high variation of the intestinal microbial species composition in Atlantic cod caught at a single location. BMC Microbiol. 2013;13:248.

23. Bakke McKellep AM, Penn MH, Salas PM, Refstie S, Sperstad S, Landsverk T, et al. Effects of dietary soyabean meal, inulin and oxytetracycline on intestinal microbiota and epithelial cell stress, apoptosis and proliferation in the teleost Atlantic salmon (Salmo salar L.). Brit J Nutr. 2007;97:699–713.

24. Kim DH, Brunt J, Austin B. Microbial diversity of intestinal contents and mucus in rainbow trout (Oncorhynchus mykiss). J Appl Microbiol. 2007;102:1654–64.

25. Feng JB, Luo P, Dong JD, Hu CQ. Intestinal microbiota of mangrove red snapper (Lutjanus argentimaculatus Forsskål, 1775) reared in sea cages. Aquac Res. 2011;42:1703–13.

26. Sun YZ, Yang HL, Ling ZC, Chang JB, Ye JD. Gut microbiota of fast and slow growing grouper Epinephelus coioides. Afr J Microbiol Res. 2009;3:713–20.

27. Landeira-Dabarca A, Sieiro C, Álvarez M. Change in food ingestion induces rapid shifts in the diversity of microbiota associated with cutaneous mucus of Atlantic salmon Salmo salar. J Fish Biol. 2013;82:893–906.

28. Boutin S, Sauvage C, Bernatchez L, Audet C, Derome N. Inter individual variations of the fish skin microbiota: host genetics basis of mutualism? PLoS One. 2014;9:e102649.

29. Ringø E, Bendiksen HR, Gausen SJ, Sundsfjord A, Olsen RE. The effect of dietary fatty acids on lactic acid bacteria associated with the epithelial mucosa and from faecalia of Arctic charr, Salvelinus alpinus (L.). J Appl Microbiol. 1998;85:855–64.

30. Mourente G, Good JE, Thompson KD, Bell JG. Effects of partial substitution of dietary fish oil with blends of vegetable oils, on blood leucocyte fatty acid compositions, immune function and histology in European sea bass (Dicentrarchus labrax L.). Brit J Nutr. 2007;98:770–9.

31. Seychelles LH, Audet C, Tremblay R, Lemarchand K, Pernet F. Bacterial colonization of winter flounder Pseudopleuronectes americanus fed live feed enriched with three different commercial diets. Aquacult Nutr. 2011;17:E196–206.

32. Silva FCP, Nicoli JR, Zambonino-Infante JL, Kaushik S, Gatesoupe FJ. Influence of the diet on the microbial diversity of faecal and gastrointestinal contents in gilthead sea bream (Sparus aurata) and intestinal contents in goldfish (Carassius auratus). FEMS Microbiol Ecol. 2011;78:285–96.

33. Geraylou Z, Souffreau C, Rurangwa E, De Meester L, Courtin CM, Delcour JA, et al. Effects of dietary arabinoxylan-oligosaccharides (AXOS) and endogenous probiotics on the growth performance, non-specific immunity and gut microbiota of juvenile Siberian sturgeon (Acipenser baerii). Fish Shellfish Immunol. 2013;35:766–75.

34. Guppy M, Withers P. Metabolic depression in animals: physiological perspectives and biochemical generalizations. Biol Rev. 1999;74:1–40.

35. Bickler PE, Buck LT. Hypoxia tolerance in reptiles, amphibians, and fishes: life with variable oxygen availability. Annu Rev Physiol. 2007;69:145–70.

36. Marrero K, Sanchez A, Rodriguez-Ulloa A, Gonzalez LJ, Castellanos-Serra L, Paz-Lago D, et al. Anaerobic growth promotes synthesis of colonization factors encoded at the Vibrio pathogenicity island in Vibrio cholerae El Tor. Res Microbiol. 2009;160:48–56.

37. Fan FX, Liu Z, Jabeen N, Birdwell LD, Zhu J, Kan B. Enhanced interaction of Vibrio cholerae virulence regulators Tcpp and Toxr under oxygen-limiting conditions. Infect Immun. 2014;82:1676–82.

38. Hiremath G, Hyakutake A, Yamamoto K, Ebisawa T, Nakamura T, Nishiyama S, et al. Hypoxia-induced localization of chemotaxis-related signaling proteins in Vibrio cholerae. Mol Microbiol. 2015;95:780–90.

39. Waldram A, Holmes E, Wang Y, Rantalainen M, Wilson ID, Tuohy KM, et al. Top-down systems biology modeling of host metabotype-microbiome associations in obese rodents. J Proteome Res. 2009;8:2361–75.

40. Lamari F, Castex M, Larcher T, Ledevin M, Mazurais D, Bakhrouf A, et al. Comparison of the effects of the dietary addition of two lactic acid bacteria on the development and conformation of sea bass larvae, Dicentrarchus labrax, and the influence on associated microbiota. Aquaculture. 2013;376:137–45.

41. Roze T, Christen F, Amerand A, Claireaux G. Trade-off between thermal sensitivity, hypoxia tolerance and growth in fish. J Therm Biol. 2013;38:98–106.

42. Vanderplancke G, Claireaux G, Quazuguel P, Madec L, Ferraresso S, Sévère A, et al. Hypoxic episode during the larval period has long-term effects on European sea bass juveniles (Dicentrarchus labrax). Mar Biol. 2015;162:367–76.

43. Genotoul (Toulouse, France). Genome & transcriptome core facilities. http://get.genotoul.fr/fileadmin/user_upload/Presentation/poster_GenoToul_2012_english_combined.pdf. Accessed 28 Oct 2016.

44. fish gut metagenome. Intestinal microbiota of sea bass juveniles Metagenome. Long-term effects of early nutritional stimuli on growth, resistance to hypoxia, and intestinal microbiota in European sea bass, Dicentrarchus labrax, juveniles fed unconventional diets. Accession: PRJNA294963, ID: 294963. https://www.ncbi.nlm.nih.gov/bioproject/?term=PRJNA294963. Accessed 28 Oct 2016.

45. Escudie F, Auer L, Cauquil L, Vidal K, Maman S, Mariadassou M, et al. FROGS: Find Rapidly OTU with Galaxy Solution. JOBIM Conf, 6–9 July 2015, Clermont-Ferrand, France. http://bioinfo.genotoul.fr/fileadmin/user_upload/FROGS_poster_Jobim_2015.pdf. Accessed 28 Oct 2016.

46. Mahé F, Rognes T, Quince C, de Vargas C, Dunthorn M. Swarm: robust and fast clustering method for amplicon-based studies. PeerJ. 2014;2:e593.

47. Rognes T, Mahé F, xflouris. vsearch: VSEARCH version 1.0.16. http://zenodo.org/record/15524. Accessed 28 Oct 2016.

48. Hammer Ø, Harper DAT, Ryan PD. PAST: paleontological statistics software package for education and data analysis. Palaeontol Electron. 2001;4. http://palaeo-electronica.org/2001_1/past/issue1_01.htm. Accessed 28 Oct 2016.

49. Segata N, Izard J, Walron L, Gevers D, Miropolsky L, Garrett W, et al. Metagenomic biomarker discovery and explanation. Genome Biol. 2011;12:R60.

50. Galaxy/Hutlab. http://huttenhower.sph.harvard.edu/galaxy/. Accessed 28 Oct 2016.

51. Lê Cao KA, Martin PGP, Robert-Granié C, Besse P. Sparse canonical methods for biological data integration: application to a cross-platform study. BMC Bioinformatics. 2009;10:34.

52. Omics Data Integration Project. mixOmics. http://mixomics.org/. Accessed 28 Oct 2016.

15

Addition of plant-growth-promoting *Bacillus subtilis* PTS-394 on tomato rhizosphere has no durable impact on composition of root microbiome

Junqing Qiao[1], Xiang Yu[2], Xuejie Liang[1], Yongfeng Liu[1], Rainer Borriss[3,4] and Youzhou Liu[1*]

Abstract

Background: Representatives of the genus *Bacillus* are increasingly used in agriculture to promote plant growth and to protect against plant pathogens. Unfortunately, hitherto the impact of *Bacillus* inoculants on the indigenous plant microbiota has been investigated exclusively for the species *Bacillus amyloliquefaciens* and was limited to prokaryotes, whilst eukaryotic member of this community, e.g. fungi, were not considered.

Results: The root-colonizing *Bacillus subtilis* PTS-394 supported growth of tomato plants and suppressed soil-borne diseases. Roche 454 pyrosequencing revealed that PTS-394 has only a transient impact on the microbiota community of the tomato rhizosphere. The impact on eukaryota could last up to 14 days, while that on bacterial communities lasted for 3 days only.

Conclusions: Ecological adaptation and microbial community-preserving capacity are important criteria when assessing suitability of bio-inoculants for commercial development. As shown here, *B. subtilis* PTS-394 is acting as an environmentally compatible plant protective agent without permanent effects on rhizosphere microbial community.

Keywords: *Bacillus subtilis* PTS-394, Phylogenomics analysis, Root colonization, Rhizosphere Microbiota community, Roche 454 Pyrosequencing

Background

Plant growth-promoting rhizobacteria (PGPR) have potential as biocontrol agents that could replace chemical pesticides thereby reducing undesired chemical remnants in agriculture. Plant growth promotion, suppression of plant diseases and rhizosphere competence (colonization and survival on plant roots) by PGPR, especially bacilli, have been considered as critical requisites for the development of commercial products [1, 2]. To be effective, PGPRs must establish and maintain a sufficient population in the rhizosphere [3]. It is reported that the robust colonization of plant roots by beneficial microbes directly contributes to the effective biocontrol of soil-borne pathogens [4]. Root colonization and rhizosphere competence is thus a critical prerequisite for the successful use of PGPRs as biocontrol and plant growth-promoting agents [5, 6].

The rhizosphere is rich in nutrients due to the accumulation of plant exudates containing amino acids and sugars that provides a rich source of energy and nutrients for colonizing bacteria [7]. This forces the bacterial community to colonize the rhizoplane and rhizosphere [8]. The colonizing capability of PGPR strains has been investigated using GFP labeling and transposon-mediated random mutagenesis approaches, in *Pseudomonas* and *Bacillus* strains, in particular [9, 10]. Bacterial traits such as chemotaxis, motility, attachment, growth, and stress resistance all appear to contribute to the colonization competence of PGPRs [8].

Rhizosphere competence is essentially a process of niche competition between PGPRs and other microbes present in the vicinity of plant roots, in which resource partitioning, competitive exclusion, and co-existence can all play a part. It is reported that PGPRs colonize more efficient in

* Correspondence: shitouren88888@163.com
[1]Institute of Plant Protection, Jiangsu Academy of Agricultural Sciences, Nanjing, Jiangsu province 210014, China
Full list of author information is available at the end of the article

poorer microbial communities than in richer soils [11]. The indigenous rhizosphere microbial community can be influenced by large-scale application of PGPRs in field trials [12]. *Pseudomonas* sp. DSMZ 13134 affects the dominant bacterial community of barley roots in the tested soil system [13], for example. By contrast, as revealed by T-RFLP community fingerprinting, application of *Bacillus amyloliquefaciens* FZB42 did not change the composition of rhizosphere bacterial community in a measurable extent [2]. Similar results were also found for *B. amyloliquefaciens* BNM122 in soybean [14]. Recently, persistence and the effect of FZB42 on the microbial community of lettuce were more deeply analyzed by 454-amplicon sequencing corroborating that inoculation with *B. amyloliquefaciens* has no or only transient minor effects on the microbiota in vicinity of plant roots [15]. A slow decrease in the number of inoculated bacteria was also registered. After five weeks only 55% of the initial number of FZB42 DNA was still traceable within the rhizosphere of lettuce in the field [16]. Unfortunately, till now, the effect of Bacillus PGPR on the non-bacterial members of indigenous plant microbiota has not been analyzed. Moreover, no other representatives of the *B. subtilis* species complex than *B. amyloliquefaciens*, the main source for commercial biofertilizer and biocontrol agents, have been proven for their impact on plant rhizosphere.

Bacillus subtilis PTS-394, isolated from the rhizosphere of tomato and fully sequenced in 2014 [17], has been shown to suppress tomato soil-borne diseases caused by *Fusarium oxysporum* and *Ralstonia solanacearum* which is the main obstacle to continuous cropping of tomato *in* greenhouse cultivation [18]. In the present study, the colonization behavior and plant growth-promoting capability of PTS-394 were investigated in laboratory and greenhouse experiments. Additionally, the effect of PTS-394 on the rhizosphere microbial community was studied using Roche 454 pyrosequencing of 16S rRNA and ITS partial sequences. The aim was to improve our understanding of the ecological consequences of microbial inoculants and of rhizosphere microbiota ecology.

Methods
Strains and growth conditions
Bacillus subtilis PTS-394 and the GFP-tagged strain of *B. subtilis* PTS-394G containing the plasmid pGFP22 were isolated and constructed by our laboratory [19]. The *Bacillus* strains were grown in YPG medium containing 0.5% yeast extract, 0.5% peptone and 0.5% glucose. 5 μg/ml of chloramphenicol was added when necessary for *B. subtilis* PTS-394G.

The software of Phylogenomic analysis
Phylogenomic analysis of strain PTS-394 was performed taking advantage of availability of the whole genome

sequence [17]. JSpeciesWS (http://jspecies.ribohost.com/jspeciesws/) was used to determine ANIb (average nucleotide identity based on BLAST+) and ANIm (average nucleotide identity based on MUMmer) values by pairwise genome comparisons. Correlation indexes of their Tetra-nucleotide signatures (TETRA) were determined by using the JSpeciesWS software (Richter et al., 2015). Electronic DNA-DNA hybridization (dDDH) is useful to mimic the wet-lab DDH and can be used for genome-based species delineation and genome-based subspecies delineation [20, 21] and was applied to finally defined PTS-394 whether belong to *B. subtilis* subsp. *subtilis*. Three formulas are applicable for the calculation: Formula: 1 (HSP length / total length), formula: 2 (identities / HSP length) and formula 3 (identities / total length). Formula 2, which is especially appropriate to analyze draft genomes, was used.

The colonization and plant growth promotion of PTS-394 on agar plates
B. subtilis PTS-394 was shaken at 180 rpm at 28 °C for 36 h, centrifuged at 5000 rpm for 10 min, then the cell pellets were suspended in sterile distilled water and adjusted to an OD_{600} of 0.01, 0.1, 1.0, 1.5, and 3.0 (OD_{600} = 1.0, ~5 × 10^7 CFU/mL). Tomato (tomato cultivar 'Moneymaker') seeds were surface-sterilized using sodium hypochlorite (3%, v/v) for 5 min, washed three times with sterile water and soaked in the above cell suspensions at 25 °C for 24 h. Seeds were sown into individual 9 cm diameter MS agar plates [22] at a density of five seeds per plate. Each treatment was performed on 10 plants and was replicated three times. Plates were incubated in a growth chamber with a daytime temperature of 25 °C and a night temperature of 18 °C, 75% humidity and a 12:12 h light: dark cycle with a light intensity of 8000 lux. At 14 days after sowing, plant fresh weight and the amount of PTS-394 on the root surface were measured as previously described [23].

The colonization and plant growth promotion of PTS-394 in pots
The dynamic colonization of PTS-394 on the tomato rhizoplane and its plant growth-promoting ability were investigated by pot experiments, and the GFP-labeled strain (PTS-394G) was used for colonization analysis. Suspensions of *B. subtilis* PTS-394 and PTS-394G with an OD_{600} = 1.0 were used. In both experiments, tomato seeds were surface-disinfected and sown into nursery soil containing a mixture of vermiculite and organic manure (1:1, w/w) for germination. Tomato seedlings at the 4-leaf stage were transplanted into pots filled with a mixture of vermiculite, rice field soil and organic manure (1:2:1, w/w). 20 mL of bacterial cell suspension was added to each plant and cultivation was continued in a

greenhouse under natural conditions with temperatures ranging from 18 to 30 °C. Each treatment was performed on 30 plants and replicated three times in a completely randomized block design. In the colonization experiment, tomato root samples were collected from three PTS-394G plants at 0, 1, 2, 3, 5, 7, 9, 11, 14, and 21 days after transplantation and the amount of PTS-394 on the root surface was determined, as described previously. Meanwhile, tomato roots were examined at 1, 3, 7 and 14 days using fluorescent microscopy in order to detect colonization with PTS-394G. At 30 days after treatment with PTS-394, the fresh weight of the root and the plant height were measured in order to evaluate the growth-promoting effect of PTS-394.

Rhizosphere soil collection and DNA extraction
During the pot experiment described above, rhizosphere soil from untreated and PTS-394-treated pot-grown tomato plants was collected from three plants at 1, 3, 7, 9 and 14 days after treatment. Rhizosphere soil total DNA

was extracted using the PowerSoil DNA Isolation Kit (MO BIO Laboratories, Inc., Carlsbad, CA, USA) and the quality was estimated using a NanoDrop spectrophotometer (ND-1000, NanoDrop Technologies, Wilmington, USA).

PCR amplification and Roche 454 pyrosequencing
For analysis of bacteria diversity, partial 16S rRNA genes were amplified using primers 27F (AGAGTTTGATCMT GGCTCAG) and 533R (TTACCGCGGCTGCTGGCAC). Primer pair ITS1 (TCCGTAGGTGAACCTGCGG) and ITS4 (TCCTCCGCTTATTGATATGC) were was used to amplify partial ITS sequences of eukaryotes. Amplifications were performed using the following program: initial denaturation at 95 °C for 2 min, 25 (for 16S rRNA) or 33 (for ITS) cycles of denaturation at 94 °C for 30 s, annealing at 55 °C for 30 s, extension at 72 °C for 30 s, and a final extension at 72 °C for 5 min. Emulsion PCR was performed with the emPCRAmp-Lib L Kit (Roche) and PCR amplicons were pyrosequenced by Shanghai Majorbio Bio-pharm Technology Co. Ltd. (Shanghai, China) using a

Table 1 Taxonomic relationship of *B. subtilis* PTS-394 based on whole genome analysis

query	accession	Z-score	ANIb	ANIm	dDDH (formula 2)	Probability >70%	>79%	G + C
Bacillus subtilis PTS-394	AWXG00000000							
reference								
Bacillus subtilis subtilis ASM74047v1	JPVW01000000	**0.99976**	**98.56**	**98.85**	**90.10% [87.8–92.1%]**	**95.85%**	**65.49%**	0.00%
Bacillus subtilis QH-1	AZQS00000000	**0.99976**	**98.59**	**98.95**	**91.10% [88.9–92.9%]**	**96.15%**	**66.94%**	0.02%
Bacillus subtilis subtilis 168(T)	NC_000964.3	**0.99929**	**98.46**	**98.85**	**90.10% [87.8–92%]**	**95.84%**	**65.47%**	0.18%
Bacillus subtilis subtilis MP9	APMW00000000.1	**0.99969**	**98.41**	**98.62**	**87.20% [84.7–89.4%]**	**94.76%**	**60.77%**	0.09%
Bacillus subtilis B4143	JXLQ00000000	**0.99969**	**98.38**	**98.65**	**88.00% [85.5–90.1%]**	**95.06%**	**61.99%**	0.03%
Bacillus subtilis subtilis RO-NN-1	CP002906.1	**0.99953**	97.84	98.07	**82.90% [80–85.4%]**	**92.55%**	53.1%	0.17%
Bacillus subtilis spizizenii W23	NC_014479.1	0.99895	92.59	92.87	49.30% [46.7–52%]	17.32%	3.61%	0.19%
Bacillus subtilis spizizenii ATCC 6633	ADGS00000000	0.99901	92.49	92.87	49.20% [46.6–51.8%]	17%	3.54%	0.13%
B.subtilis inaquosorum KCTC13429(T)	AMXN00000000	0.99832	92.37	93.05	49.90% [47.3–52.6%]	18.92%	3.93%	0.01%
Bacillus tequilensis KCTC 13622(T)	AYTO00000000	0.99741	91.29	91.83	44.90% [42.3–47.4%]	8%	1.76%	0.16%
Bacillus vallismortis DV1-F-3	AFSH00000000	0.99729	90.47	91.10	42.50% [40–45%]	4.85%	1.14%	0.06%
Bacillus mojavensis KCTC 3706(T)	AYTL00000000	0.99748	86.81	87.58	32.40% [30–34.9%]	0.27%	0.1%	0.03%
Bacillus mojavensis RRC 101	ASJT00000000	0.99731	86.74	87.64	32.40% [30–34.9%]	0.27%	0.1%	0.03%
Bacillus atrophaeus 1942	CP002207.1	0.98582	79.19	83.88	22.00% [19.8–24.5%]	0%	0%	0.48%
Bacillus amyloliquefaciens DSM 7(T)	FN597644.1	0.95303	76.39	83.88	20.40% [18.2–22.8%]	0%	0%	2.39%
Bacillus amyloliquefaciens UCMB-5036	NC_020410.1	0.94889	76.33	84.19	20.50% [18.3–22.9%]	0%	0%	2.9%
Bacillus amyloliquefaciens UCMB-5113	NC_022081.1	0.94919	76.32	84.12	20.50% [18.2–22.9%]	0%	0%	3.01%
Bacillus amyloliquefaciens FZB42(T)	NC_009725.1	0.9508	76.29	84.15	20.50% [18.3–22.9%]	0%	0%	2.78%
Bacillus amyloliquefaciens UCMB-5033	NC_022075.1	0.95049	76.26	84.09	20.40% [18.2–22.8%]	0%	0%	2.49%
Bacillus methylotrophicus SK19.001	AOFO00000000	0.95157	76.21	84.25	20.40% [18.1–22.8%]	0%	0%	2.47%
Bacillus methylotrophicus JS25R	CP009679.1	0.9523	76.16	84.21	20.40% [18.2–22.8%]	0%	0%	2.71%
Bacillus siamensis KCTC 13613(T)	AJVF00000000	0.95296	76.16	84.39	20.60% [18.3–23%]	0%	0%	2.64%

Thresholds supporting subspecies delineation (intraspecific Tetra-nucleotide signature correlation index (Z-score): > 0.999, ANIb and ANIm: >98%, and dDDH: >79%) are indicated in bold letters

Roche 454 GS FLX instrument and Titanium reagents. Pyrosequencing generated 84,579 and 157,116 raw ITS and 16S rDNA reads, respectively.

Processing of 454 sequencing data

Sequences were assigned to different samples according to sample-specific barcodes and processed using QIIME (version 1.17). Data were selected based on the following criteria: (i) an almost perfect match with barcode and primers; (ii) a length of at least 200 nucleotides (barcodes and primers excluded); (iii) an average quality score of >25, with no ambiguous bases or homopolymers longer than six bp, and without any primer mismatches. After this procedure, 67,520 and 140,099 high-quality sequences were obtained for ITS and 16S rDNA reads, respectively. The number of clean reads per samples are listed in Additional file 1: Table S1 and Additional file 2: Table S2.

Operational Taxonomic Units (OTUs) were clustered using a 97% similarity cutoff by UPARSE (version 7.1 http://drive5.com/uparse/) and chimeric sequences were identified and removed using UCHIME. The phylogenetic affiliation of each 16S rRNA gene and ITS sequence were analyzed by RDP Classifier (http://rdp.cme.msu.edu/)

against the SILVA (SSU111) 16S/18S rRNA database. Taxonomic assignment from phylum level to strain level based on hits, and these were used to plot abundance graphs.

Statistical analysis

Plant fresh weight, height and number of colonizing bacteria were analyzed using Microsoft Excel. Calculation of variance (ANOVA) and mean comparison between treatments was carried out based on the Tukey's test at the 0.05 probability level using SPSS version 19 (IBM Corporation, New York, USA). Principal component analysis (PCA) was performed to analyze the effect of PTS-394 and sampling time on the microbiota community at the OTU level using the R-forge community ecology package (vegan 2.0 was used to generate the PCA Figures).

Results

Phylogenomics of *B. subtilis* PTS-394

Phylogenomic analysis of strain PTS-394 was performed taking advantage of availability of the whole draft genome sequence. ANIb and ANIm values in comparison to the *B. subtilis* type strain 168 were determined by using the JSpecies program package. In order to finally decide,

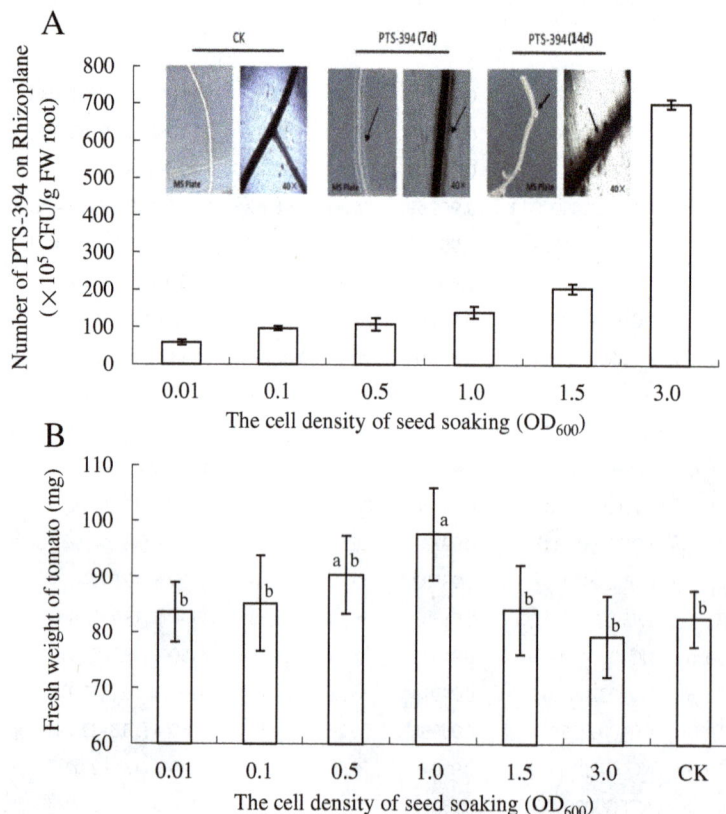

Fig. 1 Effect of *B. subtilis* PTS-394 on root colonization and tomato growth on MS agar. **a**, PTS-394 cell number (CFU) detected on the tomato rhizoplane after treatment with suspensions of different cell densities (histogram), and pattern of colonization by PTS-394 (arrows indicate the PTS-394 cells around the tomato root). **b**, Growth promotion effect of PTS-394: Fresh weight of tomato seedlings after soaking with suspensions of different cell densities (histogram). Different letters indicate significant difference at the 0.05 level by Duncan's new multiple range test

whether PTS-394 belong to *B. subtilis* subsp. *subtilis*, electronic DNA-DNA hybridization (dDDH) was applied. For calculating dDDH three different formulas can be applied (see Methods), but only results obtained with the recommended formula 2 were used in our analysis. The nearest neighbors of PTS-394 found by using different phylogenomic analyses were representatives of *B. subtilis* subsp. *subtilis* including the type strain *B. subtilis* 168 (Table 1). Tetra results (tetranucleotide signature correlation index) determined for PTS-394 when compared with *B. subtilis* subsp. *subtilis* 168(T) were in subspecies range (>0,999). The same comparison revealed ANI values far above the recommended species threshold of 0.96% [24] and dDDH values clearly exceeding subspecies delineation (>79%, [21]). By contrast, phylogenomic comparison with other members of the *B. subtilis* species complex including other subspecies of *B. subtilis*, *B. tequilensis*, *B. vallismortis*, *B. atrophaeus*, *B.amyloliquefaciens*, *B. velezensis*, *B. methylotrophicus*, and *B. siamensis* did yield values sufficient for species delineation (Table 1). We conclude that strain PTS-394 is a representative of the plant-associated taxon *B. subtilis* subsp. *subtilis*.

PTS-394 is able to colonize roots and to promote growth of tomato plants

The effect of *B. subtilis* PTS-394 on tomato growth and root colonization was evaluated using MS agar plates. After 7 and 14 days growth, PTS-394 cells became clearly visible around the tomato primary root (Fig. 1a). Density of PTS-394 on the tomato rhizoplane ranged from 10^6 to 10^7 CFU per gram of fresh root, and was found enhanced with increasing cell density of the soaking suspension (Fig. 1a). These results indicated successful root colonization by PTS-394, which is a key requirement for biocontrol action and plant growth-promoting activity. Seedling fresh weight data showed that PTS-394 exhibited impressive growth-promoting activity at a cell density (OD_{600}) of 0.5 and 1.0 in the soaking suspension, yielding 1.08×10^7 and 1.4×10^7 CFU/g of fresh root on the tomato rhizoplane, respectively (Fig. 1b). These treatments increased plant growth by 9.46% and 18.36%, respectively, compared to controls. However, tomato growth was slightly inhibited when the PTS-394 cell number exceeded 7×10^7 CFU/g of fresh root, suggesting that too high cell numbers at rhizoplane might negatively affect plant growth by that bacterium.

Evaluation of root colonization and tomato growth promotion by PTS-394 in pot experiments

Pot experiments were performed to investigate the colonization capability of PTS-394G at different time intervals. Initially, rate of root colonization decreased sharply before steadily increasing over time (Fig. 2a). Almost 2×10^6 CFU per gram of root were detected immediately

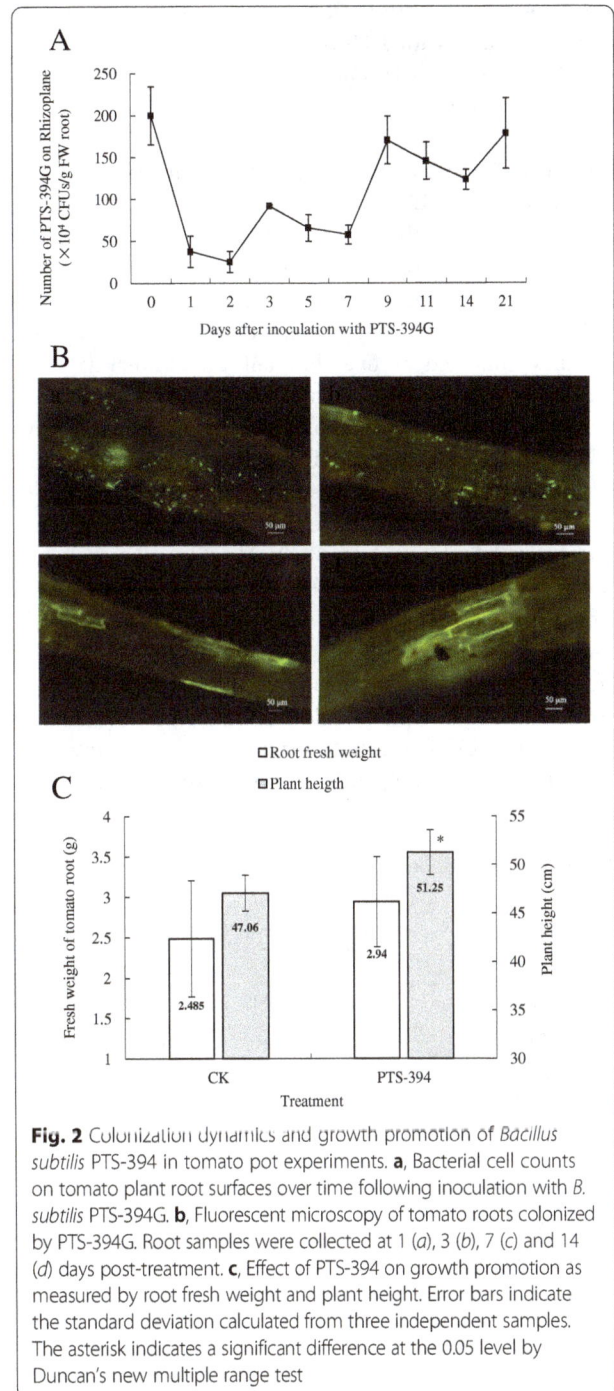

Fig. 2 Colonization dynamics and growth promotion of *Bacillus subtilis* PTS-394 in tomato pot experiments. **a,** Bacterial cell counts on tomato plant root surfaces over time following inoculation with *B. subtilis* PTS-394G. **b,** Fluorescent microscopy of tomato roots colonized by PTS-394G. Root samples were collected at 1 (*a*), 3 (*b*), 7 (*c*) and 14 (*d*) days post-treatment. **c,** Effect of PTS-394 on growth promotion as measured by root fresh weight and plant height. Error bars indicate the standard deviation calculated from three independent samples. The asterisk indicates a significant difference at the 0.05 level by Duncan's new multiple range test

after inoculation; 48 h after inoculation this number dropped to 3.7×10^5 CFU/g root. During next days PTS-394G cell number increased steadily and reached 1.7×10^6 CFU/g root nine days after inoculation. Between days 9–21, the PTS-394G cell number ranged between 1×10^6 to 2×10^6 CFU/g root, and fluorescent microscopy revealed that colonization occurred on the root surface, where PTS-394 cells formed micro-colonies or biofilms 7 days after treatment (Fig. 2b). The average plant height and root weight were 51.25 cm and 2.94 g, respectively, after treatment with PTS-394, which represented an increase of 8.90% and 18.30% compared with untreated control plants (Fig. 2c).

Impact of PTS-394 on bacterial members of the rhizosphere microbiota

Determination of the composition of the bacterial community in vicinity of plant roots was performed by metagenome sequencing (see Methods). A total of 157,116 raw 16S rRNA sequences were obtained from 10 soil samples collected at five different time intervals from PTS-394-treated and untreated plants. The number of clean sequences per sample ranged from 11,773 to 15,879, with an average of 14,010. At 97% sequence similarity, these clean sequences represented a total of 19,231 OTUs by a two-stage clustering (TSC) algorithm and ranged from 4054 to 5343. The Shannon index of diversity was determined for all samples, and ranged from 7.41 to 7.94. The Good's Coverage per sample ranged from 0.77 to 0.83. The statistical indexes and richness estimates per sample are summarized in Additional file 1: Table S1. Global Alignment for Sequence Taxonomy (GAST) was used for taxonomic assignment of 16S rRNA sequences, and 31 phyla were identified by

pooling sequences from all samples, with nineteen core phyla present in all ten samples (Table 2). The three most abundant core phyla were *Proteobacteria*, *Bacteroidetes* and *Actinobacteria*, which were present in all samples, except the one collected on the first day after treatment with PTS-394.

Principal component analysis (PCA) was used to analyze the effect of PTS-394 and sampling time on the structure of the bacterial community in the rhizosphere based on OTUs. PC1 and PC2 together accounted for more than 75% of the variation (Fig. 3a). There was a clear shift in bacterial community between 1 and 14 days after inoculation with PTS-394. Treated or untreated samples collected on day 1 were distinct from each other and from the other time points, however control and treated samples could not be distinguished at later stages. These results indicated that (i) changes in the bacterial community profile becomes visible soon after inoculation with *B. subtilis*, and (ii) application of PTS-394 affected only transiently the root bacteriome. As expected, a sudden rise of *Firmicutes* was registered due to the large quantity of *B. subtilis* PTS-394 cells applied. However, in course of the experiment the number of *Firmicutes* was steadily decreasing and became similar with the control 14 days after inoculation (Table 2).

Table 2 Relative abundance of nineteen bacterial core phyla present in rhizosphere bacterial community determined by 16S sequencing

Core phylum	Relative abundance (%) in Control treatment					Relative abundance (%) in PTS-394 treatment				
	1 d	3 d	7 d	9 d	14 d	1 d	3 d	7 d	9 d	14 d
Proteobacteria	37.94	38.43	40.11	37.13	41.58	36.56	39.18	38.29	41.37	39.22
Bacteroidetes	13.85	13.36	12.24	14.26	11.19	11.48	12.27	11.30	11.56	12.45
Actinobacteria	9.70	9.43	9.73	11.96	10.67	7.92	7.84	11.20	8.85	8.52
Chloroflexi	8.08	8.36	8.42	7.72	6.01	5.96	6.46	7.82	6.54	7.97
Acidobacteria	5.70	7.17	6.77	6.79	5.77	5.89	7.65	6.19	7.10	6.27
Gemmatimonadetes	6.12	5.28	5.38	4.37	5.36	5.83	6.01	6.12	6.59	6.29
Firmicutes	2.52	2.78	2.49	2.81	3.92	13.78	6.68	3.85	3.58	3.29
Planctomycetes	5.80	5.21	5.18	4.91	3.93	3.58	4.25	4.66	4.22	3.05
Cyanobacteria	0.58	0.50	0.44	0.87	0.86	0.45	0.47	0.55	0.64	1.18
Verrucomicrobia	0.39	0.49	0.49	1.06	0.84	0.46	0.40	0.50	0.59	0.56
Nitrospirae	0.50	0.83	0.51	0.72	0.47	0.43	0.58	0.42	0.61	0.48
Chlorobia	0.48	0.40	0.41	0.28	0.32	0.36	0.40	0.55	0.51	1.09
Deinococcus-Thermus	0.35	0.55	0.32	0.30	0.29	0.30	0.35	0.36	0.42	0.33
Fusobacteria	0.42	0.53	0.39	0.47	0.21	0.33	0.28	0.24	0.26	0.21
Armatimonadetes	0.28	0.29	0.28	0.24	0.43	0.22	0.30	0.37	0.35	0.21
Fibrobacteres	0.07	0.08	0.09	0.09	0.18	0.05	0.07	0.09	0.09	0.31
Elusimicrobia	0.04	0.05	0.04	0.09	0.15	0.03	0.08	0.06	0.06	0.05
Thermotogae	0.05	0.04	0.05	0.04	0.02	0.04	0.01	0.04	0.01	0.03
Spirochaetes	0.03	0.03	0.05	0.01	0.02	0.04	0.03	0.03	0.01	0.02

The assay was performed 1, 3, 7, 9, and 14 days after inoculation with PTS-394. Control experiments without PTS-394 were performed at same time points

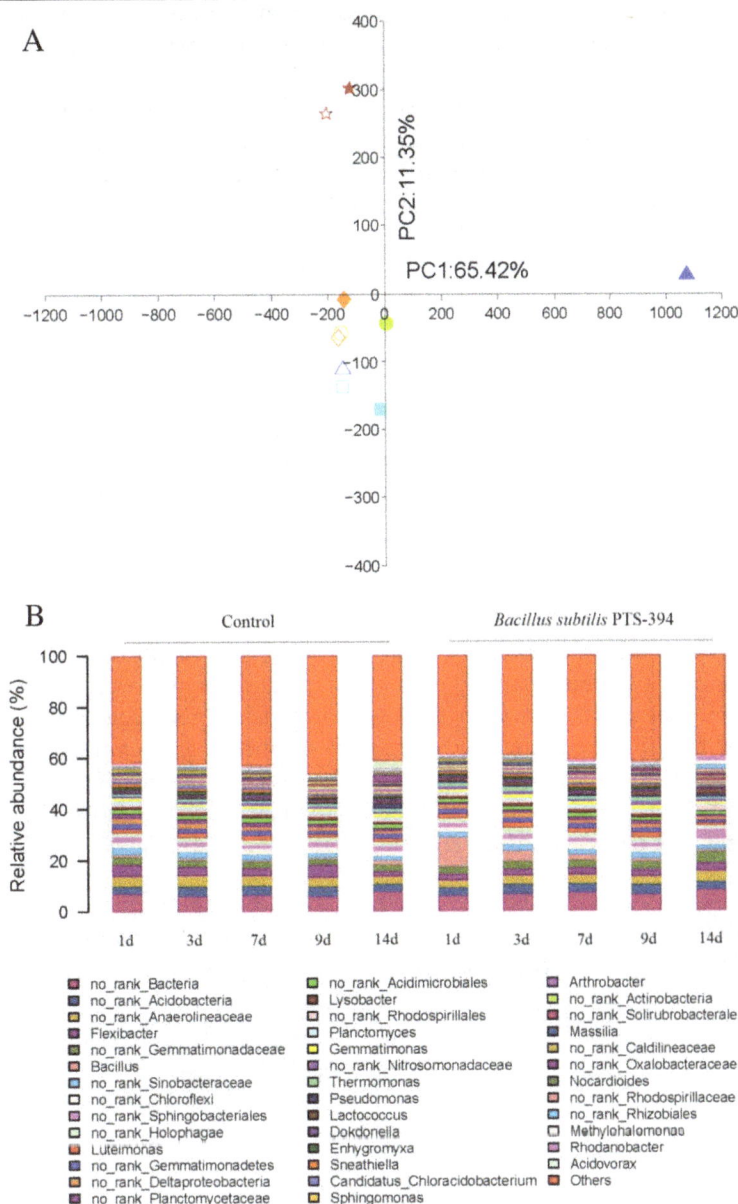

Fig. 3 Effect of *Bacillus subtilis* PTS-394 on the rhizosphere bacterial community. **a**, Principal component analysis was performed based on OTUs from16S rRNA partial sequence. The same shape and color represent the same sampling time. △, □, ○, ◇, ☆ represent 1, 3, 7, 9 and 14 days after transplantation. Filled data points represent plants treated with PTS-394 and open data points represent control plants. **b**, relative abundance of 40 core groups (genera) present in the rhizosphere bacterial community

BLAST analysis revealed that the relative abundance of 40 different groups (based on genus level under inclusion of some uncertain groups) was temporarily affected after inoculation (Fig.3b). A total of 19 groups were altered by PTS-394 treatment, occurrence of 10 groups was enhanced, including *Thermomonas, Pseudomonas, Lactococcus* and *Rhodanobacter*, while nine groups were suppressed, including *Flexibacter, Planctomyces* and *Sneathiella*. However, as in *Firmicutes* (see above), these groups were only temporarily affected and 14 days after inoculation no significant differences to the control were registered. The relative

abundance and variation in the composition of these 19 groups is presented in Additional file 3: Figure S1.

Impact of PTS-394 on eukaryotic members of the rhizosphere microbiota

Presence of eukaryotic groups in vicinity of plant roots was determined by using 18S rRNA sequences in a similar way as described for bacteria. In total, 84,579 raw ITS sequences were obtained and the number of clean sequences per sample varied ranged from3, 897 to 8402, with an average of 6752. A total of 1491 Operational

Taxonomic Units (OTUs) were obtained at a distance of 0.03 using the TSC algorithm. The Shannon index of diversity was determined for 10 samples, and ranged from 0.28 to 4.15. The Good's coverage per sample ranged from 0.96 to 0.99. Although there was some fluctuation of reads over time, analysis of these data corroborated their credibility. The statistical indexes and richness estimates of per sample are summarized in Additional file 2: Table S2. The eukaryotic OTUs could be arranged into three kingdoms (*Fungi*, *Metazoa* and *Viridiplantae*) along with the *no rank Eukaryota* group, a group of *unclassified Eukaryota* and a group of *unclassified* organisms. Altogether, thirteen phyla were identified. Seven core phyla were detected in every sample. The most abundant phyla were *Streptophyta*, *Chlorophyta Ascomycota* and *Basidiomycota* (Table 3). After the application of PTS-394, the composition of *Fungi*, *Viridiplantae*, the *no rank Eukaryota* group and the *unclassified* group were altered, compared to the untreated control. The relative abundance of *Fungi*, of the *no rank Eukaryota* group, and of the group of *unclassified* organisms was lower than that of the control at the same sample time. However, the relative abundance of *Viridiplantae* (*Solanoideae*) was found higher than in the control.

All OTUs were subjected to PCA analysis and PC1 and PC2 together accounted for more than 90% of the variation (Fig. 4a). A significant temporal change in the community profile of Eukaryota is visible in the biplot. Profiles of control and PTS-394-treated samples collected at identical time points revealed slight differences from day 1 until day 9 after treatment with PTS-394, but could not be distinguished 14 days after inoculation. These results are in accordance to those found for the rhizosphere bacteriome, and indicate that (i) PTS-394 had a transient influence on composition of the eukaryotic members of the rhizosphere microbiome, but (ii) the community recovered to its original state 14 days after treatment. BLAST analysis revealed that the relative abundance of 29 groups (based on the eukaryota genus level and including some uncertain groups) was transiently affected after inoculation with PTS-394 (Fig. 4b). A total of 19 groups were altered by PTS-394 treatment; five groups were enhanced, including *Solanoideae*, *Chlamydomonas* and *Chloromonas*, and 14 groups were decreased. The variation in these 19 groups is presented in Additional file 4: Figure S2.

PCA on OTUs classified as fungi revealed a temporal variation in the fungal community in dependence of PTS-394 inoculation (Fig. 5). The Control and the PTS-394-treated samples were found different during the first four time points, suggesting that PTS-394 inoculation affects the fungal community transiently, similar to its effect on the bacterial community. Sequence homology and analysis of the relative abundance of OTUs showed that abundance of fungi was generally suppressed during

Table 3 Relative abundance of core kingdoms/phylum present in rhizosphere eukaryota community determined by ITS sequencing

Core Kingdom	Relative abundance (%) in Control treatment					Relative abundance (%) in PTS-394 treatment				
	1 d	3 d	7 d	9 d	14 d	1 d	3 d	7 d	9 d	14 d
Fungi	16.24	20.54	5.64	36.52	1.11	17.22	12.10	5.91	9.33	1.53
Viridiplantae	29.79	24.93	81.81	39.59	97.76	58.22	79.23	92.29	81.40	97.37
Metazoa	0.39	0.62	0.05	0.98	0.17	0.63	0.15	0.43	1.09	0.14
no_rank_Eukaryota group	20.37	21.83	6.21	20.69	0.20	22.27	7.93	1.08	7.04	0.94
unclassified_Eukaryota group	0.18	0.20	0.08	0.16	0.00	0.06	0.04	0.03	0.01	0.01
Unclassified group	33.03	31.87	6.21	2.05	0.76	1.59	0.55	0.27	1.12	0.00
Core phylum										
Streptophyta	19.87	17.29	74.15	30.06	97.43	18.17	64.02	90.46	70.60	97.32
Chlorophyta	9.91	7.59	7.64	9.39	0.33	40.04	15.20	1.83	10.80	0.05
Ascomycota	2.17	3.98	1.30	8.42	0.13	2.26	1.44	0.59	0.67	0.63
Basidiomycota	1.94	1.83	0.83	2.10	0.17	1.60	4.36	0.94	1.13	0.16
Porifera	0.37	0.49	0.05	0.79	0.06	0.19	0.12	0.42	0.78	0.01
Chytridiomycota	0.18	0.42	0.20	0.58	0.03	0.40	0.19	0.11	0.23	0.07
Nematoda	0.03	0.09	0.00	0.15	0.12	0.41	0.03	0.01	0.11	0.10
no_rank_Eukaryota group	20.37	21.82	6.19	20.68	0.20	22.26	7.93	1.05	7.04	0.94
no_rank_Fungi group	11.19	12.94	3.13	18.86	0.67	12.44	5.75	4.18	6.99	0.66
unclassified_Eukaryota group	0.18	0.20	0.08	0.16	0.00	0.06	0.04	0.03	0.01	0.01
unclassified_Fungi group	0.71	1.30	0.18	6.48	0.10	0.44	0.28	0.08	0.14	0.01

The assay was performed 1, 3, 7, 9, and 14 days after inoculation with PTS-394, Control experiments without PTS-394 were performed at same time points

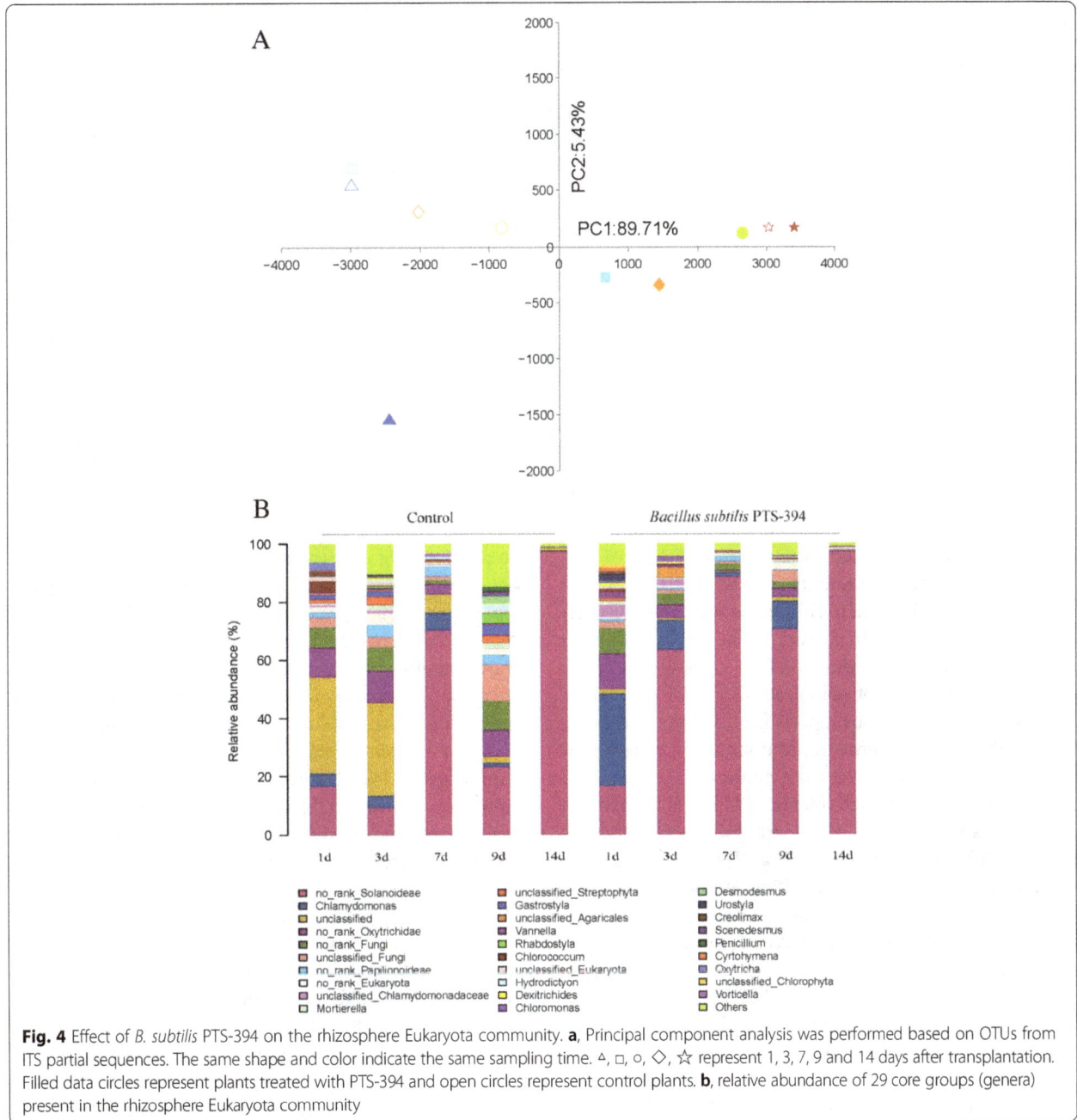

Fig. 4 Effect of *B. subtilis* PTS-394 on the rhizosphere Eukaryota community. **a**, Principal component analysis was performed based on OTUs from ITS partial sequences. The same shape and color indicate the same sampling time. △, □, ○, ◇, ☆ represent 1, 3, 7, 9 and 14 days after transplantation. Filled data circles represent plants treated with PTS-394 and open circles represent control plants. **b**, relative abundance of 29 core groups (genera) present in the rhizosphere Eukaryota community

plant development, however by PTS-394 this process was slightly enhanced (Table 3). As expected, the relative abundance of *Fusarium oxysporum* was inhibited after treatment with PTS-394 indicating antagonistic activity of PTS-394 against *F. oxysporum* (Additional file 5: Figure S3).

Discussion

Plant growth promotion and biocontrol activities are important features of commercial agents used in sustainable agriculture. To date members of the genus *Bacillus* are preferred for preparing bioformulations with beneficial impact on plant growth and health [1]. Especially,

representatives of the *B. subtilis* species complex are known for their beneficial action on plants, especially *B. amyloliquefaciens* and *B. subtilis* [25]. These *Bacilli* stimulate plant growth: (1) directly, by increasing nutrients through the production of phytohormones, siderophores, organic acids involved in P-solubilisation and/or fixation of nitrogen [26]; and (2) indirectly, by producing antagonistic substances or by inducing the plant resistance against pathogens [7]. Here, we found that plant root-colonizing *B. subtilis* subsp. *subtilis* PTS-394G is able to persist on roots and to promote growth of tomato plants. Moreover, the plant pathogen *Fusarium*

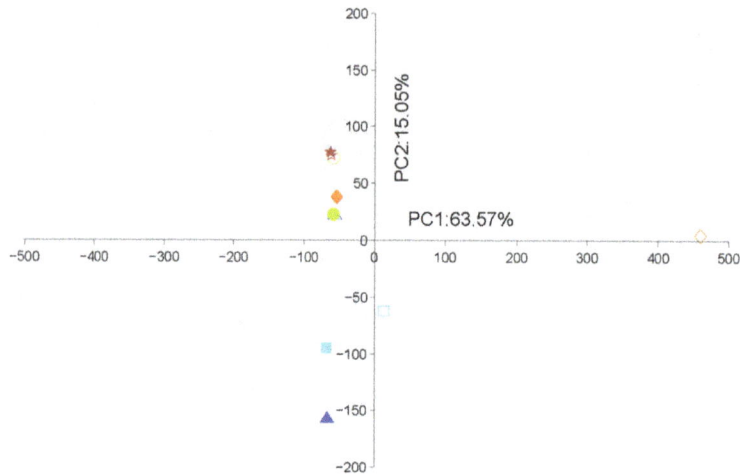

Fig. 5 Effect of *B. subtilis* PTS-394 on the rhizosphere fungi community. Principal component analysis was performed based on OTUs from fungi ITS partial sequences. The same shape and color represent the same sampling time. △, □, ○, ◇, ☆ represent 1, 3, 7, 9 and 14 days after transplantation. Filled circles represent plants treated with PTS-394 and open circles represent control plants

oxysporum was suppressed in presence of PTS-394. Laboratory MS plate experiments showed that the amount of PTS-394G on the tomato rhizoplane was approximately 10^7 CFU per gram of fresh root, while the cell number was 10^6 CFU in pot experiments. This suggests that sterile environment favors root colonization of PTS-394, which is consistent with previous reports [11, 27]. It is interesting to note that the plant growth promoting effect exerted by PTS-394 was dependent on the amount of *B. subtilis* cells used for inoculating tomato plants. Whilst cell numbers of 1.08×10^7 and 1.4×10^7 CFU/g fresh root supported plant growth, higher cell numbers such as 7×10^7 CFU/g fresh root, led to a slight growth inhibition. Without genetic experiments it remains questionable whether this different pattern of growth stimulation could be explained by a concentration dependent effect of the phytohormone indole-3-acetic acid (IAA). The PGPR *B. amyloliquefaciens* FZB42 is able to produce IAA and it is possible that higher concentrations of IAA exerts an inhibitory effect on plant growth [1]. In fact, *ysnE*, a gene encoding a putative IAA acetyltransferase, which is involved in IAA synthesis pathway, was detected in the genome of PTS-394. Our results demonstrate that, besides *B. amyloliquefaciens* and *B. pumilus*, plant-associated representatives of the *B. subtilis* subsp. *subtilis* taxon are promising candidates for developing of efficient biocontrol and biofertilizer agents. In this context it is important to know, whether addition of PTS-394 has an effect on composition of rhizosphere microbial community.

Here, the effects of PTS-394 on the entire rhizosphere microbiota community was investigated by taxonomic profiling of metagenome sequences. In accordance with the results obtained for *B. amyloliquefaciens* FZB42 [2, 15, 16], only a transient effect on composition of

the root microbiome was detected after adding the PGPR bacterium. Similarly, independent of its mode of application, applying *B. amyloliquefaciens* BNM122 to soy bean plants did not shift the composition of rhizosphere bacterial community in a measurable extent [14], suggesting that inoculation of crops with different representatives of the genus *Bacillus* has no durable impact on the bacteria living in vicinity of plant roots. By contrast, external addition of Gram-negative PGPR, such as *Pseudomonas* spp. [15, 28], *Enterobacter cowanii* [29], and *Sinorhizobium meliloti* [30] alter composition of root bacterial community in a similar extent as reported for fungal plant pathogens, e.g. *Ralstonia solani* [2].

Whilst the impact of PGPRs on the rhizosphere bacterial community is fairly investigated, relatively little is known about the effect of PGPRs on the overall eukaryotic community [8]. Analysis of 18S ribosomal ITS sequences showed that PTS-394 influences Eukaryota in the rhizosphere microbial community. However, similar to the effects exerted by rhizosphere bacteria, the impact of PTS-394 on Eukaryota was only transient, although longer-lasting than its effect on bacterial community.

The rhizosphere contains amino acids and sugars nutrients due to the accumulation of root exudates that provides a rich source of energy and nutrients for rhizosphere microbiome. This determines composition of the microbiome community [7, 8]. After application of PTS-394, relative abundance of *Fungi, Viridiplantae*, the *no rank Eukaryota* group and the *unclassified* group was similar as in the control, but changes in their relative abundance compared to control were registered. Here, a significant increase of *Viridiplantae* relative abundance was observed which might be due to enhanced root development caused by PTS-394 as seen by enhanced number of *streptophyta* counts largely corresponding to tomato cells. *Chlorophyte*, another group of the *Viridiplantae*, the relative abundance was decreased significantly in the PTS-394 treatment, which correspond to the variation trend of main genus *Chlamydomonas*. In addition, PGPR have the ability to secrete several antagonistic compounds, such as antifungal acting lipopeptide antibiotics. Nihorimbere et al. reported that surfactin, iturin and fengycin were detected when *Bacillus amyloliquefaciens* S499 colonized tomato rhizosphere [31]. We found that suspensions of *B. subtilis* PTS-394 contained lipopeptides and polyketides. Whilst PTS-394 colonizes plant root, lipopeptides and polyketides might be secreted by the strain into the environment. These secondary compounds will either directly inhibit members of the microbiome community or stimulate the ISR response in plants. The suppressing effect of PTS-394 on plant pathogen *Fusarium oxysporum* in the rhizosphere is a typical example (Additional file 5: Figure S3). We assume that the transient effect of *Bacillus subtilis* PTS-394 on composition of the rhizosphere microbiome is caused by a sudden rise of *Bacillus* cells after inoculation, which is soon compensated by the indigenous microbial community. In case of eukaryota this restoration process is somewhat delayed due to their longer generation time compared to bacteria.

Conclusions

The impact of PGPRs on the overall rhizosphere community should be considered as important criteria when assessing their suitability for commercial development. Here we found that plant-growth-promoting and biocontrol *Bacillus subtilis* subsp. *subtilis* PTS-394 is not aggressive towards the indigenous tomato rhizosphere microbiota including their eukaryotic representatives, and has only a transient impact on the composition of the community. This, together with its beneficial effect on plant growth and health, makes PTS-394 to a promising candidate for developing a successful PGPR-based bioagent.

Additional files

Additional file 1: Table S1. Statistical indexes and richness estimates of the rhizosphere bacterial sequence data.

Additional file 2: Table S2. Statistical indexes and richness estimates of the rhizosphere eukaryote sequence data.

Additional file 3: Figure S1. Variation trends in the abundance of 19 bacterial genera or groups following treatment with *Bacillus subtilis* PTS-394 (part 1 includes 10 groups stimulated by PTS-394, part 2 includes nine groups suppressed by PTS-394).

Additional file 4: Figure S2. Variation trends in the abundance of 19 *Eukaryota* genera or groups following treatment with *Bacillus subtilis* PTS-394 (part 1 includes five groups stimulated by PTS-394, part 2 includes 14 groups suppressed by PTS-394).

Additional file 5: Figure S3. The variation trends of Relative abundance of *Fusarium oxysporum* following treatment with Bacillus subtilis PTS-394

Abbreviations
ANIb: Average nucleotide identity based; ANIm: Average nucleotide identity based; dDDH: Electronic DNA-DNA hybridization.; OTUs: Operational taxonomic units; PCA: Principal component analysis; PGPR: Plant growth promoting rhizobacteria; TETRA: Tetra-nucleotide signatures; YPG medium: Yeast peptone glucose medium

Acknowledgments
Authors wish to thank all the staff of The Rice disease and Biocontrol of Plant disease Laboratory, Institute of Plant Protection, Jiangsu Academy of Agricultural Sciences.

Funding
This work was supported by Jiangsu Academy of Agricultural Innovation funds [CX (15) 1044], the National Natural Science Foundation of China (NSFC 31201556), and the Natural Science Foundation of Jiangsu Province (BK2012373).

Authors' contributions
JQ, YouL, YongL and RB: conceived and designed experiments in the study; JQ, XY and XL: performed all experimental procedures; JQ, RB and YouL draft the manuscript. All authors read and approved the manuscript.

Competing interests
The authors declares that they have no competing interests.

Author details
[1]Institute of Plant Protection, Jiangsu Academy of Agricultural Sciences, Nanjing, Jiangsu province 210014, China. [2]Suqian institute, Jiangsu Academy of Agricultural Sciences, Suqian, Jiangsu province 223831, China. [3]Institut für Agrarwissenschaften/Phytomedizin, Humboldt Universität zu Berlin, 14195 Berlin, Germany. [4]Nord Reet UG, 17489 Greifswald, Germany.

References
1. Borriss R. 'Use of plant-associated Bacillus strains as biofertilizers and biocontrol agents.' in Bacteria in Agrobiology. In: Maheshwari D, editor. Plant growth responses. Heidelberg: Springer; 2011. p. 41–76.
2. Chowdhury SP, Dietel K, Rändler M, Schmid M, Junge H, Borriss R, et al. Effects of Bacillus amyloliquefaciens FZB42 on lettuce growth and health under pathogen pressure and its impact on the rhizosphere bacterial community. PLoS One. 2013;8(7):e68818. doi:10.1371/journal.pone.0068818.
3. Compant S, Duffy B, Nowak J, Clement C, Barka EA. Use of plant growth-promoting bacteria for biocontrol of plant diseases: principles, mechanisms of action, and future prospects. Appl Environ Microbiol. 2005;71(9):4951–9.
4. Weller DM. Pseudomonas biocontrol agents of soilborne pathogens: looking back over 30 years. Phytopathology. 2007;97(2):250–6.

5. Fan B, Borriss R, Bleiss W, Wu X. Gram-positive rhizobacterium Bacillus amyloliquefaciens FZB42 colonizes three types of plants in different patterns. J Microbiol (Seoul, Korea). 2012;50(1):38–44.

6. Ongena M, Jacques P. Bacillus lipopeptides: versatile weapons for plant disease biocontrol. Trends Microbiol. 2008;16(3):115–25.

7. Beneduzi A, Ambrosini A, Passaglia LM. Plant growth-promoting rhizobacteria (PGPR): Their potential as antagonists and biocontrol agents. Genet Mol Biol. 2012;35(4 (suppl):1044–51.

8. Bulgarelli D, Schlaeppi K, Spaepen S, Themaat EVLV, Schulze-Lefert P. Structure and Functions of the Bacterial Microbiota of Plants. Annu Rev Plant Biol. 2013;64(1):807–38.

9. Fan B, Chen XH, Budiharjo A, Bleiss W, Vater J, Borriss R. Efficient colonization of plant roots by the plant growth promoting bacterium Bacillus amyloliquefaciens FZB42, engineered to express green fluorescent protein. J Biotechnol. 2011;151(4):303–11.

10. Lugtenberg B, Kamilova F. Plant-growth-promoting rhizobacteria. Annu Rev Microbiol. 2009;63:541–56.

11. Fliessbach A, Winkler M, Lutz MP, Oberholzer HR, Mader P. Soil amendment with *Pseudomonas fluorescens* CHA0: lasting effects on soil biological properties in soils low in microbial biomass and activity. Microb Ecol. 2009; 57(4):611–23.

12. Zhu W, NW, Yu X, Wang W. Effects of the Biocontrol Agent *Pseudomonas fluorescens* 2P24 on Microbial Community Diversity in the Melon Rhizosphere. Sci Agric Sin. 2010;43(7):1389–96.

13. Buddrus-Schiemann K, Schmid M, Schreiner K, Welzl G, Hartmann A. Root colonization by Pseudomonas sp. DSMZ 13134 and impact on the indigenous rhizosphere bacterial community of barley. Microb Ecol. 2010; 60(2):381–93.

14. Correa OS, Montecchia MS, Berti MF, Ferrari MCF, Pucheu NL, Kerber NL, et al. Bacillus amyloliquefaciens BNM122, a potential microbial biocontrol agent applied on soybean seeds, causes a minor impact on rhizosphere and soil microbial communities. Appl Soil Ecol. 2009;41(2):185–94.

15. Erlacher A, Cardinale M, Grosch R, Grube M, Berg G. The impact of the pathogen Rhizoctonia solani and its beneficial counterpart Bacillus amyloliquefaciens on the indigenous lettuce microbiome. Front Microbiol. 2014;5:175.

16. Kröber M, Wibberg D, Grosch R, Eikmeyer F, Verwaaijen B, Chowdhury SP, et al. Effect of the strain Bacillus amyloliquefaciens FZB42 on the microbial community in the rhizosphere of lettuce under field conditions analyzed by whole metagenome sequencing. Front Microbiol. 2014;5:252.

17. Qiao J, Liu Y, Liang X, Hu Y, Du Y. Draft Genome Sequence of Root-Colonizing Bacterium Bacillus sp. Strain PTS-394. Genome Announc. 2014;2(1):e00038–14. doi:10.1128/genomeA.00038-14.

18. Liu Y, Chen Z, Liang X, Zhu J. Screening, evaluation and identification of antagonistic bacteria against Fusarium oxysporum f. sp. lycopersici and Ralstonia solanacearum. Chin J Biol Control. 2012;28:101–8.

19. Liu Y, Liang X, Qiao J, Zhang R, Chen Z. Bacillus subtilis PTS-394 labeled by green fluorescent protein and its colonization. Acta Phytophlacica Sinica. 2014;41(4):416–2.

20. Meier-Kolthoff JP, Auch AF, Klenk HP, Göker M. Genome sequence-based species delimitation with confidence intervals and improved distance functions. BMC Bioinformatics. 2012;14(1):1–14.

21. Meier-Kolthoff JP, Hahnke RL, Petersen J, Carmen S, Victoria M, Anne F, et al. Complete genome sequence of DSM 30083(T), the type strain (U5/41(T)) of *Escherichia coli*, and a proposal for delineating subspecies in microbial taxonomy. Stand Genomic Sci. 2014;9(1):1–19.

22. Murashige TSF. A revised medium for rapid growth and bioassays with tobacco tissue cultures. Physiol Plant. 1962;15:473–9.

23. Qiao J, Liu Y, Xia Y, Mu S, Chen Z. Root colonization by Bacillus amyloliquefaciens B1619 and its impact on the microbial community of tomato rhizosphere. Acta Phytophylacica Sinica. 2013;40(6):507–11.

24. Colston SM, Fullmer MS, Beka L, Lamy B, Gogarten JP, Graf J. Bioinformatic Genome Comparisons for Taxonomic and Phylogenetic Assignments Using Aeromonas as a Test Case. MBio. 2013;5(6).

25. Fritze D. Taxonomy of the genus bacillus and related genera: the aerobic endospore-forming bacteria. Phytopathology. 2004;94(11):1245–8.

26. Kumar A, Guleria S, Mehta P, Walia A, Chauhan A, Shirkot CK. Plant growth-promoting traits of phosphate solubilizing bacteria isolated from *Hippophae rhamnoides* L. (Sea-buckthorn) growing in cold desert Trans-Himalayan Lahul and Spiti regions of India. Acta Physiol Plant. 2015;37(3):1–12.

27. Kaymak HC. Plant Growth and Health Promoting Bacteria, vol. 18. Berlin: Springer Berlin Heidelberg; 2011.

28. Blouin BS, La BB, Elizabeth L, Weller DM, McSpadden GBB. Minimal changes in rhizobacterial population structure following root colonization by wild type and transgenic biocontrol strains. FEMS Microbiol Ecol. 2004;49(2):307–18.

29. Götz M, Gomes NCM, Dratwinski A, Costa R, Berg G, Peixoto R, et al. Survival of gfp-tagged antagonistic bacteria in the rhizosphere of tomato plants and their effects on the indigenous bacterial community. FEMS Microbiol Ecol. 2006;56(2):207–18.

30. Miethling R, Wieland G, Backhaus H, Tebbe CC. Variation of Microbial Rhizosphere Communities in Response to Crop Species, Soil Origin, and Inoculation with Sinorhizobium meliloti L33. Microb Ecol. 2000;40(1):43–56.

31. Nihorimbere V, Cawoy H, Seyer A, Brunelle A, Thonart P, Ongena M. Impact of rhizosphere factors on cyclic lipopeptide signature from the plant beneficial strain B acillus amyloliquefaciens S499. FEMS Microbiol Ecol. 2012; 79(1):176–91.

16

Enrichment dynamics of *Listeria monocytogenes* and the associated microbiome from naturally contaminated ice cream linked to a listeriosis outbreak

Andrea Ottesen[1]* [iD], Padmini Ramachandran[1], Elizabeth Reed[1], James R. White[2], Nur Hasan[3], Poorani Subramanian[3], Gina Ryan[1], Karen Jarvis[4], Christopher Grim[4], Ninalynn Daquiqan[4], Darcy Hanes[4], Marc Allard[1], Rita Colwell[3], Eric Brown[1] and Yi Chen[1]

Abstract

Background: Microbiota that co-enrich during efforts to recover pathogens from foodborne outbreaks interfere with efficient detection and recovery. Here, dynamics of co-enriching microbiota during recovery of *Listeria monocytogenes* from naturally contaminated ice cream samples linked to an outbreak are described for three different initial enrichment formulations used by the Food and Drug Administration (FDA), the International Organization of Standardization (ISO), and the United States Department of Agriculture (USDA). Enrichment cultures were analyzed using DNA extraction and sequencing from samples taken every 4 h throughout 48 h of enrichment. Resphera Insight and CosmosID analysis tools were employed for high-resolution profiling of 16S rRNA amplicons and whole genome shotgun data, respectively.

Results: During enrichment, other bacterial taxa were identified, including *Anoxybacillus*, *Geobacillus*, *Serratia*, *Pseudomonas*, *Erwinia*, and *Streptococcus* spp. Surprisingly, incidence of *L. monocytogenes* was proportionally greater at hour 0 than when tested 4, 8, and 12 h later with all three enrichment schemes. The corresponding increase in *Anoxybacillus* and *Geobacillus* spp.indicated these taxa co-enriched in competition with *L. monocytogenes* during early enrichment hours. *L. monocytogenes* became dominant after 24 h in all three enrichments. DNA sequences obtained from shotgun metagenomic data of *Listeria monocytogenes* at 48 h were assembled to produce a consensus draft genome which appeared to have a similar tracking utility to pure culture isolates of *L. monocytogenes*.

Conclusions: All three methods performed equally well for enrichment of *Listeria monocytogenes*. The observation of potential competitive exclusion of *L. mono* by *Anoxybacillus* and *Geobacillus* in early enrichment hours provided novel information that may be used to further optimize enrichment formulations. Application of Resphera Insight for high-resolution analysis of 16S amplicon sequences accurately identified *L. monocytogenes*. Both shotgun and 16S rRNA data supported the presence of three slightly variable genomes of *L. monocytogenes*. Moreover, the draft assembly of a consensus genome of *L. monocytogenes* from shotgun metagenomic data demonstrated the potential utility of this approach to expedite trace-back of outbreak-associated strains, although further validation will be needed to confirm this utility.

(Continued on next page)

* Correspondence: Andrea.Ottesen@fda.hhs.gov
[1]Office of Regulatory Science, Center for Food Safety and Applied Nutrition, Food and Drug Administration, 5001 Campus Drive, College Park, MD 20740, USA
Full list of author information is available at the end of the article

(Continued from previous page)

Keywords: *Listeria monocytogenes*, Enrichment, Ice cream, Microbiota, Co-enriching bacteria, 16S rRNA, Shotgun metagenomics, Next-generation sequencing, NGS, ISO, FDA, USDA, Buffered *Listeria* enrichment broth (BLEB), Half-Fraser broth (HFB), Fraser broth (FB), University of Vermont modified broth (UVM)

Background

Optimization of enrichment methods to culture target pathogens from complex environmental, food and clinical samples is an ongoing challenge. Traditionally, samples are incubated in nonselective and/or selective enrichment broths and then plated onto selective media. Enrichment methods for specific foodborne pathogens will benefit from an improved understanding of the taxonomic diversity and relative abundance of microbiota that co-culture during enrichment. Here, we use culture independent next generation sequencing (NGS) to characterize the microbiome at four hour intervals using three different enrichment methods used for recovery of *Listeria monocytogenes* from naturally contaminated ice cream.

L. monocytogenes was first reported in 1926 by Murray, Webb and Swann as the causative agent of illness in >rabbits and guinea pigs in a laboratory breeding unit [1, 2]. Although it was long suspected that food might be a mode of transmission for human listeriosis, it was not until after 1980 that several outbreaks conclusively linked *L. monocytogenes* to foods including, coleslaw, milk, cheese, meat, pâté, and jellied pork tongues [1, 3–6]. Listeriosis outbreaks in the United States over the past five years have been associated with contaminated cheeses [7], stone fruits [8], ice cream [9], cantaloupes [10] and caramel apples [11]. Results of the study presented here were obtained from bacteriological analysis of samples from the 2010 to 2015 listeriosis outbreak with several case-patients linked to milkshakes made from contaminated ice cream. Analysis of *L. monocytogenes* in ice cream samples manufactured in the implicated production line provided information about the prevalence and level of *L. monocytogenes* [12, 13] and activity of *L. monocytogenes* in milkshakes prepared from the ice cream [14]. These analyses did not identify *Listeria* species other than *L. monocytogenes* in the ice cream samples [12].

A variety of enrichment media and methods have been developed for detection of *L. monocytogenes*. The three commonly employed methods examined in this study are as follows:

1) **BLEB** as described in the U.S. Food and Drug Administration (FDA) *Bacteriological Analytical Manual* (BAM): 4 h incubation at 30 °C in buffered *Listeria* enrichment broth (BLEB) without antibiotics, followed by 44 h incubation at 30 °C in BLEB with antibiotics [15]; 2) **HFB-FB** as described in the International

Organization of Standardization 11290–1: 24 h enrichment in Half-Fraser broth (HFB) at 30 °C followed by 24 h in Fraser broth (FB) at 37 °C [16]; and 3) **UVM-FB** as described in the U.S. Department of Agriculture (USDA) Microbiological Laboratory Guidebook (MLG): 24 h incubation at 30 °C in University of Vermont modified broth (UVM) for 24 h, followed by 24 h in FB at 37 °C [17].

Once *L. monocytogenes* is detected and isolated from a food or environmental source, pulsed-field gel electrophoresis (PFGE) has been a widely applied method for subtyping isolates. PFGE has been the gold standard for the FDA and the Centers for Disease Control and Prevention (CDC) for foodborne outbreak investigations for over 15 years. Recently however, whole genome sequencing (WGS) has been employed to improve resolution of PFGE for identifying closely related strains. Currently, both PFGE and WGS require isolation of confirmed cultures, which typically requires at least 5 to 7 days for *L. monocytogenes*.

L. monocytogenes from the same lot of ice-cream examined here was previously enumerated using an Most Probable Number (MPN) method [12]. All samples from the lot tested positive for *L. monocytogenes* with a geometric mean of 3.35 MPN $^{-1}$g. To complement enumeration data, we used a metagenomic approach to examine how *L. monocytogenes* and other members of the microbiota of the naturally contaminated ice cream responded to three commonly used enrichment methods (BLEB (FDA), HFB-FB (USDA), UVM-FB (ISO)). Ribosomal RNA amplicons and metagenomic sequence data from time-points every 4 h during a 48 h period were used to describe the taxonomic composition of co-enriching bacterial taxa and provide data to explore whether a hybrid culture/metagenomic approach can be used to source track target pathogens before they are fully isolated.

Results

Dynamics of *L. monocytogenes* and co-occurring bacteria

16S rRNA amplicon sequencing revealed that proportional abundances of *L. monocytogenes* remained low (0–10%) until 24 h of enrichment, even though each enrichment method employed selective antimicrobials (Fig. 1). After 24 h of selective enrichment, relative abundances of *L. monocytogenes* increased at each successive time point until 40 h, at which time relative abundances ranged from 90%

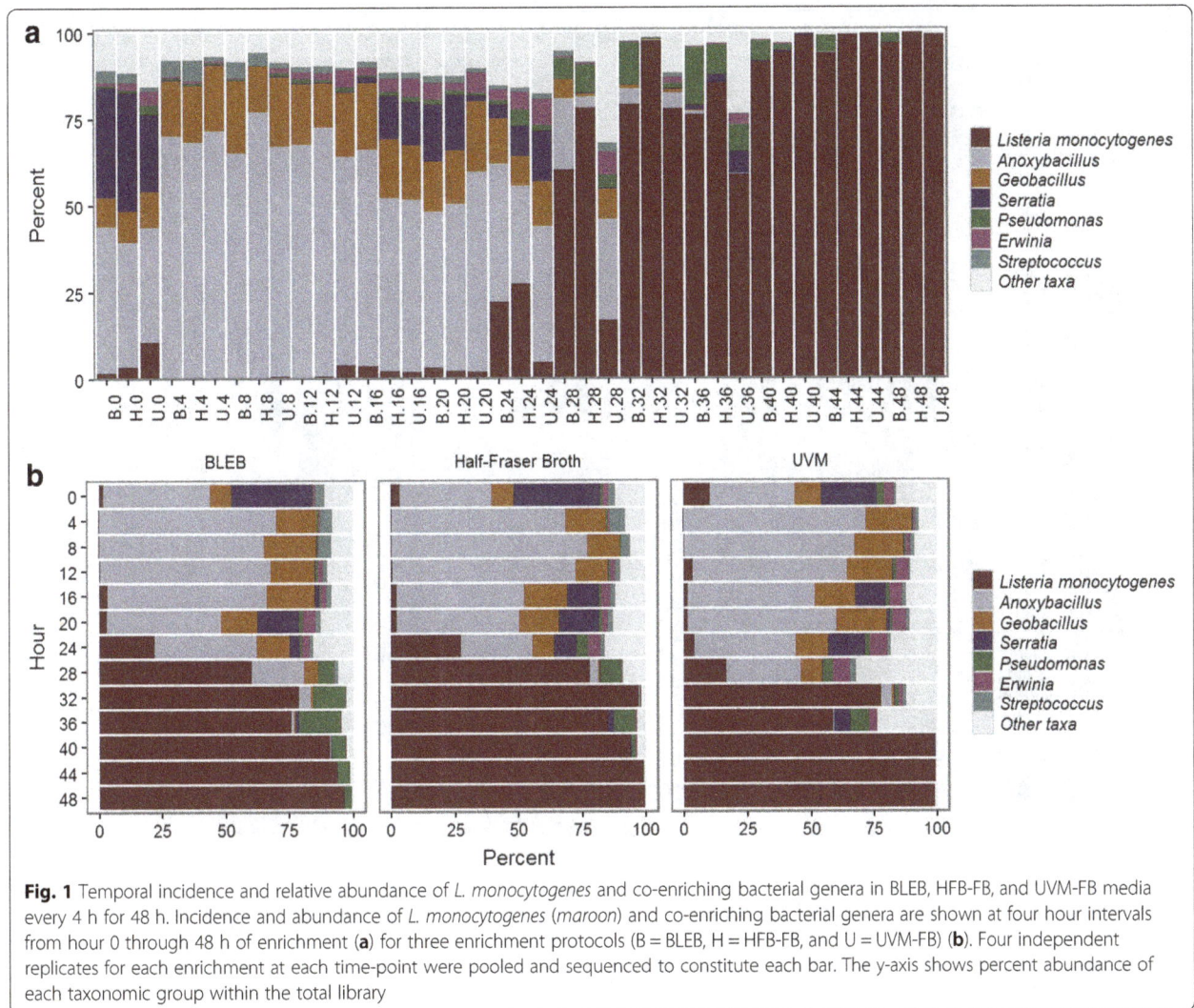

Fig. 1 Temporal incidence and relative abundance of *L. monocytogenes* and co-enriching bacterial genera in BLEB, HFB-FB, and UVM-FB media every 4 h for 48 h. Incidence and abundance of *L. monocytogenes* (*maroon*) and co-enriching bacterial genera are shown at four hour intervals from hour 0 through 48 h of enrichment (**a**) for three enrichment protocols (B = BLEB, H = HFB-FB, and U = UVM-FB) (**b**). Four independent replicates for each enrichment at each time-point were pooled and sequenced to constitute each bar. The y-axis shows percent abundance of each taxonomic group within the total library

to near 100% of the microbial community. Interestingly, *L. monocytogenes* was found to be more abundant at hour 0 than at the three subsequent time points. BLEB and HFB-FB enrichment resulted in higher proportional abundances (although not statistically significant) of *L. monocytogenes* at 24 to 36 h than UVM-FB, while all enrichment yielded equivalent results at later time points.

An examination of other community members, using 16S rRNA gene amplicons, revealed a predominance of *Anoxybacillus*, followed by *Serratia*, *Geobacillus*, and *Streptococcus* species (Fig. 1). During enrichment for all three methods (BLEB, UVM-FB, HFB-FB), Bacillaceae genera, *Anoxybacillus* and *Geobacillus* increased rapidly from a combined relative abundance of approximately 45% at hour 0 to almost 90% at hours 4, 8 and 12 (Fig. 1b). Taxonomy based on shotgun sequencing also supported the presence of *Anoxybacillus* and *Geobacillus* species (Fig. 2). Species of both genera are reported as moderately thermophilic and in this study appeared

to have an advantage over *L. monocytogenes* during early incubation at 30 °C. Additionally, shotgun metagenomic data suggested the presence of two other thermophiles, *Thermus parvatiensis* and *T. thermophilus* (Fig. 2). *Anoxybacillus* spp. have an optimum growth temperature (OGT) ranging from 50 to 62°C and their close relatives, *Geobacillus* spp., have a slightly higher OGT of 55 to 65°C [18]; *Thermus* spp. have an OGT ranging from 50 to 82°C [19, 20].

Another interesting observation of microbial dynamics during enrichment was the change in relative abundance of *Serratia* during the first 24 h of enrichment which appeared to mirror that of *Listeria* (Fig. 1). However, after 24 h of enrichment *Serratia* was outcompeted by other members of the microbial community, mainly *Listeria monocytogenes* in all three enrichment broths. Assuming that hour 0 (initiation of the experiment) resembled the mixed culture microbiota in ice cream, *Serratia* was a well-represented constituent of

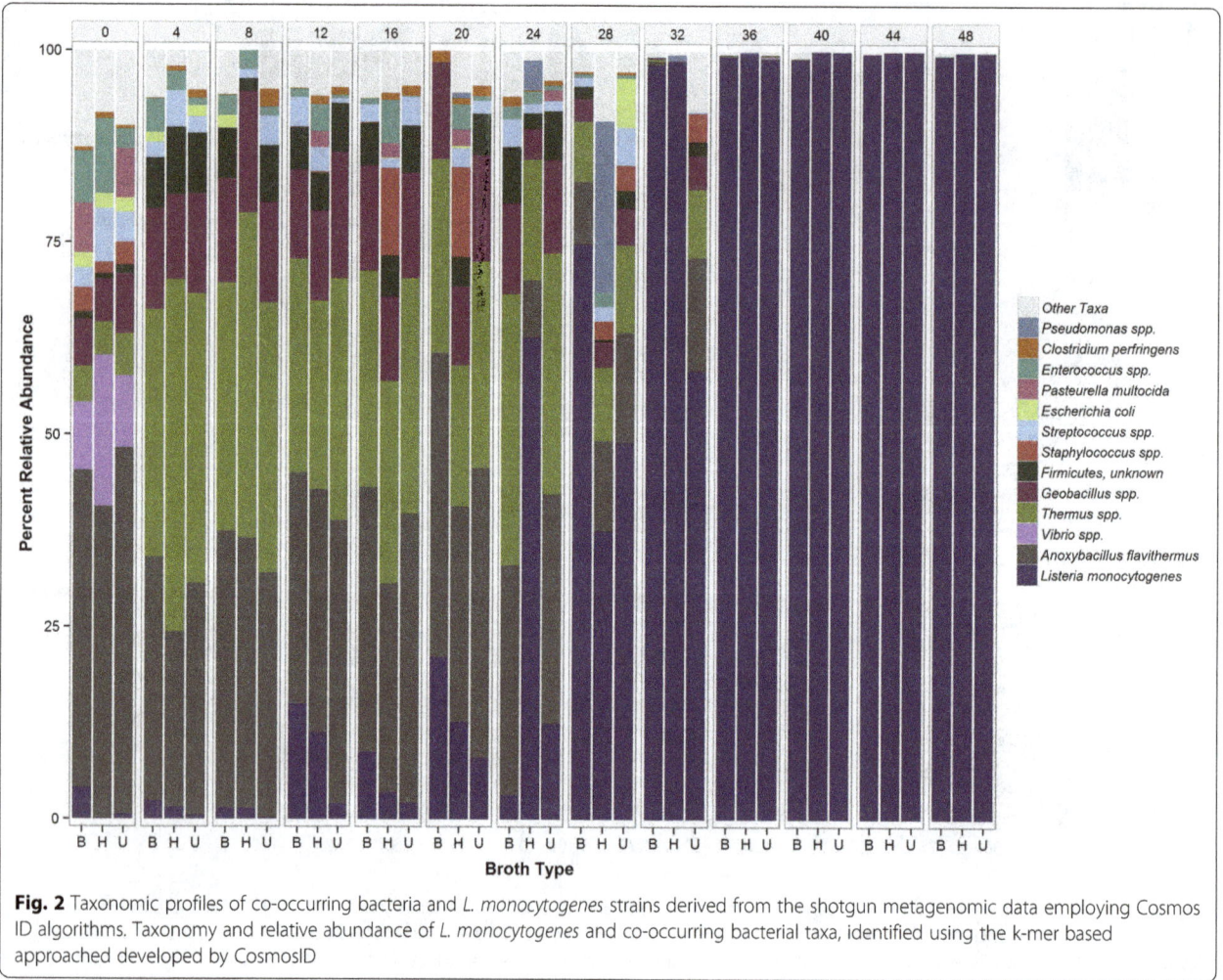

Fig. 2 Taxonomic profiles of co-occurring bacteria and *L. monocytogenes* strains derived from the shotgun metagenomic data employing Cosmos ID algorithms. Taxonomy and relative abundance of *L. monocytogenes* and co-occurring bacterial taxa, identified using the k-mer based approached developed by CosmosID

that microbiota, comprising approximately 20–30%. As observed for *L. monocytogenes*, *Anoxybacillus* and *Geobacillus* spp. outcompeted *Serratia* spp. during the 4 to 12 h time points. In terms of other taxa present in the ice cream microbiome, relative abundances of *Streptococcus* and *Erwinia* remained consistent during the first 24 h of enrichment until *L. monocytogenes* began to outcompete the rest of the community, while the relative abundances of *Pseudomonas* species increased after 24 h and decreased after 40 h of enrichment. In the absence of selective antimicrobial agents, we observed a completely different microbial community dynamics during the 48 h of non-selective enrichment (Additional file 2: Figure S2). At 0 to 8 h, we observed a very similar pattern to that in selective enrichments; however, after 16 h of non-selective enrichment, *Bacillus* species dominated the ice cream microbiome, comprising ~80% of the total population. After 24 h, *Lactococcus* emerged and became the slightly dominant species over *Bacillus* for the remainder of the non-selective enrichment.

L. monocytogenes genome coverage by shotgun data

Though 16S rRNA gene sequencing data provided a detailed description of the ice cream microbiome and resulted in the detection of *L. monocytogenes*, a deeper sequence analysis utilizing shotgun metagenomics was necessary to characterize strains of *L. monocytogenes*. Analysis of these samples revealed that near 100% genome coverage can be achieved as early as 24 h of enrichment (Table 1). *L. monocytogenes*-specific sequence reads constituted 0.20 to 0.76% of the metagenome, varying according to enrichment broths employed. At 24 h, we achieved 23 to 97% coverage of the *Listeria* genome at a 7.5× to 12× depth of coverage with 15 to 35 M total metagenomic sequence reads. A similar trend was also observed after 28 h, however, shotgun metagenomic reads from this time point yielded slightly higher genome coverages (44 to 98%) and the depth of coverage ranged from 14.5× to 95×. The highest coverage among these two time points and three enrichment schemes was achieved in BLEB ice cream enrichments at hour 28, which had 45 M total sequence reads, (>95× depth) with

Table 1 Shotgun data for potential target assemblies

Broth Type-Hour	Listeria Genome size (bp)	Size of the genome covered	Listeria read representations	Total sample reads	Predicted depth (x)	% Genome coverage
U-24 h	3109342	738261.44	59165	30102152	12.02	23.74
H-24 h	3109342	2280235.96	115521	15195488	7.6	73.34
B-24 h	3109342	3023679.63	239719	35204116	11.89	97.25
U-28 h	3109342	1360440.77	132749	81801390	14.64	43.75
H-28 h	3109342	3029431.91	419418	20818100	20.77	97.43
B-28 h	3109342	2981548.04	1884038	46165270	94.78	95.89

B BLEB, *H* HFB-FB, *U* UVM-FB

Shotgun sequence data are shown for hours 24 and 28 for all three enrichments. The target *L. monocytogenes* genome size is 3,109,342 bases

near complete genome coverage (>97%), even though the UVM-FB ice cream enrichments contained 82 M total reads (15× depth, 44% genome coverage) (Table 1).

Three putative strains of *L. monocytogenes*

Interestingly three variants of 16S rRNA gene sequences were observed in *L. monocytogenes* populations from the selective enrichments. Variable nucleotides occurred in the V2 region of the 16S rRNA gene (at positions 159 and 174, *E. coli*) with one variant comprised of AC nucleotides at those positions, a second with GT, and a third, the most abundant, possessing GC (Fig. 3, Table 2). Analysis of 68 closed genomes, available from the PATRIC database, revealed that the majority of *L. monocytogenes* genomes encoded six copies of the 16S rRNA gene (85.3%), with the remaining genomes encoding five (11.8%), or four (5.9%) copies. Six 16S rRNA gene variant sequence motifs were identified, with 96% of the 68 genomes harboring at least one of the variants identified in this study (GC, GT, or AC). Interestingly, among the 68 closed genomes, two sub-variants of type GC were identified (type GC.2 and GC.3), characterized by changes to nucleotides in the G and C allele positions. These minor variants were found in five *L. monocytogenes* lineage I genomes, with type GC.3 comprising the

sole 16S rRNA gene sequence variant present in three of these five genomes. Type GC was present in the four lineage III genomes analyzed. Overall, the distribution of major 16S rRNA gene sequence variants was not significantly associated with lineage.

Analysis of intra-genomic distribution of 16S rRNA gene variants (i.e. variant types representing the complete complement of 16S rRNA gene copies within a single genome) revealed the majority of *L. monocytogenes* genomes encoded a single 16S rRNA gene variant type (79.1%), though genomes containing multiple variant types (20.6%) were common. Overall, types GC or AC were most prevalent in genomes encoding a single variant, 51.5 and 20.6%, respectively; while types GC and GT represented the dominant type in mixed 16S rRNA gene variant genomes, 19.1 and 17.6%, respectively. Interestingly, type GT was predominantly found in mixed variant genomes, and accounted for the sole variant in only two genomes, whereas type AC was present in a single mixed variant genome (Table 2).

It is noteworthy that the AC 16S rRNA gene variant, which was the least abundant of the three types up to hour 40, and less common among the reference set of 68 closed genomes, overtook type GT at hours 44 and 48 across all of the media employed in the study. Type

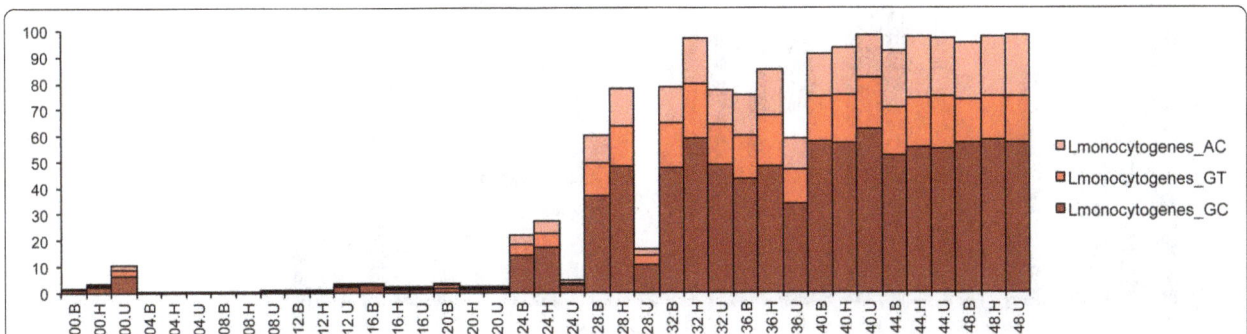

Fig. 3 Incidence and abundance of *L. monocytogenes* 16S rRNA gene sequence variants. Relative abundance of three *Listeria monocytogenes* 16S rRNA gene sequence variant types: AC, GT, and GC. Variants occurred at positions 159 and 174 (*E. coli* positions NCBI accession?) within the 16S rRNA gene. The y-axis shows the percent abundance of each *L. monocytogenes* 16S rRNA gene sequence variant within the total microbial community at each time point

Table 2 Characterization and distribution of 16S rRNA variants present in closed *L. monocytogenes* genomes

16S variants across genomes (n = 68)					16S variant within genomes						
						Variant by copy number (% total genomes)					
Variant	Genomes[a] (%)	Lineage (intra-lineal prevalence)				6	5b	4	3	2	1
		I (n = 38)	II (n = 25)	III (n = 4)	Individual variant						
GC	48 (70.6)	26 (68.4)	18 (72)	4 (100)	GC	31 (0.46)	9 (0.13)	5 (0.07)	2 (0.03)	1 (0.01)	n.d.
GT	14 (20.6)	9 (23.7)	5 (20)	n.d.	GT	2 (0.03)	n.d.	n.d.	n.d.	3 (0.04)	9 (0.13)
AC	15 (22.1)	8 (21.1)	6 (24)	n.d.	AC	12 (0.18)	2 (0.03)	n.d.	n.d.	1 (0.01)	n.d.
GC.2	1 (1.5)	5 (13.2)	n.d.	n.d.	GC.2	n.d.	n.d.	n.d.	1 (0.01)	n.d.	n.d.
GC.3	5 (7.4)	1 (2.6)	n.d.	n.d.	GC.3	3 (0.04)	n.d.	1 (0.01)	n.d.	n.d.	1 (0.01)
					Total	48 (0.71)	11 (0.16)	6 (0.09)	3 (0.04)	5 (0.07)	10 (0.15)

[a]Several genomes encoded multiple (different) 16S variants
[b]Six genomes carried only five 16S copies, with types GC and AC found in three and two genomes, respectively

GC remained the most abundant throughout all time points (Fig. 3). This observation was corroborated by analysis of shotgun datasets against a *Listeria* specialty database, suggesting the possibility that of three potentially distinct *L. monocytogenes* strains were present. CosmosID identified strain-specific biomarkers (AACA-BABA, AACABB, and AACABD) in the metagenomes, which was evidence for three putative *L. monocytogenes* variants (Fig. 4). Furthermore, we observed an interesting strain interplay in terms of their abundance over time where strain AACABD was most abundant during the first 20 h of enrichment and became the least abundant at later time points, due to the rapid upsurge of strain AACABABA, beginning at 28 h (Fig. 4). This

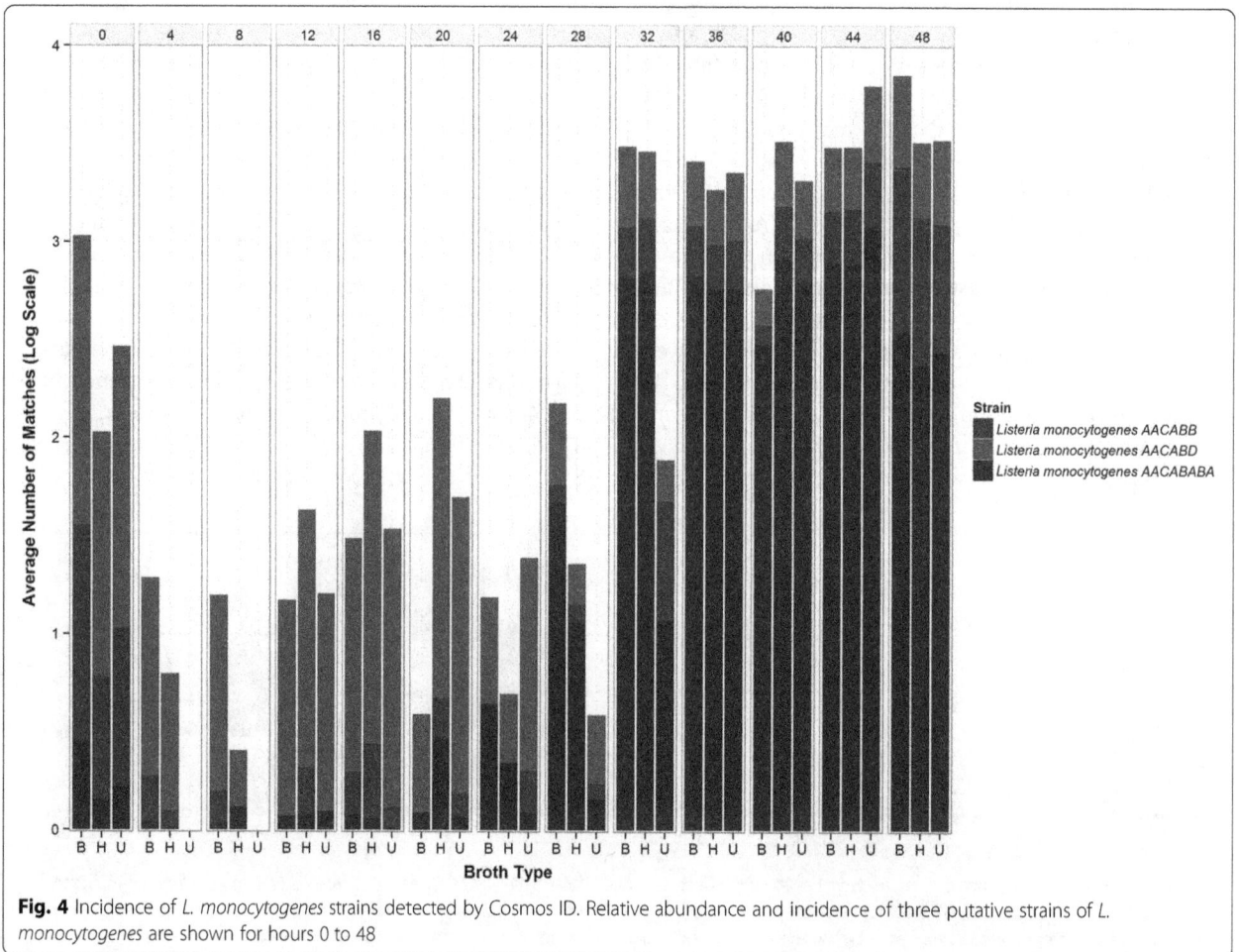

Fig. 4 Incidence of *L. monocytogenes* strains detected by Cosmos ID. Relative abundance and incidence of three putative strains of *L. monocytogenes* are shown for hours 0 to 48

same phenomenon was observed in the 16S data which suggests we may be documenting something biologically relevant.

Discussion

Every food has its own innate or imparted microbiome that will respond to enrichment conditions according to complex eco-physiology. Frequently, antagonistic micro-organisms are co-enriched along with target pathogens [21]. For example, *Paenibacillus* spp.,which are capable of inhibiting and killing *Salmonella* [22, 23], were also enriched using protocols outlined in the FDA BAM to recover *Salmonella* from tomatoes [21, 24] and cilantro [25]. Thus, continued optimization of reference enrichment protocols is still needed. In the case of *L. monocytogenes,* additional selective agents as well as changes in media formulations may improve efficiency of recovery; however, the situation may be more challenging in this case due to relatedness of *L. monocytogenes* to its co-enriching Bacilli relatives. The increase (almost doubling) of *Anoxybacillus* and *Geobacillus* spp. during the first 8 h of enrichment of the ice cream microbiome suggested these bacteria were able to adapt more readily to the environmental changes inherent to this study (Fig. 1).

Length of lag phase and growth rate typically depend on specific environmental parameters, as well as fitness of the bacterial cells. Many factors have been reported to play a role in length of lag time and/or growth rate for a given bacterial species, including nutritional content, pH, physical environment, inorganic nutrients, temperature, rate of temperature change, water activity, gas atmosphere, inhibitors, spore germination, and initial cell levels, as well as fitness, age, size, and health of individual cells [13, 26, 27]. Although lag time and/or growth rate modeling has been proposed to describe what should occur within a specific set of parameters, such results were variable [26–29]. Hence, it is evident that relationships between growth environments and lag time and growth rate are complex [26], especially for testing foodborne pathogens in food and environmental samples. Additionally, the effect of co-occurring bacterial species in enrichments has not been extensively described or considered in the literature since the analytic tools to do so have not been available and it is not always possible to obtain a sufficient number of naturally contaminated samples that are 100% positive for the target pathogen with relatively homogeneous levels of contamination [12].

The sequence depth and genome coverage achieved in the ice cream metagenomes was sufficient to generate draft *L. monocytogenes* genomes after 24–28 h of enrichment, which is considerably shorter than the 96 h typically required to enrich and isolate viable bacteria from food samples. Near complete genome coverage would facilitate high-resolution phylogenetic analysis, identification of virulence and antibiotic resistance factors, as well as subtyping. Thus this approach might be applicable for rapid and accurate microbial forensics in future work but remains to be more extensively validated.

Type GC represented the predominant allele and was present among the majority of strains belonging to lineage I (808/914 total lineage I strains examined), and was the representative pattern in 30 of the outbreak isolates sequenced in FDA labs. Type GC was also heterogeneously distributed within and between multilocus sequence typing (MLST) clonal groups, designated as clonal complex (CC), for strains within lineage II (361/914 total lineage II strains) Type GC sequences were also present in three lineage III genomes. Type GT and type AC were rare among lineage I (19/914) and lineage II (1/914) strains, respectively. The GC type was significantly prevalent among the human clinical strains from both lineages. The two variable positions (159 and 174) have been shown to contribute to the structure of helix 8, located in the 5′ major domain of the 16S rRNA gene (*E. coli*), which forms direct contact with nearby helices (e.g., helix 6) in the small ribosomal subunit [30]. As correct 16S rRNA gene folding is important for proper ribosome assembly, translational kinetics, and fidelity [31, 32] it is intriguing to consider how specific allelic changes within this region may affect *L. monocytogenes* growth under infection-associated conditions.

Conclusions

The ability to rapidly and accurately identify the etiologic agent of a foodborne outbreak from a variety of food matrices is critical for preserving public health. Culture-based recovery methods have been used for over 100 years and many have been optimized for recovery of major bacterial and fungal pathogens. We demonstrated that shotgun metagenomic sequencing facilitated the characterization of enrichment microbiota dynamics and identified the presence of competitive species co-occurring during enrichment. As *L. monocytogenes* was enriched, so were other bacterial species. The taxonomy and dynamics of co-enriching species are likely unique to each food commodity and may play a significant, as yet unidentified, role in the recovery and growth of target pathogens. Lag times and growth rates for certain pathogens may be significantly influenced by those co-enriching species "native" to the food commodity or its production environment. Thus, the approach taken in this study, namely to use a culture independent method to describe culture dependent dynamics of *L. monocytogenes* recovery and growth, provides insight and makes it possible to improve both sequence- and culture-based *L. monocytogenes* detection methods. The validation that

Resphera Insight can accurately identify *L. monocytogenes* using 16S rRNA amplicons is quite useful. As is the likelihood that draft assemblies of *L. monocytogenes* from shotgun metagenomic data may provide equivalent trace-back utility to WGS genomes - although more work will be needed to confirm this utility.

Methods
Ice cream sampling scheme
Ice cream samples (80–85 g /scoop) were aseptically transferred to Whirlpak bags, allowed to stand at room temperature for 20 to 30 min to fully melt. Analytical units were set up as follows: 25 g portions were added to 225 ml of enrichment broth, and stomached for 1 min. Four scoops were used for each of three enrichment protocols.

Enrichment methods
BLEB (FDA BAM)
Ice cream samples were incubated at 30 °C for 48 h in buffered *Listeria* enrichment broth (BLEB) (Cat. CM0897, Thermo Scientific Inc., Waltham, MA). Selective supplements (acriflavin hydrochloride 10 mg/l, nalidixic acid 40 mg/l and cycloheximide 50 mg/l, Cat. SR0149, Oxoid, UK) were added after 4 h of incubation.

UVM-FB (USDA)
Ice cream samples were incubated at 30 °C for 24 h in University of Vermont modified broth (UVM) after which, 0.1 ml of each culture was transferred to Fraser broth (FB) and incubated at 37 °C for 24 h. Selective agents were added to UVM and FB prior to enrichment.

HFB-FB (ISO)
Ice cream samples were incubated at 30 °C for 24 h in Half-Fraser broth (HFB) after which, 0.1 ml of culture was transferred to FB and incubated at 37 °C for 24 h. Selective agents were added to HFB and FB prior to the enrichment.

For each of the enrichment protocols, 4 ml was taken from each sample at hour 0. After that, every 4 h during the 48 h incubation, enrichments were stomached for 1 min and 4 ml samples taken from each of the four replicates of the three enrichment mixes. Samples were immediately frozen at –20 °C for subsequent DNA analysis.

DNA extraction
Genomic DNA was extracted using DNeasy Blood and Tissue kit (Cat No. 69506, Qiagen, Germantown, MD, USA) following the protocol for Gram-positive bacteria with minor modification: 1.5 ml of the culture was pelleted (5000 × g, 15 min) and the pellet resuspended in 200 ml of enzymatic lysis buffer containing 20 mM Tris-HCl (pH-8.0), 2 mM Sodium EDTA, 1.2% Triton X- 100, 20 mg/ml of lysozyme. The samples were subsequently

incubated for 60 min at 37 °C. DNA was extracted following manufacturer's protocol.

Shotgun library preparation
Sequencing libraries were prepared using the Truseq Nano prep kit (Illumina, SanDiego, CA, USA), according to the manufacturer's specifications.

16S rRNA gene amplification
PCR for 16S rRNA gene sequencing targeted the first 330 bases of the V1-V2 region, using the following PCR primers 27 F1 TCGTCGGCAGCGTCAGATGTGTAT AAGAGACAGAGAGTTTGATCMTGGCTCAG and 336R1 GTCTCGTGGGCTCGGAGATGTGTATAAGA-GACAGACTGCTGCSYCCCGTAGGAGTCT.

PCR cycling conditions consisted of an initial denaturation at 94 °C for 2 min, followed by 25 cycles of 94 °C for 40 s, 56 °C for 15 s, 68 °C for 40 s, and a final extension at 68 °C for 5 min. The amplicons were indexed using Illumina Nextera indexing primers following the manufacturer's instructions (Illumina, San Diego, CA, USA).

Sequencing
16S rRNA gene amplicon sequencing was performed on an Illumina MiSeq using the 500 cycle V2 chemistry and cartridges. Shotgun data was obtained by sequencing on an Illumina NextSeq, using V2 chemistry (2 by 150).

Bioinformatic analyses
16S rRNA amplicon sequence analysis
Raw paired-end reads from the MiSeq platform were merged into consensus fragments by FLASH [33] and subsequently filtered for quality (max error rate 1%) and length (minimum 300 bp) using Trimmomatic [34] and QIIME [35, 36]. Spurious hits to the PhiX control genome were identified using BLASTN and removed. Passing sequences were trimmed of primers, evaluated for chimeras with UCLUST (de novo mode) [37], and screened for chloroplast and mitochondrial contaminants using the RDP classifier [38] with a threshold of 0.5. Sequences were further evaluated for unknown contaminants using a sensitive BLASTN search against the GreenGenes 16S database [39]. High-quality 16S sequences were submitted for high-resolution taxonomic profiling using Resphera Insight (Baltimore, MD, www.respherabio.com). Sequence counts were rarefied to 5000 sequences for each independent replicate for downstream analyses.

Subtyping of *L. monocytogenes* 16S rRNA gene sequence fragments
To evaluate the potential for multiple cultured strains in the enrichments, all 16S rRNA gene sequences assigned unambiguously to *L. monocytogenes* by Resphera Insight

were aligned by PYNAST [36] to a smaller template multiple sequence alignment (MSA) of 5000 randomly selected sequences from the same set (generated by MUSCLE [40]). The full PYNAST MSA was filtered for positions with > 10% gaps and passing positions submitted for entropy calculation [41]. MSA Positions with measured entropy > 0.7 were utilized to assign putative strain membership.

Validation of Resphera Insight

To perform an external validation of the species level accuracy of Resphera Insight, Resphera Biosciences and Center for Food Safety and Applied Nutrition (CFSAN) scientists collaborated to interrogate 1695 whole-genome shotgun datasets from the GenomeTrakr Project (NCBI Project ID PRJNA183844) designated as *L. monocytogenes* isolates. Raw paired-end sequences were filtered for quality and length, followed by merging of overlapping sequences using FLASH [33]. Merged reads were screened for 16S rRNA fragments using Bowtie2 [42] against a broad database of 16S rRNA gene sequences, with additional BLAST-based filtering to confirm location specific query matches to a reference *L. monocytogenes* 16S rRNA gene (*L. monocytogenes* strain 07PF0776; NCBI accession NR_102780.1).

Passing sequences were submitted to Resphera Insight for high-resolution taxonomic identification. The primary measure of performance was the *Diagnostic True Positive rate* (DTP), defined as the percentage of reads with an unambiguous assignment to *L. monocytogenes* and differences in accuracy associated with changes in read length and gene position were evaluated. We also computed *Sensitivity* (SN), defined as percentage of reads with an unambiguous or ambiguous assignment to *L. monocytogenes*.

Overall, across all 1695 isolates, for fragments ≥200 bp originating within 16S reference gene positions 1–36, Resphera Insight achieved a mean diagnostic true positive rate of 99.43% and a mean sensitivity of 99.94%, with a misassignment rate of 0.06%. We observed increased DTP rates associated with increasing fragment lengths and a loss of species level resolution 3′ to 16S reference gene position 200 (Additional file 1: Figure S1). For fragments <200 bp, we also found a loss of DTP resolution, as more assignments became ambiguous, but sensitivity was not reduced.

To establish benchmark false positive rates of *L. monocytogenes*, we simulated amplicon fragments from our primer region for eight closely related *Listeria* species (*L. fleischmannii, L. grayi, L. innocua, L. ivanovii, L. rocourtiae, L. seeligeri, L. weihenstephanensis, L. welshimeri*). A total of 10,000 sequences per species were simulated, lengths 250–500 bp, with a random nucleotide error rate of 0.5%. Overall, the average sensitivity of detection of

these organisms was 99.9%, with a misclassification rate (assignments to *L. monocytogenes*) of 0.07%.

Shotgun metagenomic data analysis

Unassembled metagenomic sequencing reads were directly analyzed by Genius bioinformatics software package (CosmosID Inc., Rockville, MD), described elsewhere [43, 44] to achieve identification at the species, subspecies, and/or strain level and quantification of relative abundance. Briefly, the system utilizes curated genome databases and a high performance data-mining algorithm to rapidly disambiguate millions of metagenomic sequence reads into discrete microbial taxa. The pipeline has two separable comparators. The first consists of a pre-computation phase and a per-sample computation. The input to the pre-computation phase is a curated reference microbial database and its output is a whole genome phylogeny tree, together with sets of fixed length n-mer fingerprints (biomarkers) that are uniquely identified to distinct nodes of the tree. The second per-sample, computational phase searches the millions of sequence reads against the fingerprint sets. The resulting statistics are analyzed to give fine-grain composition and relative abundance estimates at all nodes of the tree. Overall classification precision is maintained through aggregation statistics.

Furthermore, CosmosID and CFSAN collaboratively developed a specialized *Listeria* database including genome sequences of all *L. monocytogenes* available at the time of this analysis. As for the pre-computational phase describe above, the *L. monocytogenes* database was organized as a phylogenetic tree with thousands of unique and shared biomarkers specific to distinct clades, branches, nodes and leaves of the tree along with specific alphabetical fingerprints reflective of their phylogenetic hierarchies. All metagenomic datasets were analyzed against this *L. monocytogenes* database to investigate the potential presence of multiple strains of *L. monocytogenes* in the ice cream samples. Predicted % genome coverages of *L. monocytogenes* were estimated based on the mean of % total match – a statistics derived from CosmosID algorithm that approximates genome coverage. Predicted depths of coverage were calculated as $X = $ (# of *Listeria* reads X Read length) / (*L. monocytogenes* genome size X Predicted % genome coverage).

Additional files

> **Additional file 1: Figure S1.** Resphera Insight validation results across 1695 *L. monocytogenes* isolates. (A) Overall, Resphera Insight achieved a mean diagnostic true positive rate of 99.43% and a mean sensitivity of 99.94% with a mis-assignment rate of 0.06%. (B) Evaluation of DTP by read length and gene position. Among reads ≥200 bp covering the first

> 200 bp of the 16S gene, Resphera Insight achieves up to 99.7% DTP rates. For sequences <200 bp, we find reduced resolution overall.
>
> **Additional file 2: Figure S2.** Enrichment protocol without antibiotics did not promote *Listeria* identification.

Abbreviations

BLEB: Buffered *Listeria* Enrichment broth; FB: Fraser broth; HFB: Half-Fraser broth; *L. monocytogenes*: *Listeria monocytogenes*; NGS: Next generation sequencing; UVM: University of Vermont modified broth

Acknowledgements

Not Applicable.

Funding

The work was supported by the Office of Regulatory Science at the Center for Food Safety and Applied Nutrition.

Authors' contributions

AO, YC, PR, LR: designed and coordinated the study, carried out laboratory work and wrote the manuscript; JRW, NH, PS, GR, PR performed bioinformatic analyses and edited the manuscript; KJ, ND, CG: carried out laboratory work and edited the manuscript; RC, EB, MA, DH provided scientific advisement and edited the manuscript. All authors read and approved the final manuscript.

Competing interests

CFSAN authors worked collaboratively with CosmosID and Resphera Biosciences (commercial bioinfomatic enterprises) to design and apply analyses described in this work.

Endnotes

Not applicable.

Author details

[1]Office of Regulatory Science, Center for Food Safety and Applied Nutrition, Food and Drug Administration, 5001 Campus Drive, College Park, MD 20740, USA. [2]Resphera Biosciences, 1529 Lancaster Street, Baltimore, MD 21231, USA. [3]CosmosID, 155 Gibbs Street, Rockville, MD 20850, USA. [4]Office of Applied Research and Safety Assessment, Center for Food Safety and Applied Nutrition, Food and Drug Administration, 8301 Muirkirk Road, Laurel, MD 20708, USA.

References

1. Low JC, Donachie W. A Review of Listeria monocytogenes and Listeriosis. Vet J. 1997;153(1):9–29.
2. Murray EGD, Webb RA, Swann MBR. A disease of rabbits characterised by a large mononuclear leucocytosis, caused by a hitherto undescribed bacillus Bacterium monocytogenes (n.sp.). J Pathol Bacteriol. 1926;29(4):407–39.
3. Jones D. Foodborne listeriosis. Lancet. 1990;336(8724):1171–4.
4. Schlech III WF, Lavigne PM, Bortolussi RA, Allen AC, Haldane EV, Wort AJ, Hightower AW, Johnson SE, King SH, Nicholls ES. Epidemic listeriosis—evidence for transmission by food. N Engl J Med. 1983;308(4):203–6.
5. Junttila J, Brander M. Listeria monocytogenes septicemia associated with consumption of salted mushrooms. Scand J Infect Dis. 1989;21(3):339–42.
6. Chen Y, Gonzalez-Escalona N, Hammack TS, Allard M, Strain EA, Brown EW. Core genome multilocus sequence typing for the identification of globally distributed clonal groups and differentiation of outbreak strains of Listeria monocytogenes. Appl Environ Microbiol. 2016;82:6258–72.
7. Multistate outbreak of listeriosis linked to Roos foods dairy products (Final Update). [http://www.cdc.gov/listeria/outbreaks/cheese-02-14]. Accessed 14 Nov 2016.
8. Jackson BR, Salter M, Tarr C, Conrad A, Harvey E, Steinbock L, Saupe A, Sorenson A, Katz L, Stroika S, et al. Notes from the field: listeriosis associated with stone fruit-United States, 2014. Morb Mortal Wkly Rep. 2015;64(10):282–3.
9. Multistate outbreak of listeriosis linked to Blue Bell creameries products (Final Update). [http://www.cdc.gov/listeria/outbreaks/ice-cream-03-15]. Accessed 14 Nov 2016.
10. McCollum JT, Cronquist AB, Silk BJ, Jackson KA, O'Connor KA, Cosgrove S, Gossack JP, Parachini SS, Jain NS, Ettestad P, et al. Multistate outbreak of listeriosis associated with cantaloupe. N Engl J Med. 2013;369(10):944–53.
11. Multistate outbreak of listeriosis linked to commercially produced, prepackaged caramel apples made from Bidart Bros. apples (Final Update). [http://www.cdc.gov/listeria/outbreaks/caramel-apples-12-14]. Accessed 14 Nov 2016.
12. Chen Y, Burall L, Macarisin D, Pouillot R, Strain E, De Jesus A, Laasri AM, Wang H, Ali A, Tatavarthy A, et al. Prevalence and level of Listeria monocytogenes in ice cream linked to a listeriosis outbreak in the United States. J Food Prot. 2016;79:1828–32. Accepted.
13. Chen Y, Pouillot R, Burall LS, Strain EA, Van Doren JM, De Jesus AJ, Laasri A, Wang H, Ali L, Tatavarthy A. Comparative evaluation of direct plating and most probable number for enumeration of low levels of Listeria monocytogenes in naturally contaminated ice cream products. Int J Food Microbiol. 2017;241:15–22.
14. Chen Y, Allard E, Wooten A, Hur M, Sheth I, Lassri A, Hammack TS, Macarisin D. Recovery and growth potential of Listeria monocytogenes in temperature abused milkshakes prepared from naturally contaminated ice cream linked to a listeriosis outbreak. Front Microbiol. 2016;7:764.
15. Food and Drug Administration Bacteriological Analytical Manual Chapter 10, Detection and enumeration of Listeria monocytogenes in foods. http://www.fda.gov/Food/FoodScienceResearch/LaboratoryMethods/ucm071400.htm. Accessed 14 Nov 2016.
16. Standardisation IOf. ISO 11290–1 Microbiology of Food and Animal Feeding Stuffs - Horizontal Method for Detection and Enumeration of Listeria Monocytogenes - Part 1: Detection Method. 1996.
17. Laboratory Guidebook . Isolation and Identification of Listeria monocytogenes from Red Meat, Poultry and Egg Products, and Environmental Samples. [http://www.fsis.usda.gov/wps/wcm/connect/1710bee8-76b9-4e6c-92fc-fdc290dbfa92/MLG-8.pdf?MOD=AJPERES]. Accessed 14 Nov 2016.
18. Goh KM, Gan HM, Chan K-G, Chan GF, Shahar S, Chong CS, Kahar UM, Chai KP. Analysis of Anoxybacillus genomes from the aspects of lifestyle adaptations, prophage diversity, and carbohydrate metabolism. PLoS One. 2014;9(3):e90549.
19. Dwivedi V, Kumari K, Gupta SK, Kumari R, Tripathi C, Lata P, Niharika N, Singh AK, Kumar R, Nigam A, et al. Thermus parvatiensis RLT sp. nov., isolated from a hot water spring, located atop the Himalayan ranges at Manikaran, India. Indian J Microbiol. 2015;55(4):357–65.
20. Ohtani N, Tomita M, Itaya M. An extreme thermophile, thermus thermophilus, is a polyploid bacterium. J Bacteriol. 2010;192(20):5499–505.
21. Pettengill JB, McAvoy E, White JR, Allard M, Brown E, Ottesen A. Using metagenomic analyses to estimate the consequences of enrichment bias for pathogen detection. BMC Res Notes. 2012;5(1):378.
22. Allard S, Enurah A, Strain E, Millner P, Rideout SL, Brown EW, Zheng J. In situ evaluation of Paenibacillus alvei in reducing carriage of Salmonella enterica serovar Newport on whole tomato plants. Appl Environ Microbiol. 2014;80(13):3842–9.
23. Luo Y, Wang C, Allard S, Strain E, Allard MW, Brown EW, Zheng J. Draft genome sequences of Paenibacillus alvei A6-6i and TS-15. Genome Announc. 2013;1:5. e00673-00613.
24. Ottesen AR, Gonzalez A, Bell R, Arce C, Rideout S, Allard M, Evans P, Strain E, Musser S, Knight R, et al. Co-enriching microflora associated with culture based methods to detect salmonella from tomato phyllosphere. PLoS One. 2013;8(9):e73079.
25. Jarvis KG, White JR, Grim CJ, Ewing L, Ottesen AR, Beaubrun JJ, Pettengill JB, Brown E, Hanes DE. Cilantro microbiome before and after nonselective pre-enrichment for Salmonella using 16S rRNA and metagenomic sequencing. BMC Microbiol. 2015;15(1):160.

26. Robinson TP, Ocio MJ, Kaloti A, Mackey BM. The effect of the growth environment on the lag phase of Listeria monocytogenes. Int J Food Microbiol. 1998;44(1):83–92.

27. Whiting RC, Bagi LK. Modeling the lag phase of Listeria monocytogenes. Int J Food Microbiol. 2002;73(2–3):291–5.

28. Delignette-Muller ML, Baty F, Cornu M, Bergis H. Modelling the effect of a temperature shift on the lag phase duration of Listeria monocytogenes. Int J Food Microbiol. 2005;100(1–3):77–84.

29. Francois K, Devlieghere F, Smet K, Standaert AR, Geeraerd AH, Van Impe JF, Debevere J. Modelling the individual cell lag phase: effect of temperature and pH on the individual cell lag distribution of Listeria monocytogenes. Int J Food Microbiol. 2005;100(1–3):41–53.

30. Noller HF. RNA structure: reading the ribosome. Science. 2005;309(5740):1508–14.

31. Kitahara K, Yasutake Y, Miyazaki K. Mutational robustness of 16S ribosomal RNA, shown by experimental horizontal gene transfer in Escherichia coli. Proc Natl Acad Sci. 2012;109(47):19220–5.

32. Roy-Chaudhuri B, Kirthi N, Culver GM. Appropriate maturation and folding of 16S rRNA during 30S subunit biogenesis are critical for translational fidelity. Proc Natl Acad Sci. 2010;107(10):4567–72.

33. Magoc T, Salzberg SL. FLASH: fast length adjustment of short reads to improve genome assemblies. Bioinformatics (Oxford, England). 2011;27(21):2957–63.

34. Bolger AM, Lohse M, Usadel B. Trimmomatic: a flexible trimmer for Illumina sequence data. Bioinformatics. 2014;30(15):2114–20.

35. Kuczynski J, Stombaugh J, Walters WA, Gonzalez A, Caporaso JG, Knight R. Using QIIME to analyze 16S rRNA gene sequences from microbial communities. Curr Protoc Bioinformatics. 2011;Chapter 10:Unit 10 17.

36. Caporaso JG, Kuczynski J, Stombaugh J, Bittinger K, Bushman FD, Costello EK. QIIME allows analysis of high-throughput community sequencing data. Nat Methods. 2010;7:335–6.

37. Edgar RC, Haas BJ, Clemente JC, Quince C, Knight R. UCHIME improves sensitivity and speed of chimera detection. Bioinformatics. 2011;27:2194–200.

38. Cox-Foster DL, Conlan S, Holmes EC, Palacios G, Evans JD, Moran NA, Quan P-L, Briese T, Hornig M, Geiser DM, et al. A metagenomic survey of microbes in honey bee colony collapse disorder. Science. 2007;318(5848):283–7.

39. McDonald D, Price MN, Goodrich J, Nawrocki EP, DeSantis TZ, Probst A, Andersen GL, Knight R, Hugenholtz P. An improved Greengenes taxonomy with explicit ranks for ecological and evolutionary analyses of bacteria and archaea. ISME J. 2012;6(3):610–8.

40. Edgar RC. MUSCLE: multiple sequence alignment with high accuracy and high throughput. Nucleic Acids Res. 2004;32(5):1792–7.

41. Eren AM, Maignien L, Sul WJ, Murphy LG, Grim SL, Morrison HG, Sogin ML. Oligotyping: differentiating between closely related microbial taxa using 16S rRNA gene data. Methods Ecol Evol. 2013;4:12.

42. Langmead B, Salzberg SL. Fast gapped-read alignment with Bowtie 2. Nat Methods. 2012;9(4):357–9.

43. Hasan NA, Young BA, Minard-Smith AT, Saeed K, Li H, Heizer EM, McMillan NJ, Isom R, Abdullah AS, Bornman DM, et al. Microbial community profiling of human saliva using shotgun metagenomic sequencing. PLoS One. 2014;9(5):e97699.

44. Lax S, Smith DP, Hampton-Marcell J, Owens SM, Handley KM, Scott NM, Gibbons SM, Larsen P, Shogan BD, Weiss S. Longitudinal analysis of microbial interaction between humans and the indoor environment. Science. 2014;345(6200):1048–52.

The process-related dynamics of microbial community during a simulated fermentation of Chinese strong-flavored liquor

Yanyan Zhang[1,2], Xiaoyu Zhu[1], Xiangzhen Li[1], Yong Tao[1,3]* ⓘ, Jia Jia[1] and Xiaohong He[1]

Abstract

Background: Famous Chinese strong-flavored liquor (CSFL) is brewed by microbial consortia in a special fermentation pit (FT). However, the fermentation process was not fully understood owing to the complicate community structure and metabolism. In this study, the process-related dynamics of microbial communities and main flavor compounds during the 70-day fermentation process were investigated in a simulated fermentation system.

Results: A three-phase model was proposed to characterize the process of the CSFL fermentation. (i) In the early fermentation period (1–23 days), glucose was produced from macromolecular carbohydrates (e.g., starch). The prokaryotic diversity decreased significantly. The *Lactobacillaceae* gradually predominated in the prokaryotic community. In contrast, the eukaryotic diversity rose remarkably in this stage. *Thermoascus, Aspergillus, Rhizopus* and unidentified *Saccharomycetales* were dominant eukaryotic members. (ii) In the middle fermentation period (23–48 days), glucose concentration decreased while lactate acid and ethanol increased significantly. Prokaryotic community was almost dominated by the *Lactobacillus*, while eukaryotic community was mainly comprised of *Thermoascus, Emericella* and *Aspergillus*. (iii) In the later fermentation period (48–70 days), the concentrations of ethyl esters, especially ethyl caproate, increased remarkably.

Conclusions: The CSFL fermentation could undergo three stages: saccharification, glycolysis and esterification. *Saccharomycetales, Monascus,* and *Rhizopus* were positively correlated to glucose concentration ($P < 0.05$), highlighting their important roles in the starch saccharification. The *Lactobacillaceae, Bacilli, Botryotinia, Aspergillus,* unidentified *Pleosporales* and *Capnodiales* contributed to the glycolysis and esterification, because they were positively correlated to most organic acids and ethyl esters ($P < 0.05$). Additionally, four genera, including *Emericella, Suillus, Mortierella* and *Botryotinia,* that likely played key roles in fermentation, were observed firstly. This study observed comprehensive dynamics of microbial communities during the CSFL fermentation, and it further revealed the correlations between some crucial microorganisms and flavoring chemicals (FCs). The results from this study help to design effective strategies to manipulate microbial consortia for fermentation process optimization in the CSFL brew practice.

Keywords: Chinese strong-flavored liquor, Microbial community, Dynamics, Flavoring chemicals

* Correspondence: taoyong@cib.ac.cn
[1]Key Laboratory of Environmental and Applied Microbiology, Chinese Academy of Sciences & Environmental Microbiology Key Laboratory of Sichuan Province, Chengdu Institute of Biology, Chinese Academy of Sciences, Chengdu 610041, People's Republic of China
[3]Chengdu Institute of Biology, Chinese Academy of Sciences, Chengdu 610041, People's Republic of China
Full list of author information is available at the end of the article

Background

Chinese strong-flavored liquor (CSFL) is a typical representatives of Chinese liquor, accounting for about 70% of Chinese liquor market share [1]. The CSFL is produced by the Chinese classic solid-state fermentation, which involves a spontaneous process with simultaneous saccharification and fermentation [2]. The procedure details include mixing pre-culture starter (Daqu) [3] and pulverizing grains (e.g., sorghum, corn, wheat and rice) [4], filling the mixture into the fermentation pit (FT, hereafter, unless otherwise indicated) under the ground and sealing it with mud (Additional file 1: Figure S1). Daqu is a traditional fermentation starter, which is produced in an open environment from non-sterilized raw materials, e.g. raw wheat, barley and/or pea. It is reported that *Lactobacillus*, *Bacillus*, *Aspergillus*, and some non-*Saccharomyces* genera (*Saccharomycopsis*, *Pichia*) are dominant microbes in different types of Daqu [3]. After fermenting for 60–70 days, the fermented grains (also called Zaopei) are taken out from the FT and are mixed a number of fresh pulverizing grains,and are distilled to gain the CSFL. After that, the steamed grains (a mixture of Zaopei and fresh grains) are reused by mixing Daqu for fermentation again [1, 5]. The microorganisms play critical roles for the production of the CSFL because they can convert carbohydrates (e.g., starch, sucrose and glucose) into ethanol [4, 6–9]. In addition, microbes also produce various flavoring compounds, such as lactic acid, butyric acid, caproic acid, and ethyl caproate [10–12]. In particular, caproic acid and ethyl caproate are defining flavoring substances that determines the quality of the CSFL to a large degree [1].

High CSFL quality is attributed to the dynamics of microbial community and their metabolisms in the fermentation process. Previously, the CSFL fermentation microbiota have been studied using cultivation-dependent and -independent approaches using denaturing gradient gel electrophoresis (DGGE) and clone library analysis of the 16S rRNA gene [4, 13, 14]. However, big discrepancies in microbial compositions existed among previous investigations. This may be attributable to differences in sampling time (different stages of fermentation), and laboratory techniques employed to characterize the community structure. In addition, most of the previous studies on the CSFL fermentation microbiota using traditional cultural and molecular methods cannot provide details of the phylogenetic compositions and process-related changes of microbial community [4, 9, 13, 15–17]. However, understanding the process-related dynamics of microbial community is important to design effective strategies to manipulate microbial consortia for fermentation process optimization in the CSFL brew practice. The next generation sequencing technique provided powerful tools to reveal the microbial community dynamics in the complicated environments [6].

In this study, we investigated the process-related dynamic of microbial communities (bacteria, archaea and fungi) and metabolites during different fermentation stages (1, 10, 23, 34, 48, 59, and 70 days) using MiSeq-sequencing targeting 16S rRNA and ITS genes, respectively, and identify the correlations between key microbial taxa and the flavoring compounds of the CSFL.

Methods

Sampling

The production of CSFL undergoes anaerobic fermentation in the FT under the ground. Sampling from real fermentation pit would disrupt fermentation process. Thus, we used batch experiments to simulate CSFL fermentation using glass bottles of 3.5 L (Additional file 1: Figure S2) as laboratory reactors. The fermentative samples (a mixture of Zaopei, Daqu, and fresh grains including sorghum, corn, wheat and rice, same as the real sample in the distiller) were collected from a well-known distillery, located in Mianzhu city, Sichuan province, China. A total of 21 bottles were filled with the fermentative samples described above, and sealed with mud,frosted-glass stopper and plastic sheets (Additional file 1: Figure S2), and cultivated at 30 °C for 70 days. During different fermentation stages of 1, 10, 23, 34, 48, 59, and 70 days, three parallel bottles were sacrificed for sampling each time. Samples were stored at −80 °C for the further use.

Chemical and physical property analysis

The pH was measured using pH meter in the suspension liquid after sample centrifugation, with a 1:5 ratio of sample to deionized water. The moisture was measured using a gravimetric approach by drying samples between 103 °C–105 °C for 48 h after sampling. For the detection of organic acid, such as lactic acid, acetic acid, butyric acid, and caproic acid, as well as glucose and ethanol content, 5 g of sample was vortex mixed with 25 mL of deionized water, centrifuged and filtered through a 0.22 μm MCE filter, and the metabolite contents were quantified using HPLC (Agilent 1260,USA). The operating condition of HPLC was as follows: Hi-Plex H HPLC column (300 × 6.5 mm), refractive index detector (RID detector), 5 mM H_2SO_4 as mobile phase with the velocity of 0.6 mL/min, and column temperature was 55 °C. For the detection of esters, such as ethyl acetate, ethyl caproate, and ethyl lactate, 5 g of sample was vortex mixed with 25 mL of ethanol, centrifuged and filtered through a 0.22 μm filter, and esters contents were determined with GC (Agilent 7890A,USA). The operating condition of GC was as following: Agilent DB-WAX column (30 m × 530 μm × 1 μm), fame ionization detector (FID detector), 40 mL/min of H_2 flow rate, 300 mL/min of air flow rate, and N_2 as the carrier gas with the velocity of 15 mL/min.

DNA extraction, PCR amplification and MiSeq sequencing

The genomic DNA was extracted from a total of 21 samples taken from seven fermentation stages using Soil DNA Kit (Omega Bio-tek, Inc.) following the manufacturer's protocol. The DNA quality and quantity were determined by NanoDrop 2000 (Thermo, USA). For prokaryotes, the V4 hypervariable region of the 16S rRNA genes was amplified using universal primer 515F and 909R [18]. For eukaryotes, the ITS2 region of fungal rRNA gene was amplified using universal primer ITS4 and ITS7 [19]. Primer 515F and ITS4 were added with barcodes. PCR conditions were described in detail previously [20]. The amplified PCR products were analyzed through a 1%(wt/vol) agarose gel and purified using a PCR purification kit (GE0101–50, TSINGKE). The concentrations of PCR purified products were assessed by NanoDrop 2000 (Thermo, USA). Subsequently, purified amplicons of all samples were equally pooled for constructing a PCR amplicon library, according to the protocols of the Illumina TruSeq. DNA sample preparation LT kit (San Diego, CA, USA), and then subjected to sequencing using the Illumina MiSeq platform at the Environmental Genomic Platform of the Chengdu Institute of Biology, CAS.

Sequencing data analysis

Sequencing data analysis was performed by QIIME Pipeline Version 1.7.0 [21]. The raw sequences were sorted with their unique barcodes. Sequences with low quality, read length below 200 bp as well as average base quality score less than 30, were filtered out. Chimera sequences were removed utilizing Uchime algorithm [22].

Sequences were clustered into operational taxonomic units (OTUs) at a 97% identity threshold. Each sample was rarefied to the same number of reads (10,568 reads for 16S rRNA gene and 4937 reads for ITS gene, respectively) for both alpha-diversity (chao1 estimator of richness, observed species and Shannon's index) and beta-diversity (PCoA, UniFrac) analyses. Taxonomy was assigned using the Ribosomal Database Project classifier (http://rdp.cme.msu.edu/).

Statistical analysis

The changes of microbial community during fermentation were evaluated by principal coordinates analysis (PCoA, UniFrac). The PerMANOVA was performed with R to present the statistical significance among datasets based on the weighted PCoA scores. One-way analysis of variance (ANOVA) was conducted to compare the differences of microbial communities among intra-group and inter-groups. Pearson's correlation analysis was performed to determine the correlations between variables. Phylogenetic analysis (maximum likelihood algorithm) of OTUs with reference sequences was performed using MEGA6 version 6 [23]. Canonical correspondence analysis (CCA) was

conducted using CANOCO 5.0 (Microcomputer Power, Ithaca, NY) to confirm the correlations between community structures and environmental variables.

Nucleotide sequence accession number

The original sequencing data are available at the European Nucleotide Archive at accession no. PRJEB19772 (http://www.ebi.ac.uk/ena/data/view/PRJEB19772).

Results

Chemical and physical properties during fermentation

Chemical and physical properties of Zaopei at different fermentation stages were shown in Table 1. During the early stage (1–23 days), glucose increased quickly, and reached a peak of 28.43 mg/g on day 23. During this period, no organic acids and ethyl esters were produced, and pH (pH 3.4) maintained constant. Ethanol concentration began to increase slightly. During the middle stage (23–48 days), glucose declined sharply. Whereas, lactic acid and ethanol began to be produced constantly. Acetic acid, propionic acid and ethyl esters changed little. The pH decreased to 3.2 with the increase of lactic acid production. During the late stages (48–70 days), glucose decreased further. Lactic acid and ethanol concentrations increased clearly, and reached up to 36.87 mg/g and 9.25 mg/g on day 70 (Table 1). Considerably, the ethyl lactate (876.82 µg/g), ethyl acetate (138.85 µg/g) and ethyl caproate (96.14 µg/g) were produced significantly at the end of fermentation (Table 1).

Microbial community structure and diversity

For 16S rRNA gene sequences, we resample to 10,568 reads per sample. The rarefaction curves reached the saturation plateau and the Good's coverages among samples were more than 95% (Additional file 2: Table S1, and Additional file 1: Figure S3). The OTU numbers of samples ranged from 192 to 976 based on the cutoff of 97% identity. As shown in Table 2, the Shannon diversity index and Chao1 estimator of richness during 1–10 days were significantly higher than those at other stages ($P < 0.05$, Table 2). PCoA analysis based on weighted UniFrac method showed that there were three clusters (Fig. 1a). The 1-day samples scattered randomly, while the 10-day samples closely clustered together. The samples from 23 to 70 days formed another cluster. PerMANOVA analysis demonstrated that there were no significant differences among prokaryotic communities of 23 to 70-day samples ($p > 0.05$), but significant different from prokaryotic communities in early days (1 to 10-day).

For eukaryotic community, 157 to 582 OTUs were observed for all samples based on 97% similarity as a cutoff (Table 2). The Good's coverages among samples were more than 90% (Additional file 2: Table S2). In contrast to the prokaryotic community, eukaryotic Chao 1 estimator,

Table 1 Chemical and physical properties of samples at different fermentation stages

Sample time (day)	1	10	23	34	48	59	70
pH	3.39 ± 0.01^a	3.40 ± 0.01^a	3.42 ± 0.03^a	3.35 ± 0.04^{ab}	3.25 ± 0.02^{bc}	3.21 ± 0.04^c	3.24 ± 0.01^c
Moisture (%)	56.78 ± 0.00^a	56.02 ± 0.00^a	57.25 ± 0.00^a	57.67 ± 0.01^a	58.58 ± 0.01^{bc}	58.57 ± 0.00^{bc}	60.23 ± 0.02^c
Glucose (mg/g)	7.27 ± 1.14^a	19.68 ± 6.60^b	28.43 ± 0.81^c	17.17 ± 2.57^b	6.73 ± 1.15^a	4.58 ± 0.36^a	5.62 ± 1.13^a
Lactic acid (mg/g)	22.23 ± 0.76^a	20.59 ± 1.31^a	21.84 ± 1.27^a	27.15 ± 1.41^b	29.39 ± 1.12^b	34.75 ± 0.22^c	36.87 ± 1.13^c
Acetic acid (mg/g)	1.27 ± 0.03^a	1.20 ± 0.08^a	1.36 ± 0.11^a	1.68 ± 0.15^b	2.16 ± 0.06^c	2.61 ± 0.05^d	2.73 ± 0.32^d
Propanic acid (mg/g)	3.68 ± 0.06^a	3.41 ± 0.02^b	3.61 ± 0.12^{ab}	3.56 ± 0.11^{ab}	4.51 ± 0.22^c	4.45 ± 0.14^c	4.56 ± 0.09^c
Ethanol (mg/g)	0.28 ± 0.16^a	0.54 ± 0.23^a	2.00 ± 0.68^{ab}	3.84 ± 1.38^b	5.80 ± 0.17^{bc}	7.65 ± 0.40^c	9.25 ± 2.63^c
Ethyl acetate (ug/g)	56.9 ± 2.42^a	59.5 ± 3.74^{ab}	70.7 ± 4.10^b	79.7 ± 9.37^b	92.1 ± 1.66^b	127.8 ± 10.20^c	138.8 ± 14.24^c
Ethyl caproate (ug/g)	ND*	ND	ND	ND	ND	93.3 ± 5.64^a	96.1 ± 5.80^a
Ethyl lactate (ug/g)	441.6 ± 32.96^a	358.6 ± 18.37^a	400.9 ± 28.31^a	41.72 ± 69.04^a	481.8 ± 39.85^a	730.25 ± 25.42^b	876.82 ± 99.47^c

*ND: not detected. All data are presented as means ±standard deviations ($n = 3$). Values with different letters in a row mean significant differences at $P < 0.05$ determined by ANOVA

Shannon index and observed OTU numbers firstly increased, then fluctuated with the fermentation process. During the whole fermentation, the succession of the eukaryotic communities was slightly different from that of prokaryotic community. PCoA analysis showed samples from 1 to 10 days formed a cluster, and samples from 23rd day formed another cluster. Samples from 48 to 59 days clustered together again. However, samples from 70th day scattered (Fig. 1b).

Microbial community compositions

At phylum level, there were six dominant prokaryotic phyla observed throughout whole fermentation process: *Firmicutes, Proteobacteria, Actinobacteria, Bacteroidetes, Euryarchaeota* and *Cyanobacteria. Firmicutes* was the most abundant phylum, accounting for average 54.2% to 99.9% of the total prokaryotic community during the whole fermentation process. During 1–10 days, *Firmicutes* abundance were 54.19 to 83.61% of the whole communities, and all other phyla approximately occupied 45.81% to 16.49% of the microbiota, including *Proteobacteria* (6.15–12.35%), *Actinobacteria* (1.73–0.97%), *Bacteroidetes*

(12.28–1.17%), *Euryarchaeota* (21.99–0.89%). As fermentation proceeded, the *Firmicutes* abundance significantly increased ($P < 0.05$) up to more than 99% during 23 and 70 days (Fig. 2a). a large proportion (>90%) of prokaryotic reads failed to be classified to the genus level. The sequence analysis on OTU level was carried out by NCBI BLAST server and RDP CLASSIFIER. As shown on Additional file 2: Table S2, prokaryotic community was dominated by eight OTUs including OTU17 (23.98–45.19% of total 16S rRNA reads), OTU27 (16.46–28.07%), OTU 211 (5.81–8.64%), OTU 125 (3.56–6.40), OTU 4 (1.51–1.78%), OTU 218 (1.15–1.38%), OTU 216 (0.0–1.53%), OTU 48 (0.0–1.23%). These OTUs accounted for 60.8–93.4% of prokaryotic community. Most of these OTUs showed high similarity (>95%) with uncultured Lactobacillus sp. clone 16S ribosomal RNA gene, but low similarity (<93%) with isolates (members of genus Lactobacillus) in NCBI's GenBank. Moreover, these OTU reads could not be classified to the genus level by RDP CLASSIFIER. Therefore, community composition was analyzed at family level. A total of 17 abundant families (abundance > 1%) were detected. Among them, 11 families

Table 2 Microbial diversity indices calculated based on the cutoff of 97% identity of 16S rRNA gene or ITS region

Sample time (day)	Chao1		Observed species		Shannon index	
	16S rRNA gene	ITS gene	16S rRNA gene	ITS gene	16S rRNA gene	ITS gene
1	1462 ± 155^a	278.27 ± 52.17^a	976 ± 84^a	170 ± 12^a	7.04 ± 0.26^a	3.84 ± 0.25^a
10	1414 ± 248^a	346.14 ± 21.58^{ab}	789 ± 54^b	197 ± 8^{ab}	5.49 ± 0.19^b	3.99 ± 0.11^{ab}
23	566 ± 78^b	609.34 ± 214.43^b	212 ± 9^c	343 ± 136^b	2.56 ± 0.07^c	5.17 ± 1.04^{bc}
34	683 ± 138^b	825.28 ± 230.83^c	192 ± 4^c	479 ± 100^c	2.52 ± 0.04^c	6.14 ± 1.01^c
48	559 ± 119^b	498.68 ± 125.07^{ab}	198 ± 7^c	276 ± 62^{ab}	2.56 ± 0.06^c	4.05 ± 0.57^{ab}
59	631 ± 97^b	560.67 ± 24.61^c	239 ± 5^c	308 ± 34^{ab}	2.70 ± 0.07^c	4.33 ± 0.31^{ab}
70	784 ± 270^b	777.76 ± 218.55^c	333 ± 200^c	462 ± 106^c	3.07 ± 0.83^c	5.96 ± 0.81^c

*All data were presented as means ± standard deviations. Values with different letters in a column mean significant difference at $p < 0.05$ tested by one-way ANOVA Duncan's test

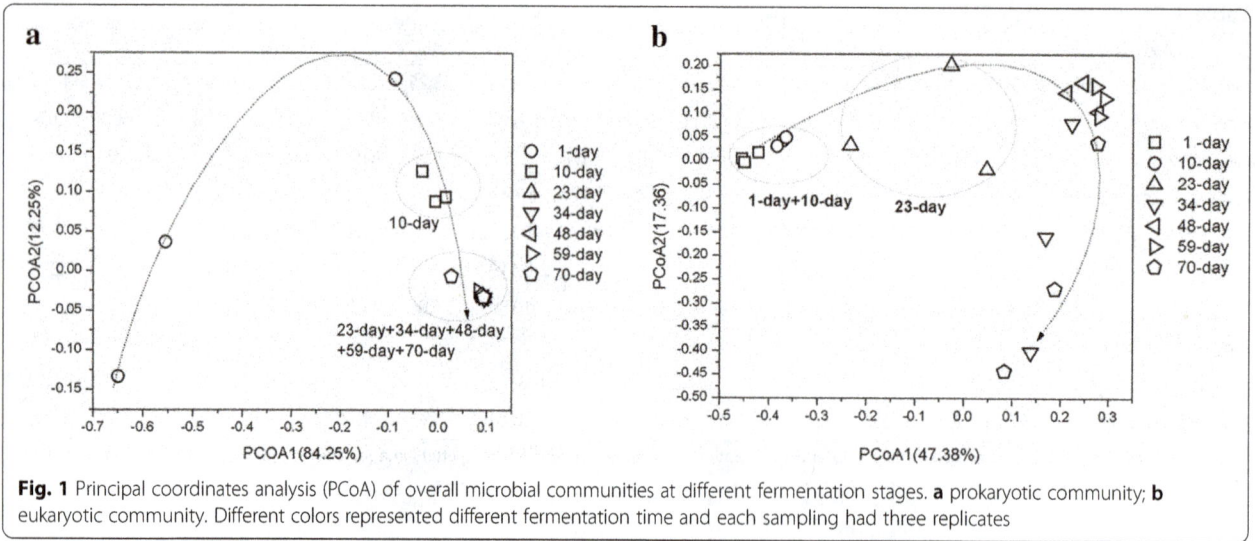

Fig. 1 Principal coordinates analysis (PCoA) of overall microbial communities at different fermentation stages. **a** prokaryotic community; **b** eukaryotic community. Different colors represented different fermentation time and each sampling had three replicates

were affiliated to phylum *Firmicutes*, including *Lactobacillaceae, Ruminococcaceae, Tissierellaceae, Bacillaceae, Clostridiaceae, Syntrophomonadaceae, Planococcaceae,* unclassified *Bacilli, Leuconostocaceae, Streptococcaceae,* and unclassified *Lactobacillales.* In particular, *Lactobacillaceae* and unclassified *Bacilli* almost dominated the microbiota during the middle and later fermentation stages (Fig. 4a). Three families were affiliated to *Proteobacteria,* including *Xanthomonadaceae, Pseudomonadaceae* and *Moraxellaceae*; Two families were affiliated to Euryarchaeota, including *Methanosarcinaceae* and *Methanobacteriaceae*; One family was affiliated to *Porphyromonadaceae* (phylum *Bacteroidetes*). To figure out the compositions and succession of microbial community more specifically, a heatmap of prokaryotic OTUs was performed (Additional file 1: Figure S4). We identified 27 representative OTUs (abundance >1%). Only two OTUs (OTU17, OTU27) were shared by all samples with their

abundances from 3.69% to 45.63%. Both of them increased rapidly on 23rd day followed by stabilizing generally in the mid-late period. The OTU17 and OTU27 revealed low similarities with their closest phylogenetic neighbour (members of genus Lactobacillus, Additional file 2: Table S2). A large proportion of diverse OTUs just appeared in early period, however most of them decreased quickly in the mid-late period and remained in a low abundance. After day 23, the entire prokaryotic population was almost covered by 8 OTUs (OTU17, OTU27, OTU 211, OTU 125, OTU 4, OTU 218, OTU 216, and OTU 48), all of which were affiliated to *Lactobacillaceae* (Additional file 2: Table S3, and Additional file 1: Figure S5).

Six phyla were observed in eukaryotic community, including *Ascomycota, Zygomycota, Basidiomycota, Chytridiomycota, Glomeromycota* and *Rozellomycota.* Among them, *Ascomycota, Zygomycota* and *Basidiomycota* occurred throughout the entire fermentation process, while

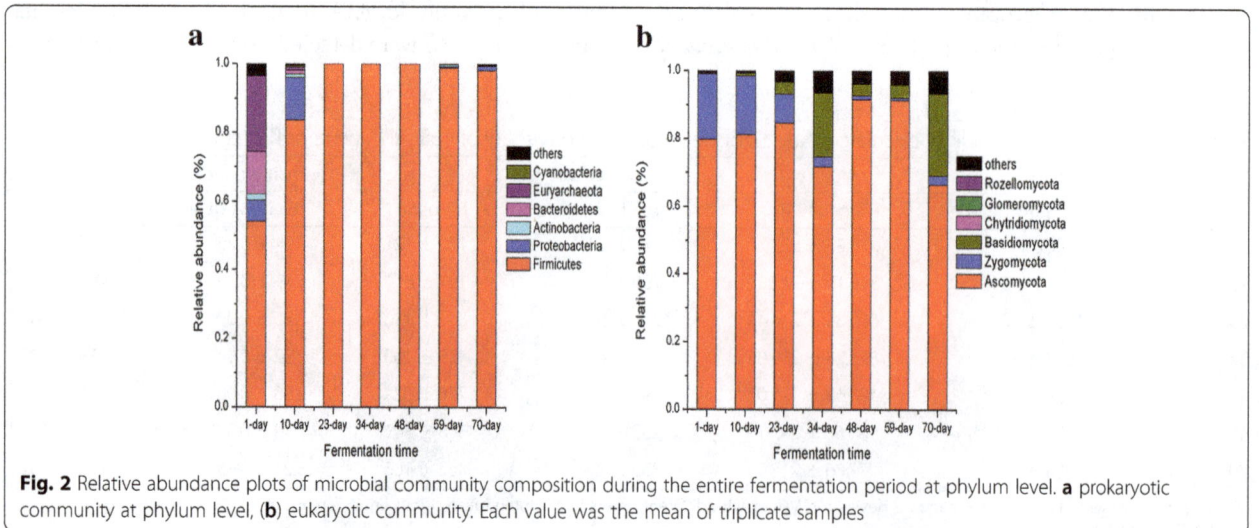

Fig. 2 Relative abundance plots of microbial community composition during the entire fermentation period at phylum level. **a** prokaryotic community at phylum level, (**b**) eukaryotic community. Each value was the mean of triplicate samples

Ascomycota predominated at the average relative abundance of 80.9%. During the early stages (1 to 10-day), *Zygomycota* was a subdominant group (17.36–19.26%), but it sharply decreased with the fermentation process up to 2.61% on the 70th day. In contrast, *Basidiomycota* abundance increased from 0.33% to 24.27% during the fermentation (Fig. 2b). At genus level, 13 abundant genera (abundance >1%) affiliated to three phyla were observed, including ten *Ascomycota* genera (*Thermoascus, Aspergillus, Emericella, Monascus, Candida*, unidentified *Pleosporales*, unidentified *Capnodiales*, unidentified *Saccharomycetales, Botryotinia* and *Pichia*), one genera (*Suillus*) belonging to *Basidiomycota*, and two genera (*Mortierella, Rhizopus*) belonging to *Zygomycota*. The *Thermoascus, Aspergillus* and *Emericella* were defined as the core genera because of their presence in whole stage, especially *Thermoascus* and *Aspergillus* with their abundances over 10% on average (Fig. 3b). Further, 16 dominant OTUs (abundance >1%) were observed. The OTU 130 and OTU 6, affiliated to *Candida* and *Aspergillus*, respectively, were shared by all the samples with abundances from 27.07% to 57.98%. The OTU 130 that was closely related to *Candida humilis*, dominated the entire process accounting for an average of 22.15% (Additional file 2: Table S4). Remarkably, three OTUs (OTU 2056, OTU 756, and OTU 4385) only appeared in late periods (59 to 70-day), and they were affiliated to *Botrytis, Mortierella* and *Cladophialophora*, respectively (Additional file 2: Table S4 and Additional file 1: Figure S6).

Correlations between microbial communities and flavoring chemicals

Canonical correspondence analysis (CCA) was conducted to reveal the correlations between microbial community and flavoring chemicals (FCs). For prokaryotes (Fig. 4), the first two axises explained 87.08% of the variation in community composition. The pH and ethanol content

were the two most influential environmental variables. The organic acid and esters were key FCs that significantly correlated with communities during 23 to 70-day. The pH was closely correlated with the community composition with higher community diversity in early period (1 to 10-day). *Lactobacillaceae* and unclassified Bacilli were positively correlated with the FCs. At OTU level, eight prokaryotic members (OTU 4,OTU 48, OTU 17,OTU 27,OTU 211,OTU 218,OTU 125, and OTU 216) showed positive correlations with FCs (Additional file 1: Figure S7), especially ethyl acetate concentration ($p < 0.05$). These OTUs were assigned to *Lactobacillaceae* (Additional file 2: Table S3). For eukaryotes, both axes explained 52.08% of the variation in community composition, which is less than that in prokaryotic communities. Organic acid and esters levels were mainly positively correlated with eukaryotic communities in late period (34 to 70-day). The pH mainly correlated with the community composition in early period (1 to 10-day) with abundant unidentified *Saccharomycetales, Rhizopus* and *Pichia*. Figure 5 showed that unclassified *Capnodiales* was positively correlated with FCs ($p < 0.01$), and *Monascus* positively correlated with the production of glucose ($p < 0.05$). At OTU level, three eukaryotic OTU 756 (*Mortierella*), OTU 2056 (Botrytis), OUT 4385 (*Cladophialophora*) showed positive correlations with FCs ($p < 0.01$) (Additional file 2: Table S4, and Additional file 1: Figure S8).

Discussion

Generally, the CSFL fermentation was achieved in strictly anaerobic FT under the ground, involving vastly complicated metabolic reactions and microbial community. It is difficult for sampling from underground pit during the fermentation process. Thus, microbial community in the FT is generally considered as a "black box". Previous studies usually collect a limited number of samples at several points, or at starting point and end point [6, 13, 16, 17],

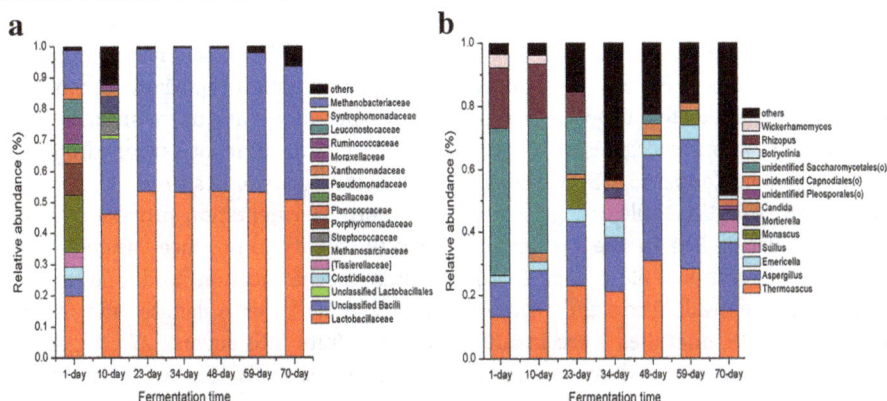

Fig. 3 Relative abundance plots of microbial community composition during the entire fermentation period. **a** prokaryotic community at family level, (**b**) eukaryotic community at genus level. Each value was the mean of triplicate samples

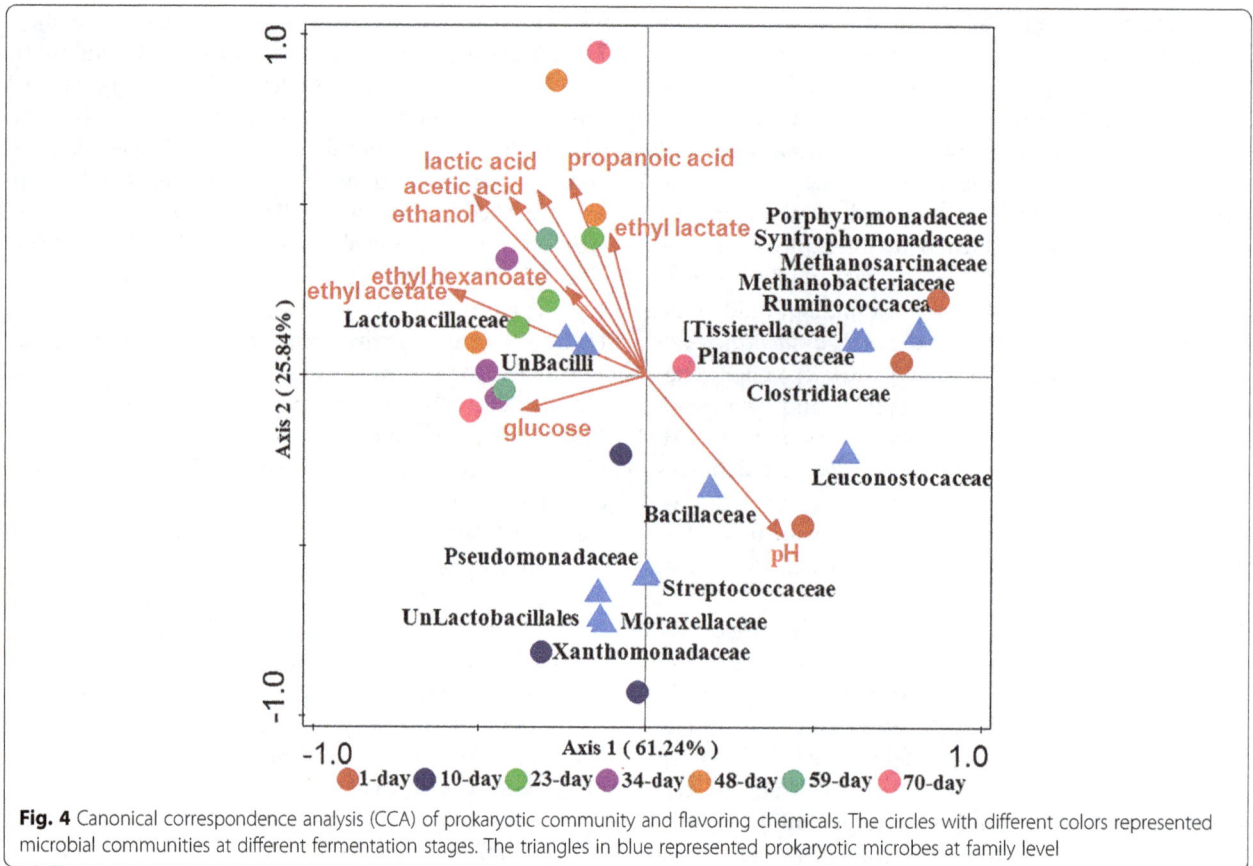

Fig. 4 Canonical correspondence analysis (CCA) of prokaryotic community and flavoring chemicals. The circles with different colors represented microbial communities at different fermentation stages. The triangles in blue represented prokaryotic microbes at family level

which leads to the lack of understanding of the process-related dynamics of microbial community during the CSFL fermentation. In this study, the dynamics of microbial communities during CSFL fermentation was investigated by simulating fermentation using batch experiments to facilitate sampling. The final contents of lactic acid and ethanol in this study were similar to those of real fermentation process [24]. The main esters related with liquor's quality were produced and the dynamics of microbial communities showed certain succession patterns. Notably, caproic acid and ethyl caproate are the important flavoring substances of CSFL, but caproic acid was not detected while ethyl caproate was detected until the late stages (59–70 days). Generally, caproic acid is mainly produced by microbes in Pit Mud (PM), while ethyl caproate is produced via the esterification by microorganisms mainly originated from Daqu starter in Zaopei [5, 25]. Compare to the real fermentation pit (Additional file 1: Figure S1), there was only a small number of PM in the bottom of in-vitro fermentation vessel (Additional file 1: Figure S2), and there was no PM on the wall inside the reactor. It was thus speculated that PM microbes were insufficient to produce detectable caproic acid in simulated fermentation, especially during the early and middle stage. During the late stage, PM microbes could produce a small amount of caproic acid that was esterified into ethyl

caproate as soon as possible. Therefore, low concentration of ethyl caproate was detected on day 59 and 70, while caproic acid was not detected. To faithfully reproduce the real CSFL fermentation, more work is needed, such as the improvement of in-vitro fermentation vessel and preculture of Pit mud. Especially, It indicated that many novel microbes and their functions remain elusive.

The *Lactobacillaceae* and unclassified *Bacilli* were dominant and occurred throughout the entire fermentation, which are consistent with previous reports based on traditional molecular methods, e.g., DGGE, 16S rRNA gene clone library and PLFA [4, 26, 27]. The *Lactobacillaceae* could produce lactic acid from glucose or starch by homolactic fermentation [4]. Many members affiliated to *Bacilli* could produce various hydrolases for the liquefaction and saccharification of carbohydrates [26, 27]. Two core eukaryotic genera (*Aspergillus*, *Thermoascus*) existed in the entire fermentation process in our study. *Aspergillus* had the ability to produce various hydrolytic enzymes for starch saccharification, and *Thermoascus* could produce protease such as xylanase and α-amylase to degrade carbohydrate into sugars [27, 28]. Additionally, four new genera (*Emericella*, *Suillus*, *Mortierella* and *Botryotinia*) were observed in fermentation, which showed positive correlations with organic acid and ethyl esters (Fig. 5). Cao et al. [29] reported that *Emericella* existed in wheat

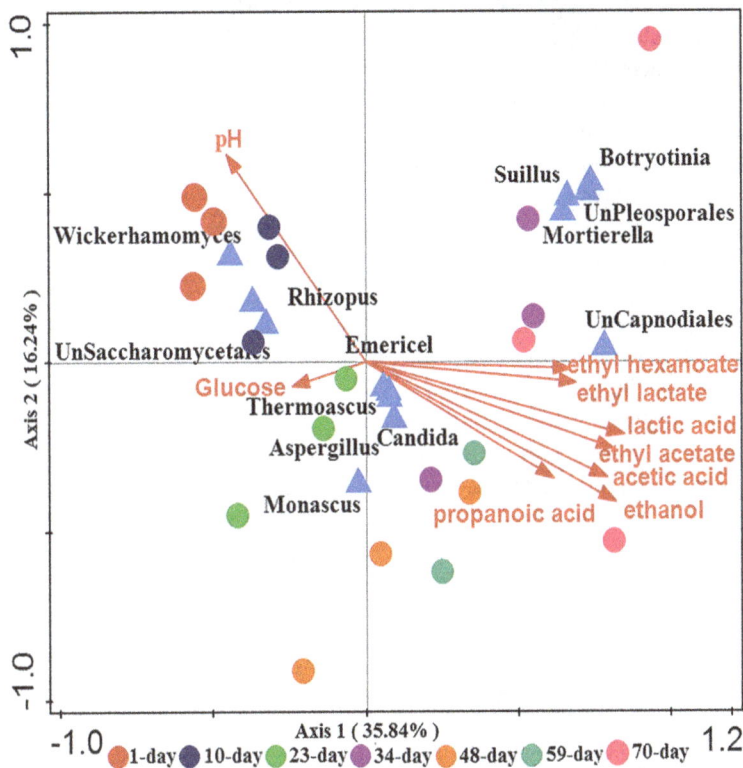

Fig. 5 Canonical correspondence analysis (CCA) of eukaryotic community and flavoring chemicals. The circles in different colors represented microbial communities at different fermentation stages. The triangles in blue represented eukaryotic microbes at genus level

Qu used for wheat Daqu fermentation. *Suillus* could decompose complex organic matter substrates, such as hemicellulose, cellulose and components of needles [30]. *Mortierella* was reported to have the ability to produce polyunsaturated fatty acid by degrading rice bran in solid substrate fermentation [31]. *Botryotinia* was reported to produce rhamnogalacturonan hydrolase, cell-wall-degrading enzymes and other low-molecular-weight compounds such as oxalic acid [32, 33]. However, their definite functions in CSFL fermentation still remain elusive. Ethyl esters are crucial factors that determine the quality of the CSFL [12]. Three microbial taxa (*Monascus, Candida and Pichia*) were reported to involve in the formation of ethyl esters [10, 17, 25, 34]. Microbial community showed distinct succession in the fermentation. In early fermentation period (1 to 23-day), the microbial communities were mainly dominated by *Lactobacillus* and eight eukaryotic genera (*Thermoascus, Aspergillus, Emericella, Monascus, Candida,* unidentified *Saccharomycetales, Rhizopus, and Pichia*). Glucoses were produced and reached a peak due to the degradation effects on macromolecular carbohydrates by above microbes [4, 6, 9, 13, 15, 17]. Meanwhile, to our knowledge, Daqu is an important saccharifying and fermenting agent [35]. Filamentous fungi (e.g. *Rhizopus, Aspergillus*), yeasts (e.g. *Saccharomyces, Candida*) and bacteria (e.g. acetic

acid bacteria, lactic acid bacteria), are considered to be the functional populations in Daqu, which are responsible for lyase production and polysaccharide degradation [36]. Thus, this stage could be described as "the stage of saccharification". In the middle fermentation period (23 to 48-day), *Lactobacillus* converts sugars into lactic acid, as the precursor to form ethyl lactate, and lead to the decrease of pH. Ethanol begins to be generated via the glycolysis by some bacterial and fungal microbes [7, 37, 38]. *Lactobacillus* absolutely predominates the microbial community, and eukaryotic community is dominated by *Thermoascus, Aspergillus, Emericella, Candida,* unidentified *Saccharomycetales*. During this period, lactic acid and ethanol are the main fermentation products. It could be described as "the stage of glycolysis stage". At late fermentation period (48 to 70-day), ethyl esters are produced, especially ethyl caproate increases remarkably. In this stage, organic acids are transformed into esters by microbes, such as *Clostridium* and *Pichia* via esterification between alcohol and organic acids [9, 11, 36, 39]. Low pH results into the predominant of *Lactobacillus*. Eukaryotic populations fad significantly due to the unfavorable environmental conditions. The period could be described as "the stage of esterification". The dynamics of microbial community in the CSFL fermentation process provide various metabolites which constitute unique flavor of Chinese liquor.

Conclusion

This study comprehensively revealed a dynamic of microbial communities including prokaryotes and eukaryotes during the CSFL fermentation. The overall fermentation presents three phases: saccharification, glycolysis and esterification stage. During the fermentation, *Lactobacillaceae* was the most abundant prokaryotic taxon, and *Thermoascus, Aspergillus* and *Emericella* dominated entire eukaryotic communities during the whole fermentation. *Lactobacillaceae, Bacilli, Botryotinia, Aspergillus*, unidentified *Pleosporales* and *Capnodiales* were positively correlated to the production of FCs. *Emericella, Suillus, Mortierella* and *Botryotinia* were firstly observed in CSFL fermentation. This study provide deep theoretical basis to design effective strategies to manipulate microbial consortia for better controlling CSFL production systems and improving liquor quality in the brew practice.

Additional files

Additional file 1: Figure S1. Schematic diagram of the real CSFL fermentation pit. **Figure S2.** Diagram of glass bottles used for simulating fermentation experiments. **Figure S3.** The rarefaction curves of sequencing depths. **Figure S4.** Heatmaps of prokaryotic and eukaryotic communities. **Figure S5.** Phylogenetic analysis (maximum likelihood algorithm) result of prokaryotic communities. **Figure S6.** Phylogenetic analysis (maximum likelihood algorithm) result of eukaryotic communities. **Figure S7.** Canonical correspondence analysis (CCA) of prokaryotic OTUs and flavoring chemicals. **Figure S8.** Canonical correspondence analysis (CCA) of eukaryotic OTUs and flavoring chemicals.

Additional file 2: Table S1. Good coverages of prokaryotic Sequencing (16S rRNA gene). **Table S2.** Good coverages of eukaryotic Sequencing (ITS gene). **Table S3.** The OTU BLAST result based on 16S rRNA gene. **Table S4.** The OTU BLAST result based on ITS region.

Abbreviations

ANOVA: Analysis of variance; CCA: Canonical correspondence analysis; CSFL: Chinese strong-flavored liquor; FCs: Flavoring chemicals; FT: Fermentation pit; GC: gas chromatography; HPLC: High performance liquid chromatography; OTU: Operational taxonomic unit; PCoA: Principal coordinate analysis; PCR-DGGE: Polymerase chain reaction-denaturing gradient gel electrophoresis; PerMANOVA: Permutational multivariate analysis of variance; PLFA: Phospholipid fatty acid analysis; PM: Pit mud; qPCR: Real-time Quantitative polymerase chain reaction; RDP: Ribosomal Database Project

Acknowledgements

Not applicable

Authors' contribution

YZ performed the experiments, and wrote the manuscript. YT conceived and designed the study, and revised the manuscript. XL revised the manuscript. JJ and XH participated in the planning and coordination of the study. All authors read and approved the final manuscript.

Funding

This work was supported by the Natural Science Foundation of China (Nos. 31,470,020 and 31,770,090), the Open-foundation project of Key Laboratory of Environmental and Applied Microbiology, CAS (KLCAS-2017-02 and KLCAS-2016-04), Sichuan Science and Technology Support Program (2016JY0219 and 2016JZ0010), Project of Resources Service Network, CAS (ZSYS-004), 973 Project (No. 2013CB733502) and China Biodiversity Observation Networks (Sino BON).

Competing interests

The authors declare that they have no competing interests.

Author details

[1]Key Laboratory of Environmental and Applied Microbiology, Chinese Academy of Sciences & Environmental Microbiology Key Laboratory of Sichuan Province, Chengdu Institute of Biology, Chinese Academy of Sciences, Chengdu 610041, People's Republic of China. [2]University of Chinese Academy of Sciences, Beijing 100049, People's Republic of China. [3]Chengdu Institute of Biology, Chinese Academy of Sciences, Chengdu 610041, People's Republic of China.

References

1. Tao Y, Li JB, Rui JP, Xu ZC, Zhou Y, Hu XH, Wang X, Liu MH, Li DP, Li XZ. Prokaryotic communities in pit mud from different-aged cellars used for the production of Chinese strong-flavored liquor. Appl Environ Microbiol. 2014;80(7):2254–60.
2. Lu XW, Wu Q, Zhang Y, Xu Y. Genomic and transcriptomic analyses of the Chinese Maotai-flavored liquor yeast MT1 revealed its unique multi-carbon co-utilization. BMC Genomics. 2015;16(1):1064.
3. Wang P, Wu Q, Jiang XJ, Wang ZQ, Tang JL, Xu Y. Bacillus licheniformis affects the microbial community and metabolic profile in the spontaneous fermentation of Daqu starter for Chinese liquor making. Int J Food Microbiol. 2017;250:59–67.
4. Zhang WX, Qiao ZW, Shigematsu T, Tang YQ, Hu C, Morimura S, Kida K. Analysis of the bacterial community in Zaopei during production of Chinese Luzhou-flavor liquor. J Inst Brew. 2005;111(2):215–22.
5. Hu XL, Du H, Ren C, Xu Y. Illuminating anaerobic microbial community and cooccurrence patterns across a quality gradient in Chinese liquor fermentation pit muds. Appl Environ Microbiol. 2016;82(8):2506–15.
6. Sun WN, Xiao HZ, Peng Q, Zhang QG, Li XX, Han Y. Analysis of bacterial diversity of Chinese Luzhou-flavor liquor brewed in different seasons by Illumina Miseq sequencing. Ann Microbiol. 2016;66(3):1293–301.
7. Li XR, Ma EB, Yan LZ, Meng H, Du XW, Zhang SW, Quan ZX. Bacterial and fungal diversity in the traditional Chinese liquor fermentation process. Int J Food Microbiol. 2011;146(1):31–7.
8. Tamang JP, Watanabe K, Holzapfel WH. Review: diversity of microorganisms in global fermented foods and beverages. Front Microbiol. 2016;7:377.
9. Zhang WX, Qiao ZW, Tang YQ, Hu C, Sun Q, Morimura S, Kida K. Analysis of the fungal community in Zaopei during the production of Chinese Luzhou-flavour liquor. J Inst Brew. 2007;113(1):21–7.
10. Liang HP, Li WF, Luo QC, Liu CL, Wu ZY, Zhang WX. Analysis of the bacterial community in aged and aging pit mud of Chinese Luzhou-flavour liquor by combined PCR-DGGE and quantitative PCR assay. J Sci Food Agric. 2015;95(13):2729–35.
11. Ding XF, Wu CD, Huang J, Zhou RQ. Changes in volatile compounds of Chinese Luzhou-flavor liquor during the fermentation and distillation process. J Food Sci. 2015;80(11):2373–81.
12. Fan WL, Qian MC. Characterization of aroma compounds of Chinese "Wuliangye" and "Jiannanchun" liquors by aroma extract dilution analysis. J Agric Food Chem. 2006;54(7):2695–704.
13. Zheng J, Wu CD, Huang J, Zhou RQ, Liao XP. Spatial distribution of bacterial communities and related biochemical properties in Luzhou-flavor liquor-fermented grains. J Food Sci. 2014;79(12):2491–8.
14. Luo QC, Liu CL, Wu ZY, Wang HY, Li WF, Zhang KH, Huang D, Zhang J, Zhang WX. Monitoring of the prokaryotic diversity in pit mud from a Luzhou-flavour liquor distillery and evaluation of two predominant archaea using qPCR assays. J Inst Brew. 2014;120(3):253–61.
15. Zheng J, Liang R, Zhang LQ, Wu CD, Zhou RQ, Liao XP. Characterization of microbial communities in strong aromatic liquor fermentation pit muds of different ages assessed by combined DGGE and PLFA analyses. Food Res Int. 2013;54(1):660–6.

16. Wang Q, Zhang H, Liu X. Microbial community composition associated with Maotai liquor fermentation. J Food Sci. 2016;81(6):1485–94.

17. Shi S, Zhang L, Wu ZY, Zhang WX, Deng Y, Zhong FD, Li JM. Analysis of the fungi community in multiple-and single-grains Zaopei from a Luzhou-flavor liquor distillery in western China. World J Microbiol Biotechnol. 2011;27(8):1869–74.

18. Lin Q, Vrieze JD, Li JB, Li XZ. Temperature affects microbial abundance, activity and interactions in anaerobic digestion. Bioresour Technol. 2016;209:228–36.

19. Ihrmark K, Bodeker ITM, Cruz-Martinez K, Friberg H, Kubartova A, Schenck J, Strid Y, Stenlid J, Brandstrom-Durling M, Clemmensen KE, et al. New primers to amplify the fungal ITS2 region-evaluation by 454-sequencing of artificial and natural communities. FEMS Microbiol Ecol. 2012;82(3):666–77.

20. Li X, Rui J, Mao Y, Yannarell A, Mackie R. Dynamics of the bacterial community structure in the rhizosphere of a maize cultivar. Soil Biol Biochem. 2014;68:392–401.

21. Caporaso JG, Kuczynski J, Stombaugh J, Bittinger K, Bushman FD, Costello EK, Fierer N, Peña AG, Goodrich JK, Gordon JI, et al. QIIME allows analysis of high-throughput community sequencing data. Nat Methods. 2010;7(5):335–6.

22. Edgar RC, Haas BJ, Clemente JC, Quince C, Knight R. UCHIME improves sensitivity and speed of chimera detection. Bioinformatics. 2011;27(16):2194–200.

23. Tamura K, Stecher G, Peterson D, Filipski A, Kumar S. MEGA6: Molecular evolutionary genetics analysis version 6.0. Mol Biol Evol. 2013;30(12):2725–9.

24. Zhao D, Qiao ZW, Peng ZY. Investigation on the microflora in fermented grains & the evolution of its ecological factors during the fermentation of Luzhou-flavor liquor (in Chinese). Liquor-making Sci Technol. 2007;7:37–9.

25. Li P, Liang HB, Lin WT, Feng F, Luo LX. Microbiota dynamics associated with environmental conditions and potential roles of cellulolytic communities in traditional Chinese cereal starter solid-state fermentation. Appl Environ Microbiol. 2015;81(15):5144–56.

26. Ding XF, Wu CD, Zhang LQ, Zheng J, Zhou RQ. Characterization of eubacterial and archaeal community diversity in the pit mud of Chinese Luzhou-flavor liquor by nested PCR-DGGE. World J Microbiol Biotechnol. 2014;30(2):605–12.

27. Gou M, Wang HZ, Yuan HW, Zhang WX, Tang YQ, Kida K. Characterization of the microbial community in three types of fermentation starters used for Chinese liquor production. J Inst Brew. 2015;121(4):620–7.

28. Bertoldo C, Antranikian G. Starch-hydrolyzing enzymes from thermophilic archaea and bacteria. Curr Opin Chem Biol. 2002;6(2):151–60.

29. Cao Y, Chen JY, Xie GF, Lu J. Study on the factors of fungal community formation during the fermentation course of wheat Qu. J Food Sci Biotechnol (in Chinese). 2008;27(5):95–101.

30. Durall DM, Todd AW, Trappe JM. Decomposition of C-14- labeled substrates by ectomycorrhizal fungi in association with Douglas-fir. New Phytol. 1994; 127(4):725–9.

31. Jang HD, Lin YY, Yang S. Polyunsaturated fatty acid production with Mortierella Alpina by solid substrate fermentation. Bot Bul Acad Sin. 2000;41(1):41–8.

32. Fu J, Prade R, Mort A. Expression and action pattern of Botryotinia fuckeliana (Botrytis Cinerea) rhamnogalacturonan hydrolase in Pichia Pastoris. Carbohydr Res. 2001;330(1):73–81.

33. Williamson B, Tudzynsk B, Tudzynski P, Kan van JAL. Botrytis Cinerea: the cause of grey mould disease. Mol Plant Pathol. 2007;8(5):561–80.

34. Liu GY, Lu SH, Huang DY, Wu YY. Ethyl caproate synthesis by extracellular lipase of monascus fulginosus. Chinese J Biotechnol (in Chinese). 1995;3:288–90.

35. Li P, Lin WF, Liu X, Wang XW, Gan X, Luo LX, Lin WT. Effect of bioaugmented inoculation on microbiota dynamics during solid-state fermentation of Daqu starter using autochthonous of bacillus, Pediococcus, Wickerhamomyces and Saccharomycopsis. Food Microbiol. 2017;61:83–92.

36. Li P, Liang H, Lin W-T, Feng F, Luo LX. Microbiota dynamics associated with environmental conditions and potential roles of cellulolytic communities in traditional Chinese cereal starter solid-state fermentation. Appl Environ Microbiol. 2015;81(15):5144–56.

37. Torija MJ, Beltran G, Novo M, Poblet M, Guillamon JM, Mas A, Rozes N. Effects of fermentation temperature and saccharomyces species on the cell fatty acid composition and presence of volatile compounds in wine. Int J Food Microbiol. 2003;85(1–2):127–36.

38. Hierro N, Esteve-Zarzoso B, Mas A, Guillamon JM. Monitoring of saccharomyces and hanseniaspora populations during alcoholic fermentation by real-time quantitative PCR. FEMS Yeast Res. 2007;7(8):1340–9.

39. Zheng XW, Rezaei MR, Nout MJR, Han B-Z. Daqu— a traditional Chinese liquor fermentation starter. J Inst Brew. 2011;117(1):82–90.

PERMISSIONS

The contributors of this book come from diverse backgrounds, making this book a truly international effort. This book will bring forth new frontiers with its revolutionizing research information and detailed analysis of the nascent developments around the world.

We would like to thank all the contributing authors for lending their expertise to make the book truly unique. They have played a crucial role in the development of this book. Without their invaluable contributions this book wouldn't have been possible. They have made vital efforts to compile up to date information on the varied aspects of this subject to make this book a valuable addition to the collection of many professionals and students.

This book was conceptualized with the vision of imparting up-to-date information and advanced data in this field. To ensure the same, a matchless editorial board was set up. Every individual on the board went through rigorous rounds of assessment to prove their worth. After which they invested a large part of their time researching and compiling the most relevant data for our readers.

The editorial board has been involved in producing this book since its inception. They have spent rigorous hours researching and exploring the diverse topics which have resulted in the successful publishing of this book. They have passed on their knowledge of decades through this book. To expedite this challenging task, the publisher supported the team at every step. A small team of assistant editors was also appointed to further simplify the editing procedure and attain best results for the readers.

Apart from the editorial board, the designing team has also invested a significant amount of their time in understanding the subject and creating the most relevant covers. They scrutinized every image to scout for the most suitable representation of the subject and create an appropriate cover for the book.

The publishing team has been an ardent support to the editorial, designing and production team. Their endless efforts to recruit the best for this project, has resulted in the accomplishment of this book. They are a veteran in the field of academics and their pool of knowledge is as vast as their experience in printing. Their expertise and guidance has proved useful at every step. Their uncompromising quality standards have made this book an exceptional effort. Their encouragement from time to time has been an inspiration for everyone.

The publisher and the editorial board hope that this book will prove to be a valuable piece of knowledge for researchers, students, practitioners and scholars across the globe.

LIST OF CONTRIBUTORS

Aleš Ulrych, Nela Holečková, Jana Goldová, Linda Doubravová, Oldřich Benada, Olga Kofroňová, Petr Halada and Pavel Branny
Institute of Microbiology, v.v.i., Academy of Sciences of the Czech Republic, Vídeňská 1083, 142 20 Prague, Czech Republic

Zhen Li, Ailyn Pérez-Osorio, Kaye Eckmann and William A. Glover
Washington State Department of Health, Public Health Laboratories, Shoreline, Washington, USA

Eric W. Brown, Yi Chen, Yu Wang and Marc W. Allard
Center for Food Safety and Applied Nutrition,Food and Drug Administration, College Park, MD, USA

Tekalign Kejela
Department of Biology, Faculty of Natural and Computational Sciences, Mettu University, Mettu, Ethiopia
BRD school of Biosciences, Sardar Patel University, Vallabh Vidyanagar 388120, India

Vasudev R. Thakkar and Parth Thakor
BRD School of Biosciences, Sardar Patel University, Vadtal Road, Satellite Campus, Post Box No.39, Vallabh Vidyanagar 388120, Gujarat, India

Subhashini Sivasamboo and Arbakariya B. Ariff
Department of Microbiology, Faculty of Biotechnology and Biomolecular Sciences, Universiti Putra Malaysia, 43400 UPM Serdang, Selangor, Malaysia

Sahar Abbasiliasi and Shuhaimi Mustafa
Department of Microbiology, Faculty of Biotechnology and Biomolecular Sciences, Universiti Putra Malaysia, 43400 UPM Serdang, Selangor, Malaysia
Bioprocessing and Biomanufacturing Research Centre, Faculty of Biotechnology and Biomolecular Sciences, Universiti Putra Malaysia, 43400 UPM Serdang, Selangor, Malaysia

Joo Shun Tan
School of Industrial Technology, Universiti Sains Malaysia, 11800 George Town, Penang, Malaysia

Fatemeh Bashokouh and Faezeh Vakhshiteh
Institute of Bioscience, Universiti Putra Malaysia, 43300 Serdang, Selangor, Malaysia

Tengku Azmi Tengku Ibrahim
Institute of Bioscience, Universiti Putra Malaysia, 43300 Serdang, Selangor, Malaysia
Faculty of Veterinary Medicine, Universiti Putra Malaysia, 43400 UPM Serdang, Selangor, Malaysia

Ali Kassim, Zul Premji and Gunturu Revathi
Aga Khan University Hospital, Nairobi, Kenya

Claudia Daubenberger
Swiss Tropical and Public Health Institute, Basel, Switzerland

Valentin Pflüger
Mabritec AG, Riehen, Switzerland

Joseph C. Bryant, Ridge C. Dabbs, Katie L. Oswalt, Lindsey R. Brown and Justin A. Thornton
Department of Biological Sciences, Mississippi State University, 295 E Lee Blvd., Harned Hall, Rm 219, Mississippi State, MS 39762, USA

Jason W. Rosch
Department of Infectious Diseases, St. Jude Children's Research Hospital, Memphis, TN, USA

Keun S. Seo
Department of Basic Sciences, College of Veterinary Medicine, Mississippi State University, Mississippi State, MS, USA

Janet R. Donaldson
Department of Biological Sciences, University of Southern Mississippi, Hattiesburg, MS, USA

Larry S. McDaniel
Department of Microbiology and Immunology, University of Mississippi Medical Center, Jackson, MS, USA

Gammadde Hewa Ishan Maduka Wickramasinghe, Pilimathalawe Panditharathna Attanayake Mudiyanselage Samith Indika Rathnayake, Naduviladath Vishvanath Chandrasekharan and Mahindagoda Siril Samantha Weerasinghe
Department of Chemistry, Faculty of Science, University of Colombo, Colombo, Sri Lanka

Ravindra Lakshman Chundananda Wijesundera
Department of Plant Sciences, Faculty of Science, University of Colombo, Colombo, Sri Lanka

Wijepurage Sandhya Sulochana Wijesundera
Department of Biochemistry and Molecular Biology, Faculty of Medicine, University of Colombo, Kynsey Road, Colombo 08, Sri Lanka

Il-Kyu Park and Dong-Hyun Kang
Department of Food and Animal Biotechnology, College of Agricultural Biotechnology, Center for Food and Bioconvergence, and Institute of GreenBio Science & Technology, Research Institute for Agricultural and Life Sciences, Seoul National University, Seoul 08826, Korea

Jae-Won Ha
Department of Food and Biotechnology, College of Engineering, Food & Bio-industry Research Center, Hankyong National University, Anseong-si 17579, Korea

Julia U. Brandt, Friederike-Leonie Born, Frank Jakob and Rudi F. Vogel
Technische Universität München, Lehrstuhl für Technische Mikrobiologie, Gregor-Mendel-Straße 4, 85354 Freising, Germany

Fengjiao Hu, Qiaoxing Wu, Shuang Song, Ruiping She, Yue Zhao, Yifei Yang, Fang Du, Majid Hussain Soomro and Ruihan Shi
Department of Veterinary Pathology and Public Health, Key Laboratory of Zoonosis of Ministry of Agriculture College of Veterinary Medicine, China Agricultural University, Beijing 100193, China

Meikun Zhang
Beijing Huadu Broiler Corporations, Beijing 102211, China

Yuhong Su, Yunhe Xu, Donghui Shi, Haidi Xiao and Yumin Tian
Department of Animal Husbandry & Veterinary Medicine, Liaoning Medical University, Jinzhou, Liaoning 121000, China

Huixin Yang
Department of Veterinary Medicine, Nanjing Agricultural University, Nanjing, Jiangsu 210095, China

Lili Zhang
Department of Food Science, Liaoning Medical University, Jinzhou, Liaoning, China

Nagaraju Indugu, Bonnie Vecchiarelli, Linda D. Baker, James D. Ferguson and Dipti W. Pitta
Department of Clinical Studies, School of Veterinary Medicine, New Bolton Center, University of Pennsylvania, Kennett Square, PA 19348, USA

Jairam K. P. Vanamala
Department of Food Science, Pennsylvania State University, University Park, State College, PA 16802, USA
Penn State Hershey Cancer Institute, Hershey, PA 17033, USA

André Strauss, Penelope Heather Dobrowsky, Thando Ndlovu, Brandon Reyneke and Wesaal Khan
Department of Microbiology, Faculty of Science, Stellenbosch University, Private Bag X1, Stellenbosch 7602, South Africa

François-Joël Gatesoupe
NUMEA, INRA, Univ. Pau & Pays Adour, 64310 Saint Pée sur Nivelle, France
PFOM/ARN, Ifremer, Centre de Bretagne, CS 10070, 29280 Plouzané, France

Christine Huelvan, Nicolas Le Bayon, Hervé Le Delliou, Lauriane Madec, Olivier Mouchel, Patrick Quazuguel, David Mazurais and José-Luis Zambonino-Infante
Ifremer, UMR 6539 (LEMAR), PFOM/ARN, Centre de Bretagne, CS 10070, 29280 Plouzané, France

Junqing Qiao, Xuejie Liang, Yongfeng Liu and Youzhou Liu
Institute of Plant Protection, Jiangsu Academy of Agricultural Sciences, Nanjing, Jiangsu province 210014, China

Xiang Yu
Suqian institute, Jiangsu Academy of Agricultural Sciences, Suqian, Jiangsu province 223831, China

Rainer Borriss
Institut für Agrarwissenschaften/Phytomedizin, Humboldt Universität zu Berlin, 14195 Berlin, Germany
Nord Reet UG, 17489 Greifswald, Germany

Andrea Ottesen, Padmini Ramachandran, Elizabeth Reed, Gina Ryan, Eric Brown, Yi Chen and Marc Allard
Office o f R egulatory S cience, C enter f or Food Safety and Applied Nutrition, Food and Drug Administration, 5001 Campus Drive, College Park, MD 20740, USA

Nur Hasan, Poorani Subramanian and Rita Colwell
CosmosID, 155 Gibbs Street, Rockville, MD 20850, USA

Karen Jarvis, Christopher Grim, Ninalynn Daquiqan and Darcy Hanes
Office of Applied Research and Safety Assessment, Center for Food Safety and Applied Nutrition, Food and Drug Administration, 8301 Muirkirk Road, Laurel, MD 20708, USA

Xiaoyu Zhu, Xiangzhen Li, Jia Jia and Xiaohong He
Key Laboratory of Environmental and Applied Microbiology, Chinese Academy of Sciences & Environmental Microbiology Key Laboratory of Sichuan Province, Chengdu Institute of Biology, Chinese Academy of Sciences, Chengdu 610041, People's Republic of China

James R. White
Resphera Biosciences, 1529 Lancaster Street, Baltimore, MD 21231, USA

Yanyan Zhang
Key Laboratory of Environmental and Applied Microbiology, Chinese Academy of Sciences & Environmental Microbiology Key Laboratory of Sichuan Province, Chengdu Institute of Biology, Chinese Academy of Sciences, Chengdu 610041, People's Republic of China
University of Chinese Academy of Sciences, Beijing 100049, People's Republic of China

Yong Tao
Key Laboratory of Environmental and Applied Microbiology, Chinese Academy of Sciences & Environmental Microbiology Key Laboratory of Sichuan Province, Chengdu Institute of Biology, Chinese Academy of Sciences, Chengdu 610041, People's Republic of China
Chengdu Institute of Biology, Chinese Academy of Sciences, Chengdu 610041, People's Republic of China

Index

www.ingramcontent.com/pod-product-compliance
Lightning Source LLC
Chambersburg PA
CBHW082025190326
41458CB00010B/3281